INTERNATIONAL SERIES OF MONOGRAPHS IN
AERONAUTICS AND ASTRONAUTICS

DIVISION I: SOLID AND STRUCTURAL MECHANICS

EDITORS: W. S. HEMP AND R. L. BISPLINGHOFF

VOLUME 9

VARIATIONAL METHODS IN ELASTICITY AND PLASTICITY

Prof. Rockey

VARIATIONAL METHODS IN ELASTICITY AND PLASTICITY

SECOND EDITION

KYUICHIRO WASHIZU

Professor of Aeronautics and Astronautics, University of Tokyo

PERGAMON PRESS

Oxford · New York · Toronto · Sydney · Braunschweig

Pergamon Press Ltd., Headington Hill Hall, Oxford
Pergamon Press Inc., Maxwell House Fairview Park, Elmsford, New York 10523
Pergamon of Canada Ltd., 207 Quay West, Toronto 1
Pergamon Press (Aust.) Pty. Ltd., 19a Boundary Street, Rushcutters Bay, N.S.W. 2011, Australia
Pergamon Press GmbH., Burgplatz 1, Braunschweig 3300. West Germany

---

First edition 1968
Second edition 1975

Library of Congress Cataloging in Publication Data

Washizu, Kyuichiro, 1921–
Variational methods in elasticity and plasticity.

(International series of monographs in aeronautics and astronautics. Division I: solid and structural mechanics, v. 9)
Includes bibliographies.
1. Elasticity. 2. Plasticity. 3. Calculus of variations. I. Title.
QA931. W33 1974 620.1'123 74–8861
ISBN 0–08–017653–4

Printed in Great Britain at The Pitman Press, Bath

FOREWORD

THE variational principle and its application to many branches of mechanics including elasticity and plasticity has had a long history of development. However, the importance of this principle has been high-lighted in recent years by developments in the use of finite element methods which have been widely employed in structural analysis since the pioneering work by M. J. Turner *et al.* appeared in Vol. 23, No. 9 issue of the *Journal of Aeronautical Sciences* in 1956. It has been shown repeatedly since that time that the variational principle provides a powerful tool in the mathematical formulation of the finite element approach. Conversely, the rapid development of the finite element method has given much stimulus to the advancement of the variational principle and new forms of the principle have been developed during the past decade as outlined in Section 1 of Appendix I of the present book.

The first edition of Professor Washizu's book, entitled *Variational Methods in Elasticity and Plasticity* and published in 1968, was well received by engineers, teachers and students working in solid and structural mechanics. Its publication was timely, because it coincided with a period of rapid growth of application of the finite element method. The principle features of the first edition was that of providing a systematic way of deriving variational principles in elasticity and plasticity, of transforming one variational principle to another and of providing a systematic basis for the mathematical formulation of the finite element method. The book was widely used and referenced frequently in literature related to the finite element method.

Now, Professor Washizu has prepared a revised edition which adds a new Appendix I. The new appendix introduces an outline of variational principles which are used frequently as a basis for mathematical formulations in elasticity and plasticity including those new variational principles developed in connection with the finite element method. As in the case of the first edition, Appendix I is written in the clear, concise and elegant style for which Professor Washizu is so widely known. The revised edition should form an extremely valuable addition to the libraries and reference shelves of all who are interested in solid and structural mechanics.

R. L. BISPLINGHOFF
National Science Foundation,
Washington D.C.

ACKNOWLEDGEMENTS

THE author feels extremely honored and wishes to express his deepest gratitude to Dr. R. L. Bisplinghoff, Deputy Director of National Science Foundation, for having given the Foreword to the revised edition of this book. The author would like to express his deepest appreciation to Professor T. H. H. Pain of the Massachusetts Institute of Technology and Professor R. H. Gallagher of Cornell University for having given valuable comments to the manuscript for the new appendix. Dr. Oscar Orringer of the Massachusetts Institute of Technology collaborated again with the author in correcting the writing of the manuscript of the new appendix. Moreover, the author should remember that he has been given numerous comments, criticisms and encouragements from the reader since the publication of the first edition of this book. The author would like to express his sincere appreciation to all of these people, without whose encouragement and collaboration, this revised edition couldn't be realized.

K. WASHIZU

CONTENTS

INTRODUCTION

THE calculus of variations is a branch of mathematics, wherein the stationary property of a function of functions, namely, a functional, is studied. Thus, the object of the calculus of variations is not to find extrema of a function of a finite number of variables, but to find, among the group of admissible functions, the one which makes the given functional stationary.† A well-established example is to find, among the admissible curves joining two points in the prescribed space, that curve on which the distance between the points is a minimum. The problem of finding a curve which encloses a given area with minimum peripheral length is another typical example.

The calculus of variations has a wide field of application in mathematical physics. This is due to the fact that a physical system often behaves in a manner such that some functional depending on its behavior assumes a stationary value. In other words, the equations governing the physical phenomenon are often found to be stationary conditions of some variational problem. Fermat's principle in optics may be mentioned as a typical example. It states that a ray of light travels between two points along the path which requires the least time. This leads immediately to the conclusion that a ray of light travels in a straight line in any homogeneous medium.

Mechanics is one of the fields of mathematical physics, wherein the variational technique has been extensively investigated. We shall take a problem of a system of particles as an example and review the derivation of its variational formulations.‡

First, we shall consider the problem of a system of particles in static equilibrium under external and internal forces. It is well known that the basis of variational formulation is the principle of virtual work,†† which may be stated as follows: *Assume that the mechanical system is in equilibrium under applied forces and prescribed geometrical contraints. Then, the sum of all the virtual work, denoted by $\delta' W$, done by the external and internal forces existing in the system in any arbitrary infinitesimal virtual displacements satisfying the prescribed geometrical constraints is zero:*

$$\delta' W = 0. \qquad (1)‡‡$$

The principle may be stated alternatively in the following manner: *If $\delta' W$ vanishes for any arbitrary infinitesimal virtual displacements satisfying the*

† For details of the calculus of variations, see Refs. 1 through 8 (see pp. 6–7).

‡ For details of the variational methods in mechanics, see Refs. 2, 9, 10 and 11.

†† This principle is also called the principle of virtual displacements.

‡‡ $\delta' W$ is not a variation of some state function W, but denotes merely the total virtual work.

prescribed geometrical constraints, the mechanical system is in equilibrium. Thus, the principle of virtual work is equivalent to the equations of equilibrium of the system. However, the former has a much wider field of application to the formulation of mechanics problems than the latter. When all the external and internal forces are derived from a potential function U, which is a function of the coordinates of the system of particles,† such that

$$\delta' W = -\delta U, \tag{2}$$

the principle of virtual work leads to the establishment of the principle of stationary potential energy: *Among the set of all admissible configurations, the state of equilibrium is characterized by the stationary property of the potential energy U:*

$$\delta U = 0. \tag{3}$$

The above formulation may be extended to the dynamical problem of a system of particles subject to time-dependent applied forces and geometrical constraints. By the use of d'Alembert's principle which states that the system can be considered to be in equilibrium if inertial forces are taken into account, the principle of virtual work of the dynamical problem can be derived in a manner similar to the static problem case, except that terms representing the virtual work done by the inertial forces are now included. The principle thus obtained is integrated with respect to time t between two limits $t = t_1$ and $t = t_2$. Through integration by parts and by the use of the convention that virtual displacements vanish at the limits, we finally obtain the following principle of virtual work for the dynamical problem:

$$\delta \int_{t_1}^{t_2} T\,dt + \int_{t_1}^{t_2} \delta' W\,dt = 0, \tag{4}$$

where T is the kinetic energy of the system. Since Lagrange's equations of motion of the system may be derived from the principle of virtual work thus obtained, it is evident that the principle is extremely useful for obtaining the equations of motion of a system of particles with geometrical constraints.

When it is further assured that all the external and internal forces are derived from a potential function U, which is defined in the same manner as Eq. (2) and is a function of coordinates and the time,‡ we obtain Hamilton's principle, which states that *among the set of all admissible configurations of the system, the actual motion makes the quantity*

$$\int_{t_1}^{t_2} (T - U)\,dt \tag{5}$$

† Forces of this category are called conservative forces.

‡ If U is time-independent, the forces are called conservative. In Ref. 2, the name "monogenic" is given to forces derivable from a scalar quantity which is in the most general case a function of coordinates and velocities of the particles and the time.

stationary, provided the configuration of the system is prescribed at the limits $t = t_1$ *and* $t = t_2$. Hamilton's principle may be stated mathematically as follows:

$$\delta \int_{t_1}^{t_2} L\, dt = 0, \tag{6}$$

where $L = T - U$ is the Lagrangian function of the system. It is well known that Hamilton's principle can be transformed by the use of Legendre's transformation into a new and equivalent principle, and that Lagrange's equations of motion are reduced to the so-called canonical equations. Transformations of Hamilton's principle were extensively investigated and an elegant theory known as canonical transformation was established.

The main object of this book is to derive the principle of virtual work and related variational principles in elasticity and plasticity in a systematic way.† We shall formulate these principles in a manner similar to the development in the problem of a system of particles. The outline is as follows: we define a problem involving a solid body in static equilibrium under body forces plus mechanical and geometrical boundary conditions prescribed on the surface of the body. To begin with, we derive the principle of virtual work. This principle is equivalent to the equations of equilibrium and the mechanical boundary conditions of the solid body, and is derived for small displacement theory as well as finite displacement theory.‡ Within the realm of small displacement theory we obtain another principle which will be called the principle of complementary virtual work.†† It is worthy of special mention that the principles of virtual work and complementary virtual work are invariant under coordinate transformations and that they hold independently of the stress–strain relations of the material of the body. However, the stress–strain relations should be taken into account for the formulations of variational principles, and the theories of elasticity and plasticity should be treated separately.

The variational method finds one of the most fruitful fields of application in the small displacement theory of elasticity. When the existence of a strain energy function is assured and the external forces are assumed to be kept unchanged during displacement variation, the principle of virtual work leads to the establishment of the principle of minimum potential energy. The variational principle is generalized by the introduction of Lagrange multipliers to yield a family of variational principles which includes the Hellinger–

† For variational principles in elasticity and plasticity, see Refs. 11 through 20.

‡ In the small displacement theory, the displacements are assumed so small as to allow linearizations of all governing equations of the solid body except the stress–strain relations. Consequently, the equations of equilibrium, the strain–displacement relations and the boundary conditions are reduced to linearized forms in small displacement theory.

†† This principle is also called the principle of virtual stress, the principle of virtual force or the principle of virtual changes in the state of stress.

Reissner principle, the principle of minimum complementary energy and so forth.

On the other hand, the principle of complementary virtual work leads to the establishment of the principle of minimum complementary energy when the stress–strain relations assure the existence of a complementary energy function and the geometrical boundary conditions are assumed to be kept unchanged during stress variation. The principle of minimum complementary energy is generalized by the introduction of Lagrange multipliers to yield the Hellinger–Reissner principle, the principle of minimum potential energy and so forth. It is seen that these two approaches to the formulation of the variational principles are reciprocal and equivalent to each other as far as the small displacement theory of elasticity is concerned.

In the finite displacement theory of elasticity, the principle of virtual work leads to the establishment of the principle of stationary potential energy when the existence of a strain energy function of the body material and potential functions of the external forces is assured. Once the principle of stationary potential energy is thus established, it can be generalized through the use of Lagrange multipliers.

The above technique is extended to dynamical elastic body problems by taking inertial forces into account. Thus, we derive the principle of virtual work for the dynamical problem with the introduction of the concept of kinetic energy. The principle of virtual work is then transformed into a variational principle under the assumption of the existence of a strain energy function and potential functions of the external forces. The newly obtained variational principle may be thought of as Hamilton's principle extended to the dynamical elastic body problem, and it can be generalized through the use of Lagrange multipliers.

The variational principle of an elasticity problem provides the governing equations of the problem as stationary conditions and, in that sense, is equivalent to the governing equations. However, the variational formulation has several advantages. First, the functional which is subject to variation usually has a definite physical meaning and is invariant under coordinate transformation. Consequently, once the variational principle has been formulated in one coordinate system, governing equations expressed in another coordinate system can be obtained by first writing the invariant quantity in the new coordinate system and then applying variational procedures. For example, once the variational principle has been formulated in the rectangular Cartesian coordinate system, governing equations expressed in cylindrical or polar coordinate systems can be obtained through the above technique. It may be observed that this property makes the variational method extremely powerful for the analysis of structures.

Second, the variational formulation is helpful in carrying out a common mathematical procedure, namely, the transformation of a given problem into an equivalent problem that can be solved more easily than the original.

In a variational problem with subsidiary conditions, the transformation is achieved by the Lagrange multiplier method, a very useful and systematic tool. Thus, we may derive a family of variational principles which are equivalent to each other.

Third, variational principles sometimes lead to formulae for upper or lower bounds of the exact solution of the problem under consideration. As will be shown in Chapter 6, upper and lower bound formulae for the torsional rigidity of a bar are provided by simultaneous use of two variational principles. Another example is an upper bound formula, derived from the principle of stationary potential energy, for the lowest frequency of free vibrations of an elastic body.

Fourth, when a problem of elasticity cannot be solved exactly, the variational method often provides an approximate formulation for the problem which yields a solution compatible with the assumed degree of approximation. Here, the variational method provides not only approximate governing equations, but also suggestions on approximate boundary conditions. Since it is almost impossible to obtain the exact solution of an elasticity problem except in a few special cases, we must be satisfied with approximate solutions for practical purposes. Theories of beams, plates, shells and multi-component structures are typical examples of such approximate formulations and show the power of the principle of virtual work and related variational methods. However, one should take care in relying upon the accuracy of approximate solutions thus obtained. Consider, for example, an application of the Rayleigh–Ritz method combined with the principle of stationary potential energy. The method may provide a good approximate solution for the displacements of a body if admissible functions are chosen properly. However, the accuracy of stress distribution calculated from the approximate displacements is not as reliable. This is obvious if we remember that, in the governing equations obtained by the approximate method, the exact equations of equilibrium and mechanical boundary conditions have been replaced by their weighted means and that the accuracy of an approximate solution decreases with differentiation. Thus, the equations of equilibrium and mechanical boundary conditions are generally violated at least locally in the approximate solution. In understanding approximate solutions thus obtained, the principle of Saint-Venant is sometimes helpful. It states:[14] "*If the forces acting on a small portion of the surface of an elastic body are replaced by another statically equivalent system of forces acting on the same portion of the surface, this redistribution of loading produces substantial changes in the stresses locally, but has a negligible effect on the stresses at distances which are large in comparison with the linear dimensions of the surface on which the forces are changed.*"

Due to the author's preference, approximate governing equations of elasticity problems will be derived very frequently from the principle of virtual work rather than from the variational principle, since the former holds independently of the stress–strain relations of the body and the existence of potential functions. An approximate method of solution using the principle of virtual work will be called the generalized Galerkin's method.† As far as conservative problems in elasticity are concerned, results obtained by the combined use of the principle of virtual work and the generalized Galerkin's method are equivalent to those obtained by the combined use of the principle of stationary potential energy and the Rayleigh–Ritz method.

It is quite natural in theories of plasticity to make the principle of virtual work a basis for the establishment of variational principles. If the problem is confined to the small displacement theory, the principle of complementary virtual work may also be employed as another basis. Since stress–strain relations in the theories of plasticity are more complicated than those in the theory of elasticity, it may be expected that the establishment of a variational principle in plasticity is more difficult. Several variational principles which have been established for the theories of plasticity can be shown to be formally derivable in a manner similar to those in the theory of elasticity, although rigorous proofs should follow for showing the validity of the variational principles.

The most successful application of variational formulations in the flow theory of plasticity is the theory of limit analysis for a body consisting of material which obeys the perfectly plastic Prandtl–Reuss equation. Limit analysis concerns the determination of an eigenvalue called the collapse load of the body. Two variational principles provide upper and lower bound formulae for locating the collapse load.

Since a great many papers have been written on variational treatment of problems in elasticity and plasticity, the bibliography of this book is not intended to be complete. The author is satisfied with citing only a limited number of papers for the reader's reference. Literature such as Refs. 22 and 23 may be helpful for reviewing recent developments of the topic.

The variational method can, of course, be applied to problems other than those mentioned herein. For example, it has been applied to problems in fluid mechanics, conduction of heat and so forth. [24–26] As a recent application of engineering concern, we may add that problems of the performance of flight vehicles have been extensively treated in the literature by the optimization technique.[27]

Bibliography

1. R. Courant and D. Hilbert, *Methods of Mathematical Physics*, Vol. 1, Interscience, New York, 1953.
2. C. Lanczos, *The Variational Principles of Mechanics*, University of Toronto Press, 1949.

† This is also called the method of weighting functions. It is a special case of the approximate method of solution called the method of weighted residuals.[21]

3. O. Bolza, *Lectures on the Calculus of Variations*, The University of Chicago Press, 1946.
4. G. A. Bliss, *Lectures on the Calculus of Variations*, The University of Chicago Press, 1946.
5. C. Fox, *An Introduction to the Calculus of Variations*, Oxford University Press, London, 1950.
6. R. Weinstock, *Calculus of Variations with Application to Physics and Engineering*, McGraw-Hill, 1952.
7. P. M. Morse and H. Feshbach, *Methods of Theoretical Physics*, Vols. 1 and 2, McGraw-Hill, 1953.
8. S. G. Mikhlin, *Variational Methods in Mathematical Physics*, Pergamon Press, 1964.
9. H. Goldstein, *Classical Mechanics*, Addison-Wesley, 1953.
10. J. L. Synge and B. A. Griffth, *Principles of Mechanics*, McGraw-Hill, 1959.
11. H. L. Langhaar, *Energy Methods in Applied Mechanics*, John Wiley, 1962.
12. C. B. Biezeno and R. Grammel, *Technische Dynamik*, Springer, Berlin, 1939.
13. R. V. Southwell, *Introduction to the Theory of Elasticity*, Clarendon Press, Oxford, 1941.
14. S. Timoshenko and J. N. Goodier, *Theory of Elasticity*, McGraw-Hill, 1951.
15. N. J. Hoff, *The Analysis of Structures*, John Wiley, 1956.
16. C. E. Pearson, *Theoretical Elasticity*, Harvard University Press, 1959.
17. J. H. Argyris and S. Kelsey, *Energy Theorems and Structural Analysis*, Butterworth, 1960.
18. V. V. Novozhilov, *Theory of Elasticity*, translated by J. K. Lusher, Pergamon Press, 1961.
19. J. H. Greenberg, *On the Variational Principles of Plasticity*, Brown University, ONR, NR–041–032, March 1949.
20. R. Hill, *Mathematical Theory of Plasticity*, Oxford, 1950.
21. M. Becker, *The Principles and Applications of Variational Methods*, The Massachusetts Institute of Technology Press, 1964.
22. *Applied Mechanics Reviews*, published monthly by the American Society of Mechanical Engineers.
23. *Structural Mechanics in U.S.S.R. 1917–1957*, edited by I. M. Rabinovich. English translation edited by G. Herrmann was published by Pergamon Press in 1960.
24. J. Serrin, Mathematical Principles of Classical Fluid Mechanics, *Handbuch der Physik*, Band VII/I. Strömungsmechanik I, pp. 125–265, Springer, 1959.
25. M. A. Biot, Lagrangian Thermodynamics of Heat Transfer in Systems including Fluid Motion. *Journal of the Aeronautical Sciences*, Vol. 25, No. 5, pp. 568–77, May 1962.
26. K. Washizu, *Variational Principles in Continuum Mechanics*, University of Washington, College of Engineering, Department of Aeronautical Engineering, Report 62–2, June 1962.
27. G. Leitmann (Editor), *Optimization Techniques with Applications to Aerospace Systems*, Academic Press, 1962.

CHAPTER 1

SMALL DISPLACEMENT THEORY OF ELASTICITY IN RECTANGULAR CARTESIAN COORDINATES

1.1. Presentation of a Problem in Small Displacement Theory

In the beginning of his classical work,[1] Love states: "The Mathematical Theory of Elasticity is occupied with an attempt to reduce to calculation the state of strain, or relative displacement, within a solid body which is subject to the action of an equilibrating system of forces, or is in a state of slight internal relative motion, and with endeavours to obtain results which shall be practically important in applications to architecture, engineering, and all other useful arts in which the material of construction is solid." This seems to have been a guiding definition of the theory of elasticity.

In the first and second chapters of this book we shall deal with the small displacement theory of elasticity and derive the principle of virtual work and related variational principles for the problem of an elastic body in static equilibrium under body forces and prescribed boundary conditions.[1,2] Rectangular Cartesian coordinates (x, y, z) will be employed for defining the three-dimensional space containing the body. In the small displacement theory of elasticity displacement components, u, v, w, of a point of the body are assumed so small that we are justified in linearizing equations governing the problem. The linearized governing equations may be summarized as follows:

(a) *Stress.* The state of internal force at a point of the body is defined by nine components of stress:

$$\begin{matrix} \sigma_x & \tau_{yx} & \tau_{zx}, \\ \tau_{xy} & \sigma_y & \tau_{zy}, \\ \tau_{xz} & \tau_{yz} & \sigma_z, \end{matrix} \tag{1.1}$$

which should satisfy the equations of equilibrium:

$$\begin{aligned} \frac{\partial \sigma_x}{\partial x} + \frac{\partial \tau_{yx}}{\partial y} + \frac{\partial \tau_{zx}}{\partial z} + \bar{X} = 0, \\ \frac{\partial \tau_{xy}}{\partial x} + \frac{\partial \sigma_y}{\partial y} + \frac{\partial \tau_{zy}}{\partial z} + \bar{Y} = 0, \\ \frac{\partial \tau_{xz}}{\partial x} + \frac{\partial \tau_{yz}}{\partial y} + \frac{\partial \sigma_z}{\partial z} + \bar{Z} = 0, \end{aligned} \tag{1.2†}$$

† Throughout the present book, an overbar indicates that the barred quantity is prescribed, unless otherwise stated.

and $$\tau_{yz} = \tau_{zy}, \quad \tau_{zx} = \tau_{xz}, \quad \tau_{xy} = \tau_{yx}, \tag{1.3}$$

where $\bar{X}$, $\bar{Y}$ and $\bar{Z}$ are components of the body forces per unit volume. We shall eliminate τ_{zy}, τ_{xz} and τ_{yx} by the use of Eqs. (1.3), and specify the state of stress at a point of the body with six components $(\sigma_x, \sigma_y, \sigma_z, \tau_{yz}, \tau_{zx}, \tau_{xy})$. Then, Eqs. (1.2) become:

$$\begin{aligned} \frac{\partial \sigma_x}{\partial x} + \frac{\partial \tau_{xy}}{\partial y} + \frac{\partial \tau_{zx}}{\partial z} + \bar{X} &= 0, \\ \frac{\partial \tau_{xy}}{\partial x} + \frac{\partial \sigma_y}{\partial y} + \frac{\partial \tau_{yz}}{\partial z} + \bar{Y} &= 0, \\ \frac{\partial \tau_{zx}}{\partial x} + \frac{\partial \tau_{yz}}{\partial y} + \frac{\partial \sigma_z}{\partial z} + \bar{Z} &= 0. \end{aligned} \tag{1.4}$$

(b) *Strain.* The state of strain at a point of the body is defined by six components of strain $(\varepsilon_x, \varepsilon_y, \varepsilon_z, \gamma_{yz}, \gamma_{zx}, \gamma_{xy})$.

(c) *Strain–displacement relations.* In small displacement theory the strain–displacement relations are given as follows:

$$\begin{aligned} &\varepsilon_x = \frac{\partial u}{\partial x}, \quad \varepsilon_y = \frac{\partial v}{\partial y}, \quad \varepsilon_z = \frac{\partial w}{\partial z}, \\ &\gamma_{yz} = \frac{\partial w}{\partial y} + \frac{\partial v}{\partial z}, \quad \gamma_{zx} = \frac{\partial u}{\partial z} + \frac{\partial w}{\partial x}, \quad \gamma_{xy} = \frac{\partial v}{\partial x} + \frac{\partial u}{\partial y}. \end{aligned} \tag{1.5}$$

(d) *Stress–strain relations.* In small displacement theory, the stress–strain relations are given in linear, homogeneous form:

$$\begin{bmatrix} \sigma_x \\ \sigma_y \\ \sigma_z \\ \tau_{yz} \\ \tau_{zx} \\ \tau_{xy} \end{bmatrix} = \begin{bmatrix} a_{11} & a_{12} & a_{13} & a_{14} & a_{15} & a_{16} \\ a_{21} & a_{22} & a_{23} & a_{24} & a_{25} & a_{26} \\ a_{31} & a_{32} & a_{33} & a_{34} & a_{35} & a_{36} \\ a_{41} & a_{42} & a_{43} & a_{44} & a_{45} & a_{46} \\ a_{51} & a_{52} & a_{53} & a_{54} & a_{55} & a_{56} \\ a_{61} & a_{62} & a_{63} & a_{64} & a_{65} & a_{66} \end{bmatrix} \begin{bmatrix} \varepsilon_x \\ \varepsilon_y \\ \varepsilon_z \\ \gamma_{yz} \\ \gamma_{zx} \\ \gamma_{xy} \end{bmatrix} \tag{1.6}$$

The coefficients of these equations are called elastic constants. Among them, there exist relations of the form:

$$a_{rs} = a_{sr}, \quad (r, s = 1, 2, \ldots, 6). \tag{1.7}$$

Eqs. (1.6) may be inverted to yield:

$$\begin{bmatrix} \varepsilon_x \\ \varepsilon_y \\ \varepsilon_z \\ \gamma_{yz} \\ \gamma_{zx} \\ \gamma_{xy} \end{bmatrix} = \begin{bmatrix} b_{11} & b_{12} & b_{13} & b_{14} & b_{15} & b_{16} \\ b_{21} & b_{22} & b_{23} & b_{24} & b_{25} & b_{26} \\ b_{31} & b_{32} & b_{33} & b_{34} & b_{35} & b_{36} \\ b_{41} & b_{42} & b_{43} & b_{44} & b_{45} & b_{46} \\ b_{51} & b_{52} & b_{53} & b_{54} & b_{55} & b_{56} \\ b_{61} & b_{62} & b_{63} & b_{64} & b_{65} & b_{66} \end{bmatrix} \begin{bmatrix} \sigma_x \\ \sigma_y \\ \sigma_z \\ \tau_{yz} \\ \tau_{zx} \\ \tau_{xy} \end{bmatrix}, \tag{1.8}$$

where

$$b_{rs} = b_{sr}, \quad (r, s = 1, 2, \ldots, 6). \tag{1.9}$$

For an isotropic material, the number of the independent elastic constants reduces to 2, and the stress–strain relations are given by:

$$\begin{aligned}
\sigma_x &= 2G\left[\varepsilon_x + \frac{\nu}{1-2\nu}(\varepsilon_x + \varepsilon_y + \varepsilon_z)\right], \quad \tau_{yz} = G\gamma_{yz},\\
\sigma_y &= 2G\left[\varepsilon_y + \frac{\nu}{1-2\nu}(\varepsilon_x + \varepsilon_y + \varepsilon_z)\right], \quad \tau_{zx} = G\gamma_{zx},\\
\sigma_z &= 2G\left[\varepsilon_z + \frac{\nu}{1-2\nu}(\varepsilon_x + \varepsilon_y + \varepsilon_z)\right], \quad \tau_{xy} = G\gamma_{xy},
\end{aligned} \qquad (1.10)\dagger$$

or, inversely,

$$\begin{aligned}
E\varepsilon_x &= \sigma_x - \nu(\sigma_y + \sigma_z), \quad G\gamma_{yz} = \tau_{yz},\\
E\varepsilon_y &= \sigma_y - \nu(\sigma_z + \sigma_x), \quad G\gamma_{zx} = \tau_{zx},\\
E\varepsilon_z &= \sigma_z - \nu(\sigma_x + \sigma_y), \quad G\gamma_{xy} = \tau_{xy}.
\end{aligned} \qquad (1.11)\dagger$$

(e) *Boundary conditions.* The surface of the body can be divided into two parts from the viewpoint of the boundary conditions: the part S_1 over which boundary conditions are prescribed in terms of external forces and the part S_2 over which boundary conditions are prescribed in terms of displacements. Obviously $S = S_1 + S_2$. Denoting the components of the prescribed external forces per unit area of the boundary surface by $\bar{X}_\nu$, $\bar{Y}_\nu$ and $\bar{Z}_\nu$, the mechanical boundary conditions are given by

$$X_\nu = \bar{X}_\nu, \quad Y_\nu = \bar{Y}_\nu, \quad Z_\nu = \bar{Z}_\nu \quad \text{on} \quad S_1, \qquad (1.12)$$

where

$$\begin{aligned}
X_\nu &= \sigma_x l + \tau_{xy} m + \tau_{zx} n,\\
Y_\nu &= \tau_{xy} l + \sigma_y m + \tau_{yz} n,\\
Z_\nu &= \tau_{zx} l + \tau_{yz} m + \sigma_z n,
\end{aligned} \qquad (1.13)$$

l, m, n being the direction cosines of the unit normal ν drawn outwards on the boundary:‡ $l = \cos(x, \nu)$, $m = \cos(y, \nu)$ and $n = \cos(z, \nu)$. On the other hand, denoting the components of the prescribed displacements by $\bar{u}$, $\bar{v}$ and $\bar{w}$, the geometrical conditions are given by

$$u = \bar{u}, \quad v = \bar{v}, \quad w = \bar{w} \quad \text{on} \quad S_2. \qquad (1.14)$$

Thus, we obtain all the governing equations of the elasticity problem in small displacement theory: the equations of equilibrium (1.4), the strain–displacement relations (1.5) and the stress–strain relations (1.6) in the interior V of the body, and the mechanical and geometrical boundary conditions, (1.12) and (1.14), on the surface S of the body. These conditions show that we have 15 unknowns, namely, 6 stress components, 6 strain components and 3 displacement components in the 15 equations (1.4), (1.5) and (1.6).

† Young's modulus E, Poisson's ratio ν and the modulus of rigidity G are related by the equation $E = 2G(1 + \nu)$. Thus there are only two independent elastic constants.

‡ It is noted that the symbol ν is used in the present book to denote the Poisson's ratio as well as the unit normal drawn outwards on the boundary.

Our problem is then to solve these 15 equations under the boundary conditions (1.12) and (1.14). Since all the governing equations have linear forms, the law of superposition can be applied in solving the problem. Thus, we obtain linear relationships between the prescribed quantities such as the applied load on S_1 and resulting quantities such as stress and displacement caused in the body.

1.2. Conditions of Compatibility

We observe from Eqs. (1.5) that when a continuum deforms, the six strain components ($\varepsilon_x, \varepsilon_y, \varepsilon_z, \gamma_{yz}, \gamma_{zx}, \gamma_{xy}$) cannot behave independently, but must be derived from three functions u, v and w as shown. This statement can be expressed in a different way as follows: Let the continuum under consideration be separated into a large number of infinitesimal rectangular parallelepiped elements before deformation. Assume that each element is given six strain components ($\varepsilon_x, \ldots, \gamma_{xy}$) of arbitrary magnitude. Then, trials to reassemble the elements again into a continuum are assumed to be made. In general, such trials cannot be successful. Some relations should exist between the magnitude of the strain components for a reassemblage to be successful. Thus, a problem will arise which may be stated as follows: *What are the necessary and sufficient conditions for the elements to be reassembled into a continuous body?*

The necessary and sufficient conditions that the six strain components can be derived from three single-valued function as given in Eqs. (1.5) are called *the conditions of compatibility*. It is shown in Refs. 1 through 5, for example, that the conditions of compatibility are given in a matrix form as,

$$[R] = \begin{bmatrix} R_x & U_z & U_y \\ U_z & R_y & U_x \\ U_y & U_x & R_z \end{bmatrix} = 0, \tag{1.15}$$

where

$$\begin{aligned}
R_x &= \frac{\partial^2 \varepsilon_z}{\partial y^2} + \frac{\partial^2 \varepsilon_y}{\partial z^2} - \frac{\partial^2 \gamma_{yz}}{\partial y\, \partial z}, \\
R_y &= \frac{\partial^2 \varepsilon_x}{\partial z^2} + \frac{\partial^2 \varepsilon_z}{\partial x^2} - \frac{\partial^2 \gamma_{zx}}{\partial z\, \partial x}, \\
R_z &= \frac{\partial^2 \varepsilon_y}{\partial x^2} + \frac{\partial^2 \varepsilon_x}{\partial y^2} - \frac{\partial^2 \gamma_{xy}}{\partial x\, \partial y}, \\
U_x &= -\frac{\partial^2 \varepsilon_x}{\partial y\, \partial z} + \frac{1}{2}\frac{\partial}{\partial x}\left(-\frac{\partial \gamma_{yz}}{\partial x} + \frac{\partial \gamma_{zx}}{\partial y} + \frac{\partial \gamma_{xy}}{\partial z}\right), \\
U_y &= -\frac{\partial^2 \varepsilon_y}{\partial z\, \partial x} + \frac{1}{2}\frac{\partial}{\partial y}\left(\frac{\partial \gamma_{yz}}{\partial x} - \frac{\partial \gamma_{zx}}{\partial y} + \frac{\partial \gamma_{xy}}{\partial z}\right), \\
U_z &= -\frac{\partial^2 \varepsilon_z}{\partial x\, \partial y} + \frac{1}{2}\frac{\partial}{\partial z}\left(\frac{\partial \gamma_{yz}}{\partial x} + \frac{\partial \gamma_{zx}}{\partial y} - \frac{\partial \gamma_{xy}}{\partial z}\right).
\end{aligned} \tag{1.16}$$

The proof that the conditions (1.15) are necessary follows immediately from Eqs. (1.5) by direct differentiation. The proof that they are sufficient is rather lengthy and is not given here. The interested reader is advised to read the cited references.

It is noted at the end of this section that there exist identities between $R_x, R_y, \ldots, U_z$:

$$\begin{aligned} \frac{\partial R_x}{\partial x} + \frac{\partial U_z}{\partial y} + \frac{\partial U_y}{\partial z} &= 0, \\ \frac{\partial U_z}{\partial x} + \frac{\partial R_y}{\partial y} + \frac{\partial U_x}{\partial z} &= 0, \\ \frac{\partial U_y}{\partial x} + \frac{\partial U_x}{\partial y} + \frac{\partial R_z}{\partial z} &= 0. \end{aligned} \tag{1.17}$$

These identities can be proved easily by direct calculations. They show that the quantities R_x, R_y, ..., and U_z are not mutually independent, and that the conditions of compatibility (1.15) can be replaced by[6]

$$R_x = R_y = R_z = 0 \quad \text{in} \quad V, \tag{1.18a}$$

and

$$R_x = R_y = R_z = U_x = U_y = U_z = 0 \quad \text{on } S, \tag{1.18b}$$

or alternatively

$$U_x = U_y = U_z = 0 \quad \text{in } V, \tag{1.19a}$$

and

$$R_x = R_y = R_z = U_x = U_y = U_z = 0 \quad \text{on } S. \tag{1.19b}$$

1.3. Stress Functions

We know from Eqs. (1.4) that when the body forces are absent, the equations of equilibrium can be written as:

$$\begin{aligned} \frac{\partial \sigma_x}{\partial x} + \frac{\partial \tau_{xy}}{\partial y} + \frac{\partial \tau_{zx}}{\partial z} &= 0, \\ \frac{\partial \tau_{xy}}{\partial x} + \frac{\partial \sigma_y}{\partial y} + \frac{\partial \tau_{yz}}{\partial z} &= 0, \\ \frac{\partial \tau_{zx}}{\partial x} + \frac{\partial \tau_{yz}}{\partial y} + \frac{\partial \sigma_z}{\partial z} &= 0. \end{aligned} \tag{1.20}$$

These equations are satisfied identically when stress components are expressed in terms of either Maxwell's stress functions χ_1, χ_2 and χ_3 defined by

$$\begin{aligned}
\sigma_x &= \frac{\partial^2 \chi_3}{\partial y^2} + \frac{\partial^2 \chi_2}{\partial z^2}, & \tau_{yz} &= -\frac{\partial^2 \chi_1}{\partial y\, \partial z}, \\
\sigma_y &= \frac{\partial^2 \chi_1}{\partial z^2} + \frac{\partial^2 \chi_3}{\partial x^2}, & \tau_{zx} &= -\frac{\partial^2 \chi_2}{\partial z\, \partial x}, \\
\sigma_z &= \frac{\partial^2 \chi_2}{\partial x^2} + \frac{\partial^2 \chi_1}{\partial y^2}, & \tau_{xy} &= -\frac{\partial^2 \chi_3}{\partial x\, \partial y},
\end{aligned} \tag{1.21}$$

or Morera's stress functions ψ_1, ψ_2 and ψ_3 defined by

$$\begin{aligned}
\sigma_x &= \frac{\partial^2 \psi_1}{\partial y\, \partial z}, & \tau_{yz} &= -\frac{1}{2}\frac{\partial}{\partial x}\left(-\frac{\partial \psi_1}{\partial x} + \frac{\partial \psi_2}{\partial y} + \frac{\partial \psi_3}{\partial z}\right), \\
\sigma_y &= \frac{\partial^2 \psi_2}{\partial z\, \partial x}, & \tau_{zx} &= -\frac{1}{2}\frac{\partial}{\partial y}\left(\frac{\partial \psi_1}{\partial x} - \frac{\partial \psi_2}{\partial y} + \frac{\partial \psi_3}{\partial z}\right), \\
\sigma_z &= \frac{\partial^2 \psi_3}{\partial x\, \partial y}, & \tau_{xy} &= -\frac{1}{2}\frac{\partial}{\partial z}\left(\frac{\partial \psi_1}{\partial x} + \frac{\partial \psi_2}{\partial y} - \frac{\partial \psi_3}{\partial z}\right).
\end{aligned} \tag{1.22}$$

It is interesting to note that, when these two kinds of stress functions are combined such that

$$\begin{aligned}
\sigma_x &= \frac{\partial^2 \chi_3}{\partial y^2} + \frac{\partial^2 \chi_2}{\partial z^2} - \frac{\partial \psi_1}{\partial y\, \partial z}, \ldots, \ldots, \\
\tau_{yz} &= -\frac{\partial^2 \chi_1}{\partial y\, \partial z} + \frac{1}{2}\frac{\partial}{\partial x}\left(-\frac{\partial \psi_1}{\partial x} + \frac{\partial \psi_2}{\partial y} + \frac{\partial \psi_3}{\partial z}\right), \ldots, \ldots,
\end{aligned} \tag{1.23}$$

the expressions (1.16) and (1.23) have similar forms.

In a two-dimensional stress problem, where the equations of equilibrium are

$$\frac{\partial \sigma_x}{\partial x} + \frac{\partial \tau_{xy}}{\partial y} = 0, \qquad \frac{\partial \tau_{xy}}{\partial x} + \frac{\partial \sigma_y}{\partial y} = 0, \tag{1.24}$$

the so-called Airy stress function defined by

$$\sigma_x = \frac{\partial^2 F}{\partial y^2}, \quad \sigma_y = \frac{\partial^2 F}{\partial x^2}, \quad \tau_{xy} = -\frac{\partial^2 F}{\partial x\, \partial y} \tag{1.25}$$

satisfies the equations of equilibrium identically.

1.4. Principle of Virtual Work

In this section we shall derive the principle of virtual work for the problem defined in Section 1.1. We consider the body in equilibrium under prescribed body forces and boundary conditions, and denote the stress components by σ_x, σ_y, ..., and τ_{xy}. Obviously,

$$\frac{\partial \sigma_x}{\partial x} + \frac{\partial \tau_{xy}}{\partial y} + \frac{\partial \tau_{zx}}{\partial z} + \bar{X} = 0, \ldots, \ldots, \quad \text{in } V, \tag{1.26}$$

and

$$X_\nu - \bar{X}_\nu = 0, \ldots, Z_\nu - \bar{Z}_\nu = 0 \quad \text{on } S_1. \tag{1.27}$$

Now, the body is assumed to execute an arbitrary set of infinitesimal virtual displacements δu, δv and δw from this equilibrating configuration.

Then, we have

$$-\iiint_V \left[\left(\frac{\partial \sigma_x}{\partial x} + \frac{\partial \tau_{xy}}{\partial y} + \frac{\partial \tau_{zx}}{\partial z} + \bar{X}\right)\delta u + (\ldots)\,\delta v + \left(\frac{\partial \tau_{zx}}{\partial x} + \frac{\partial \tau_{yz}}{\partial y} + \frac{\partial \sigma_z}{\partial z} + \bar{Z}\right)\delta w\right] dV + \iint_{S_1} [(X_\nu - \bar{X}_\nu)\,\delta u + (\ldots)\,\delta v + (Z_\nu - \bar{Z}_\nu)\,\delta w]\,dS = 0, \tag{1.28}$$

where $dV = dx\,dy\,dz$ and dS are the elementary volume and the elementary area of the surface of the body, respectively.

Here, we shall choose the arbitrary set of virtual displacements such that the geometrical boundary conditions on S_2 are not violated. Namely, they are so chosen as to satisfy the following equations:

$$\delta u = 0, \quad \delta v = 0, \quad \delta w = 0 \quad \text{on } S_2. \tag{1.29}$$

Then, by the use of geometrical relations

$$dy\,dz = \pm l\,dS, \quad dz\,dx = \pm m\,dS, \quad dx\,dy = \pm n\,dS \tag{1.30}$$

which hold on the boundary, and through integrations by parts such that

$$\iiint_V \frac{\partial \sigma_x}{\partial x}\,\delta u\,dx\,dy\,dz = \iint_S \sigma_x l\,\delta u\,dS - \iiint_V \sigma_x \frac{\partial \delta u}{\partial x}\,dx\,dy\,dz, \tag{1.31†}$$

we may transform Eq. (1.28) into

$$\iiint_V (\sigma_x\,\delta\varepsilon_x + \sigma_y\,\delta\varepsilon_y + \sigma_z\,\delta\varepsilon_z + \tau_{yz}\,\delta\gamma_{yz} + \tau_{zx}\,\delta\gamma_{zx} + \tau_{xy}\,\delta\gamma_{xy})\,dV - \iiint_V (\bar{X}\,\delta u + \bar{Y}\,\delta v + \bar{Z}\,\delta w)\,dV - \iint_{S_1} (\bar{X}_\nu\,\delta u + \bar{Y}_\nu\,\delta v + \bar{Z}_\nu\,\delta w)\,dS = 0 \tag{1.32}$$

† This is an application of the divergence theorem of Gauss expressed by the equation

$$\iiint_V \left(\frac{\partial L}{\partial x} + \frac{\partial M}{\partial y} + \frac{\partial N}{\partial z}\right) dx\,dy\,dz = \iint_S (Ll + Mm + Nn)\,dS.$$

where
$$\delta\varepsilon_x = \frac{\partial \delta u}{\partial x}, \quad \delta\varepsilon_y = \frac{\partial \delta v}{\partial y}, \quad \delta\varepsilon_z = \frac{\partial \delta w}{\partial z},$$
$$\delta\gamma_{yz} = \frac{\partial \delta w}{\partial y} + \frac{\partial \delta v}{\partial z}, \quad \delta\gamma_{zx} = \frac{\partial \delta u}{\partial z} + \frac{\partial \delta w}{\partial x}, \quad \delta\gamma_{xy} = \frac{\partial \delta v}{\partial x} + \frac{\partial \delta u}{\partial y}. \tag{1.33}$$

This is the principle of virtual work for the problem defined in Section 1.1. The principle holds for arbitrary infinitesimal virtual displacements satisfying the prescribed geometrical boundary conditions.*

Next, we shall consider what kind of relations will be obtained if the principle of virtual work is required to hold for any admissible virtual displacements. Reversing the above development, we may obtain Eq. (1.28) from Eq. (1.32). Since δu, δv and δw are chosen arbitrarily in V and on S_1, all the coefficients in Eq. (1.28) are required to vanish. Thus, we have another statement of the principle of virtual work: *Introduction of the strain–displacement relations (1.5) and the geometrical boundary conditions (1.14) into the principle of virtual work yields the equations of equilibrium (1.4) and mechanical boundary conditions (1.12).* Consequently, once the strain–displacement relations have been derived, the equations of equilibrium may be obtained from the principle of virtual work. It is worthy of special mention that the principle of virtual work holds irrespective of the material stress–strain relations.

1.5. Approximate Method of Solution Based on the Principle of Virtual Work

An approximate method of solution can be formulated by employing the principle of virtual work.[5] This approach will be called the generalized Galerkin method.† The first step of the method is to assume that the displacement components u, v and w can be expressed approximately as follows:

$$\begin{aligned} u(x, y, z) &= u_0(x, y, z) + \sum_{r=1}^{n} a_r u_r(x, y, z), \\ v(x, y, z) &= v_0(x, y, z) + \sum_{r=1}^{n} b_r v_r(x, y, z), \\ w(x, y, z) &= w_0(x, y, z) + \sum_{r=1}^{n} c_r w_r(x, y, z), \end{aligned} \tag{1.34‡}$$

where u_0, v_0 and w_0 are so chosen that
$$u_0 = \bar{u}, \quad v_0 = \bar{v}, \quad w_0 = \bar{w} \quad \text{on} \quad S_2, \tag{1.35}$$

* For a physical interpretation of the principle, see Appendix J.

† This is a generalization of the so-called Galerkin method which requires that approximate displacements of Eqs. (1.34) are chosen to satisfy not only the geometrical boundary conditions on S_2, but also by substitution of the stress–strain relations the mechanical boundary conditions on S_1. For Galerkin's method, see Refs. 5, 7 through 11, for instance.

‡ It is noted that the number of the terms under the three summation signs need not be equal to each other. In other words, some terms among u_r, v_r and w_r may be missing.

and u_r, v_r, w_r; $\gamma = 1, 2, \ldots, n$ are linearly independent functions which satisfy the conditions

$$u_r = 0, \quad v_r = 0, \quad w_r = 0, \ (r = 1, 2, \ldots, n) \quad \text{on } S_2. \tag{1.36}$$

The constants a_r, b_r and c_r are arbitrary. We then have:

$$\delta u = \sum_{r=1}^{n} \delta a_r u_r \quad \delta v = \sum_{r=1}^{n} \delta b_r v_r, \quad \delta w = \sum_{r=1}^{n} \delta c_r w_r. \tag{1.37}$$

Introducing Eqs. (1.34) into the principle (1.32), we have

$$\sum_{r=1}^{n} [L_r \, \delta a_r + M_r \, \delta b_r + N_r \, \delta c_r) = 0, \tag{1.38}$$

where

$$L_r = \iiint_V \left(\sigma_x \frac{\partial u_r}{\partial x} + \tau_{xy} \frac{\partial u_r}{\partial y} + \tau_{zx} \frac{\partial u_r}{\partial z} - \bar{X} u_r \right) dV - \iint_{S_1} \bar{X}_\nu u_r \, dS,$$

$$M_r = \iiint_V \left(\tau_{xy} \frac{\partial v_r}{\partial x} + \sigma_y \frac{\partial v_r}{\partial y} + \tau_{yz} \frac{\partial v_r}{\partial z} - \bar{Y} v_r \right) dV - \iint_{S_1} \bar{Y}_\nu v_r \, dS,$$

$$N_r = \iiint_V \left(\tau_{zx} \frac{\partial w_r}{\partial x} + \tau_{yz} \frac{\partial w_r}{\partial y} + \sigma_z \frac{\partial w_r}{\partial z} - \bar{Z} w_r \right) dV - \iint_{S_1} \bar{Z}_\nu w_r \, dS. \tag{1.39}$$

Since δa_r, δb_r and δc_r are arbitrary, we obtain the following equations:

$$L_r = 0, \quad M_r = 0, \quad N_r = 0, \quad (r = 1, 2, \cdots, n). \tag{1.40}$$

We note that the expressions (1.39) are transformed via integration by parts into,

$$L_r = -\iiint_V \left(\frac{\partial \sigma_x}{\partial x} + \frac{\partial \tau_{xy}}{\partial y} + \frac{\partial \tau_{zx}}{\partial z} + \bar{X} \right) u_r \, dV + \iint_{S_1} (X_\nu - \bar{X}_\nu) \, u_r \, dS,$$

$$M_r = -\iiint_V \left(\frac{\partial \tau_{xy}}{\partial x} + \frac{\partial \sigma_y}{\partial y} + \frac{\partial \tau_{yz}}{\partial z} + \bar{Y} \right) v_r \, dV + \iint_{S_1} (Y_\nu - \bar{Y}_\nu) \, v_r \, dS,$$

$$N_r = -\iiint_V \left(\frac{\partial \tau_{zx}}{\partial x} + \frac{\partial \tau_{yz}}{\partial y} + \frac{\partial \sigma_z}{\partial z} + \bar{Z} \right) w_r \, dV + \iint_{S_1} (Z_\nu - \bar{Z}_\nu) \, w_r \, dS. \tag{1.41}$$

The second step is to calculate the stress components in terms of Eqs. (1.34) by the use of Eqs. (1.5) and the stress–strain relations. Here we assume isotropy of the material to obtain the following stress–displacement relations:

$$\sigma_x = 2G \left\{ \frac{\partial u_0}{\partial x} + \sum_{r=1}^{n} a_r \frac{\partial u_r}{\partial x} + \frac{\nu}{1 - 2\nu} \left[\frac{\partial u_0}{\partial x} + \frac{\partial v_0}{\partial y} + \frac{\partial w_0}{\partial z} + \sum_{r=1}^{n} \left(a_r \frac{\partial u_r}{\partial x} + b_r \frac{\partial v_r}{\partial y} + c_r \frac{\partial w_r}{\partial z} \right) \right] \right\},$$

$$\sigma_y = \cdots, \quad \cdots. \tag{1.42}$$

Introducing Eq. (1.42) into Eq. (1.40), we have a set of $3n$ simultaneous linear equations with respect to the $3n$ unknowns a_r, b_r and c_r; $r = 1, 2, \ldots, n$. By solving these equations, values of a_r, b_r and c_r are determined. By substituting the constants thus determined into the expressions (1.34), an approximate solution for the displacement components is obtained.

By a proper choice of the functions $u_0, v_0, w_0, u_r, v_r, w_r$; $r = 1, 2, \ldots, n$, and the number n, it is possible to obtain good approximate solutions for the deformation of the body. However, the accuracy of the stresses calculated by the use of Eqs. (1.42), employing the values of a_r, b_r and c_r thus determined, is in general not as good. This is obvious if we remember that we have replaced the equilibrium conditions (1.4) and the mechanical boundary conditions (1.12) by the $3n$ weighted expressions shown in Eqs. (1.41), and that the accuracy of an approximate solution decreases with differentiation. The equations of equilibrium as well as the mechanical boundary conditions are generally violated, at least locally, in the approximate solution.

The accuracy of the approximate solution may be improved by increasing the number of terms n. If Eqs. (1.34) represent the set of all admissible functions when n tends to infinity, we may hope that the approximate solution will approach close to the exact solution for a sufficiently large n, and tend to it when the number of terms increases without limit. However, experience and intuition are required if one wishes to obtain an accurate approximation while retaining only a small number of terms in Eq. (1.34).

Modifications of the above method are frequently employed. For example, we might choose

$$\begin{aligned} u(x, y, z) &= \sum_{m=0}^{n} u_m(x, y)\, g_m(z), \\ v(x, y, z) &= \sum_{m=0}^{n} v_m(x, y)\, g_m(z), \\ w(x, y, z) &= \sum_{m=0}^{n} w_m(x, y)\, g_m(z), \end{aligned} \tag{1.43}$$

where $g_m(z)$; $m = 0, 1, 2, \ldots, n$ are prescribed functions of z, while u_m, v_m and w_m are undetermined. Equations governing u_m, v_m and w_m are derived from the principle of virtual work. We shall cite frequent examples of this method in Chapters 7, 8 and 9.

1.6. Principle of Complementary Virtual Work

Within the realm of small displacement theory we can formulate another principle which is complementary to the principle of virtual work in defining the problem presented in Section 1.1. We consider the body in equilibrium under the prescribed body forces and boundary conditions, and denote the strain and displacement components by $\varepsilon_x, \ldots, \gamma_{xy}$ and u, v, w, respectively. Obviously,

$$\varepsilon_x - \frac{\partial u}{\partial x} = 0, \dots, \qquad \gamma_{xy} - \frac{\partial u}{\partial y} - \frac{\partial v}{\partial x} = 0 \quad \text{in} \quad V, \tag{1.44}$$

$$u - \bar{u} = 0, \dots, \quad w - \bar{w} = 0 \quad \text{on} \quad S_2. \tag{1.45}$$

Now, the body is assumed to take an arbitrary set of infinitesimal virtual variations of the stress components $(\delta\sigma_x, \delta\sigma_y, \dots, \delta\tau_{xy})$ from this equilibrating configuration. Then we have

$$\iiint_V \left[\left(\varepsilon_x - \frac{\partial u}{\partial x}\right)\delta\sigma_x + \left(\varepsilon_y - \frac{\partial v}{\partial y}\right)\delta\sigma_y + \cdots + \left(\gamma_{xy} - \frac{\partial u}{\partial y} - \frac{\partial v}{\partial x}\right)\delta\tau_{xy}\right] dV$$

$$+ \iint_{S_2} [(u - \bar{u})\,\delta X_\nu + (v - \bar{v})\,\delta Y_\nu + (w - \bar{w})\,\delta Z_\nu]\, dS = 0, \tag{1.46}$$

which, via integrations by parts, is tranformed into:

$$\iiint_V \left[\varepsilon_x\,\delta\sigma_x + \varepsilon_y\,\delta\sigma_y + \cdots + \gamma_{xy}\,\delta\tau_{xy} + \left(\frac{\partial\delta\sigma_x}{\partial x} + \frac{\partial\delta\tau_{xy}}{\partial y} + \frac{\partial\delta\tau_{zx}}{\partial z}\right)u\right.$$

$$\left. + (\cdots)\,v + (\cdots)\,w\right] dV - \iint_{S_1} (u\delta X_\nu + v\delta Y_\nu + w\delta Z_\nu)\, dS$$

$$- \iint_{S_2} (\bar{u}\,\delta X_\nu + \bar{v}\,\delta Y_\nu + \bar{w}\delta Z_\nu) dS = 0. \tag{1.47}$$

Here, we shall choose the arbitrary set of virtual stresses such that the equations of equilibrium and the mechanical boundary conditions are not violated. Namely, they are so chosen as to satisfy the following equations:

$$\begin{aligned} \frac{\partial\delta\sigma_x}{\partial x} + \frac{\partial\delta\tau_{xy}}{\partial y} + \frac{\partial\delta\tau_{zx}}{\partial z} &= 0, \\ \frac{\partial\delta\tau_{xy}}{\partial x} + \frac{\partial\delta\sigma_y}{\partial y} + \frac{\partial\delta\tau_{yz}}{\partial z} &= 0, \\ \frac{\partial\delta\tau_{zx}}{\partial x} + \frac{\partial\delta\tau_{yz}}{\partial y} + \frac{\partial\delta\sigma_z}{\partial z} &= 0, \end{aligned} \tag{1.48}$$

in the interior of the body V and

$$\begin{aligned} \delta X_\nu &= \delta\sigma_x l + \delta\tau_{xy} m + \delta\tau_{zx} n = 0, \\ \delta Y_\nu &= \delta\tau_{xy} l + \delta\sigma_y m + \delta\tau_{yz} n = 0, \\ \delta Z_\nu &= \delta\tau_{zx} l + \delta\tau_{yz} m + \delta\sigma_z n = 0, \end{aligned} \tag{1.49}$$

on S_1. Then, Eq. (1.47) reduces to

$$\iiint_V (\varepsilon_x\,\delta\sigma_x + \varepsilon_y\,\delta\sigma_y + \cdots + \gamma_{xy}\,\delta\tau_{xy})\, dV$$

$$- \iint_{S_2} (\bar{u}\,\delta X_\nu + \bar{v}\,\delta Y_\nu + \bar{w}\,\delta Z_\nu)\, dS = 0. \tag{1.50}$$

The formula (1.50) will be called the principle of complementary virtual work. The principle holds for arbitrary infinitesimal virtual stress variations satisfying the equations of equilibrium and prescribed mechanical boundary conditions. It is seen that the principle of complementary virtual work has a form which is complementary to the principle of virtual work given by Eq. (1.32).

Next, we shall consider what conditions result if the principle of complementary virtual work is required to hold for an arbitrary set of admissible virtual stress variations. For such a formulation the Lagrange multiplier method provides a systematic tool.† We shall treat Eqs. (1.48) and (1.49) as constraints and employ the displacements u, v and w as the Lagrange multipliers associated with these conditions. Thus, reversing the above development, we obtain Eq. (1.46) from Eq. (1.50). Since the quantities $\delta\sigma_x$, $\delta\sigma_y$, ..., $\delta\tau_{xy}$ have been made independent of each other by introduction of Lagrange multipliers, all the coefficients in Eq. (1.46) are required to vanish. This leads to another statement of the principle of complementary virtual work: *Introduction of the equations of equilibrium* (*1.4*) *and the mechanical boundary conditions* (*1.12*) *into the principle of complementary virtual work yields the strain–displacement relations* (*1.5*) *and the geometrical boundary conditions* (*1.14*). Consequently, once the equations of equilibrium have been derived in the small displacement theory, the strain–displacement relations may be obtained from the principle of complementary virtual work. It is worthy of special mention that the principle of complementary virtual work holds irrespective of the material stress–strain relations.

1.7. Approximate Method of Solution Based on the Principle of Complementary Virtual Work

An approximate method of solution can be formulated by employing the principle of complementary virtual work. This approach is similar to the one mentioned in Section 1.5 and may also be called the generalized Galerkin method. For the sake of simplicity, we shall consider a two-dimensional elasticity problem of a simply connected body.‡ The side boundary of the body is cylindrical with the generating line parallel to the

† For Lagrange multiplier method, see Chapter 4 of Ref. 12, and Chapters 2 and 5 of Ref. 13.

‡ The two-dimensional elasticity problem defined here is a good approximation to the so-called plane stress problem of a thin isotropic plate with traction-free upper and lower surfaces. In a plane stress problem we assume $\sigma_z = 0$ and obtain $E\varepsilon_z = -\nu(\sigma_x + \sigma_y)$.[2] On the other hand, this two-dimensional elasticity problem can be shown to be mathematically equivalent to a plane strain problem of an isotropic body, by replacing E and ν in Eqs. (1.51) with $E'[= E/(1 - \nu^2)]$ and $\nu'[= \nu/(1 - \nu)]$, respectively, and employing the assumptions $\varepsilon_z = 0$ and $\sigma_z = -\nu(\sigma_x + \sigma_y)$.[2]

z-axis, and the deformation of the body is assumed independent of z. The stress components σ_z, τ_{zx} and τ_{yz} are assumed to vanish. The remaining stress components σ_x, σ_y and τ_{xy} are assumed to be functions of (x, y) only, and related to the strain components as follows:

$$E\varepsilon_x = \sigma_x - \nu\sigma_y, \quad E\varepsilon_y = -\nu\sigma_x + \sigma_y, \quad G\gamma_{xy} = \tau_{xy}, \tag{1.51}$$

where

$$\varepsilon_x = \frac{\partial u}{\partial x}, \quad \varepsilon_y = \frac{\partial v}{\partial y}, \quad \gamma_{xy} = \frac{\partial u}{\partial y} + \frac{\partial v}{\partial x}. \tag{1.52}$$

Under assumption of absence of body forces, the equations of equilibrium then reduce to Eqs. (1.24), which suggests the use of the Airy stress function defined by Eqs. (1.25).

The boundary conditions on the side surface must be prescribed independently of z, and are assumed to be given, for the sake of simplicity, in terms of external forces only, namely

$$X_\nu = \bar{X}_\nu, \quad Y_\nu = \bar{Y}_\nu \tag{1.53}$$

on the side boundary C, where

$$X_\nu = \sigma_x l + \tau_{xy} m, \quad Y_\nu = \tau_{xy} l + \sigma_y m. \tag{1.54}$$

In the above l and m are the direction cosines of the outward normal ν to the boundary C. If the contour of the side boundary C is given parametrically in terms of the arc length s measured along C, such that

$$x = x(s), \quad y = y(s), \tag{1.55}$$

we have

$$l = dy/ds, \quad m = -dx/ds. \tag{1.56}$$

The arc length s is measured as shown in Fig. 1.1. By introducing the Airy stress function and Eqs. (1.56) into Eqs. (1.54), we obtain X_ν and Y_ν in

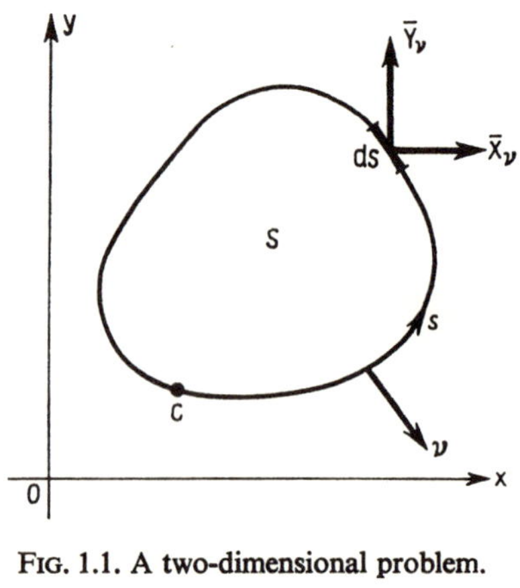

FIG. 1.1. A two-dimensional problem.

terms of F:

$$X_\nu = \frac{\partial^2 F}{\partial y^2}\frac{dy}{ds} + \frac{\partial^2 F}{\partial x\,\partial y}\frac{dx}{ds} = \frac{d}{ds}\left(\frac{\partial F}{\partial y}\right),$$
$$Y_\nu = -\frac{\partial^2 F}{\partial x\,\partial y}\frac{dy}{ds} - \frac{\partial^2 F}{\partial x^2}\frac{dx}{ds} = -\frac{d}{ds}\left(\frac{\partial F}{\partial x}\right). \tag{1.57}$$

We shall assume an expression for the stress function of the following form:

$$F(x, y) = F_0(x, y) + \sum_{r=1}^{n} a_r F_r(x, y), \tag{1.58}$$

where F_0 and F_r are chosen so that

$$\frac{d}{ds}\left(\frac{\partial F_0}{\partial y}\right) = \bar{X}_\nu, \qquad -\frac{d}{ds}\left(\frac{\partial F_0}{\partial x}\right) = \bar{Y}_\nu,$$
$$\frac{d}{ds}\left(\frac{\partial F_r}{\partial y}\right) = 0, \qquad -\frac{d}{ds}\left(\frac{\partial F_r}{\partial x}\right) = 0, \quad (r = 1, 2, \ldots, n) \tag{1.59}$$

on the boundary C, and a_r; $r = 1, 2, \ldots, n$ are arbitrary constants. The equations (1.59) suggest that both $\partial F_r/\partial x$ and $\partial F_r/\partial y$ are constant along C. Since the addition to the stress function F of a function $ax + by + c$, where a, b and c are arbitrary constants, is immaterial as far as the simply connected body is concerned, we may take

$$F_r = 0, \qquad \frac{\partial F_r}{\partial x} = 0, \qquad \frac{\partial F_r}{\partial y} = 0 \quad \text{on} \quad C, \ (r = 1, 2, \ldots, n) \tag{1.60}$$

without loss of generality.

Introduction of Eq. (1.58) into Eqs. (1.25) results in the following expressions for the stress components:

$$\sigma_x = \frac{\partial^2 F}{\partial y^2} = \frac{\partial^2 F_0}{\partial y^2} + \sum_{r=1}^{n} a_r \frac{\partial^2 F_r}{\partial y^2},$$
$$\sigma_y = \frac{\partial^2 F}{\partial x^2} = \frac{\partial^2 F_0}{\partial x^2} + \sum_{r=1}^{n} a_r \frac{\partial^2 F_r}{\partial x^2}, \tag{1.61}$$
$$\tau_{xy} = -\frac{\partial^2 F}{\partial x\,\partial y} = -\frac{\partial^2 F_0}{\partial x\,\partial y} - \sum_{r=1}^{n} a_r \frac{\partial^2 F_r}{\partial x\,\partial y}.$$

A set of admissible virtual stress variation is then given by

$$\delta\sigma_x = \sum_{r=1}^{n} \frac{\partial^2 F_r}{\partial y^2}\delta a_r, \quad \delta\sigma_y = \sum_{r=1}^{n} \frac{\partial^2 F_r}{\partial x^2}\delta a_r,$$
$$\delta\tau_{xy} = -\sum_{r=1}^{n} \frac{\partial^2 F_r}{\partial x\,\partial y}\delta a_r. \tag{1.62}$$

Substituting Eq. (1.62) into the principle (1.50), and remembering that all the surface boundary conditions are given in terms of forces only, we have

$$\sum_{r=1}^{n} L_r\,\delta a_r = 0, \tag{1.63}$$

where
$$L_r = \iint_S \left(\varepsilon_x \frac{\partial^2 F_r}{\partial y^2} + \varepsilon_y \frac{\partial^2 F_r}{\partial x^2} - \gamma_{xy} \frac{\partial^2 F_r}{\partial x \, \partial y} \right) dx \, dy. \tag{1.64}$$

In Eq. (1.64), the length of the body in the direction of the z-axis is taken as unity, and the integrations extend through the region of the body in the (x, y) plane. Since the variations of the constants, δa_r, are arbitrary, we obtain the following equations:
$$L_r = 0, \quad (r = 1, 2, \ldots, n). \tag{1.65}$$
We note that by the use of Eqs. (1.60) and via integrations by parts, the expression (1.64) is transformed into
$$L_r = \iint_S \left(\frac{\partial^2 \varepsilon_x}{\partial y^2} + \frac{\partial^2 \varepsilon_y}{\partial x^2} - \frac{\partial^2 \gamma_{xy}}{\partial x \, \partial y} \right) F_r \, dx \, dy. \tag{1.66}$$

By the use of Eqs. (1.51) and (1.61), Eqs. (1.65) can be reduced to n simultaneous equations with respect to a_r; $r = 1, 2, \ldots, n$. By solving these equations, values of a_r are determined. Substituting the value of a_r thus determined into Eqs. (1.61), we obtain an approximate solution for the stresses. By judicious choice of $F_0, F_1, \ldots, F_n$, approximate solutions of considerable accuracy may be obtained. The factors which govern the accuracy of the approximate solution are similar to those mentioned at the end of Section 1.5.

It is noted here that the strains calculated from the approximate stress solution and the stress–strain relations do not satisfy, in general, the conditions of compatibility, unless the number n is increased without limit. For example, as the expression (1.66) shows, Eqs. (1.65) are weighted means, and consequently, approximations to the condition of compatibility for the two-dimensional problem. Although we have taken a two-dimensional problem as an example, the extension to three dimensions is straightforward.

1.8. Relations between Conditions of Compatibility and Stress Functions†

We have observed in Section 1.4 that the equations of equilibrium can be obtained from the principle of virtual work (1.32). In view of the development in Section 1.4, we might ask what kind of relations will be obtained if the conditions of compatibility (1.15), instead of u, v and w, are introduced into the principle (1.32) by the use of Lagrange multipliers. The body forces will be assumed absent throughout the present discussion.

We shall employ Eqs. (1.18a) as the field conditions of compatibility and write the principle of virtual work (1.32) as follows:
$$\begin{aligned} \iiint_V [\sigma_x \, \delta\varepsilon_x + \sigma_y \, \delta\varepsilon_y + \cdots + \tau_{xy} \, \delta\gamma_{xy} \\ - \chi_1 \, \delta R_x - \chi_2 \, \delta R_y - \chi_3 \, \delta R_z] \, dV \\ + (\text{surface terms}) = 0, \end{aligned} \tag{1.67}$$

† Refs. 14 through 18.

where χ_1, χ_2 and χ_3 are the Lagrange multipliers. After some calculation, including partial integrations, Eq. (1.67) is transformed into:

$$\iiint_V \left\{\left[\sigma_x - \left(\frac{\partial^2 \chi_3}{\partial y^2} + \frac{\partial^2 \chi_2}{\partial z^2}\right)\right] \delta\varepsilon_x + \left[\sigma_y - \left(\frac{\partial^2 \chi_1}{\partial z^2} + \frac{\partial^2 \chi_3}{\partial x^2}\right)\right] \delta\varepsilon_y + \cdots + \left[\tau_{xy} + \frac{\partial^2 \chi_3}{\partial x \partial y}\right] \delta\gamma_{xy}\right\} dV$$

$$+ \text{surface terms} = 0. \tag{1.68}$$

Therefore, since the quantities $\delta\varepsilon_x$, $\delta\varepsilon_y \cdots$, and $\delta\gamma_{xy}$ are arbitrary, we have

$$\sigma_x = \frac{\partial^2 \chi_3}{\partial y^2} + \frac{\partial^2 \chi_2}{\partial z^2}, \quad \cdots, \quad \tau_{xy} = -\frac{\partial^2 \chi_3}{\partial x \partial y}, \tag{1.69}$$

thus proving that the Lagrange multipliers χ_1, χ_2 and χ_3 are Maxwell's stress functions. A similar procedure employing Eqs. (1.19a) as the field conditions of compatibility leads to Morera's stress functions. The present method of finding stress functions is applicable to any problem where the principle of virtual work and conditions of compatibility have been formulated.

On the other hand, we have observed in Section 1.6 that the strain–displacement relations may be obtained from the principle of complementary virtual work if the equations of equilibrium have been derived. Now, we shall inquire what conditions result if stress functions are used in place of the equations of equilibrium and Lagrange multipliers in conjunction with the principle of complementary virtual work.

We shall employ as an example Maxwell's stress functions defined by Eqs. (1.21). The principle (1.50) can now be written as follows:

$$\iiint_V \left[\varepsilon_x \left(\frac{\partial^2\, \delta\chi_3}{\partial y^2} + \frac{\partial^2\, \delta\chi_2}{\partial z^2}\right) + \cdots - \gamma_{xy} \frac{\partial^2\, \delta\chi_3}{\partial x\, \partial y}\right] dx\, dy\, dz$$

$$+ (\text{surface terms}) = 0. \tag{1.70}$$

After some calculation, including partial integration, Eq. (1.70) is transformed into

$$\iiint_V \left[\left(\frac{\partial^2 \varepsilon_z}{\partial y^2} + \frac{\partial^2 \varepsilon_y}{\partial z^2} - \frac{\partial^2 \gamma_{yz}}{\partial y\, \partial z}\right)\delta\chi_1 + \left(\frac{\partial^2 \varepsilon_x}{\partial z^2} + \frac{\partial^2 \varepsilon_z}{\partial x^2} - \frac{\partial^2 \gamma_{zx}}{\partial z\, \partial x}\right)\delta\chi_2\right.$$

$$\left. + \left(\frac{\partial^2 \varepsilon_y}{\partial x^2} + \frac{\partial^2 \varepsilon_x}{\partial y^2} - \frac{\partial^2 \gamma_{xy}}{\partial x\, \partial y}\right)\delta\chi_3\right] dx\, dy\, dz + (\text{surface terms}) = 0. \tag{1.71}$$

Since $\delta\chi_1$, $\delta\chi_2$ and $\delta\chi_3$ are arbitrary, we have

$$R_x = R_y = R_z = 0, \tag{1.72}$$

and conclude that Eq. (1.71) provides Eqs. (1.18a) as the field conditions of compatibility. A similar procedure employing Morera's stress functions leads to the conditions of compatibility given by Eqs. (1.19a).

The reader has already seen in Section 1.7 that the employment of Airy stress function in the principle of complementary virtual work leads to the condition of compatibility for the two-dimensional problem.

It is noted here that for a multiply connected body, such as a body with several holes, formulation via the principle of complementary virtual work combined with stress functions provides other geometrical conditions, the so-called *conditions of compatibility in the large.*[19, 20] A simple example of these conditions will be illustrated in Section 6.3. In Chapter 10 we shall show that the conditions of compatibility in the large play an essential part in the theory of structures.

1.9. Some Remarks

We have observed in Sections 1.4 and 1.6 that the principles of virtual work and complementary virtual work are complementary to each other in defining the elasticity problem. Here, we consider extensions of these principles.

It has been assumed in deriving the principle of virtual work that the virtual displacements are so chosen as to satisfy Eqs. (1.29). This restriction may be removed to obtain an extension of the principle of virtual work as follows:

$$\iiint_V (\sigma_x \,\delta\varepsilon_x + \sigma_y \,\delta\varepsilon_y + \cdots + \tau_{xy}\,\delta\gamma_{xy})\, dV - \iiint_V (\bar{X}\,\delta u + \bar{Y}\,\delta v + \bar{Z}\,\delta w)\, dV - \iint_{S_1} (\bar{X}_\nu\,\delta u + \bar{Y}_\nu\,\delta v + \bar{Z}_\nu\,\delta w)\, dS - \iint_{S_2} (X_\nu\,\delta u + Y_\nu\,\delta v + Z_\nu\,\delta w)\, dS = 0. \tag{1.73}$$

On the other hand, we have assumed in deriving the principle of complementary virtual work that the virtual variation of the stress components are so chosen as to satisfy Eqs. (1.48) and (1.49). These restrictions may be removed to obtain an extension of the principle of complementary virtual work as follows:

$$\iiint_V (\varepsilon_x \,\delta\sigma_x + \varepsilon_y \,\delta\sigma_y + \cdots + \gamma_{xy}\,\delta\tau_{xy})\, dV - \iiint_V (u\,\delta X + v\,\delta Y + w\,\delta Z)\, dV - \iint_{S_1} (u\,\delta X_\nu + v\,\delta Y_\nu + w\,\delta Z_\nu)\, dS - \iint_{S_2} (\bar{u}\,\delta X_\nu + \bar{v}\,\delta Y_\nu + \bar{w}\,\delta Z_\nu)\, dS = 0, \tag{1.74}$$

where δX, δY and δZ are given by

$$\frac{\partial\delta\sigma_x}{\partial x}+\frac{\partial\delta\tau_{xy}}{\partial y}+\frac{\partial\delta\tau_{zx}}{\partial z}+\delta X=0,$$
$$\frac{\partial\delta\tau_{xy}}{\partial x}+\frac{\partial\delta\sigma_y}{\partial y}+\frac{\partial\delta\tau_{yz}}{\partial z}+\delta Y=0, \tag{1.75}$$
$$\frac{\partial\delta\tau_{zx}}{\partial x}+\frac{\partial\delta\tau_{yz}}{\partial y}+\frac{\partial\delta\sigma_z}{\partial z}+\delta Z=0.$$

In view of the above developments, we find that these principles are special cases of the following divergence theorem:

$$\begin{aligned}\iiint_V (\sigma_x\varepsilon_x + \sigma_y\varepsilon_y + \cdots + \tau_{xy}\gamma_{xy})\,dV \\ = \iiint_V (\bar{X}u + \bar{Y}v + \bar{Z}w)\,dV \\ + \iint_{S_1} (X_\nu u + Y_\nu v + Z_\nu w)\,dS \\ + \iint_{S_2} (X_\nu u + Y_\nu v + Z_\nu w)\,dS,\end{aligned} \tag{1.76}$$

where $(\sigma_x, \sigma_y, \ldots, \tau_{xy})$ are an arbitrary set of stress components which satisfy the equations of equilibrium (1.4), and (X_ν, Y_ν, Z_ν) are derived from the stress components by the use of Eqs. (1.13), while (u, v, w) are an arbitrary set of displacement components, and $(\varepsilon_x, \varepsilon_y, \ldots, \gamma_{xy})$ are derived from these displacement components by the use Eqs. (1.5). The proof of the theorem (1.76) is given in a manner similar to those mentioned in Sections 1.4 and 1.6. It should be noted here that the sets $(\sigma_x, \sigma_y, \ldots, \tau_{xy})$ and $(\varepsilon_x, \varepsilon_y, \ldots, \gamma_{xy};\ u, v, w)$ are independent of each other. Namely, no relations are assumed to exist between these two sets. The divergence theorem has a wide field of application in continuum mechanics. We find that this theorem constitutes a basis for the unit displacement method and the unit load method† which play important roles in the analysis of structures.[11]

We note that continuity of stresses as well as displacements is assumed for the derivation of the divergence theorem. If some discontinuity exists in stresses and/or displacements, Eq. (1.76) should contain additional terms. For example, consider that the stress components $(\sigma_x, \ldots, \tau_{xy})$ are continuous, while the displacement components (u, v, w) are discontinuous across an interface $S_{(12)}$ which divides the body V into two parts $V_{(1)}$ and $V_{(2)}$.

Then, a term

$$\iint_{S_{(12)}} \{X_\nu[u] + Y_\nu[v] + Z_\nu[w]\}\,dS \tag{1.77}$$

† This method is also called the dummy load method.[10]

should be added to the righthand side of Eq. (1.76), where (X_ν, Y_ν, Z_ν) are defined on the surface $S_{(12)}$ with unit normal ν drawn from $V_{(1)}$ to $V_{(2)}$, and the square brackets denote the jumps of u, v and w across the surface: $[u] = u_{(1)} - u_{(2)}$, $[v] = v_{(1)} - v_{(2)}$, $[w] = w_{(1)} - w_{(2)}$. A similar care should be taken when the stress components show discontinuity.

Bibliography

1. A. E. H. Love, *A Treatise on the Mathematical Theory of Elasticity*, Cambridge University Press, 4th edition, 1927.
2. S. Timoshenko and J. N. Goodier, *Theory of Elasticity*, McGraw-Hill, 1951.
3. S. Moriguti, Fundamental Theory of Dislocation of Elastic Bodies (in Japanese), *Ōyo Sugaku Rikigaku*, Vol. 1, No. 2, pp. 87–90, 1947.
4. C. Pearson, *Theoretical Elasticity*, Harvard University Press, 1959.
5. V. V. Novozhilov, *Theory of Elasticity*, Translated by J. K. Lusher, Pergamon Press, 1961.
6. K. Washizu, A Note on the Conditions of Compatibility, *Journal of Mathematics and Physics*, Vol. 36, No. 4, pp. 306–12, January 1958.
7. W. J. Duncan, *Galerkin's Method in Mechanics and Differential Equations*, Aeronautical Research Committee, Report and Memoranda No. 1798, 1937.
8. C. Biezeno and R. Grammel, *Technische Dynamik*, Springer-Verlag, 1939.
9. L. Collatz, *Numerische Behandlung von Differentialgleichungen*, Springer-Verlag, 1951.
10. N. J. Hoff, *The Analysis of Structures*, John Wiley, 1956.
11. J. H. Argyris and S. Kelsey, *Energy Theorems and Structural Analysis*, Butterworth, 1960.
12. R. Courant and D. Hilbert, *Methods of Mathematical Physics*, Vol. 1, Interscience, New York, 1953.
13. C. Lanczos, *The Variational Principles of Mechanics*, University of Toronto Press, 1949.
14. R. V. Southwell, Castigliano's Principle of Minimum Strain Energy, *Proceedings of the Royal Society*, Vol. 154, No. 881, pp. 4–21, March 1936.
15. R. V. Southwell, Castigliano's Principle of Minimum Strain Energy and Conditions of Compatibility for Strains, *S. Timoshenko 60th Aniversary Volume*, pp. 211–17, 1938.
16. W. S. Dorn and A. Schild, A Converse to the Virtual Work Theorem for Deformable Solids, *Quarterly of Applied Mathematics*, Vol. 14, No. 2, pp. 209–13, July 1956.
17. C. Truesdell, General Solution for the Stresses in a Curved Membrane, *Proceedings of the National Academy of Science, Washington*, Vol. 43, No. 12, pp. 1070–2, December 1957.
18. C. Truesdell, Invariant and Complete Stress Functions for General Continua, *Archives for Rational Mechanics and Analysis*, Vol. 4, No. 1, pp. 1–29, November 1959.
19. S. Moriguti, On Castigliano's Theorem in Three-Dimensional Elastostatics (in Japanese), *Journal of the Society of Applied Mechanics of Japan*, Vol. 1, No. 6, pp. 175–80, 1948.
20. Y. C. Fung, *Foundations of Solid Mechanics*, Prentice-Hall Inc., 1965.
21. W. Prager and P. G. Hodge Jr., *Theory of Perfectly Plastic Solids*, John Wiley & Sons, 1951.

CHAPTER 2

VARIATIONAL PRINCIPLES IN THE SMALL DISPLACEMENT THEORY OF ELASTICITY

2.1. Principle of Minimum Potential Energy

We shall treat variational principles in the small displacement theory of elasticity in the present chapter. In this section the principle of minimum potential energy will be derived from the principle of virtual work established in Section 1.4.

First, it is observed that we can derive a state function $A(\varepsilon_x, \varepsilon_y, \ldots, \gamma_{xy})$ from the stress–strain relations (1.6), such that

$$\delta A = \sigma_x \,\delta\varepsilon_x + \sigma_y \,\delta\varepsilon_y + \cdots + \tau_{xy} \,\delta\gamma_{xy}, \tag{2.1}$$

where

$$\begin{aligned} 2A = {} & (a_{11}\varepsilon_x + a_{12}\varepsilon_y + \cdots + a_{16}\gamma_{xy})\,\varepsilon_x \\ & + \cdots \\ & + (a_{61}\varepsilon_x + a_{62}\varepsilon_y + \cdots + a_{66}\gamma_{xy})\,\gamma_{xy}. \end{aligned} \tag{2.2}$$

For the stress–strain relations of an isotropic material, namely Eqs. (1.10), we have

$$\begin{aligned} A = {} & \frac{Ev}{2(1+v)(1-2v)}(\varepsilon_x + \varepsilon_y + \varepsilon_z)^2 + G(\varepsilon_x^2 + \varepsilon_y^2 + \varepsilon_z^2) \\ & + \frac{G}{2}(\gamma_{yz}^2 + \gamma_{zx}^2 + \gamma_{xy}^2). \end{aligned} \tag{2.3}$$

We shall refer to A as the strain energy function.† From physical considerations which will be given in Chapter 3, we may assume the strain energy function to be a positive definite function of the strain components. This assumption involves some relations of inequality among the elastic constants.[1] For later convenience we introduce a notation $A(u, v, w)$ to indicate that the strain energy function is expressed in terms of the displacement components by introduction of the strain–displacement relations (1.5). For

† The quantity A is also called the strain energy per unit volume or the strain energy density.

example, we have

$$A(u, v, w) = \frac{E\nu}{2(1+\nu)(1-2\nu)}\left(\frac{\partial u}{\partial x}+\frac{\partial v}{\partial y}+\frac{\partial w}{\partial z}\right)^2 + G\left[\left(\frac{\partial u}{\partial x}\right)^2+\left(\frac{\partial v}{\partial y}\right)^2+\left(\frac{\partial w}{\partial z}\right)^2\right] + \frac{G}{2}\left[\left(\frac{\partial v}{\partial z}+\frac{\partial w}{\partial y}\right)^2+\left(\frac{\partial w}{\partial x}+\frac{\partial u}{\partial z}\right)^2+\left(\frac{\partial u}{\partial y}+\frac{\partial v}{\partial x}\right)^2\right], \quad (2.4)$$

for an isotropic material.

When the existence of the strain energy function is thus assured, the principle of virtual work (1.32) can be transformed into:

$$\delta \iiint_V A(u, v, w)\, dV - \iiint_V (\bar{X}\,\delta u + \bar{Y}\,\delta v + \bar{Z}\,\delta w)\, dV - \iint_{S_1} (\bar{X}_\nu\,\delta u + \bar{Y}_\nu\,\delta v + \bar{Z}_\nu\,\delta w)\, dS = 0. \quad (2.5)$$

This expression is useful in application to elasticity problems in which external forces are not derivable from potential functions.

Next, we shall assume that the body forces and surface forces are derivable from potential function $\Phi(u, v, w)$ and $\Psi(u, v, w)$ such that

$$-\delta\Phi = \bar{X}\,\delta u + \bar{Y}\,\delta v + \bar{Z}\,\delta w, \quad (2.6)$$

$$-\delta\Psi = \bar{X}_\nu\,\delta u + \bar{Y}_\nu\,\delta v + \bar{Z}_\nu\,\delta w. \quad (2.7)$$

Then, the principle (2.5) can be transformed into

$$\delta\Pi = 0, \quad (2.8)$$

where

$$\Pi = \iiint_V [A(u, v, w) + \Phi(u, v, w)]\, dV + \iint_{S_1} \Psi(u, v, w)\, dS, \quad (2.9)$$

is the total potential energy. The principle (2.8) states that *among all the admissible displacements u, v and w which satisfy the prescribed geometrical boundary conditions, the actual displacements make the total potential energy stationary.*

Hereafter, we shall confine our elasticity problem by assuming that the body forces $(\bar{X}, \bar{Y}, \bar{Z})$, the surface forces $(\bar{X}_\nu, \bar{Y}_\nu, \bar{Z}_\nu)$ and the surface displacements $(\bar{u}, \bar{v}, \bar{w})$ are prescribed, and kept unchanged in magnitudes and directions during variation. Then, potential energy functions are derived for these forces as follows:

$$-\Phi = \bar{X}u + \bar{Y}v + \bar{Z}w, \quad (2.10)$$

$$-\Psi = \bar{X}_\nu u + \bar{Y}_\nu v + \bar{Z}_\nu w, \quad (2.11)$$

and we have a variational principle called the principle of minimum potential energy: *Among all the admissible displacement functions, the actual displacements make the total potential energy*

$$\Pi = \iiint_V A(u, v, w)\, dV - \iiint_V (\bar{X}u + \bar{Y}v + \bar{Z}w)\, dV$$
$$- \iint_{S_1} (\bar{X}_\nu u + \bar{Y}_\nu v + \bar{Z}_\nu w)\, dS, \tag{2.12}$$

an absolute minimum.

For the proof of the principle of minimum potential energy, let the displacement components of the actual solution and a set of admissible, arbitrarily chosen displacement components be denoted by u, v, w and u^*, v^*, w^*, respectively, and put $u^* = u + \delta u$, $v^* = v + \delta v$ and $w^* = w + \delta w$. We then have

$$\Pi(u^*, v^*, w^*) = \Pi(u, v, w) + \delta\Pi + \delta^2\Pi, \tag{2.13}$$

where $\delta\Pi$ and $\delta^2\Pi$ are the first and second variations of the total potential energy. The first and second variations are respectively linear and quadratic in δu, δv, δw and their derivatives, namely,

$$\delta\Pi = \iiint_V \left[\sigma_x\left(\frac{\partial \delta u}{\partial x}\right) + \cdots + \tau_{xy}\left(\frac{\partial \delta u}{\partial y} + \frac{\partial \delta v}{\partial x}\right) - (\bar{X}\,\delta u + \cdots + \bar{Z}\,\delta w)\right] dV$$
$$- \iint_{S_1} (\bar{X}_\nu\,\delta u + \cdots + \bar{Z}_\nu\,\delta w)\, dS, \tag{2.14}$$

$$\delta^2\Pi = \iiint_V A(\delta u, \delta v, \delta w)\, dV, \tag{2.15}$$

where $\sigma_x, \ldots,$ and τ_{xy} are the stress components of the actual solution. Since $\delta u = \delta v = \delta w = 0$ on S_2, and the stress components belong to the actual solution, we find that the first variation, Eq. (2.14), vanishes:

$$\delta\Pi = 0. \tag{2.16}$$

Furthermore, since A is a positive definite function, we must have

$$\delta^2\Pi \geqq 0, \tag{2.17}$$

where the equality sign holds only when all the strain components which are derived from δu, δv and δw are zero. Consequently, we obtain

$$\Pi_{\text{admissible}} \geqq \Pi_{\text{actual}}. \tag{2.18}$$

Since no restrictions have been made on the magnitudes of δu, δv and δw in the above proof, we conclude that the total potential energy is made an absolute minimum for the actual solution.

2.2. Principle of Minimum Complementary Energy

It will now be shown that another variational principle can be derived from the principle of complementary virtual work, Eq. (1.50). We observe that a state function $B(\sigma_x, \sigma_y, \ldots, \tau_{xy})$ may be derived from the stress–strain relations (1.8), such that

$$\delta B = \varepsilon_x \, \delta\sigma_x + \varepsilon_y \, \delta\sigma_y + \cdots + \gamma_{xy} \, \delta\tau_{xy}, \tag{2.19}$$

where

$$\begin{aligned} 2B = & (b_{11}\sigma_x + b_{12}\sigma_y + \cdots + b_{16}\tau_{xy})\, \sigma_x \\ & + \cdots \\ & + (b_{61}\sigma_x + b_{62}\sigma_y + \cdots + b_{66}\tau_{xy})\, \tau_{xy}. \end{aligned} \tag{2.20}$$

For the stress–strain relations of an isotropic material, namely Eqs. (1.11), we have

$$\begin{aligned} B = \frac{1}{2E} \{ & (\sigma_x + \sigma_y + \sigma_z)^2 \\ & + 2(1+\nu)(\tau_{yz}^2 + \tau_{zx}^2 + \tau_{xy}^2 - \sigma_y\sigma_z - \sigma_z\sigma_x - \sigma_x\sigma_y)\}. \end{aligned} \tag{2.21}$$

We shall refer to B as the complementary energy function.† It is obvious that, the strain energy function A defined by Eq. (2.2) is equal to the complementary energy function B defined by Eq. (2.20) and that, if the former is positive definite, so is the latter. When the existence of the complementary energy function is thus assured, the principle of complementary virtual work can be transformed into:

$$\delta \iiint_V B(\sigma_x, \sigma_y, \ldots, \tau_{xy})\, dV - \iint_{S_2} (\bar{u}\, \delta X_\nu + \bar{v}\, \delta Y_\nu + \bar{w}\, \delta Z_\nu)\, dS = 0. \tag{2.22}$$

Employing the assumption that the quantities $\bar{u}$, $\bar{v}$ and $\bar{w}$ are kept unchanged during variation, we can derive from Eq. (2.22) a variational principle called the principle of minimum complementary energy: *Among all the sets of admissible stresses σ_x, σ_y, ..., and τ_{xy}, which satisfy the equations of equilibrium and the prescribed mechanical boundary conditions on S_1, the set of actual stress components makes the total complementary energy Π_c defined by*

$$\Pi_c = \iiint_V B(\sigma_x, \sigma_y, \ldots, \tau_{xy})\, dV - \iint_{S_2} (\bar{u}X_\nu + \bar{v}Y_\nu + \bar{w}Z_\nu)\, dS, \tag{2.23}$$

an absolute minimum.

For the proof, we denote the stress components of the actual solution and a set of admissible, arbitrarily chosen stress components by $\sigma_x, \sigma_y, \ldots, \tau_{xy}$ and $\sigma_x^*, \sigma_y^*, \ldots, \tau_{xy}^*$, respectively and put $\sigma_x^* = \sigma_x + \delta\sigma_x$, $\sigma_y^* = \sigma_y + \delta\sigma_y, \ldots,$ $\tau_{xy}^* = \tau_{xy} + \delta\tau_{xy}$. Then, in a manner similar to the development in the preceding section, we find that the first variation of the total complementary energy vanishes for the actual solution and that, since B is a positive definite function, the second variation of the total complementary energy is non-

† The quantity B is also called the complementary energy per unit volume, complementary energy density or the stress energy per unit volume.

negative. Thus, we are assured of the validity of the principle of minimum complementary energy.†

We observe that the arguments of A are strain components, while those of B are stress components. For the linear stress–strain relations, Eqs. (1.6) and (1.8), B is equal to A and has the same physical meaning: the strain energy stored in a unit volume of the elastic body. It should be noted, however, that when stress–strain relations are nonlinear, B defined by Eq. (2.19) is different from A defined by Eq. (2.1). For example, in the simple case of a bar in tension, we have

$$\sigma_x = \sigma_x(\varepsilon_x), \quad \varepsilon_x = \varepsilon_x(\sigma_x). \tag{2.24}$$

The functions A and B are then given by

$$A = \int_0^{\varepsilon_x} \sigma_x \, d\varepsilon_x, \quad B = \int_0^{\sigma_x} \varepsilon_x \, d\sigma_x. \tag{2.25}$$

These are illustrated in Fig. 2.1 by the shaded area OPS and unshaded area OSR, respectively. It is seen that A and B are complementary to each other in respresenting the area $OPSR$, namely $A + B = \sigma_x \varepsilon_x$.

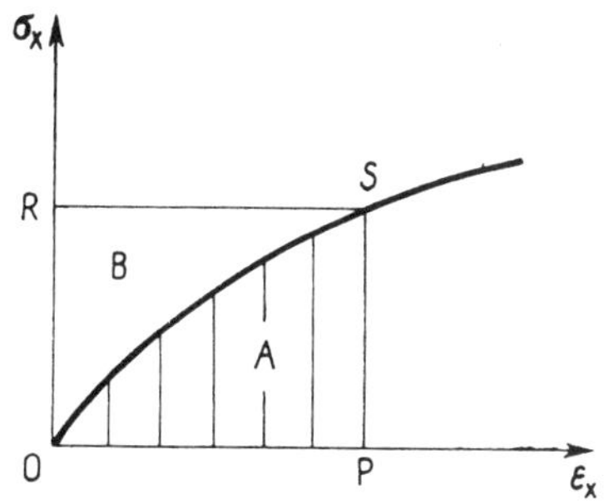

FIG. 2.1. Strain and complementary energies in a uniaxial tension.

2.3. Generalization of the Principle of Minimum Potential Energy

In the present section we shall consider a generalization of the principle of minimum potential energy. To begin with, we shall summarize the steps by which the principle of minimum potential energy has been obtained

† For an elasticity problem in which the part S_2 of the boundary is held rigidly fixed, namely $\bar{u} = \bar{v} = \bar{w} = 0$, the functional Π_c reduces to

$$\Pi_c = \iiint_V B(\sigma_x, \sigma_y, \ldots, \tau_{xy}) \, dV$$

to yield the principle of least work.[2]

from the principle of virtual work. We have assumed that: (1) it is possible to derive a positive definite state function $A(\varepsilon_x, \varepsilon_y, \ldots, \gamma_{xy})$ from the given stress–strain relations; (2) the above strain components satisfy the conditions of compatibility, that is, they can be derived from u, v and w as in the relationships of Eqs. (1.5); (3) the displacement components u, v and w thus defined satisfy the geometrical boundary conditions (1.14), and (4) the body forces and surface forces can be derived from potential functions Φ and Ψ as given by Eqs. (2.10) and (2.11). The principle of minimum potential energy then asserts that, on the basis of the above assumptions, the actual deformation can be obtained from the minimizing conditions of the functional Π defined by Eq. (2.12).

We shall now show that the subsidiary conditions stated in the assumptions (2) and (3) above can be put into the framework of the variational expression by introducing Lagrange multipliers,† and the principle of minimum potential energy can be generalized.[4,5] By the introduction of nine Lagrange multipliers $\sigma_x, \sigma_y, \ldots, \tau_{xy}$ and p_x, p_y, p_z defined in V and on S_2, respectively, the generalized principle can be expressed as follows: *The actual solution can be given by the stationary conditions of a functional Π_{I} defined as*††

$$
\begin{aligned}
\Pi_{\text{I}} = & \iiint_V [A(\varepsilon_x, \varepsilon_y, \ldots, \gamma_{xy}) - (\bar{X}u + \bar{Y}v + \bar{Z}w)]\, dV \\
& - \iiint_V \Big[\left(\varepsilon_x - \frac{\partial u}{\partial x}\right)\sigma_x + \left(\varepsilon_y - \frac{\partial v}{\partial y}\right)\sigma_y + \left(\varepsilon_z - \frac{\partial w}{\partial z}\right)\sigma_z \\
& + \left(\gamma_{yz} - \frac{\partial w}{\partial y} - \frac{\partial v}{\partial z}\right)\tau_{yz} + \left(\gamma_{zx} - \frac{\partial u}{\partial z} - \frac{\partial w}{\partial x}\right)\tau_{zx} \\
& + \left(\gamma_{xy} - \frac{\partial v}{\partial x} - \frac{\partial u}{\partial y}\right)\tau_{xy}\Big]\, dV - \iint_{S_1} (\bar{X}_\nu u + \bar{Y}_\nu v + \bar{Z}_\nu w)\, dS \\
& - \iint_{S_2} [(u - \bar{u})\, p_x + (v - \bar{v})\, p_y + (w - \bar{w})\, p_z]\, dS.
\end{aligned} \tag{2.26}
$$

The independent quantities subject to variation in the functional (2.26) are eighteen in number, namely, $\varepsilon_x, \varepsilon_y, \ldots, \gamma_{xy}$; u, v, w; $\sigma_x, \sigma_y, \ldots, \tau_{xy}$; p_x, p_y and p_z with no subsidiary conditions. On taking variations with respect

† See Chapter 4, § 9 of Ref. 3 for the Lagrange multiplier method and involutory transformations. See also Appendix A.

†† It should be noted that once Lagrange multipliers have been employed, the phrase "minimizing conditions" used in the principle of minimum potential energy must be replaced by "stationary conditions".

to these quantities, we have

$$\delta\Pi_{\mathrm{I}} = \iiint_V \Big[\Big(\frac{\partial A}{\partial \varepsilon_x} - \sigma_x\Big)\delta\varepsilon_x + \cdots + \Big(\frac{\partial A}{\partial \gamma_{xy}} - \tau_{xy}\Big)\delta\gamma_{xy}$$

$$- \Big(\varepsilon_x - \frac{\partial u}{\partial x}\Big)\delta\sigma_x - \cdots - \Big(\gamma_{xy} - \frac{\partial v}{\partial x} - \frac{\partial u}{\partial y}\Big)\delta\tau_{xy}$$

$$- \Big(\frac{\partial \sigma_x}{\partial x} + \frac{\partial \tau_{xy}}{\partial y} + \frac{\partial \tau_{zx}}{\partial z} + \bar{X}\Big)\delta u - \cdots - (\cdots)\,\delta w\Big]\,dV$$

$$+ \iint_{S_1} [(X_\nu - \bar{X}_\nu)\,\delta u + \cdots + (Z_\nu - \bar{Z}_\nu)\,\delta w]\,dS$$

$$- \iint_{S_2} [(u - \bar{u})\,\delta p_x + \cdots + (w - \bar{w})\,\delta p_z]\,dS$$

$$+ \iint_{S_2} [(X_\nu - p_x)\,\delta u + \cdots + (Z_\nu - p_z)\,\delta w]\,dS, \tag{2.27}$$

and the stationary conditions are shown to be

$$\sigma_x = a_{11}\varepsilon_x + a_{12}\varepsilon_y + \cdots + a_{16}\gamma_{xy}, \ldots \quad \text{in} \quad V, \tag{2.28}$$

$$\varepsilon_x = \frac{\partial u}{\partial x}, \ldots, \gamma_{xy} = \frac{\partial v}{\partial x} + \frac{\partial u}{\partial y} \quad \text{in} \quad V, \tag{2.29}$$

$$\frac{\partial \sigma_x}{\partial x} + \frac{\partial \tau_{xy}}{\partial y} + \frac{\partial \tau_{zx}}{\partial z} + \bar{X} = 0, \ldots \quad \text{in} \quad V, \tag{2.30}$$

$$X_\nu = \bar{X}_\nu, \ldots, \quad Z_\nu = \bar{Z}_\nu \quad \text{on} \quad S_1, \tag{2.31}$$

$$u = \bar{u}, \ldots, \quad w = \bar{w} \quad \text{on} \quad S_2, \tag{2.32}$$

$$p_x = X_\nu, \ldots, \quad p_z = Z_\nu \quad \text{on} \quad S_2. \tag{2.33}$$

It is seen that Eqs. (2.28) and (2.33) determine physical meanings of the Lagrange multipliers σ_x, σ_y, ..., τ_{xy}, p_x, p_y and p_z, and that the relationships for Π_{I} to be stationary are the equations which define the elasticity problem stated in Section 1.1. If Eqs. (2.29) and (2.32) are taken as subsidiary conditions, Π_{I} is reduced again to Π defined by Eq. (2.12).

We may obtain another expression of the variational principle in which the Lagrange multipliers p_x, p_y and p_z have been eliminated. For this purpose, we may require the coefficients of δu, δv and δw in the integral term on S_2 of the expression (2.27) to vanish. Thus, by the use of Eqs. (2.33) we may transform the functional (2.26) into

$$\Pi_{\mathrm{II}} = \iiint_V \Big[A(\varepsilon_x, \varepsilon_y, \ldots, \gamma_{xy}) - (\bar{X}u + \bar{Y}v + \bar{Z}w)$$

$$- \left(\varepsilon_x - \frac{\partial u}{\partial x}\right)\sigma_x - \cdots - \left(\gamma_{xy} - \frac{\partial v}{\partial x} - \frac{\partial u}{\partial y}\right)\tau_{xy}\Big]\, dV$$

$$- \iint_{S_1} (\bar{X}_\nu u + \bar{Y}_\nu v + \bar{Z}_\nu w)\, dS$$

$$- \iint_{S_2} [(u - \bar{u})\, X_\nu + (v - \bar{v})\, Y_\nu + (w - \bar{w})\, Z_\nu]\, dS, \tag{2.34}$$

or, through integrations by parts, into

$$\Pi_{\mathrm{III}} = - \iiint_V \Big[[\sigma_x\varepsilon_x + \sigma_y\varepsilon_y + \cdots + \tau_{xy}\gamma_{xy} - A(\varepsilon_x, \varepsilon_y, \ldots, \gamma_{xy})$$

$$+ \left(\frac{\partial \sigma_x}{\partial x} + \frac{\partial \tau_{xy}}{\partial y} + \frac{\partial \tau_{zx}}{\partial z} + \bar{X}\right) u + (\cdots)\, v + (\cdots)\, w\Big]\, dV$$

$$+ \iint_{S_1} [(X_\nu - \bar{X}_\nu)\, u + (Y_\nu - \bar{Y}_\nu)\, v + (Z_\nu - \bar{Z}_\nu)\, w]\, dS$$

$$+ \iint_{S_2} (X_\nu\bar{u} + Y_\nu\bar{v} + Z_\nu\bar{w})\, dS. \tag{2.35}$$

The independent quantities subject to variation in the functional (2.34) or (2.35) are 15 in number, namely, $\varepsilon_x, \varepsilon_y, \ldots, \gamma_{xy}$; u, v, w, $\sigma_x, \sigma_y, \ldots,$ and τ_{xy} with no subsidiary conditions. On taking variations of these 15 quantities, we find that the stationary conditions are given by Eqs. (2.28) through (2.32).

2.4. Derived Variational Principles

It will be shown in the present section that the Hellinger–Reissner principle and the principle of minimum complementary energy can be interpreted as special cases of the generalized principle (2.26). Let the coefficients of $\delta\varepsilon_x$, $\delta\varepsilon_y, \ldots, \delta\gamma_{xy}$ in the expression of $\delta\Pi_I$ be required to vanish. This means that $\varepsilon_x, \varepsilon_y, \ldots$ and γ_{xy} are no longer independent, but must instead be determined in a new formulation by the conditions (2.28), namely,

$$\begin{aligned} \varepsilon_x &= b_{11}\sigma_x + \cdots + b_{16}\tau_{xy}, \\ &\cdots\cdots\cdots\cdots\cdots\cdots \\ \gamma_{xy} &= b_{61}\sigma_x + \cdots + b_{66}\,\tau_{xy}. \end{aligned} \tag{2.36}$$

By the use of Eqs. (2.36), the strain components can be eliminated from the functional (2.26) to yield another functional of the principle, Π_R, as follows:

$$\Pi_R = \iiint_V \Big[\sigma_x \frac{\partial u}{\partial x} + \sigma_y \frac{\partial v}{\partial y} + \cdots + \tau_{xy} \left(\frac{\partial v}{\partial x} + \frac{\partial u}{\partial y} \right)$$
$$- B(\sigma_x, \sigma_y, \ldots, \tau_{xy}) - (\bar{X}u + \bar{Y}v + \bar{Z}w) \Big] dV$$
$$- \iint_{S_1} (\bar{X}_\nu u + \bar{Y}_\nu v + \bar{Z}_\nu w)\, dS$$
$$- \iint_{S_2} [(u - \bar{u})\, p_x + (v - \bar{v})\, p_y + (w - \bar{w})\, p_z]\, dS, \qquad (2.37)\dagger$$

where the quantity B is defined, as the above derivation shows, by

$$B = \sigma_x \varepsilon_x + \sigma_y \varepsilon_y + \cdots + \tau_{xy} \gamma_{xy} - A, \qquad (2.38)$$

in which the strain components are eliminated by the introduction of the stress–strain relationships (2.36). Since we have

$$\delta B = \sigma_x\, \delta\varepsilon_x + \cdots + \tau_{xy}\, \delta\gamma_{xy} + \varepsilon_x\, \delta\sigma_x + \cdots + \gamma_{xy}\, \delta\tau_{xy} - \delta A$$
$$= \varepsilon_x\, \delta\sigma_x + \cdots + \gamma_{xy}\, \delta\tau_{xy}, \qquad (2.39)$$

with the aid of Eq. (2.1), it is seen that the quantity B defined by Eq. (2.38) is the complementary energy function defined by (2.19). The functional (2.37) is equivalent to those in the Hellinger–Reissner principle.[6,7]

Because of the elimination of the strain components, the number of the independent quantities subject to variation in the functional Π_R is reduced to 12: $u, v, w; \sigma_x, \sigma_y, \ldots, \tau_{xy}; p_x, p_y, p_z$ with no subsidiary conditions. On taking variations of these quantities, we find that the stationary conditions are

$$\frac{\partial u}{\partial x} = b_{11}\sigma_x + \cdots + b_{16}\tau_{xy},$$
$$\cdots\cdots\cdots\cdots\cdots\cdots\cdots\cdots, \qquad (2.40)$$
$$\frac{\partial v}{\partial x} + \frac{\partial u}{\partial y} = b_{61}\sigma_x + \cdots + b_{66}\tau_{xy},$$

together with Eqs. (2.30) through (2.33).

The functional (2.37) may also be written via integrations by parts in the following form:

$$-\Pi_R^* = \iiint_V \Big[B(\sigma_x, \sigma_y, \ldots, \tau_{xy}) + \left(\frac{\partial \sigma_x}{\partial x} + \frac{\partial \tau_{xy}}{\partial y} + \frac{\partial \tau_{zx}}{\partial z} + \bar{X} \right) u$$
$$+ \left(\frac{\partial \tau_{xy}}{\partial x} + \frac{\partial \sigma_y}{\partial y} + \frac{\partial \tau_{yz}}{\partial z} + \bar{Y} \right) v + \left(\frac{\partial \tau_{zx}}{\partial x} + \frac{\partial \tau_{yz}}{\partial y} + \frac{\partial \sigma_z}{\partial z} + \bar{Z} \right) w \Big] dV$$
$$- \iint_{S_1} [(X_\nu - \bar{X}_\nu)\, u + (Y_\nu - \bar{Y}_\nu)\, v + (Z_\nu - \bar{Z}_\nu)\, w]\, dS$$
$$- \iint_{S_2} [X_\nu \bar{u} + Y_\nu \bar{v} + Z_\nu \bar{w}]\, dS, \qquad (2.41)$$

† This is a special case of the Legendre transformation in the calculus of vanation. The unique inverse relations of Eqs. (2.28) should exist for the transformation to be justified.

where Eqs. (2.33) have been used for the elimination of p_x, p_y and p_z. The quantities subjected to variation in the functional (2.41) are u, v, w; σ_x, σ_y, ... and τ_{xy} with no subsidiary conditions.

We shall now impose further restrictions on the number of independent functions in the generalized functional. All the coefficients of $\delta\varepsilon_x$, $\delta\varepsilon_y$, ..., $\delta\gamma_{xy}$; δu, δv and δw in the expression of $\delta\Pi_{\mathrm{I}}$ are required to vanish: thus the strains and displacements are eliminated by the use of Eqs. (2.28), (2.30), (2.31) and (2.33) to transform the functional Π_{I} into a functional $\tilde{\Pi}_c$ defined by

$$\tilde{\Pi}_c = -\iiint_V B(\sigma_x, \sigma_y, \ldots, \tau_{xy})\, dV + \iint_{S_2} (X_\nu \bar{u} + Y_\nu \bar{v} + Z_\nu \bar{w})\, dS, \tag{2.42}$$

where the quantities subject to variation are σ_x, σ_y, ... and τ_{xy} under the subsidiary conditions (2.30) and (2.31). Taking into account the positive definiteness of the function B, we may state this new principle as follows: *Of all the admissible functions* σ_x, σ_y, ... and τ_{xy} *which satisfy Eqs.* (*2.30*) *and* (*2.31*), *the stress components of the actual solution make the functional* $\tilde{\Pi}_c$ *an absolute maximum.* We observe that the principle (2.42) is equivalent to the principle of minimum complementary energy derived in Section 2.2. In reversing the above development, we find that the functions u, v, w in the functional (2.41) play the role of introducing the subsidiary conditions (2.30) and (2.31) into the variational expression.

We have seen that, in the expression for Π, admissible functions are chosen to satisfy the conditions of compatibility, Eqs. (1.5), and the geometrical boundary conditions on S_2, Eqs. (1.14), while in the expression of $\tilde{\Pi}_c$, admissible functions are chosen to satisfy the equations of equilibrium, Eqs. (1.4), and the mechanical boundary conditions on S_1, Eqs. (1.12). Consequently, Π and $\tilde{\Pi}_c$ are complementary to each other in defining the elasticity problem. The transformation of Π into $\tilde{\Pi}_c$ is known as Friedrichs' transformation;[3,8] the actual solution characterized by the minimum property of Π is also given by the maximum property of $\tilde{\Pi}_c$.

Thus far, it has been shown that once the principle of minimum potential energy has been established from the principle of virtual work, it can be generalized by the introduction of Lagrange multipliers to yield a family of variational principles which include the Hellinger–Reissner principle, the principle of minimum complementary energy and so forth. The avenue of this formulation is shown diagramatically in Table 2.1.

The principle of minimum complementary energy was derived in Section 2.2 from the principle of complementary virtual work. It is easily verified that the principle of minimum potential energy can be derived from the principle of minimum complementary energy by reversing the development in the present and preceding sections. The equivalence between these two approaches is quite obvious as far as the small displacement theory of elasticity is concerned. However, we shall emphasize the avenue of approach

TABLE 2.1. VARIATIONAL PRINCIPLES IN THE SMALL DISPLACEMENT THEORY OF ELASTICITY

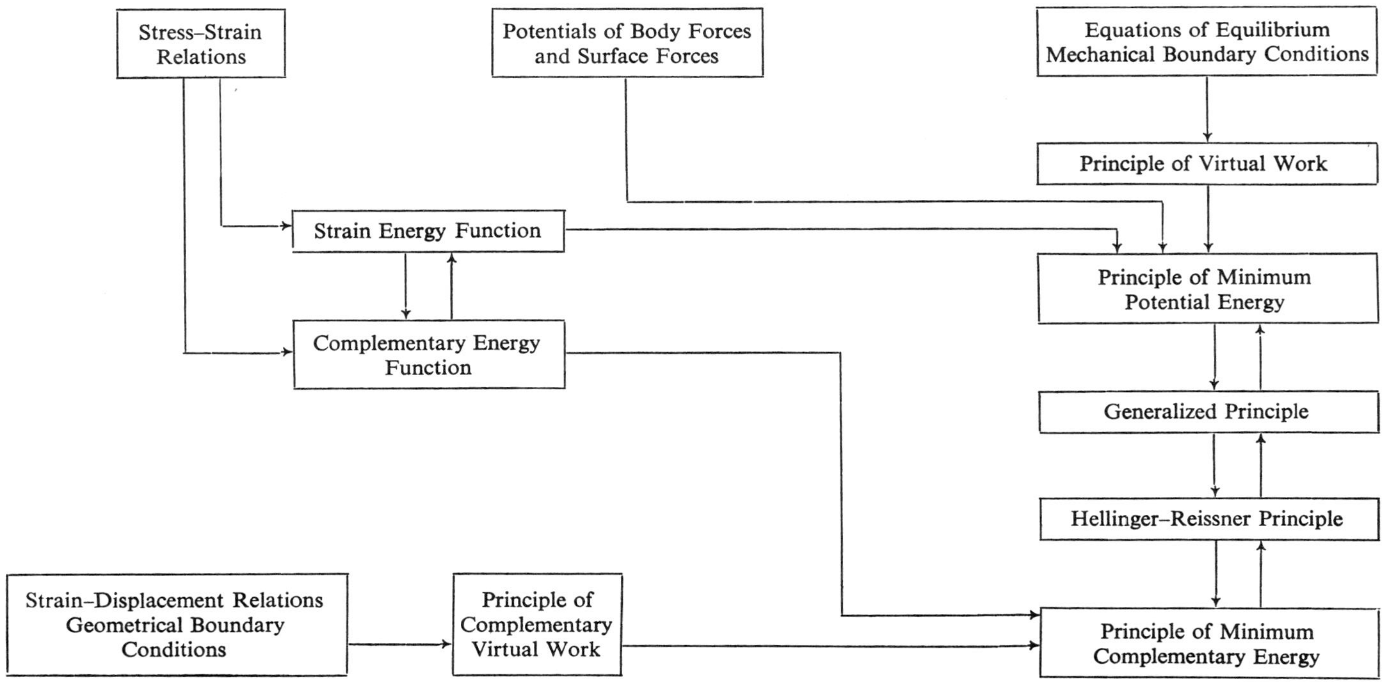

which leads from the principle of virtual work to the principle of minimum potential energy and other related variational principles, because this choice seems advantageous for a systematic treatment of problems in solid mechanics.

It is noted here that these variational principles can be applied to an elastic body consisting of several different materials, if the stress–strain relations of each material assure the existence of a strain energy or complementary energy function. For example, if the body consists of n different materials, and the strain energy function of i-th material is denoted by A_i, the principle of minimum potential energy may be formulated by replacing $\iiint_V A\,dV$ with $\sum_{i=1}^{n} \iiint_{V_i} A_i\,dV$. The continuity of displacement components on the interface between the various materials must be satisfied if neither slipping nor tearing is assumed. Similar statements can be made concerning the other variational principles. It is also noted here that several other related variational principles in elasticity have been proposed in Refs. 9, 10 and 11.

2.5. Rayleigh–Ritz Method—(1)

It has been shown that the elasticity problem in small displacement theory can be formulated by variational methods under the assumption that the three functions A, Φ and Ψ exist. The exact differential equations and boundary conditions defining the problem are then given by the stationary property of the total potential energy and related functionals. However, one of the greatest advantages of the variational procedure is its usefulness in obtaining approximate solutions. The so-called Rayleigh–Ritz method is the best established technique for obtaining approximate solutions through the use of the variational method.† We shall illustrate the Rayleigh–Ritz method with two examples.

Let us first consider the principle of minimum potential energy applied to the elasticity problem of Section 1.5. Let us assume a set of admissible displacement functions u, v and w as given by Eqs. (1.34), (1.35) and (1.36). Introducing Eqs. (1.34) into Eq. (2.12) and carrying out the volume and surface integrals, we can express Π in terms of a_r, b_r and c_r $(r = 1, 2, \ldots, n)$. The Rayleigh–Ritz method determines the values of these constants by requiring $\delta\Pi = 0$, which, in the present case, becomes:

$$\frac{\partial \Pi}{\partial a_r} = 0, \quad \frac{\partial \Pi}{\partial b_r} = 0, \quad \frac{\partial \Pi}{\partial c_r} = 0, \quad (r = 1, 2, \ldots, n). \tag{2.43}$$

The Eqs. (2.43) lead to a set of $3n$ simultaneous linear algebraic equations in which the $3n$ unknowns are a_r, b_r and c_r $(r = 1, 2, \ldots, n)$. It is observed that the $3n$ equations thus obtained are equivalent to those obtained in Section 1.5.

† Refs. 2, 3 and 12 through 17.

Next, let us consider the principle of minimum complementary energy applied to the two-dimensional problem of Section 1.7. Noting that the stresses expressed by Eq. (1.61) constitute a set of admissible functions, we shall substitute them into

$$\Pi_c = \iint \frac{1}{2E} [(\sigma_x + \sigma_y)^2 + 2(1 + \nu)(\tau_{xy}^2 - \sigma_x\sigma_y)] \, dx \, dy, \tag{2.44}$$

which, after integration, can be written in terms of a_r $(r = 1, 2, \ldots, n)$. The Rayleigh–Ritz method asserts that the stationary property of the exact solution can be satisfied approximately by requiring

$$\frac{\partial \Pi_c}{\partial a_r} = 0 \quad (r = 1, 2, \ldots, n). \tag{2.45}$$

The n simultaneous equations thus obtained determine values of a_r $(r = 1, 2, \ldots, n)$, which, when substituted into Eqs. (1.61), provide approximate solutions for the stress components. We also observe that the n equations thus derived are equivalent to those obtained in Section 1.7.

Thus, we see that the Rayleigh–Ritz method leads to formulations equivalent to those of the approximate methods developed in Sections 1.5 and 1.7, as far as the elasticity problem of the small displacement theory is concerned. However, each has its own advantages and disadvantages in applications to problems outside the elasticity problem. The approximate methods are valid irrespective of the stress–strain relations employed and potentials of the external forces, but the proof that the approximate solutions converge to the exact solution with increasing n is usually difficult. On the other hand, the stress–strain relations, body forces and surface forces must assure the existence of the state functions A, B, Φ and Ψ for the formulation of the variational expressions, if the Rayleigh–Ritz method is to be used. However, the convergence proof is less difficult, especially when the maximum or minimum property of the variational expressions has been established.

When boundary value problems of elasticity can be solved only approximately, it is desirable to obtain upper and lower bounds of the exact solution. However, this requirement is seldom answered, because bounds are usually much more difficult to obtain than approximate solutions. Trefftz proposed a method of deriving upper and lower bound formulae for the torsional rigidity of a bar by simultaneous use of the principles of minimum potential and complementary energy (see Ref. 18 and Section 6.5). Since his paper was published, many papers on this and related subjects have appeared in the field of elasticity. Among them, the concept of function space devised by W. Prager and J. L. Synge may be mentioned as a notable contribution.[(19)]

In function space a set of stress components $(\sigma_x, \sigma_y, \ldots, \tau_{xy})$, related to a set of strain components $(\varepsilon_x, \varepsilon_y, \ldots, \gamma_{xy})$ by Eqs. (1.6), is considered as a vector. Denoting two arbitrary vectors by **N** and **N*** and their components of stress and strain by $(\sigma_x, \ldots, \tau_{xy})$, $(\varepsilon_x, \ldots, \gamma_{xy})$ and $(\sigma_x^*, \ldots, \tau_{xy}^*)$, $(\varepsilon_x^*, \ldots, \gamma_{xy}^*)$, respectively, we define the scalar product of the two vectors in function space by

$$(\mathbf{N}, \mathbf{N}^*) = \iiint_V (\sigma_x \varepsilon_x^* + \sigma_y \varepsilon_y^* + \cdots + \tau_{xy}\gamma_{xy}^*) \, dV, \tag{2.46}$$

the integral being taken throughout the body. Since the strain energy function is a positive definite form, the following relations are obtained immediately:

$$(\mathbf{N}, \mathbf{N}) \geqq 0, \tag{2.47}$$

$$(\mathbf{N}, \mathbf{N}^*) \leqq (\mathbf{N}, \mathbf{N})^{\frac{1}{2}} (\mathbf{N}^*, \mathbf{N}^*)^{\frac{1}{2}}. \tag{2.48}$$

The function space thus defined enables us to grasp intuitively approximate methods of solution and their convergence characteristics, and to estimate the error of approximate solutions.(20) Due to the space available, the method of deriving bound formulae in function space will not be shown here. The interested reader is directed to Ref. 21 for details of the concept of function space.

2.6. Variation of the Boundary Conditions and Castigliano's Theorem

Thus far, we have derived the principle of minimum potential energy and its family under the assumption that the boundary conditions on S_1 and S_2 are kept constant during variation. Now, we shall consider variation of the boundary conditions. We assume that the problem defined in Section 1.1 has been solved and that components of the stress and strain as well as the functions A and B of the actual solution have been expressed in terms of the prescribed body forces, surface forces on S_1 and surface displacements on S_2. We denote the stress, strain and displacement components of the actual solution by $\sigma_x, \sigma_y, \ldots; \varepsilon_x, \varepsilon_y, \ldots; u, v, w$, respectively in the present section.

We shall consider first the variation of the geometrical boundary conditions. The displacement components are given infinitesimal increments $d\bar{u}$, $d\bar{v}$ and $d\bar{w}$ on S_2, while the body forces as well as the mechanical boundary conditions of S_1 remain unchanged. We assume that the incremental displacements have yielded a new configuration and denote incremental displacements caused in the body by du, dv and dw. Then we have

$$\begin{aligned} dU = &\iiint_V (\bar{X}\,du + \bar{Y}\,dv + \bar{Z}\,dw)\,dV \\ &+ \iint_{S_1} (\bar{X}_\nu\,du + \bar{Y}_\nu\,dv + \bar{Z}_\nu\,dw)\,dS \\ &+ \iint_{S_2} (X_\nu\,d\bar{u} + Y_\nu\,d\bar{v} + Z_\nu\,d\bar{w})\,dS, \end{aligned} \tag{2.49}$$

where

$$U = \iiint_V A \, dV, \tag{2.50}$$

is the strain energy of the elastic body. We have derived Eq. (2.49) in a manner similar to the developments of the divergence theorem, Eq. (1.76), remembering that

$$dA = \sigma_x \, d\varepsilon_x + \sigma_y \, d\varepsilon_y + \cdots + \tau_{xy} \, d\gamma_{xy}, \tag{2.51}$$

and observing that the stress components σ_x, ... and the incremental stain $d\varepsilon_x$, ... satisfy the equations of equilibrium and the conditions of compatibility, respectively. We shall see in Chapter 3 that Eq. (2.49) holds for finite displacement theory of elasticity as well.

The formula (2.49) is useful in determining the values of X_ν, Y_ν and Z_ν on the boundary S_2. As an example, we shall consider the truss structure consisting of two equal members of uniform cross-section shown in Fig. 2.2. Let the problem be defined such that the displacement $\bar{\delta}$ at the joint is prescribed and the resulting force P is to be obtained. We denote the lengths of members before and after deformation by l_0 and l, respectively, and the strain of the members by ε. From geometrical considerations we have $l^2 = a^2 + (b + \bar{\delta})^2$ and $l_0^2 = a^2 + b^2$, and we obtain

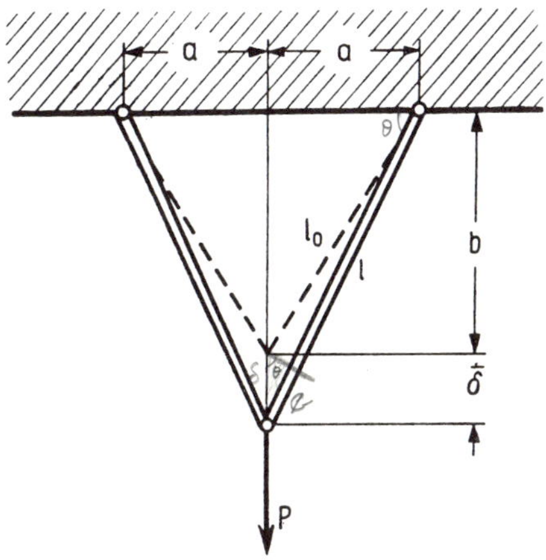

FIG. 2.2. A truss structure.

$$\varepsilon = (l - l_0)/l_0 = (b\bar{\delta})/l_0^2, \tag{2.52}$$

where higher order terms are neglected. Consequently, we have

$$U = [(\tfrac{1}{2}) \, EA_0 l_0 \varepsilon^2] \times 2 = EA_0 b^2 (\bar{\delta})^2 / l_0^3, \tag{2.53}$$

where A_0 is the cross-sectional area of the member. Applying Eq. (2.49), we obtain

$$P = \partial U / \partial \bar{\delta} = (2EA_0 b^2 / l_0^3) \, \bar{\delta}. \tag{2.54}$$

Next, we shall consider the variation of the body forces and mechanical boundary conditions. The body forces and the external forces on S_1 are given infinitesimal increments $d\bar{X}$, $d\bar{Y}$, $d\bar{Z}$ and $d\bar{X}_\nu$, $d\bar{Y}_\nu$, $d\bar{Z}_\nu$, respectively, while the geometrical boundary conditions on S_2 remain unchanged. We assume that these incremental forces yield a new configuration, and denote incremental stresses caused in the body by $d\sigma_x$, $d\sigma_y$, ... and $d\tau_{xy}$. Then, we have

$$dV = \iiint_V (u\, d\bar{X} + v\, d\bar{Y} + w\, d\bar{Z})\, dV + \iint_{S_1} (u\, d\bar{X}_\nu + v\, d\bar{Y}_\nu + w\, d\bar{Z}_\nu)\, dS + \iint_{S_2} (\bar{u}\, dX_\nu + \bar{v}\, dY_\nu + \bar{w}\, dZ_\nu)\, dS, \tag{2.55}$$

where

$$V = \iiint_V B\, dV \tag{2.56}$$

is the complementary energy of the elastic body. We have derived Eq. (2.55) in a manner similar to the development of the divergence theorem remembering that

$$dB = \varepsilon_x\, d\sigma_x + \varepsilon_y\, d\sigma_y + \cdots + \gamma_{xy}\, d\tau_{xy}, \tag{2.57}$$

and observing that the strain components ε_x, ... and the incremental stresses $d\sigma_x$, ... satisfy the conditions of compatibility and the equations of equilibrium, respectively.

The formula (2.55) is useful in determining the values of u, v and w on the boundary S_1. As an example, we consider a body which is held rigidly fixed on the boundary S_2, and is subject to n concentrated loads $\bar{P}_1$, $\bar{P}_2$, ..., $\bar{P}_n$ on the boundary S_1. For the sake of simplicity, these concentrated loads are assumed independent. In other words, any of these forces can be assumed to be given increments without interfering with those remaining. Denoting the displacement component of the point of application of the load $\bar{P}_i$ in the direction of the load by Δ_i, we have from Eq. (2.55):

$$dV = \sum_{i=1}^{n} \Delta_i\, d\bar{P}_i. \tag{2.58}$$

Since V is a function of the external forces, we have:

$$dV = \sum_{i=1}^{n} (\partial V/\partial \bar{P}_i)\, d\bar{P}_i. \tag{2.59}$$

Combining these two equations, we obtain:

$$\sum_{i=1}^{n} \left(\frac{\partial V}{\partial \bar{P}_i} - \Delta_i\right) d\bar{P}_i = 0. \tag{2.60}$$

Since the forces are assumed independent, we have

$$\frac{\partial V}{\partial \bar{P}_i} = \Delta_i \quad (i = 1, 2, \ldots, n). \tag{2.61}$$

The formula (2.55) and its family are called Castigliano's theorem—a powerful tool for analyzing problems in the small displacement theory of elasticity (see Refs. 2 and 12 through 15, for instance).

2.7. Free Vibrations of an Elastic Body

The variational principles derived so far have been for the boundary value problem of elasticity. In the last two sections of this chapter we shall consider variational formulations of a problem of free vibrations of an elastic body in small displacement theory. The problem is defined by allowing the body to be mechanically free on S_1 and geometrically fixed on S_2. Since the problem is confined to small displacement theory, all the equations defining the problem are linear, and displacements and stresses in the body behave sinusoidally with respect to time. Consequently if we denote the amplitudes of stress, strain and displacement by $\sigma_x, \ldots, \varepsilon_x, \ldots$ and u, v, w, respectively, we have for the equations of motion,

$$\begin{gathered}
\frac{\partial \sigma_x}{\partial x} + \frac{\partial \tau_{xy}}{\partial y} + \frac{\partial \tau_{zx}}{\partial z} + \lambda \varrho u = 0, \\
\frac{\partial \tau_{xy}}{\partial x} + \frac{\partial \sigma_y}{\partial y} + \frac{\partial \tau_{yz}}{\partial z} + \lambda \varrho v = 0, \\
\frac{\partial \tau_{zx}}{\partial x} + \frac{\partial \tau_{yz}}{\partial y} + \frac{\partial \sigma_z}{\partial z} + \lambda \varrho w = 0.
\end{gathered} \tag{2.62}$$

In Eqs. (2.62), $\lambda = \omega^2$ where ω is the natural circular frequency, and ϱ is the density of the material. The boundary conditions are given by

$$X_\nu = 0, \quad Y_\nu = 0, \quad Z_\nu = 0 \quad \text{on} \quad S_1, \tag{2.63}$$

and

$$u = 0, \quad v = 0, \quad w = 0 \quad \text{on} \quad S_2. \tag{2.64}$$

From Eqs. (2.62) and (2.63), we have

$$\begin{aligned}
&-\iiint\limits_V \left[\left(\frac{\partial \sigma_x}{\partial x} + \frac{\partial \tau_{xy}}{\partial y} + \frac{\partial \tau_{zx}}{\partial z} + \lambda \varrho u\right) \delta u + (\cdots)\, \delta v + (\cdots)\, \delta w\right] dV \\
&+ \iint\limits_{S_1} (X_\nu\, \delta u + Y_\nu\, \delta v + Z_\nu\, \delta w)\, dS = 0.
\end{aligned} \tag{2.65}$$

Here we chose the arbitrary set of virtual displacements δu, δv and δw such that the geometrical boundary conditions are not violated, namely, $\delta u = \delta v$

$= \delta w = 0$ on S_2. Then, we may transform Eq. (2.65) into

$$\iiint_V (\sigma_x \,\delta\varepsilon_x + \sigma_y \,\delta\varepsilon_y + \cdots + \tau_{xy} \,\delta\gamma_{xy}) \, dV - \lambda \iiint_V (u \,\delta u + v \,\delta v + w \,\delta w) \varrho \, dV = 0. \tag{2.66}$$

This is the principle of virtual work for the free vibration problem.

If the relations between the amplitudes of stress and strain are given by

$$\sigma_x = a_{11}\varepsilon_x + a_{12}\varepsilon_y + \cdots + a_{16}\gamma_{xy}, \ldots, \tag{2.67}$$

where

$$a_{ij} = a_{ji} \quad (i, j = 1, 2, \ldots, 6),$$

we are assured of the existence of the strain energy function defined by Eq. (2.2). Moreover, the body forces $\lambda\varrho u$, $\lambda\varrho v$ and $\lambda\varrho w$ are derivable from a potential function Φ defined by Eq. (2.6) such that

$$-\Phi = \tfrac{1}{2}\lambda\varrho(u^2 + v^2 + w^2). \tag{2.68}$$

Consequently, we obtain from Eq. (2.66) the principle of stationary potential energy as follows: *Among all the admissible displacement functions u, v and w which satisfy the prescribed geometrical boundary conditions, the actual displacements make the total potential energy*

$$\Pi = \iiint_V A(u, v, w) \, dV - \tfrac{1}{2}\lambda \iiint_V (u^2 + v^2 + w^2) \varrho \, dV, \tag{2.69}$$

stationary. In the functional (2.69), the quantities subject to variation are u, v and w under the subsidiary conditions (2.64), while λ is treated as a parameter not subject to variation.

The principle of stationary potential energy can be generalized through the use of Lagrange multipliers as follows:

$$\Pi_{\mathrm{I}} = \iiint_V \Big[A(\varepsilon_x, \varepsilon_y, \ldots, \gamma_{xy}) - \frac{1}{2}\lambda\varrho(u^2 + v^2 + w^2) - \left(\varepsilon_x - \frac{\partial u}{\partial x}\right)\sigma_x - \cdots - \left(\gamma_{xy} - \frac{\partial v}{\partial x} - \frac{\partial u}{\partial y}\right)\tau_{xy}\Big] dV - \iint_{S_2} (p_x u + p_y v + p_z w) \, dS, \tag{2.70}$$

where the independent quantities subject to variation are $\varepsilon_x, \ldots$; $u, \ldots$; $\sigma_x, \ldots$; $p_x, \ldots$ and p_z. The stationary conditions are shown to be Eqs. (2.67); Eqs. (2.62), (2.63) and

$$p_x = X_\nu, \ldots, \quad p_z = Z_\nu \quad \text{on} \quad S_2; \tag{2.71}$$

$$\varepsilon_x = \frac{\partial u}{\partial x}, \ldots, \quad \gamma_{xy} = \frac{\partial v}{\partial x} + \frac{\partial u}{\partial y}; \tag{2.72}$$

and Eqs. (2.64).

Several variational principles can be derived from the generalized principle.[22] Here, we shall derive a functional for the principle of stationary complementary energy. It is shown that elimination of the strain components by the use of Eqs. (2.67) and a simple calculation by the use of Eqs. (2.62), (2.63) and (2.71) lead to a transformation of the functional (2.70) as follows:

$$\Pi_c = \iiint_V B(\sigma_x, \sigma_y, \ldots, \tau_{xy})\, dV - \tfrac{1}{2}\lambda \iiint_V (u^2 + v^2 + w^2)\, \varrho\, dV, \tag{2.73}$$

where the quantities subject to variation are $u, \ldots; \sigma_x, \ldots$ and τ_{xy} under the subsidiary conditions (2.62) and (2.63), and the stationary conditions are shown to be equivalent to Eqs. (2.64) and (2.72). The functional (2.73) is an expression for the principle of stationary complementary energy of the free vibration problem. We note that another expression of the principle of stationary complementary energy can be obtained by eliminating u, v and w from the functional (2.73) by the use of Eqs. (2.62), thus expressing the functional in terms of $\sigma_x, \sigma_y, \ldots$ and τ_{xy} only.

It was shown in Refs. 23 and 24 that the principle of stationary complementary energy might be extended to eigenvalue problems such as free vibration and stability of elastic bodies. The principle was introduced and proved in Ref. 25 by E. Reissner for a problem in which loads, stresses and displacements are simple harmonic functions of time. The functional (2.73) is equivalent to that introduced by E. Reissner.

It is well established that the principle of stationary potential energy (2.69) is equivalent to finding, among admissible functions u, v and w which satisfy the prescribed geometrical boundary conditions, those which make the quotient

$$\lambda = U/T \tag{2.74}$$

stationary, where

$$U = \iiint_V A(u, v, w)\, dV, \tag{2.75}$$

$$T = \tfrac{1}{2} \iiint_V (u^2 + v^2 + w^2)\, \varrho\, dV, \tag{2.76}$$

and the stationary values of λ provide the eigenvalues of the solution. For the proof, we see that

$$\delta\lambda = \frac{\delta U}{T} - \frac{U\,\delta T}{T^2} = \frac{1}{T}(\delta U - \lambda\, \delta T), \tag{2.77}$$

where the variation is taken with respect to u, v and w. Consequently, the condition that the quotient λ is stationary is equivalent to the principle of stationary potential energy. The expression (2.74) is the Rayleigh quotient for the free vibration problem.[16, 26]

It is also well known that the principle of stationary potential energy (2.69) is equivalent to a problem of finding, among admissible functions

u, v and w which satisfy the geometrical boundary conditions, those which make U stationary under a subsidiary condition

$$T(u, v, w) - 1 = 0. \tag{2.78}$$

For the proof, we see that this problem is equivalent to obtaining the stationary conditions of a functional defined by

$$U - \lambda(T - 1), \tag{2.79}$$

where λ plays a role of a Lagrange multiplier and the variation is taken with respect to u, v, w and λ under the subsidiary conditions (2.64).

2.8. Rayleigh–Ritz Method—(2)

We have seen the variational principles established for the free vibration problem in the preceding section. When the variational expressions are thus available, the Rayleigh–Ritz method provides a powerful tool for obtaining approximate values of eigenvalues. We shall consider a free vibration problem of a beam as an example and follow the outline of the method.

We shall take a beam clamped at one end $x = 0$ and simply supported at the other end $x = l$ as shown in Fig. 7.5. The functional for the principle of stationary potential energy for the present problem is given by

$$\Pi = \tfrac{1}{2}\int_0^l EI(w'')^2\, dx - \tfrac{1}{2}\lambda \int_0^l mw^2\, dx, \tag{2.80†}$$

where EI, w and m are the bending rigidity, deflection and mass per unit span of the beam, respectively, and $(\;)' = d(\;)/dx$. In the functional (2.80), the quantity subject to variation is w under the subsidiary conditions

$$w(0) = w(l) = w'(0) = 0. \tag{2.81}$$

We denote the exact eigenvalues by

$$\lambda_i \quad (i = 1, 2, 3, \ldots), \tag{2.82}$$

in ascending order of magnitude such that $0 < \lambda_1 < \lambda_2 < \cdots$.

We may transform the functional (2.80) to yield a functional for the principle of stationary complementary energy:

$$\Pi_c = \frac{1}{2}\int_0^l \frac{M^2}{EI}\, dx - \frac{1}{2}\lambda \int_0^l mw^2\, dx, \tag{2.83†}$$

where the quantities subject to variation are M and w under the subsidiary conditions

$$M'' + \lambda mw = 0, \tag{2.84}$$

† See Section 7.4 for the derivation of these functionals.

and

$$M(l) = 0. \tag{2.85}$$

We shall consider first the Rayleigh–Ritz method applied to the principle of stationary potential energy. The well-known procedure proposed by the method is followed by choosing a set of n linearly independent admissible functions $w_i(x)$, the so-called coordinate functions, which satisfy Eqs. (2.81), and assuming w as a linear combination of these coordinate functions, namely:

$$w = \sum_{i=1}^{n} c_i w_i, \tag{2.86}$$

where $c_i (i = 1, 2, \ldots, n)$ are arbitrary constants. Substituting Eq. (2.86) into Eq. (2.80), and setting

$$\partial\Pi/\partial c_i = 0 \quad (i = 1, 2, \ldots, n), \tag{2.87}$$

we obtain a set of n homogeneous equations. The requirement that the determinant of the set must vanish for a nontrivial solution provides another algebraic equation, called the characteristic equation of the set of the form:

$$\det (m_{ij} - \lambda n_{ij}) = 0. \tag{2.88}$$

If we denote the roots of the characteristic equation (2.88) by Λ_i $(i = 1, 2, \ldots, n)$ in ascending order of magnitude, i.e. $\Lambda_1 < \Lambda_2 \cdots < \Lambda_n$, we have

$$\lambda_i \leqq \Lambda_i \quad (i = 1, 2, \ldots, n). \tag{2.89}†$$

Next, we shall consider the method applied to the principle of stationary complementary energy. This is sometimes called the modified Rayleigh–Ritz method,[13] and its outline is as follows: We choose w as given by Eq. (2.86), where the coordinate functions $w_i(x)$ are so chosen as to satisfy Eqs. (2.81). We substitute Eq. (2.86) into Eq. (2.84) and perform integrations with the boundary condition Eq. (2.85) to obtain

$$(1/\lambda)\, M = c(x - l) - \sum_{i=1}^{n} c_i \int_x^l \left[\int_\eta^l m(\xi)\, w_i(\xi)\, d\xi \right] d\eta, \tag{2.90}$$

where c is an integration constant. Substituting Eqs. (2.86) and (2.90) into the functional (2.83) and requiring that

$$\partial\Pi_c/\partial c = 0, \tag{2.91}$$

and

$$\partial\Pi_c/\partial c_i = 0, \quad i = 1, 2, \ldots, n, \tag{2.92}$$

we obtain a characteristic equation which determines approximate eigenvalues. For later convenience, these approximate eigenvalues are denoted by Λ_i^* $(i = 1, 2, \ldots, n)$ in ascending order of magnitude, i.e. $\Lambda_1^* < \Lambda_2^* < \cdots < \Lambda_n^*$.

† For the proof, see Refs. 3, 26 and 27.

We see that the inclusion of the term $c(x - l)$ in Eq. (2.90) and the requirement of Eq. (2.91) are equivalent to obtaining the exact beam deflection due to the inertial loading $\lambda m \Sigma c_i w_i$. Thus, the method is equivalent to the Grammel's method in which the exact deflection due to the inertial loading is obtained by the use of the Green's function or the so-called influence functions.[28,29] It is stated in Ref. 28 that if the same assumed modes (2.86) are employed, we have the following inequality relations:

$$\lambda_i \leqq \Lambda_i^* \leqq \Lambda_i, \quad i = 1, 2, \ldots, n. \tag{2.93}$$

Thus, the Rayleigh–Ritz method provides an upper bound for each eigenvalue. It is well established that the accuracy of approximate eigenvalues thus obtained is good and sometimes excellent if the coordinate functions are chosen properly. However, since an approximate method of solution is applied to a problem whose exact solution cannot be obtained, we can usually expect to have no information on the exact eigenvalues beforehand. Therefore, formulae providing lower bounds are indispensable for locating the exact eigenvalues.

There have been proposed several theorems for locating lower bounds of eigenvalues. Among them, the Temple–Kato theorem and Weinstein's method may be mentioned as typical. The Temple–Kato theorem provides a lower bound for the eigenvalue λ_n when the value or a lower bound of the eigenvalue λ_{n+1} is known.† This theorem often proves to be an effective tool for eigenvalue location. On the other hand, Weinstein's method employs as a basis one of Rayleigh's principles that, if the prescribed boundary conditions are partly relaxed, all the eigenvalues decrease.‡ That is, if we denote eigenvalues of a relaxed or intermediate problem by $\bar{\lambda}_i$, $i = 1, 2, \ldots$ in ascending order of magnitude such that $\bar{\lambda}_1 < \bar{\lambda}_2 < \cdots$, then we have

$$\bar{\lambda}_i \leqq \lambda_i \quad (i = 1, 2, \ldots). \tag{2.94}$$

Therefore, if we obtain exact eigenvalues of the intermediate problem, they provide lower bounds for the eigenvalues of the original problem.

The Rayleigh–Ritz method for the free vibration problem has been illustrated. It is obvious that the method also finds a field of application in other eigenvalue problems. The reader is directed to Refs. 13, 16 and 26 for further details and numerical illustrations of the Rayleigh–Ritz method applied to eigenvalue problems.

2.9. Some Remarks

We have derived some extensions of the principles of virtual work and complementary virtual work in Section 1.9. It is obvious that the first terms of Eqs. (1.73) and (1.74) may be replaced by δU and δV, respectively, for

† Refs. 30 through 35.
‡ Refs. 36 through 38.

elasticity problems, and several extensions of the principles (2.5) and (2.22) may be obtained from these equations. For example, we have

$$\delta V - \iiint_V (u\,\delta X + v\,\delta Y + w\,\delta Z)\,dV = 0, \tag{2.95}$$

for an elasticity problem, where the boundary conditions are specified such that

$$\bar{X}_\nu = \bar{Y}_\nu = \bar{Z}_\nu = 0 \quad \text{on} \quad S_1, \tag{2.96}$$

$$\bar{u} = \bar{v} = \bar{w} = 0 \qquad \text{on} \quad S_2, \tag{2.97}$$

and stress variations are so chosen as to satisfy Eqs. (1.75) and the boundary conditions

$$\delta X_\nu = \delta Y_\nu = \delta Z_\nu = 0 \quad \text{on} \quad S_1. \tag{2.98}$$

Next, a mention is made of the generalized Galerkin's method treated in Section 1.5. The object of this mention is to note that the principle (2.5), which may be used instead of the principle (1.32) for elasticity problems, suggests a modification of the generalized Galerkin's method as follows: Since

$$\delta U = \sum_{r=1}^{n} [(\partial U/\partial a_r)\,\delta a_r + (\partial U/\partial b_r)\,\delta b_r + (\partial U/\partial c_r)\,\delta c_r], \tag{2.99}$$

we find that the principle (2.5) leads to an approximate method of solution in which equations for the determination of the unknown constants a_r, b_r and c_r $(r = 1, 2, \ldots, n)$ are given by

$$L_r = 0, \quad M_r = 0, \quad N_r = 0, \quad (r = 1, 2, \ldots, n) \tag{2.100}$$

where

$$\begin{aligned} L_r &= \frac{\partial U}{\partial a_r} - \iiint_V \bar{X} u_r\,dV - \iint_{S_1} \bar{X}_\nu u_r\,dS, \\ M_r &= \frac{\partial U}{\partial b_r} - \iiint_V \bar{Y} v_r\,dV - \iint_{S_1} \bar{Y}_\nu v_r\,dS, \\ N_r &= \frac{\partial U}{\partial c_r} - \iiint_V \bar{Z} w_r\,dV - \iint_{S_1} \bar{Z}_\nu w_r\,dS. \end{aligned} \tag{2.101}$$

We shall see in Section 5.6 that this approximate method of solution is equivalent to that employed in deriving Lagrange's equation of motion of the dynamical problem. It is obvious that the principle (2.22) suggests a similar modification of the generalized Galerkin's method treated in Section 1.7.

The approximate method of solution above mentioned can also be applied to eigenvalue problems of an elastic body in which external forces are not derivable from potential functions. Such applications are usually based on the principle of virtual work, Eq. (2.5), as illustrated in Refs. 39 and 40. As an example of applications based on the principle of complementary virtual work, we may refer to E. Reissner's work for flutter calculations.[41] An examination of his paper reveals that his method may be considered as an application of the principle (2.95) if aerodynamic and inertial forces are taken as types of body forces.

Bibliography

1. A. E. H. LOVE, *A Treatise on the Mathematical Theory of Elasticity*, Cambridge University Press, 4th edition, 1927.
2. S. TIMOSHENKO and J. N. GOODIER, *Theory of Elasticity*, McGraw-Hill, 1951.
3. R. COURANT and D. HILBERT, *Methods of Mathematical Physics*, Vol. 1, Interscience, New York, 1953.
4. K. WASHIZU, *On the Variational Principles of Elasticity and Plasticity*, Aeroelastic and Structures Research Laboratory, Massachusetts Institute of Technology, Technical Report 25–18, March 1955.
5. H. C. HU, On Some Variational Principles in the Theory of Elasticity and Plasticity, *Scintia Sinica*, Vol. 4, No. 1, pp. 33–54, March 1955.
6. E. HELLINGER, Der allgemeine Ansatz der Mechanik der Kontinua, *Encyclopädie der Mathematischen Wissenschaften*, Vol. 4, Part 4, pp. 602–94, 1914.
7. E. REISSNER, On a Variational Theorem in Elasticity, *Journal of Mathematics and Physics*, Vol. 29, No. 2, pp. 90–95, July 1950.
8. K. FRIEDRICHS, Ein Verfahren der Variationsrechnung das Minimum eines Integrals als das Maximum eines anderen Ausdruckes darzustellen, *Nachrichten der Academie der Wissenschaften in Göttingen*, pp. 13–20, 1929.
9. T. IAI, A Method of Solution of Elastic Problems by the Theorem of Maximum Energy (in Japanese), *Journal of the Society of Aeronautical Science of Nippon*, Vol. 10, No. 96, pp. 276–98, April 1943.
10. E. REISSNER, On Variational Principles in Elasticity, *Proceedings of Symposia in Applied Mathematics*, Vol. 8, pp. 1–6, MacGraw-Hill, 1958.
11. P. M. NAGHDI, On a Variational Theorem in Elasticity and its Application to Shell Theory, *Journal of Applted Mechanics*, Vol. 31, No. 4, pp. 647–53, December 1964.
12. C. BIEZENO and R. GRAMMEL, *Technische Dynamik*, Springer Verlag, 1939.
13. R. L. BISPLINGHOFF, H. ASHLEY and R. L. HALFMAN, *Aeroelasticity*, Addison-Wesley, 1955.
14. N. J. HOFF, *The Analysis of Structures*, John Wiley, 1956.
15. J. H. ARGYRIS and S. KELSEY, *Energy Theorems and Structural Analysis*, Butterworth, 1960.
16. G. TEMPLE and W. G. BICKLEY, *Rayleigh's Principle and its Applications to Engineering*, Oxford University Press, 1933.
17. S. G. MIKHLIN, *Variational Methods in Mathematical Physics*, Pergamon Press, 1964.
18. E. TREFFTZ, Ein Gegenstück zum Ritzschen Verfahren, *Proceedings of the 2nd International Congress for Applied Mechanics*, Zürich, pp. 131–7, 1926.
19. W. PRAGER and J. L. SYNGE, Approximations in Elasticity Based on the Concept of Function Space, *Quarterly of Applied Mathematics*, Vol. 5, No. 3, pp. 241–69, October 1947.
20. K. WASHIZU, Bounds for Solutions of Boundary Value Problems in Elasticity, *Journal of Mathematics and Physics*, Vol. 32, No. 2–3, pp. 119–28, July–October 1953.
21. J. L. SYNGE, *The Hypercircle in Mathematical Physics*, Cambridge University Press, 1957.

22. K. Washizu, Note on the Principle of Stationary Complementary Energy Applied to Free Vibration of an Elastic Body, *International Journal of Solids and Structures*, Vol. 2, No. 1, pp. 27–35, January 1966.
23. S. Timoshenko, *Theory of Elastic Stability*, McGraw-Hill, 1936.
24. H. M. Westergaard, On the Method of Complementary Energy and Its Application to Structures Stressed Beyond the Proportional Limit, to Buckling and Vibrations, and to Suspension Bridges, *Proceedings of American Society of Civil Engineers*, Vol. 67, No. 2, pp. 199–227, February 1941.
25. E. Reissner, Note on the Method of Complementary Energy, *Journal of Mathematics and Physics*, Vol. 27, pp. 159–60, 1948.
26. L. Collatz, *Eigenwertaufgaben mit technischen Anwendungen*, Akademische Verlagsgesellschaft, Leipzig, 1949.
27. S. H. Gould, *Variational Methods for Eigenvalue Problems*, University of Toronto Press, 1957.
28. R. Grammel, Ein neues Verfahren zur Lösung technischer Eigenwertprobleme, *Ingenieur Archiv*, Vol. 10, pp. 35–46, 1939.
29. A. I. van de Vooren and J. H. Greidanus, Complementary Energy Method in Vibration Analysis, Reader's Forum, *Journal of Aeronautical Sciences*, Vol. 17, No. 7, pp. 454–5, July 1950.
30. G. Temple, The Calculation of Characteristic Numbers and Characteristic Functions, *Proceedings of London Mathematical Society*, Vol. 29, Series 2, No. 1690, pp. 257–80, 1929.
31. W. Kohn, A Note on Weinstein's Variational Method, *Physical Review*, Vol. 71, No. 12, pp. 902–4, June 1947.
32. T. Kato, On the Upper and Lower Bounds of Eigenvalues, *Journal of Physical Society of Japan*, Vol. 4, No. 4–6, pp. 334–9, July–December 1949.
33. G. Temple, The Accuracy of Rayleigh's Method of Calculating the Natural Frequencies of Vibrating Systems, *Proceedings of Royal Society*, Vol. A211, No. 1105, pp. 204–24, February 1950.
34. R. V. Southwell, Some Extensions of Rayleigh's Principle, *Quarterly Journal of Mechanics and Applied Mathematics*, Vol. 6, Part 3, pp. 257–72, October 1953.
35. K. Washizu, On the Bounds of Eigenvalues, *Quarterly Journal of Mechanics and Applied Mathematics*, Vol. 8, Part 3, pp. 311–25, September 1955.
36. A. Weinstein, Étude des spectres des équations aux derivées partielles de la théorie des plaques élastiques, *Memorial des Sciences Mathematiques*, Vol. 88, Paris, 1937.
37. N. Aronszajn and A. Weinstein, On the Unified Theory of Eigenvalues of Plates and Membranes, *American Journal of Mathematics*, Vol. 64, No. 4, pp. 623–45, December 1942.
38. J. B. Diaz, Upper and Lower Bounds for Eigenvalues, *Proceedings of Symposia in Applied Mathematics*, Vol. 8, pp. 53–78, McGraw-Hill, 1958.
39. R. L. Bisplinghoff and H. Ashley, *Principles of Aeroelasticity*, John Wiley, 1962.
40. V. V. Bolotin, *Nonconservative Problems of the Theory of Elastic Stability*, Translated by T. K. Lusher and edited by G. Herrmann, Pergamon Press, 1963.
41. E. Reissner, Complementary Energy Procedure for Flutter Calculations, Reader's Forum, *Journal of Aeronautical Sciences*, Vol. 16, No. 5, pp. 316–17, May 1949.
42. F. R. Gantmacher, *The Theory of Matrices*, Chelsea Publishing Company, 1959.
43. B. M. Fraeijs de Veubeke, Upper and Lower Bounds in Matrix Structural Analysis, in *Matrix Methods of Structural Analysis*, edited by B. M. F. de Veubeke and published by Pergamon Press, 1964.

CHAPTER 3

FINITE DISPLACEMENT THEORY OF ELASTICITY IN RECTANGULAR CARTESIAN COORDINATES

3.1. Analysis of Strain

In the present chapter we shall treat finite displacement theory of elasticity in rectangular Cartesian coordinates.† The difference between spatial variables and material variables cannot be overemphasized in the formulation of finite displacement theory. Unless otherwise stated, we shall employ the Lagrangian approach, in which the coordinates defining a point of the body before deformation are employed for locating the point during the subsequent deformation.‡

Let the rectangular Cartesian coordinates (x^1, x^2, x^3) be fixed in space and the position vector of an arbitrary point $P^{(0)}$ of the body before deformation be represented by

$$\mathbf{r}^{(0)} = \mathbf{r}^{(0)}(x^1, x^2, x^3), \tag{3.1}$$

as shown in Fig. 3.1, where the superscript $^{(0)}$ means that the quantity is referred to the state before deformation.†† Let us henceforth employ a set of the values (x^1, x^2, x^3), which the point $P^{(0)}$ occupies before deformation, as parameters which specify the material point during deformation.

The base vectors in this coordinate system are given by

$$\mathbf{i}_\lambda = \frac{\partial \mathbf{r}^{(0)}}{\partial x^\lambda} = \mathbf{r}^{(0)}_{,\lambda} \quad (\lambda = 1, 2, 3), \qquad (3.2)‡‡$$

where and throughout the present chapter, the notation $(\ )_{,\lambda}$ denotes differentiation with respect to x^λ, namely, $(\ )_{,\lambda} = \partial(\ )/\partial x^\lambda$. They are unit vectors in the directions of the coordinate axes and are mutually orthogonal:

$$\mathbf{i}_\lambda \cdot \mathbf{i}_\mu = \delta_{\lambda\mu}, \qquad (3.3)†††$$

† Refs. 1 through 6.

‡ On the other hand, coordinates associated with the deformed body are employed in the Eulerian approach.

†† Superscript indices should not be mistaken for exponents.

‡‡ A Greek index will be assigned in place of (1, 2, 3) in Chapters 3, 4 and 5.

††† The notation $\mathbf{a} \cdot \mathbf{b}$ denotes the scalar product of two vectors $\mathbf{a}$ and $\mathbf{b}$.

where $\delta_{\lambda\mu}$ is the Kronecker symbol defined by

$$\delta_{\lambda\mu} = 1 \quad (\lambda = \mu),$$
$$\delta_{\lambda\mu} = 0 \quad (\lambda \neq \mu). \tag{3.4}$$

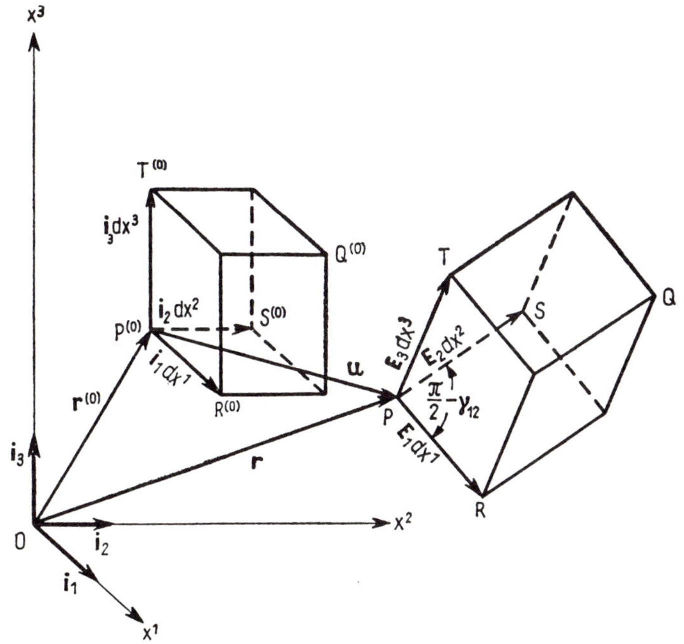

FIG. 3.1. Geometry of an infinitesimal parallelepiped.
(a) before deformation. (b) after deformation.

We shall take a point $Q^{(0)}$ in the neighborhood of the point $P^{(0)}$ and denote the coordinates of $Q^{(0)}$ by $(x^1 + dx^1, x^2 + dx^2, x^3 + dx^3)$. Then the position vector $d\mathbf{r}^{(0)}$ and the distance $ds^{(0)}$ between these two points can be expressed as

$$d\mathbf{r}^{(0)} = \mathbf{r}^{(0)}_{,\lambda}\, dx^\lambda = \mathbf{i}_\lambda\, dx^\lambda, \tag{3.5}†$$

and

$$(ds^{(0)})^2 = d\mathbf{r}^{(0)} \cdot d\mathbf{r}^{(0)} = \delta_{\lambda\mu}\, dx^\lambda\, dx^\mu, \tag{3.6}$$

† The summation convention will be employed in Chapters 3, 4 and 5. Therefore a Greek letter index which appears twice in the same term indicates summation with respect to (1, 2, 3). For example:

$$x^\lambda \mathbf{i}_\lambda = \sum_{\lambda=1}^{3} x^\lambda \mathbf{i}_\lambda = x^1\mathbf{i}_1 + x^2\mathbf{i}_2 + x^3\mathbf{i}_3.$$

$$\sigma^{\lambda\mu}e_{\lambda\mu} = \sum_{\lambda=1}^{3}\sum_{\mu=1}^{3} \sigma^{\lambda\mu}e_{\lambda\mu} = \sigma^{11}e_{11} + \sigma^{12}e_{12} + \sigma^{13}e_{13}$$
$$+ \sigma^{21}e_{21} + \sigma^{22}e_{22} + \sigma^{23}e_{23} + \sigma^{31}e_{31} + \sigma^{32}e_{32} + \sigma^{33}e_{33}.$$

respectively. For later convenience, an infinitesimal rectangular parallelepiped enclosed by the following six surfaces:

$$x^\lambda = \text{constant}, \quad x^\lambda + dx^\lambda = \text{constant} \quad (\lambda = 1, 2, 3)$$

will be fixed to the body. It has the line element $P^{(0)}Q^{(0)}$ as one of the principal diagonals and $R^{(0)}$, $S^{(0)}$, and $T^{(0)}$ as the vertices adjacent to the point $P^{(0)}$.

The body is now assumed to be deformed into a strained configuration. The points $P^{(0)}$, $Q^{(0)}$, $R^{(0)}$, $S^{(0)}$ and $T^{(0)}$ move to new positions denoted by P, Q, R, S and T, respectively, and the infinitesimal rectangular parallelepiped is deformed into a parallelepiped which, in general, is no longer rectangular. Let us denote the position vector of the point P by

$$\mathbf{r} = \mathbf{r}(x^1, x^2, x^3), \tag{3.7}$$

and introduce the lattice vectors defined by

$$\mathbf{E}_\lambda = \frac{\partial \mathbf{r}}{\partial x^\lambda} = \mathbf{r}_{,\lambda} \quad (\lambda = 1, 2, 3). \tag{3.8}$$

The sides of the parallelepiped issuing from the point P are then given by $\mathbf{E}_1\, dx^1$, $\mathbf{E}_2\, dx^2$ and $\mathbf{E}_3\, dx^3$. Consequently, the position vector $d\mathbf{r}$ and the distance ds between P and Q can be expressed by

$$d\mathbf{r} = \mathbf{r}_{,\lambda}\, dx^\lambda = \mathbf{E}_\lambda\, dx^\lambda, \tag{3.9}$$

and

$$(ds)^2 = d\mathbf{r} \cdot d\mathbf{r} = E_{\lambda\mu}\, dx^\lambda\, dx^\mu, \tag{3.10}$$

respectively, where

$$E_{\lambda\mu} = \mathbf{E}_\lambda \cdot \mathbf{E}_\mu = E_{\mu\lambda}. \tag{3.11}$$

Let us consider the geometrical meaning of $E_{\lambda\mu}$. The lengths of the infinitesimal line element $P^{(0)}R^{(0)}$ before and after deformation are

$$(ds^{(0)})^2 = (dx^1)^2 \quad \text{and} \quad (ds)^2 = E_{11}(dx^1)^2,$$

respectively. Therefore, the rate of elongation of $P^{(0)}R^{(0)}$ is given by

$$(ds - ds^{(0)})/ds^{(0)} = \sqrt{E_{11}} - 1. \tag{3.12}$$

The geometrical meanings of E_{22} and E_{33} follow similarly.

Next, consider two infinitesimal line elements $P^{(0)}R^{(0)}$ and $P^{(0)}S^{(0)}$, which are orthogonal before deformation. After deformation, these two elements move to new positions, PR and PS, the relative positions of which are given by the vectors $\mathbf{E}_1\, dx^1$ and $\mathbf{E}_2\, dx^2$, respectively. If we denote the acute angle between PR and PS by $\left(\frac{\pi}{2} - \gamma_{12}\right)$, we have

$$\mathbf{E}_1\, dx^1 \cdot \mathbf{E}_2\, dx^2 = |\mathbf{E}_1|\, |\mathbf{E}_2|\, dx^1\, dx^2 \cos\left(\frac{\pi}{2} - \gamma_{12}\right),$$

or

$$E_{12} = \sqrt{E_{11}E_{22}} \sin \gamma_{12}. \tag{3.13†}$$

This gives the geometrical meaning of E_{12}. The meanings of E_{23} and E_{31} follow similarly.

Therefore, we conclude that after deformation an infinitesimal rectangular parallelepiped is transformed into a skew parallelepiped, and the geometry of the deformation can be specified by the set of values of the quantities $E_{\lambda\mu}$ $(\lambda, \mu = 1, 2, 3)$. Consequently, we define strains of the parallelepiped by

$$e_{\lambda\mu} = \tfrac{1}{2}(E_{\lambda\mu} - \delta_{\lambda\mu}) = e_{\mu\lambda} \quad (\lambda, \mu = 1, 2, 3), \tag{3.14}$$

and employ the nine components $e_{\lambda\mu}$, under the symmetry conditions $e_{\lambda\mu} = e_{\mu\lambda}$, as the quantities which specify the strain of the parallelepiped.

Let us express the position vector of the point P as

$$\mathbf{r} = \mathbf{r}^{(0)} + \mathbf{u}, \tag{3.15}$$

where $\mathbf{u}$ is the displacement vector, whose components (u^1, u^2, u^3) are defined by

$$\mathbf{u} = u^\lambda \mathbf{i}_\lambda. \tag{3.16}$$

From Eqs. (3.8) and (3.15), we have

$$\mathbf{E}_\lambda = (\delta^\kappa_\lambda + u^\kappa_{.\lambda})\, \mathbf{i}_\kappa, \tag{3.17}$$

where δ^κ_λ is the Kronecker symbol. By the use of Eqs. (3.14) and (3.17), the strains can be calculated in terms of the displacement components as follows:

$$e_{\lambda\mu} = \tfrac{1}{2}(u^\lambda_{.\mu} + u^\mu_{.\lambda} + u^\kappa_{.\lambda} u^\kappa_{.\mu}) = e_{\mu\lambda}. \tag{3.18}$$

If u, v, w are used instead of u^1, u^2, u^3, respectively, and x, y, z in place of 1, 2, 3, or x^1, x^2, x^3, respectively, Eqs. (3.18) can be written as follows:

$$
\begin{aligned}
e_{xx} &= \frac{\partial u}{\partial x} + \frac{1}{2}\left[\left(\frac{\partial u}{\partial x}\right)^2 + \left(\frac{\partial v}{\partial x}\right)^2 + \left(\frac{\partial w}{\partial x}\right)^2\right],\\
e_{yy} &= \frac{\partial v}{\partial y} + \frac{1}{2}\left[\left(\frac{\partial u}{\partial y}\right)^2 + \left(\frac{\partial v}{\partial y}\right)^2 + \left(\frac{\partial w}{\partial y}\right)^2\right],\\
e_{zz} &= \frac{\partial w}{\partial z} + \frac{1}{2}\left[\left(\frac{\partial u}{\partial z}\right)^2 + \left(\frac{\partial v}{\partial z}\right)^2 + \left(\frac{\partial w}{\partial z}\right)^2\right],\\
2e_{yz} &= \frac{\partial w}{\partial y} + \frac{\partial v}{\partial z} + \frac{\partial u}{\partial y}\frac{\partial u}{\partial z} + \frac{\partial v}{\partial y}\frac{\partial v}{\partial z} + \frac{\partial w}{\partial y}\frac{\partial w}{\partial z} = 2e_{zy},\\
2e_{zx} &= \frac{\partial u}{\partial z} + \frac{\partial w}{\partial x} + \frac{\partial u}{\partial z}\frac{\partial u}{\partial x} + \frac{\partial v}{\partial z}\frac{\partial v}{\partial x} + \frac{\partial w}{\partial z}\frac{\partial w}{\partial x} = 2e_{xz},\\
2e_{xy} &= \frac{\partial v}{\partial x} + \frac{\partial u}{\partial y} + \frac{\partial u}{\partial x}\frac{\partial u}{\partial y} + \frac{\partial v}{\partial x}\frac{\partial v}{\partial y} + \frac{\partial w}{\partial x}\frac{\partial w}{\partial y} = 2e_{yx}.
\end{aligned} \tag{3.19‡}
$$

† $|\mathbf{a}| = (\mathbf{a} \cdot \mathbf{a})^{1/2}$.
‡ Compare with Eqs. (1.5).

We note here that, when the strain components are sufficiently small, Eqs. (3.12) and (3.13) may be linearized with respect to the strains to obtain the following approximate relations:

$$(ds - ds^{(0)})/ds^{(0)} = \sqrt{1 + 2e_{11}} - 1 \approx e_{11}, \tag{3.20}$$

$$\gamma_{12} = \sin^{-1}(E_{12}/\sqrt{E_{11}E_{22}}) \approx 2e_{12}. \tag{3.21}$$

Similar relations hold for the other strain components.

3.2. Analysis of Stress and Equations of Equilibrium

It has been shown in the preceding section that an infinitesimal rectangular parallelepiped is transformed into a skew parallelepiped after deformation. We shall now consider the equilibrium of the deformed parallelepiped.[1]

The forces acting on the deformed parallelepiped are internal forces exerted by neighboring parts of the body through the six side surfaces and body forces, as shown in Fig. 3.2. Let the internal forces acting on one of the side surfaces, the area of which before deformation was $dx^2\,dx^3$ and the

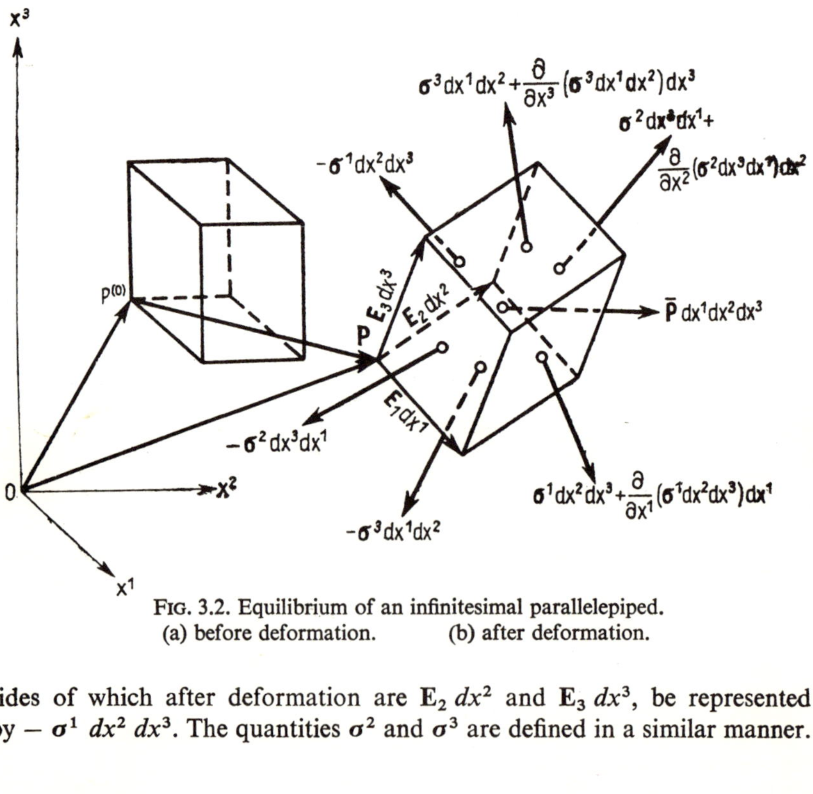

FIG. 3.2. Equilibrium of an infinitesimal parallelepiped. (a) before deformation. (b) after deformation.

sides of which after deformation are $\mathbf{E}_2\,dx^2$ and $\mathbf{E}_3\,dx^3$, be represented by $-\boldsymbol{\sigma}^1\,dx^2\,dx^3$. The quantities $\boldsymbol{\sigma}^2$ and $\boldsymbol{\sigma}^3$ are defined in a similar manner.

The internal forces acting on the six side surfaces are as follows:

$$-\boldsymbol{\sigma}^1\,dx^2\,dx^3,\quad \boldsymbol{\sigma}^1\,dx^2\,dx^3+\frac{\partial}{\partial x^1}(\boldsymbol{\sigma}^1\,dx^2\,dx^3)\,dx^1,$$

$$-\boldsymbol{\sigma}^2\,dx^3\,dx^1,\quad \boldsymbol{\sigma}^2\,dx^3\,dx^1+\frac{\partial}{\partial x^2}(\boldsymbol{\sigma}^2\,dx^3\,dx^1)\,dx^2,$$

$$-\boldsymbol{\sigma}^3\,dx^1\,dx^2,\quad \boldsymbol{\sigma}^3\,dx^1\,dx^2+\frac{\partial}{\partial x^3}(\boldsymbol{\sigma}^3\,dx^1\,dx^2)\,dx^3.$$

The body forces acting in the deformed parallelepiped will be represented by $\bar{\mathbf{P}}\,dx^1\,dx^2\,dx^3$. The force equations of equilibrium of the deformed parallelepiped are then given by

$$\boldsymbol{\sigma}^{\lambda}_{,\lambda}+\bar{\mathbf{P}}=0. \tag{3.22}$$

Let us define the components of $\boldsymbol{\sigma}^{\lambda}$ by resolving it in the directions of the lattice vectors:

$$\boldsymbol{\sigma}^{\lambda}=\sigma^{\lambda\mu}\mathbf{E}_{\mu}, \tag{3.23}†$$

as shown in Fig. 3.3. Then, the moment equation of equilibrium of the deformed parallelepiped are given by

$$(\boldsymbol{\sigma}^1\,dx^2\,dx^3)\times\mathbf{E}_1\,dx^1+(\boldsymbol{\sigma}^2\,dx^3\,dx^1)\times\mathbf{E}_2\,dx^2+(\boldsymbol{\sigma}^3\,dx^1\,dx^2)\times\mathbf{E}_3\,dx^3$$
$$=(\boldsymbol{\sigma}^{\lambda}\times\mathbf{E}_{\lambda})\,dx^1\,dx^2\,dx^3=0, \tag{3.24}‡$$

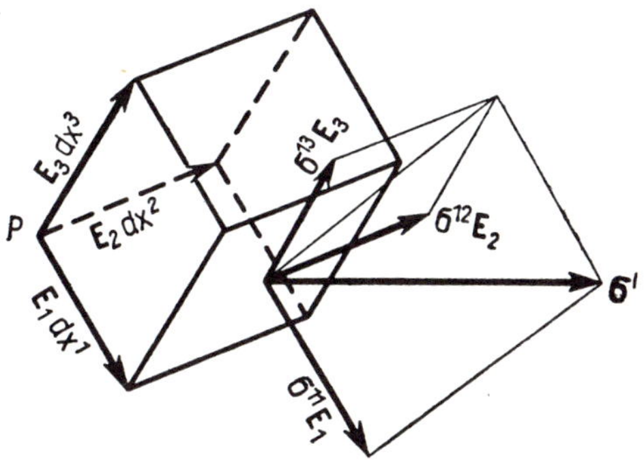

FIG. 3.3. $\boldsymbol{\sigma}^1=\sigma^{1\lambda}\,\mathbf{E}_{\lambda}$.

where higher order terms are neglected. By the use of Eq. (3.23), we obtain from Eq. (3.24) the following relations:

$$\sigma^{\lambda\mu}=\sigma^{\mu\lambda}. \tag{3.25}††$$

† The quantity $\sigma^{\lambda\mu}$ defined by Eq. (3.23) is called *pseudo-stress* or *generalized stress*. However, the familiar nomenclature *stress* will be used instead in subsequent formulations.

‡ A notation $\mathbf{a}\times\mathbf{b}$ denotes a vector product of two vectors $\mathbf{a}$ and $\mathbf{b}$.

†† It is noted that

$$\boldsymbol{\sigma}^{\lambda}\times\mathbf{E}_{\lambda}=\sigma^{\lambda\mu}\mathbf{E}_{\mu}\times\mathbf{E}_{\lambda}=(\sigma^{21}-\sigma^{12})\,\mathbf{E}_1\times\mathbf{E}_2+(\sigma^{32}-\sigma^{23})\,\mathbf{E}_2\times\mathbf{E}_3+(\sigma^{13}-\sigma^{31})\times\mathbf{E}_3\times\mathbf{E}_1,$$

and the three vectors $\mathbf{E}_1\times\mathbf{E}_2$, $\mathbf{E}_2\times\mathbf{E}_3$ and $\mathbf{E}_3\times\mathbf{E}_1$ are mutually independent.

We shall employ the nine components $\sigma^{\lambda\mu}$, under the symmetry conditions $\sigma^{\lambda\mu} = \sigma^{\mu\lambda}$, as the quantities which specify the state of stress on the infinitesimal parallelepiped.

Equation (3.22) is a vector equation. One way of expressing it in scalar form is to resolve it in the direction of $\mathbf{i}_\lambda$. By defining the components of the body force as

$$\bar{\mathbf{P}} = \bar{P}^\lambda \mathbf{i}_\lambda, \tag{3.26}$$

we obtain the following scalar equations from Eq. (3.22):

$$[(\delta^\lambda_\mu + u^\lambda_{.,\mu})\,\sigma^{\varkappa\mu}]_{.\varkappa} + \bar{P}^\lambda = 0 \quad (\lambda = 1, 2, 3). \tag{3.27}\dagger$$

We note here that $\boldsymbol{\sigma}^\lambda$ are defined per unit area and $\bar{\mathbf{P}}$ is defined per unit volume, both with respect to the undeformed state.

3.3. Transformation of the Stress Tensor

The quantities $\sigma^{\lambda\mu}$ which define the state of stress at the point P depend on the choice of coordinates. We shall now find a law of transformation for the stress tensor. Form an infinitesimal tetrahedron $P^{(0)}R^{(0)}S^{(0)}T^{(0)}$ bounded by three surfaces of the rectangular parallelepiped and an oblique surface, as shown in Fig. 3.4. If the area of the inclined surface before deform-

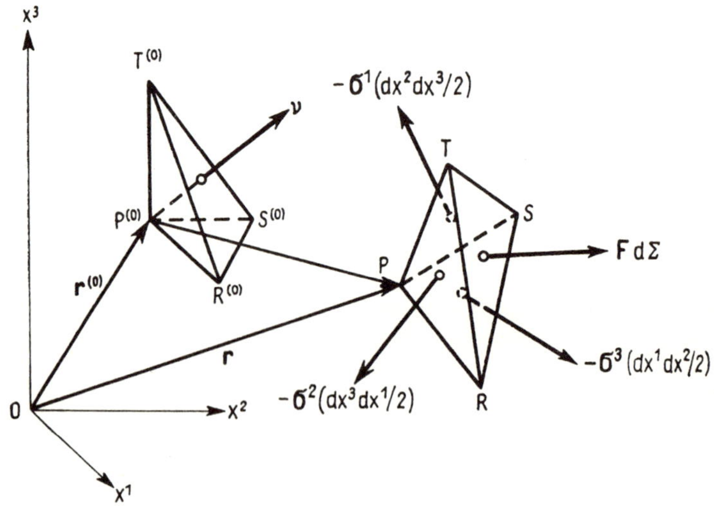

FIG. 3.4. Equilibrium of an infinitesimal tetrahedron.
(a) before deformation. (b) after deformation.

† Compare with Eqs. (1.2).

ation be denoted by $d\Sigma$ and the internal force acting on the surface RST after deformation be represented by $\mathbf{F}\, d\Sigma$, the equilibrium equation of the infinitesimal tetrahedron is

$$\mathbf{F}\, d\Sigma = \boldsymbol{\sigma}^1(dx^2\, dx^3/2) + \boldsymbol{\sigma}^2(dx^3\, dx^1/2) + \boldsymbol{\sigma}^3(dx^1\, dx^2/2). \tag{3.28}$$

From the geometry before deformation, we have

$$dx^2\, dx^3 = 2(\mathbf{i}_1 \cdot \boldsymbol{\nu})\, d\Sigma, \quad dx^3\, dx^1 = 2(\mathbf{i}_2 \cdot \boldsymbol{\nu})\, d\Sigma,$$
$$dx^1\, dx^2 = 2(\mathbf{i}_3 \cdot \boldsymbol{\nu})\, d\Sigma, \qquad (3.29)\dagger$$

where $\boldsymbol{\nu}$ is the unit normal vector drawn outwards on the inclined surface $R^{(0)}S^{(0)}T^{(0)}$ before deformation. Substituting the relations (3.29) into Eq. (3.28), we obtain

$$\mathbf{F} = (\mathbf{i}_\varkappa \cdot \boldsymbol{\nu})\, \boldsymbol{\sigma}^\varkappa. \tag{3.30}$$

This gives the direction and magnitude of the internal force $\mathbf{F}$ acting on the oblique surface. By resolving $\mathbf{F}$ in the direction of the base vectors as

$$\mathbf{F} = F^\lambda \mathbf{i}_\lambda, \tag{3.31}$$

we obtain Eq. (3.30) in scalar form:

$$F^\lambda = \sigma^{\varkappa\mu} n_\varkappa (\delta^\lambda_\mu + u^\lambda_{;\mu}), \tag{3.32}$$

where $n_\varkappa = \mathbf{i}_\varkappa \cdot \boldsymbol{\nu}$.

3.4. Stress–Strain Relations

In the present chapter we shall assume that the deformation under consideration takes place either isothermally or adiabatically, and postulate the existence of functions which define the stress in terms of the strain such that[1]

$$\sigma^{\lambda\mu} = \sigma^{\lambda\mu}(e_{11}, e_{12}, \ldots, e_{33}) \quad (\lambda, \mu = 1, 2, 3), \qquad (3.33)\ddagger$$

where the zero stress state corresponds to the zero strain state, namely, $\sigma^{\lambda\mu}(0, 0, \ldots, 0) = 0$. We also assume the existence of unique inverse functions which define the strain in terms of the stress:

$$e_{\lambda\mu} = e_{\lambda\mu}(\sigma^{11}, \sigma^{12}, \ldots, \sigma^{33}) \quad (\lambda, \mu = 1, 2, 3). \qquad (3.34)\ddagger$$

When the strain components are assumed sufficiently small, we may expand Eqs. (3.33) into power series with respect to $e_{\alpha\beta}$ and neglect higher

† See footnote of Eqs. (4.63).

‡ Only six equations are physically independent in Eqs. (3.33). However, we may write them and their inverse relations in nine equations of symmetrical form as given by Eqs. (3.33) and (3.34), respectively.

order terms to obtain the following linear stress–strain relations:

$$\sigma^{\lambda\mu} = a^{\lambda\mu\alpha\beta} e_{\alpha\beta}. \qquad (3.35)\dagger$$

It is obvious that, due to the symmetry property of the stress and strain tensors, there exist the following relations among the coefficients:

$$a^{\lambda\mu\alpha\beta} = a^{\mu\lambda\alpha\beta} = a^{\lambda\mu\beta\alpha}.$$

Eqs. (3.35) may be inverted to yield:

$$e_{\lambda\mu} = b_{\lambda\mu\alpha\beta}\,\sigma^{\alpha\beta}, \qquad (3.36)\dagger$$

where

$$b_{\lambda\mu\alpha\beta} = b_{\mu\lambda\alpha\beta} = b_{\lambda\mu\beta\alpha}.$$

When the material is isotropic, the numerical values of $a^{\lambda\mu\alpha\beta}$ must be independent of the coordinate system in which the stress and strain are defined. This leads to the conclusion that $a^{\lambda\mu\alpha\beta}$ is a fourth-order isotropic tensor and is given by[3, 5]

$$a^{\lambda\mu\alpha\beta} = \frac{E\nu}{(1+\nu)(1-2\nu)}\,\delta^{\lambda\mu}\delta^{\alpha\beta} + G(\delta^{\lambda\alpha}\delta^{\mu\beta} + \delta^{\lambda\beta}\delta^{\mu\alpha}), \qquad (3.37)$$

and Eqs. (3.35) and (3.36) reduce to

$$\sigma^{\lambda\mu} = \frac{E}{(1-2\nu)}\,e\delta_{\lambda\mu} + 2Ge'_{\lambda\mu}, \qquad (3.38)\dagger$$

and

$$e_{\lambda\mu} = \frac{(1-2\nu)}{E}\,\sigma\delta^{\lambda\mu} + \frac{1}{2G}\,\sigma^{\lambda\mu\prime}, \qquad (3.39)\dagger$$

respectively, where $E = 2(1+\nu)\,G$. The quantities $\sigma^{\lambda\mu\prime}$ and $e'_{\lambda\mu}$ are deviator stresses and deviator strains defined by $\sigma^{\lambda\mu\prime} = \sigma^{\lambda\mu} - \sigma\delta^{\lambda\mu}$ and $e'_{\lambda\mu} = e_{\lambda\mu} - e\delta_{\lambda\mu}$, respectively, where $\sigma = (1/3)\,\sigma^{\lambda\lambda} = (1/3)\,(\sigma^{11} + \sigma^{22} + \sigma^{33})$ and $e = (1/3)\,e_{\lambda\lambda} = (1/3)\,(e_{11} + e_{22} + e_{33})$.

3.5. Presentation of a Problem

With the above preliminaries, we shall now define a boundary value problem in the finite displacement theory of elasticity. Consider an elastic body subjected to the following boundary conditions and body forces:

(1) *Mechanical boundary conditions on S_1*,

$$\mathbf{F} = \bar{\mathbf{F}}, \qquad (3.40)$$

where $\mathbf{F}$ is given by Eq. (3.30) with the understanding that the vector $\boldsymbol{\nu}$ is now the unit normal drawn outwards on the boundary and $\bar{\mathbf{F}}$ is the pre-

† Compare with Eqs. (1.6), (1.8), (1.10) and (1.11), respectively.

scribed external force. Both **F** and $\bar{\mathbf{F}}$ are defined per unit area of the undeformed state. Resolving $\bar{\mathbf{F}}$ in the directions of the base vectors,

$$\bar{\mathbf{F}} = \bar{F}^{\lambda}\mathbf{i}_{\lambda}, \tag{3.41}$$

we obtain from Eq. (3.40) the following scalar equations:

$$F^{\lambda} = \bar{F}^{\lambda} \quad (\lambda = 1, 2, 3). \qquad (3.42)\dagger$$

(2) *Geometrical boundary conditions on* S_2,

$$u^{\lambda} = \bar{u}^{\lambda} \quad (\lambda = 1, 2, 3). \qquad (3.43)\dagger$$

(3) *The body forces*

$$\bar{P}^{\lambda} \quad (\lambda = 1, 2, 3). \tag{3.44}$$

Our problem is then to find the stresses and displacements existing in the deformed body by employing the stress–strain relations (3.33).

By combining Eqs. (3.18) and (3.33), we can represent $\sigma^{\lambda\mu}$ in terms of $u^{\varkappa}$. Introducing $\sigma^{\lambda\mu}$ thus represented into Eqs. (3.27) and (3.42), we obtain three simultaneous differential equations and the mechanical boundary conditions in terms of $u^{\varkappa}$. If these differential equations can be solved under the boundary conditions on S_1 and S_2, we can obtain the required equilibrium configuration. Once the displacement components $u^{\varkappa}$ have been obtained, the state of stress induced in the body may be determined from Eqs. (3.18) and (3.33).

Since the problem is nonlinear, the solution, if obtainable, generally yields nonlinear relations between the applied external loads and the resulting deformations. Some typical examples of nonlinear relations are illustrated in Figs. 3.5, 3.6, 3.7 and 3.8, where the ordinate is the applied external load and the abscissa denotes the resulting displacement of the point

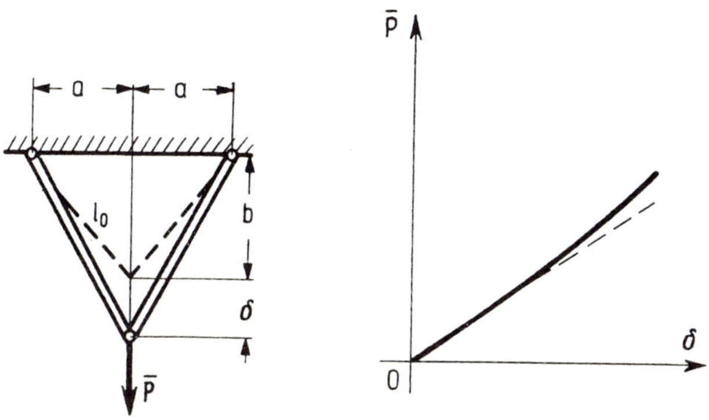

FIG. 3.5. A load–deflection curve of a truss structure.

† Compare with Eqs. (1.12) and (1.14), respectively.

of application in the direction of the external load.[7] In the example illustrated in Fig. 3.5, the slope of the curve, $d\bar{P}/d\delta$, increases with the increase of the deflection. This shows that the body is stable under the external load as long as it behaves elastically. On the other hand, in the example illustrated in Fig. 3.6, the slope $d\bar{M}/d\varphi$ decreases with increasing deformation

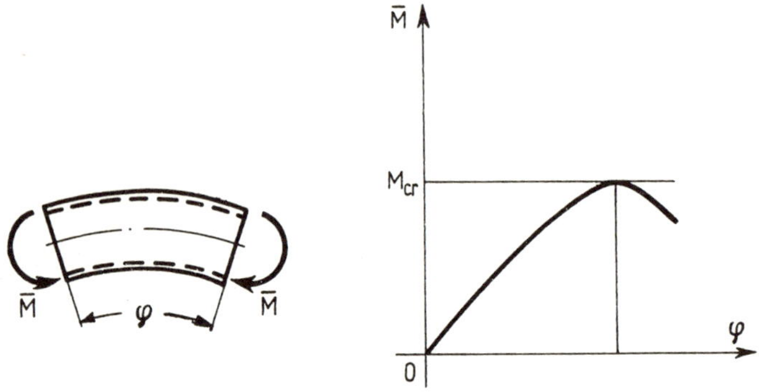

FIG. 3.6. Flattening instability of a thin cylindrical tube under bending.

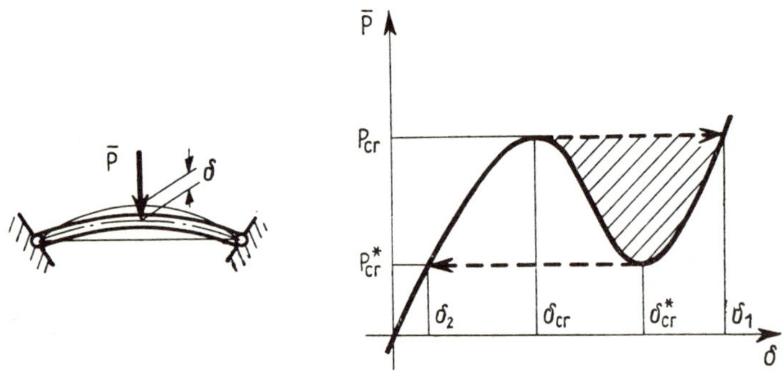

FIG. 3.7. Durchschlag or snap-through of a curved beam.

and the load $\bar{M}$ eventually reaches its maximum M_{cr}. This shows that the body ceases to be stable at M_{cr}, beyond which the body fails elastically.[8, 9] The phenomenon illustrated in Fig. 3.7, where the load–deflection curve has the S-shape characteristic, is called Durchschlag or snap-through.[10] If $\bar{P}$ is a dead load and increases from zero to P_{cr}, the deflection jumps from δ_{cr} to δ_1 with kinetic energy illustrated by the shaded area of the figure. In the case of unloading, another jump occurs at P^*_{cr}. If the load–deflection relation is as shown in Fig. 3.8, where a point of bifurcation exists at P_{cr}, the system has two states of equilibrium at loads beyond the critical load. The body prefers a stable configuration and changes its deflection suddenly from unstable to stable under the stimulus of small external disturbances. The phenomena shown in the last three figures constitute the main parts of the theory of elastic stability.[11-13]

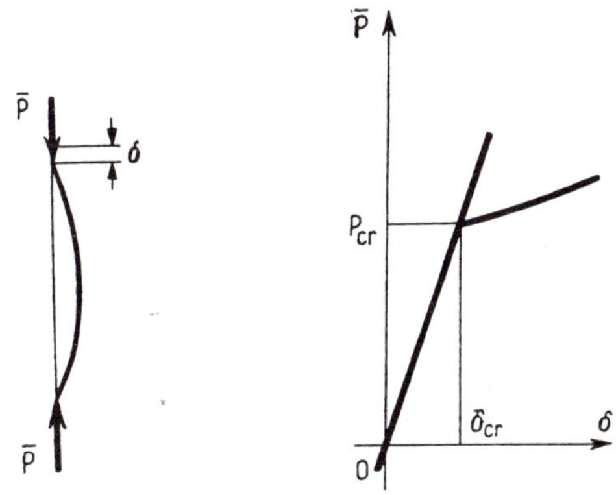

FIG. 3.8. Bifurcation of a rod under compression.

3.6. Principle of Virtual Work

In this section we shall derive the principle of virtual work of the continuous body under consideration. Assume that the body is in equilibrium under the body forces, the applied external forces on S_1 and the prescribed geometrical boundary conditions on S_2. Now, the body is assumed to execute an infinitesimal virtual displacement $\delta\mathbf{u}$ from this equilibrating configuration without violating the prescribed boundary conditions on S_2. Then, by employing the equations of equilibrium (3.22) and the mechanical boundary conditions (3.40), and remembering that $\delta\mathbf{u} = \delta\mathbf{r}$, we obtain

$$-\iiint_V (\sigma^\lambda{}_{,\lambda} + \bar{\mathbf{P}}) \cdot \delta\mathbf{r}\, dV + \iint_{S_1} (\mathbf{F} - \bar{\mathbf{F}}) \cdot \delta\mathbf{r}\, dS = 0, \qquad (3.45)$$

where $dV = dx^1\, dx^2\, dx^3$ and dS are the elementary volume of the parallelepiped and the elementary area on the surface of the body before deformation, respectively. By the use of the geometrical relations,

$$dx^2\, dx^3 = \pm n_1\, dS, \quad dx^3\, dx^1 = \pm n_2\, dS, \quad dx^1\, dx^2 = \pm n_3\, dS, \qquad (3.46)\dagger$$

which hold on the surface of the body, the first term of Eq. (3.45) may be transformed into

$$-\iiint_V \sigma^\lambda{}_{,\lambda} \cdot \delta\mathbf{r}\, dV =$$
$$-\iint_{S_1+S_2} \mathbf{F} \cdot \delta\mathbf{r}\, dS + \iiint_V \sigma^\lambda \cdot (\delta\mathbf{r})_{,\lambda}\, dV.$$

† See footnote of Eqs. (4.79).

Introducing the above into Eq. (3.45), and remembering that $\delta\mathbf{r} = 0$ on S_2, we obtain

$$\iiint_V \boldsymbol{\sigma}^\lambda \cdot (\delta\mathbf{r})_{,\lambda}\, dV - \iiint_V \bar{\mathbf{P}} \cdot \delta\mathbf{r}\, dV - \iint_{S_1} \bar{\mathbf{F}} \cdot \delta\mathbf{r}\, dS = 0. \tag{3.47}$$

The integrand $\boldsymbol{\sigma}^\lambda \cdot (\delta\mathbf{r})_{,\lambda}\, dV$ of the first term in Eq. (3.47) may be interpreted as the virtual work done, during the infinitesimal virtual displacement, by the body forces and the internal forces acting on the deformed infinitesimal parallelepiped, while the second and the third terms represent the virtual work done by the body forces and the external forces on S_1, respectively. Combination of Eqs. (3.23), (3.11), (3.25) and (3.14) yields

$$\begin{aligned} \boldsymbol{\sigma}^\lambda \cdot (\delta\mathbf{r})_{,\lambda} &= \sigma^{\lambda\mu}\mathbf{r}_{,\mu}\,(\delta\mathbf{r})_{,\lambda} \\ &= \tfrac{1}{2}\sigma^{\lambda\mu}\,\delta E_{\lambda\mu} = \sigma^{\lambda\mu}\,\delta e_{\lambda\mu}. \end{aligned}$$

Introducing the above into Eq. (3.47), we obtain

$$\iiint_V \sigma^{\lambda\mu}\,\delta e_{\lambda\mu}\, dV - \iiint_V \bar{\mathbf{P}} \cdot \delta\mathbf{r}\, dV - \iint_{S_1} \bar{\mathbf{F}} \cdot \delta\mathbf{r}\, dS = 0. \tag{3.48}$$

This is the principle of virtual work for the elasticity problem in finite displacement theory. By the use of Eqs. (3.16), (3.26) and (3.41), the principle may be expressed alternatively as follows:

$$\iiint_V (\sigma^{\lambda\mu}\,\delta e_{\lambda\mu} - \bar{P}^\lambda\,\delta u^\lambda)\, dV - \iint_{S_1} \bar{F}^\lambda\,\delta u^\lambda\, dS = 0. \tag{3.49†}$$

Reversing the above development, we obtain from the principle, the equations of equilibrium (3.27) and mechanical boundary conditions (3.42) under the assumptions of the strain-displacement relations (3.18), the symmetry of stress components (3.25), and the prescribed geometrical boundary conditions (3.43). We note that the principle holds regardless of the form of the stress–strain relations of the body.

3.7. Strain Energy Function

Let us consider an element of the body which is a rectangular parallelepiped and occupies a unit volume before deformation. When the element is subjected to deformation along a loading path and is brought into a strained state expressed by $(e_{11}, e_{12}, \ldots, e_{33})$, we may calculate an integral,

$$\int_{(0,\ldots,0)}^{(e_{11},\ldots,e_{33})} \sigma^{\lambda\mu}\, de_{\lambda\mu}, \tag{3.50}$$

† Compare with Eq. (1.32).

For consistency in tensor notation, it is better to write $\delta_{\lambda\mu}\,\bar{P}^\lambda\,\delta u^\mu$ or $\bar{P}^\lambda\,\delta u_\lambda$ instead of $\bar{P}^\lambda\,\delta u^\lambda$ where $\delta_{\lambda\mu}$ is the Kronecker symbol and $\delta u_\lambda = \delta_{\lambda\mu}\,\delta u^\mu$. However, we shall use the simpler expressions for the sake of brevity, whenever the rectangular Cartesian coordinate system is employed.

along the loading path by the use of the stress–strain relations (3.33). The value of the integral thus obtained generally depends on the loading path. However, if the value does not depend on the loading path, but depends only on the final strain, the quantity $\sigma^{\lambda\mu}\, de_{\lambda\mu}$ is called a perfect differential and the existence of a state function $A(e_{11}, e_{12}, \dots, e_{33})$ is assured such that

$$dA = \sigma^{\lambda\mu}\, de_{\lambda\mu}, \tag{3.51}$$

or equivalently,

$$\frac{\partial A}{\partial e_{\lambda\mu}} = \sigma^{\lambda\mu} \quad (\lambda, \mu = 1, 2, 3). \tag{3.52}$$

The state function thus defined is the strain energy function in the finite displacement theory of elasticity.

The present section is concerned with the conditions under which the quantity $\sigma^{\lambda\mu}\, de_{\lambda\mu}$ is a perfect differential. A detailed discussion of these conditions may be found in any book on partial differential equations. However, we may summarize them as follows:

If the stress–strain relations (3.33) satisfy the equations

$$\frac{\partial \sigma^{\lambda\mu}}{\partial e_{\alpha\beta}} = \frac{\partial \sigma^{\alpha\beta}}{\partial e_{\lambda\mu}} \quad (\lambda, \mu, \alpha, \beta = 1, 2, 3) \tag{3.53}$$

the quantity $\sigma^{\lambda\mu}\, de_{\lambda\mu}$ proves to be a perfect differential.

Consequently, if the stress–strain relations (3.35) satisfy the equations

$$a^{\lambda\mu\alpha\beta} = a^{\alpha\beta\lambda\mu}, \tag{3.54}$$

we have

$$A = \tfrac{1}{2} a^{\lambda\mu\alpha\beta} e_{\lambda\mu} e_{\alpha\beta}. \tag{3.55}\dagger$$

When the material is isotropic and the stress–strain relations given by Eq. (3.38) may be employed, we have

$$A = \frac{3E}{2(1-2\nu)} e^2 + G e'_{\lambda\mu} e'_{\lambda\mu}. \tag{3.56}\dagger$$

So far, conditions for the existence of the strain energy function have been studied mathematically. We shall now show from physical considerations that such a function really exists when an elastic body deforms either isothermally or adiabatically in a reversible process.[2, 14, 15]

We assume a unit volume of an elastic body again and call it an element. The first law of thermodynamics is applied to the element by taking the sum of energy supplied to the element during the strain's increase by $de_{\lambda\mu}$ to obtain,

$$dU_0 = d'Q + \sigma^{\lambda\mu}\, de_{\lambda\mu}, \tag{3.57}$$

† Compare with Eqs. (2.2) and (2.3), respectively.

where dU_0 and $d'Q$ are the increments of the internal energy and heat energy supplied to the element, respectively. The quantity dU_0 is the differential of the internal energy U_0, which is a single-valued function of the temperature and the instantaneous state of strain of the element. Physically speaking, the internal energy U_0 is a function of the mean position of the molecules under the intermolecular forces and the kinetic energy of the molecules about their mean positions.(16) However, Q is not a state function and $d'Q$ merely denotes an infinitesimal amount of heat energy supplied. Therefore, the special notation d' is used here to avoid confusion. The quantity $\sigma^{\lambda\mu}\, de_{\lambda\mu}$ is an infinitesimal amount of supplied mechanical energy. After these two kinds of energy have been supplied to the element, there can never be any distinction between portions of dU_0 contributed by $d'Q$ and $\sigma^{\lambda\mu}\, de_{\lambda\mu}$.

The second law of thermodynamics assures the existence of a state function S, called entropy, such that

$$d'Q = T\, dS, \tag{3.58}$$

where T is the absolute temperature of the element. Combining Eqs. (3.57) and (3.58), we obtain,

$$dU_0 = T\, dS + \sigma^{\lambda\mu}\, de_{\lambda\mu}. \tag{3.59}$$

Equation (3.57) shows that if the deformation takes place adiabatically in a reversible process, we have

$$dU_0 = \sigma^{\lambda\mu}\, de_{\lambda\mu}. \tag{3.60}$$

Consequently, the mechanical energy is stored in the element in the form of internal energy and we obtain

$$A = U_0 + \text{constant}. \tag{3.61}$$

On the other hand, if the deformation takes place isothermally in a reversible process, we have from Eq. (3.59)

$$dF_0 = \sigma^{\lambda\mu}\, de_{\lambda\mu}, \tag{3.62}$$

where

$$F_0 = U_0 - TS \tag{3.63}$$

is the Helmholtz free energy function. This shows that the mechanical energy is stored in the form of Helmholtz free energy and we obtain

$$A = F_0 + \text{constant}. \tag{3.64}$$

Therefore, it may be concluded that the quantity $\sigma^{\lambda\mu}\, de_{\lambda\mu}$ is a perfect differential for these two special cases, and the existence of the strain energy function is assured.

The differences between the assumptions of adiabatic and isothermal deformations appear, in mathematical formulations, only as differences between adiabatic and isothermal elastic constants. Generally speaking, the differences between these elastic constants have been proved by experiments to be negligible. Consequently, the strain energy function is usually assumed to exist in the theory of elasticity, although the deformation process may be somewhere between adiabatic and isothermal.

We know from experimental evidence that when the strains are sufficiently small, an element of the elastic body is stable. This requires that the strain energy function must be a positive definite function of the strain components for the small strain. Since we have also found that, when the strains are small enough, the strain energy function can be expressed by Eq. (3.55), we may conclude that the strain energy function (3.55) is a positive definite function of the strains.

3.8. Principle of Stationary Potential Energy

We have investigated in the preceding section the condition under which the strain energy function can exist. When the strain energy function is assured to exist, the principle of virtual work (3.49) can be written as follows:

$$\delta \iiint_V A(u^\lambda)\, dV - \iiint_V \bar{P}^\lambda\, \delta u^\lambda\, dV - \iint_{S_1} \bar{F}^\lambda\, \delta u^\lambda\, dS = 0, \qquad (3.65)\dagger$$

where the strain energy function $A(u^\lambda)$ is written in terms of u^λ by the use of Eqs. (3.18). The principle (3.65) is very useful in application to elasticity problems in which external forces are not derivable from potential functions.

Next, we shall assume further that the applied external forces are conservative, namely, they are derivable from potential functions $\Phi(u^\lambda)$ and $\Psi(u^\lambda)$ such that

$$\delta\Phi = -\bar{P}^\lambda\, \delta u^\lambda, \quad \delta\Psi = -\bar{F}^\lambda\, \delta u^\lambda. \qquad (3.66)$$

If the applied external forces vary neither in magnitude nor in direction during the virtual displacements, namely, if they are treated as dead loads, we may have

$$\Phi = -\bar{P}^\lambda u^\lambda, \quad \Psi = -\bar{F}^\lambda u^\lambda. \qquad (3.67)$$

Under the assumption of the existence of the strain energy function A and two potential functions Φ and Ψ, the principle of virtual work (3.49) yields the principle of stationary potential energy as follows:

$$\delta\Pi = 0, \qquad (3.68)$$

where

$$\Pi = \iiint_V [A(u^\lambda) + \Phi(u^\lambda)]\, dV + \iint_{S_1} \Psi(u^\lambda)\, dS \qquad (3.69)\ddagger$$

† Compare with Eq. (2.5).
‡ Compare with Eq. (2.9).

is called the total potential energy of the mechanical system under consideration and is a functional containing u^λ which are the independent variables subject to variation. The variables u^λ should be chosen so that they satisfy the required continuity and differentiability criteria and the boundary conditions on S_2. The principle (3.68) can be stated as follows: *Among all admissible displacement functions u^λ, the actual ones are those which render the total potential energy stationary.* Retracing the development in the latter half of Section 3.6, we can easily show that the conditions which make the total potential energy stationary provide the equations of equilibrium and the mechanical boundary conditions on S_1. However, in these equations and conditions all the stress components are expressed in terms of displacement components, and we obtain three simultaneous differential equations of equilibrium in V and three boundary condition equations on S_1 in terms of u^λ. The formulation thus obtained can also be reached by direct elimination of the stress components from the equations of equilibrium and the mechanical boundary conditions by the use of the relations (3.18) and (3.33), as observed at the end of Section 3.5.

3.9. Generalization of the Principle of Stationary Potential Energy

It is obvious that the principle of stationary potential energy established in Section 3.8 can be generalized by the use of Lagrange multipliers. Through familiar procedures, we obtain a generalized functional as follows:

$$\begin{aligned}\Pi_I = \iiint_V \{&A(e_{\lambda\mu}) + \Phi(u^\varkappa) \\ &- \sigma^{\lambda\mu}[e_{\lambda\mu} - \tfrac{1}{2}(u^\lambda_{.\mu} + u^\mu_{.\lambda} + u^\varkappa_{.\lambda}u^\varkappa_{.\mu})]\}\, dV \\ &+ \iint_{S_1} \Psi(u^\varkappa)\, dS - \iint_{S_2} p^\lambda(u^\lambda - \bar{u}^\lambda)\, dS,\end{aligned} \qquad (3.70)\dagger$$

where the independent quantities subject to variation are $e_{\lambda\mu}$, u^λ, $\sigma^{\lambda\mu}$ and p^λ with no subsidiary conditions. The functional for Reissner's principle[17] can be derived from the functional (3.70) through elimination of $e_{\lambda\mu}$ as follows:

$$\begin{aligned}\Pi_R = \iiint_V [&\tfrac{1}{2}(u^\lambda_{,\mu} + u^\mu_{,\lambda} + u^\varkappa_{,\lambda}u^\varkappa_{,\mu})\,\sigma^{\lambda\mu} \\ &- B(\sigma^{\lambda\mu}) + \Phi(u^\varkappa)]\, dV \\ &+ \iint_{S_1} \Psi(u^\varkappa)\, dS - \iint_{S_2} p^\lambda(u^\lambda - \bar{u}^\lambda)\, dS,\end{aligned} \qquad (3.71)\dagger$$

where the independent quantities subject to variation are u^λ, $\sigma^{\lambda\mu}$ and p^λ with no subsidiary conditions.

† Compare with the functionals (2.26) and (2.37), respectively.

The quantity $B(\sigma^{\lambda\mu})$ appearing in Eq. (3.71) is the complementary energy function. As the above development shows, it is defined by

$$B = \sigma^{\lambda\mu} e_{\lambda\mu} - A. \tag{3.72}$$

The stress–strain relations (3.34) are introduced into Eq. (3.72) to express B entirely in terms of the stress components. From Eqs. (3.51) and (3.72), we have

$$dB = e_{\lambda\mu}\, d\sigma^{\lambda\mu}, \tag{3.73}$$

or equivalently,

$$\frac{\partial B}{\partial \sigma^{\lambda\mu}} = e_{\lambda\mu} \quad (\lambda, \mu = 1, 2, 3). \tag{3.74}$$

We note here that if Eqs. (3.34) satisfy the equations

$$\frac{\partial e_{\lambda\mu}}{\partial \sigma^{\alpha\beta}} = \frac{\partial e_{\alpha\beta}}{\partial \sigma^{\lambda\mu}} \quad (\lambda, \mu, \alpha, \beta = 1, 2, 3), \tag{3.75}$$

the function B is assured to exist and may be determined from Eq. (3.73) independently of the function A. Consequently, if Eqs. (3.36) satisfy the equations

$$b_{\lambda\mu\alpha\beta} = b_{\alpha\beta\lambda\mu} \quad (\lambda, \mu, \alpha, \beta = 1, 2, 3), \tag{3.76}$$

we have

$$B = \tfrac{1}{2} b_{\lambda\mu\alpha\beta} \sigma^{\lambda\mu} \sigma^{\alpha\beta}. \tag{3.77}\dagger$$

When Eq. (3.39) is employed, we obtain

$$B = \frac{3(1 - 2\nu)}{2E} \sigma^2 + \frac{1}{4G} \sigma^{\lambda\mu'} \sigma^{\lambda\mu'} \tag{3.78}\dagger$$

for the isotropic material.

Under the assumptions of small displacement theory, the principle of minimum complementary energy can be expressed in terms of the stress components only, as shown in Section 2.2. However, coupling of the displacements with the stress components in finite displacement problems complicates the derivation of the principle of stationary complementary energy from Π_R; the principle can no longer be expressed purely in terms of stress components.

3.10. Energy Criterion for Stability

We know that, as long as the applied external loads are sufficiently small, we obtain linear relations between the loads and resulting deformations. However, the deformation characteristics gradually deviate from the linear relations with increasing loads. This tendency is usually pronounced in

† Compare with Eqs. (2.20) and (2.21), respectively.

slender or thin bodies, and a point is finally reached beyond which the bodies cease to be stable for some loading conditions, as shown in the examples illustrated in Figs. 3.6, 3.7 and 3.8. In the present section, we shall consider the energy criterion for determining the stability and critical load of equilibrium configurations of elastic bodies under conservative external forces.†

We assume that the equilibrium configuration of the body has been obtained for the problem defined in Section 3.5, and call it the original configuration. Then, let the original configuration be given small virtual displacements without violating the geometrical boundary conditions, thus obtaining a new configuration. If the virtual work done by the external forces does not exceed the increase of the stored strain energy, the body is considered stable. If this condition is not met for some virtual displacements, then the excess energy will appear as kinetic energy. This indicates an instability of the original configuration for the virtual displacements.

The above considerations lead to the following mathematical formulation. Let the displacement components of the original equilibrium configuration and the new configuration be denoted by u^λ and $u^\lambda + \delta u^\lambda$ respectively. Denoting the total potential energy of the original configuration and the new configuration by $\Pi(u^\lambda)$ and $\Pi(u^\lambda + \delta u^\lambda)$ respectively, we have

$$\Pi(u^\lambda + \delta u^\lambda) = \Pi(u^\lambda) + \delta\Pi + \delta^2\Pi + \delta^3\Pi + \cdots, \tag{3.79}$$

where $\delta\Pi$, $\delta^2\Pi$, $\delta^3\Pi$, ..., are the first, second, third, ... variations of the total potential energy. They are linear, quadratic, ... with respect to δu^λ and their derivatives, and their coefficients contain the displacement components of the original configuration as parameters.‡ Since the original configuration is in equilibrium, we have

$$\delta\Pi = 0. \tag{3.80}$$

With these preliminaries, we may now conclude that the stability of the original configuration in its neighborhood can be determined by the sign of the second variation $\delta^2\Pi$ as follows:

(1) *The configuration is stable if* $\delta^2\Pi > 0$ *holds for all admissible virtual displacements.*††

† Refs. 1 and 18 through 22.

‡ When all the applied external forces are dead loads and their potential functions are given by Eqs. (3.67), we have

$$\delta^2\Pi = \tfrac{1}{2}\iiint_V [(\partial^2 A/\partial e_{\lambda\mu}\,\partial e_{\alpha\beta})\,\delta e_{\lambda\mu}\,\delta e_{\alpha\beta} + \sigma^{\lambda\mu}\,\delta u^{\varkappa}_{.\lambda}\,\delta u^{\varkappa}_{.\mu}]\,dV,$$

where $\sigma^{\lambda\mu}$ are the stress components of the original configuration and

$$2\delta e_{\lambda\mu} = (\delta^{\varkappa}_{\lambda} + u^{\varkappa}_{.\lambda})\,\delta u^{\varkappa}_{.\mu} + (\delta^{\varkappa}_{\mu} + u^{\varkappa}_{.\mu})\,\delta u^{\varkappa}_{.\lambda}.$$

†† This statement, combined with the principle of minimum potential energy derived in Section 2.1, assures that no unstable configurations exist as far as problems of the small displacement theory of elasticity are concerned.

(2) *The configuration is unstable if $\delta^2\Pi < 0$ holds for at least one admissible set of virtual displacements.*

Following Trefftz,[(18)] we shall consider the determination of the lowest critical load beyond which the body ceases to be stable for the first time during the loading process. We have seen that the original configuration is stable as long as $\delta^2\Pi > 0$ holds for all admissible virtual displacements. This criterion will now be expressed in a different manner. We introduce a properly chosen functional N which is positive definite and quadratic with respect to δu^λ and their derivatives,† and seek, among admissible virtual displacements which satisfy

$$\delta u^\lambda = 0 \quad \text{on} \quad S_2, \tag{3.81}$$

those which make the quotient

$$\lambda = \delta^2\Pi/N \tag{3.82}$$

a minimum. The criterion then states: *if the minimum value of the quotient is found to be positive, the original configuration is stable.*

We know that since

$$\hat{\delta}\lambda = \hat{\delta}(\delta^2\Pi/N) = \hat{\delta}(\delta^2\Pi - \lambda N)/N, \tag{3.83}$$

where $\hat{\delta}$ indicates that the variation is taken with respect to δu^λ, the stationary condition of the quotient is given by

$$\hat{\delta}(\delta^2\Pi - \lambda N) = 0. \tag{3.84}$$

Equation (3.84) yields differential equations and mechanical boundary conditions, which, together with the geometrical boundary conditions (3.81), determine the stationary values of the quotient as eigenvalues. Consequently, the stability criterion may be expressed as follows: *if the minimum of the eigenvalues is found to be positive, the original configuration is stable.*

The above consideration leads to a conclusion for the determination of the lowest critical load: *the external load acting on the original configuration is considered critical when the minimum of the eigenvalues reaches the value zero; the variational equation*

$$\hat{\delta}(\delta^2\Pi) = 0 \tag{3.85}$$

under the subsidiary conditions (3.81) then yields governing equations which determine the lowest critical load. We note here that the governing equations determine all the critical configurations which possess at least one eigenvalue of the value zero in the eigenvalue problem derived from Eq. (3.84), and the configuration corresponding to the lowest critical load is one of the critical configurations. The lowest critical load thus determined is frequently

† The functional N, which is introduced here for the normalization of the virtual displacements, has no effect on the final result, namely, Eq. (3.85).

followed by unstable states upon further increase of the external load, but the sign of the higher-order variation must be considered for a sufficiency proof.

3.11. The Euler Method for Stability Problem

Another method of determining the critical load concerns the problem of whether there exists at least one additional, distinct equilibrium configuration in the very close neighborhood of the original configuration. If such an adjacent equilibrium configuration exists, the body may change suddenly from one equilibrium configuration to the other under the stimulus of small external disturbances. We shall formulate the stability problem by this approach, which is sometimes called *the Euler method*, for a body under nonconservative external forces.[21, 23]†

We assume the existence of a critical original configuration and develop a linearized theory which determines the adjacent configuration. We denote stresses, strains, displacements and external forces of the original and adjacent equilibrium configuration by

$$\sigma^{\lambda\mu},\ e_{\lambda\mu},\ u^{\lambda},\ \bar{P}^{\lambda},\ \bar{F}^{\lambda}$$

and

$$\sigma^{\lambda\mu}+\sigma_*^{\lambda\mu},\quad e_{\lambda\mu}+e^*_{\lambda\mu},\quad u^{\lambda}+u_*^{\lambda}\quad \bar{P}^{\lambda}+\bar{P}_*^{\lambda},\quad \bar{F}^{\lambda}+\bar{F}_*^{\lambda},$$

respectively, where

$$u_\lambda^* = 0 \quad \text{on} \quad S_2, \tag{3.86}$$

because the geometrical boundary conditions are the same for both configurations. We may derive from Eq. (3.49) the principle of virtual work of the adjacent equilibrium configuration by replacing $\sigma^{\lambda\mu}$, u^{λ}, ... by $\sigma^{\lambda\mu}+\sigma_*^{\lambda\mu}$, $u^{\lambda}+u_*^{\lambda}$, ..., respectively:

$$\iiint\limits_V (\sigma^{\lambda\mu}+\sigma_*^{\lambda\mu})\,\delta(e_{\lambda\mu}+e^*_{\lambda\mu})\,dV - \iiint\limits_V (\bar{P}^{\lambda}+\bar{P}_*^{\lambda})\,\delta u_*^{\lambda}\,dV$$
$$- \iint\limits_{S_1} (\bar{F}^{\lambda}+\bar{F}_*^{\lambda})\,\delta u_*^{\lambda}\,dS = 0, \tag{3.87}$$

where

$$2(e_{\lambda\mu}+e^*_{\lambda\mu}) = (u^{\lambda}+u_*^{\lambda})_{,\mu} + (u^{\mu}+u_*^{\mu})_{,\lambda}$$
$$+ (u^{\varkappa}+u_*^{\varkappa})_{,\lambda}(u^{\varkappa}+u_*^{\varkappa})_{,\mu}, \tag{3.88}$$

and the variation is taken with respect to u_*^{λ}. The equation (3.87) is the principle of virtual work for an incremental theory of elasticity.

† It is emphasized in Ref. 23 that the stability problem of a nonconservative system should be investigated not only by *the Euler method* which deals with the static instability, but also by *the dynamic method* which deals with the dynamic instability of small oscillations of the system about the original equilibrium configuration.

Since we are interested in a linearized theory, we may assume that $\sigma_*^{\lambda\mu}$, $\bar{P}_*^\lambda$ and $\bar{F}_*^\lambda$ are linear functions of $u_*^\varkappa$ and their derivatives. Remembering that the original configuration is in equilibrium, we may reduce the principle (3.87) into the following form:

$$\iiint_V (\sigma_*^{\lambda\mu}\,\delta\varepsilon_{\lambda\mu}^* + \sigma^{\lambda\mu}u_{*,\lambda}^\varkappa\,\delta u_{*,\mu}^\varkappa)\,dV$$

$$-\iiint_V \bar{P}_*^\lambda\,\delta u_*^\lambda\,dV - \iint_{S_1} \bar{F}_*^\lambda\,\delta u_*^\lambda\,dS = 0, \tag{3.89}$$

where

$$2\varepsilon_{\lambda\mu}^* = (\delta_\lambda^\varkappa + u^\varkappa{}_{,\lambda})\,u_{*,\mu}^\varkappa + (\delta_\mu^\varkappa + u^\varkappa{}_{,\mu})\,u_{*,\lambda}^\varkappa \tag{3.90}$$

and higher order terms are neglected. The equation (3.89) yields differential equations in V and mechanical boundary conditions on S_1:

$$[\sigma_*^{\varkappa\mu}(\delta_\mu^\lambda + u^\lambda{}_{,\mu}) + \sigma^{\varkappa\mu}u_{*,\mu}^\lambda]_{,\varkappa} + \bar{P}_*^\lambda = 0, \tag{3.91}$$

$$\sigma_*^{\varkappa\mu}n_\varkappa(\delta_\mu^\lambda + u^\lambda{}_{,\mu}) + \sigma^{\varkappa\mu}n_\varkappa u_{*,\mu}^\lambda = \bar{F}_*^\lambda. \tag{3.92}$$

Consequently, if the relations between the incremental stresses and strains are given in a linearized form by

$$\sigma_*^{\lambda\mu} = a^{\lambda\mu\alpha\beta}\varepsilon_{\alpha\beta}^*, \tag{3.93}$$

we have all the governing equations which determine the critical load and the adjacent equilibrium configuration.

When Eqs. (3.93) satisfy the symmetry relations:

$$a^{\lambda\mu\alpha\beta} = a^{\alpha\beta\lambda\mu}, \tag{3.94}$$

the principle can be written as follows:

$$\delta\left\{\tfrac{1}{2}\iiint_V [a^{\lambda\mu\alpha\beta}\varepsilon_{\lambda\mu}^*\varepsilon_{\alpha\beta}^* + \sigma^{\lambda\mu}u_{*,\lambda}^\varkappa u_{*,\mu}^\varkappa]\,dV\right\}$$

$$-\iiint_V \bar{P}_*^\lambda\,\delta u_*^\lambda\,dV - \iint_{S_1} \bar{F}_*^\lambda\,\delta u_*^\lambda\,dS = 0, \tag{3.95}$$

where Eqs. (3.90) have been substituted to express $\varepsilon_{\lambda\mu}^*$ in terms of $u_*^\varkappa$.

When the body is elastic and the external forces are conservative, we find that the principle (3.89) reduces to the principle (3.85).

The above formulations show that the critical load depends upon the relations between the incremental stresss and incremental strain measured from the original configuration, rather than the previous stress–strain relations. This suggests that the critical load problem can be treated more generally as an instability of a body with initial stresses and deformations. A stability problem of a body with initial stresses will be treated in Section 5.2 under an assumption that changes in the geometrical configuration of the body remain negligible until the instability occurs.

3.12. Some Remarks

Thus far, the principle of virtual work and related variational principles have been written for the elasticity problem in finite displacement theory. It is noted here that the approximate methods of solution such as the generalized Galerkin's method mentioned in Section 1.4 and the Rayleigh–Ritz method mentioned in Section 2.5 can similarly be applied to the elasticity problem in the finite displacement theory. It is also noted that for a variation of the geometrical boundary conditions on S_2 by $d\bar{u}^\kappa$, we have

$$dU = \iiint_V \bar{F}^\lambda \, du^\lambda \, dV + \iint_{S_1} \bar{F}^\lambda \, du^\lambda \, dS + \iint_{S_2} F^\lambda \, d\bar{u}^\lambda \, dS, \qquad (3.96)$$

which is an extension of Eq. (2.49) to the elasticity problem in the finite displacement theory.

We have derived in Section 1.2 the conditions of compatibility for the small displacement theory in rectangular Cartesian coordinates. The same kind of conditions may be derived of the finite displacement theory for the strains $e_{\lambda\mu}$ to be derivable from scalar functions u^λ. However, we shall not show them in the present chapter, but shall be satisfied with formulating the conditions for finite displacement theory in general curvilinear coordinates later in Section 4.2.

Mention is made here of the principles of complementary virtual work and minimum complementary energy. We have observed that these principles play important roles in the small displacement theory of elasticity. However, extensions of these principles to the finite displacement theory of elasticity are not found successful, since the displacements couple with the stress components as mentioned in Section 3.9.

Bibliography

1. R. Kappus, Zur Elastizitätstheorie endlicher Verschiebungen, *Zeitschrift für Angewandte Mathematik und Mechanik*, Vol. 19, pp. 271–85, October 1939 and pp. 344–61, December 1939.
2. V. V. Novozhilov, *Theory of Elasticity*, Pergamon Press, 1961.
3. A. E. Green and W. Zerna, *Theoretical Elasticity*, Oxford University Press, 1954.
4. I. S. Sokolnikoff, *Mathematical Theory of Elasticity*, McGraw-Hill, 1956.
5. C. E. Pearson, *Theoretical Elasticity*, Harvard University Press, 1959.
6. C. Truesdell, editor, *Problems of Non-linear Elasticity*, Gordon and Breach, Science Publishers, 1965.
7. C. Biezeno and R. Grammel, *Technische Dynamik*, Springer, 1939.
8. L. G. Brazier, The Flexure of Thin Cylindrical Shells and Other Thin Sections, *R. & M.* No. 1081, British Aeronautical Research Council, 1926.
9. L. G. Brazier, On the Flexure of Thin Cylindrical Shells and Other Thin Sections, *Proceedings of the Royal Society of London*, Series A, Vol. 116, pp. 104–14, November 1927.
10. C. B. Biezeno, Das Durchschlagen eines schwach gekrümmten Stabes. *Zeitschrift für Angewandte Mathematik und Mechanik*, Vol. 18, No. 1, pp. 21–30, February 1938.

11. S. Timoshenko, *Theory of Elastic Stability*, McGraw-Hill, 1936.
12. H. L. Langhaar, General Theory of Buckling, *Applied Mechanics Reviews*, Vol. 11, No. 11, pp. 585–8, November 1958.
13. M. Yoshiki *et al.*, *Handbook of Elastic Stability* (in Japanese), edited by Column Research Committee of Japan, Corona Publishing Co. Tokyo, revised edition 1960.
14. A. E. H. Love, *A Treatise on the Mathematical Theory of Elasticity*, Cambridge University Press, 4th edition 1927.
15. Y. C. Fung, *Foundations of Solid Mechanics*, Prentice-Hall, 1965.
16. R. L. Bisplinghoff, Some Structural and Aeroelastic Considerations of High Speed Flight, *Journal of the Aeronautical Sciences*, Vol. 23, No. 4, pp. 289–327, April 1956.
17. E. Reissner, On a Variational Theorem for Finite Elastic Deformations, *Journal of Mathematics and Physics*, Vol. 32, No. 2–3, pp. 129–35, July–October, 1953.
18. E. Trefftz, Über die Ableitung der Stabilitäts-Kriterien des elastischen Gleichgewichtes aus der Elastizitätstheorie endlicher Deformationen, *Proceedings of the 3rd International Congress for Applied Mechanics*, pp. 44–50, Stockholm 1930.
19. K. Marguerre, Die Behandlung von Stabilitätsproblemen mit Hilfe der energetischen Methoden, *Zeitschrift für Angewandte Mathematik und Mechanik*, Vol. 18, No. 1, pp. 57–73, February 1938.
20. K. Marguerre, Über die Anwendung der energetischen Methode auf Stabilitätsprobleme, *Jahrbuch der Deutschen Luftfahrtforschung, Flugwerk*, pp. 433–43, 1938.
21. V. V. Novozhilov, *Foundations of the Nonlinear Theory of Elasticity*, Graylock Press, 1953.
22. H. L. Langhaar, *Energy Methods in Applied Mechanics*, John Wiley, 1962.
23. V. V. Bolotin, *Nonconservative Problems of the Theory of Elastic Stability*, translated by T. K. Lusher, Pergamon Press, 1963.

CHAPTER 4

THEORY OF ELASTICITY IN CURVILINEAR COORDINATES

4.1. Geometry before Deformation

We shall devote this chapter to the theory of elasticity expressed in general curvilinear coordinates.† Let the space coordinates be defined by three parameters $(\alpha^1, \alpha^2, \alpha^3)$ before deformation. We shall employ a set of values of $(\alpha^1, \alpha^2, \alpha^3)$ which locate an arbitrary point $P^{(0)}$ of the body before deformation as parameters which specify the point during the deformation. Therefore, the position vector of the point $P^{(0)}$ before deformation is given by

$$\mathbf{r}^{(0)} = \mathbf{r}^{(0)}(\alpha^1, \alpha^2, \alpha^3). \tag{4.1}$$

The relations between the rectangular Cartesian coordinates (x^1, x^2, x^3) and the curvilinear coordinates $(\alpha^1, \alpha^2, \alpha^3)$ are usually written

$$x^\lambda = x^\lambda(\alpha^1, \alpha^2, \alpha^3) \quad (\lambda = 1, 2, 3). \tag{4.2}$$

A simple example of a curvilinear coordinates system is cylindrical coordinates, in which the relations (4.2) are

$$x^1 = r\cos\theta, \quad x^2 = r\sin\theta, \quad x^3 = z, \tag{4.3}$$

where $\alpha^1 = r$, $\alpha^2 = \theta$, $\alpha^3 = z$. By introducing the unit vector $\mathbf{i}_\lambda$ associated with the x^λ axis in the rectangular Cartesian coordinate system, we can write Eq. (4.1) for the present example as follows:

$$\mathbf{r}^{(0)} = r\cos\theta\,\mathbf{i}_1 + r\sin\theta\,\mathbf{i}_2 + z\,\mathbf{i}_3. \tag{4.4}$$

We shall summarize geometrical relations which are useful in subsequent formulations. For details of their derivations, the reader is advised to refer to books on the tensor calculus and differential geometry.‡ First, we define the covariant base vectors associated with the point $P^{(0)}$ by

$$\mathbf{g}_\lambda = \frac{\partial \mathbf{r}^{(0)}}{\partial \alpha^\lambda} = \mathbf{r}^{(0)}_{,\lambda} \tag{4.5}$$

as shown in Fig. 4.1, where and throughout the present chapter, the notation $(\ )_{,\lambda}$ denotes differentiation with respect to α^λ, namely, $(\ )_{,\lambda}$

† Refs. 1 through 6.
‡ Refs. 7 through 11.

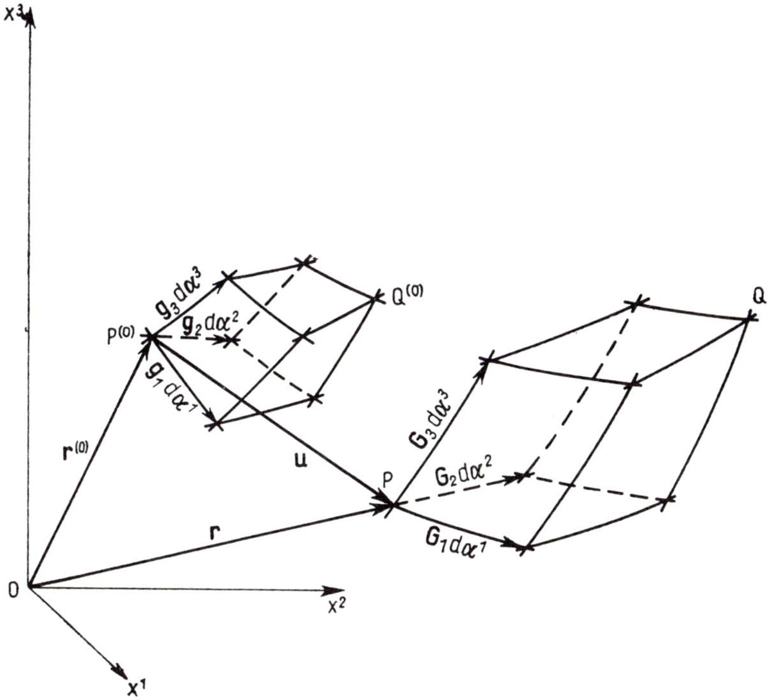

FIG. 4.1. Geometry of an infinitesimal parallelepiped.
(a) before deformation. (b) after deformation.

$= \partial(\ \ )/\partial\alpha^\lambda$. By the use of the covariant base vectors, we define the covariant metric tensor $g_{\lambda\mu}$ by

$$g_{\lambda\mu} = \mathbf{g}_\lambda \cdot \mathbf{g}_\mu = \mathbf{g}_\mu \cdot \mathbf{g}_\lambda = g_{\mu\lambda}, \tag{4.6}$$

the contravariant metric tensor $g^{\lambda\mu}$ by

$$g^{\lambda\varkappa} g_{\varkappa\mu} = \delta^\lambda_\mu, \tag{4.7}$$

where δ^λ_μ is the Kronecker symbol, and the contravariant base vector $\mathbf{g}^\lambda$ by

$$\mathbf{g}^\lambda = g^{\lambda\mu}\mathbf{g}_\mu. \tag{4.8}$$

From these relations, we obtain

$$\mathbf{g}_\lambda = g_{\lambda\mu}\mathbf{g}^\mu, \tag{4.9}$$

$$\mathbf{g}^\lambda \cdot \mathbf{g}_\mu = \delta^\lambda_\mu. \tag{4.10}$$

Next, we shall consider the derivative of the covariant base vector $\mathbf{g}_\mu$ with respect to α^ν. Since the derivative is again a vector, we may write

$$\mathbf{g}_{\mu,\nu} = \begin{Bmatrix} \lambda \\ \mu\nu \end{Bmatrix} \mathbf{g}_\lambda, \tag{4.11}$$

where the notation $\left\{ {\lambda \atop \mu\nu} \right\}$ is related to the magnitude of the components of $\mathbf{g}_{\mu,\nu}$ in the direction of $\mathbf{g}_\lambda$. Since

$$\mathbf{g}_{\mu,\nu} = (\mathbf{r}^{(0)}_{,\mu})_{,\nu} = (\mathbf{r}^{(0)}_{,\nu})_{,\mu} = \mathbf{g}_{\nu,\mu},$$

we obtain

$$\left\{ {\lambda \atop \mu\nu} \right\} = \left\{ {\lambda \atop \nu\mu} \right\}. \tag{4.12}$$

Differentiating both sides of $\mathbf{g}_\mu \cdot \mathbf{g}_\nu = g_{\mu\nu}$ with respect to $\alpha^\varkappa$, we have

$$\mathbf{g}_{\mu,\varkappa} \cdot \mathbf{g}_\nu + \mathbf{g}_\mu \cdot \mathbf{g}_{\nu,\varkappa} = g_{\mu\nu,\varkappa}.$$

Substituting (4.11) into the above, we obtain

$$\left\{ {\varrho \atop \mu\varkappa} \right\} g_{\varrho\nu} + \left\{ {\varrho \atop \nu\varkappa} \right\} g_{\varrho\mu} = g_{\mu\nu,\varkappa}. \tag{a}$$

Interchanging $\varkappa$ with ν in (a), we have

$$\left\{ {\varrho \atop \mu\nu} \right\} g_{\varrho\varkappa} + \left\{ {\varrho \atop \varkappa\nu} \right\} g_{\varrho\mu} = g_{\mu\varkappa,\nu}. \tag{b}$$

Another interchange of $\varkappa$ with μ in (a) leads to

$$\left\{ {\varrho \atop \varkappa\mu} \right\} g_{\varrho\nu} + \left\{ {\varrho \atop \nu\mu} \right\} g_{\varrho\varkappa} = g_{\varkappa\nu,\mu}. \tag{c}$$

Subtraction of (a) from the sum of (b) and (c) then yields

$$2\left\{ {\varrho \atop \mu\nu} \right\} g_{\varrho\varkappa} = g_{\varkappa\mu,\nu} + g_{\varkappa\nu,\mu} - g_{\mu\nu,\varkappa}.$$

By multiplying both sides of the above equation by $g^{\lambda\varkappa}$ and summing with respect to $\varkappa$, we finally obtain

$$\left\{ {\lambda \atop \mu\nu} \right\} = \frac{1}{2} g^{\lambda\varkappa}(g_{\varkappa\mu,\nu} + g_{\varkappa\nu,\mu} - g_{\mu\nu,\varkappa}). \tag{4.13}$$

The quantity $\left\{ {\lambda \atop \mu\nu} \right\}$ is called the Christoffel three-index symbol of the second kind. The Christoffel symbols are a measure of the curvature of the curvilinear coordinate axes, and play an important role in tensor calculus. From Eqs. (4.8) and (4.11), we obtain the derivative of the contravariant base vector $\mathbf{g}^\mu$ with respect to α^ν as follows:

$$\mathbf{g}^\mu_{,\nu} = -\left\{ {\mu \atop \lambda\nu} \right\} \mathbf{g}^\lambda. \tag{4.14}$$

Let us consider next a vector field in space and denote the vector by $\mathbf{u}(\alpha^1, \alpha^2, \alpha^3)$. We define components of the vector $\mathbf{u}$ by resolving it in the directions of $\mathbf{g}_\lambda$ at the point $P^{(0)}$ as follows:

$$\mathbf{u} = v^\lambda \mathbf{g}_\lambda, \tag{4.15}$$

where v^λ is called the contravariant component of the vector **u**. Differentiating **u** with respect to α^ν, and using Eq. (4.11), we obtain

$$\begin{aligned}\mathbf{u}_{,\nu} &= (v^\lambda \mathbf{g}_\lambda)_{,\nu} = v^\lambda{}_{,\nu}\mathbf{g}_\lambda + v^\lambda \mathbf{g}_{\lambda,\nu} \\ &= v^\lambda{}_{;\nu}\mathbf{g}_\lambda,\end{aligned} \tag{4.16}$$

where $v^\lambda{}_{;\nu}$ is called the covariant derivative of v^λ and is given by

$$v^\lambda{}_{;\nu} = v^\lambda{}_{,\nu} + \begin{Bmatrix} \lambda \\ \varrho\nu \end{Bmatrix} v^\varrho. \tag{4.17}$$

Components of the vector **u** may be expressed alternatively by resolving it in the directions of $\mathbf{g}^\lambda$ at the point $P^{(0)}$:

$$\mathbf{u} = v_\lambda \mathbf{g}^\lambda, \tag{4.18}$$

where v_λ is called the covariant component of the vector **u** and is given from Eqs. (4.9), (4.15) and (4.18) by

$$v_\lambda = g_{\lambda\mu} v^\mu. \tag{4.19}$$

Differentiating **u** with respect to α^ν, we obtain

$$\mathbf{u}_{,\nu} = (v_\lambda \mathbf{g}^\lambda)_{,\nu} = v_{\lambda;\nu}\mathbf{g}^\lambda, \tag{4.20}$$

where $v_{\lambda;\nu}$ is called the covariant derivative of v_λ and is given by

$$v_{\lambda;\nu} = v_{\lambda,\nu} - \begin{Bmatrix} \varrho \\ \lambda\nu \end{Bmatrix} v_\varrho. \tag{4.21}$$

It is defined in the theory of tensor calculus that the covariant derivative of a tensor $T^{\lambda_1\cdots\lambda_r}_{\mu_1\cdots\mu_s}$ with respect to α^ν is

$$\begin{aligned} T^{\lambda_1\cdots\lambda_r}_{\mu_1\cdots\mu_s;\nu} &= T^{\lambda_1\cdots\lambda_r}_{\mu_1\cdots\mu_s,\nu} + \sum_{i=1}^{r} T^{\lambda_1\cdots\lambda_{i-1}\varrho\lambda_{i+1}\cdots\lambda_r}_{\mu_1\cdots\cdots\cdots\cdots\mu_s} \begin{Bmatrix} \lambda_i \\ \varrho\nu \end{Bmatrix} \\ &\quad - \sum_{j=1}^{s} T^{\lambda_1\cdots\cdots\cdots\cdots\lambda_r}_{\mu_1\cdots\mu_{j-1}\varrho\mu_{j+1}\cdots\mu_s} \begin{Bmatrix} \varrho \\ \mu_j\nu \end{Bmatrix}. \end{aligned} \tag{4.22}$$

An application of this relation to vectors has been shown in Eqs. (4.17) and (4.21). As another application, we may show that the covariant derivatives of the covariant and contravariant metric tensors $g_{\lambda\mu}$ and $g^{\lambda\mu}$ vanish:

$$g_{\lambda\mu;\nu} = 0, \quad g^{\lambda\mu}{}_{;\nu} = 0. \tag{4.23}$$

It is also added that the following formula holds for the covariant derivative of a tensor product of two tensors $S^{\lambda\cdots}_{\mu\cdots}$ and $T^{\varkappa\cdots}_{\varrho\cdots}$:

$$(S^{\lambda\cdots}_{\mu\cdots}T^{\varkappa\cdots}_{\varrho\cdots})_{;\nu} = S^{\lambda\cdots}_{\mu\cdots;\nu}T^{\varkappa\cdots}_{\varrho\cdots} + S^{\lambda\cdots}_{\mu\cdots}T^{\varkappa\cdots}_{\varrho\cdots;\nu}. \tag{4.24}$$

Two more geometrical relations are noted here before proceeding to the next section. If we take a point $Q^{(0)}$ in the neighborhood of the point $P^{(0)}$ and denote the coordinates of $Q^{(0)}$ by $(\alpha^1 + d\alpha^1, \alpha^2 + d\alpha^2, \alpha^3 + d\alpha^3)$, the distance between $P^{(0)}$ and $Q^{(0)}$ can be expressed by

$$(ds^{(0)})^2 = d\mathbf{r}^{(0)} \cdot d\mathbf{r}^{(0)} = (\mathbf{g}_\lambda \, d\alpha^\lambda) \cdot (\mathbf{g}_\mu \, d\alpha^\mu)$$
$$= g_{\lambda\mu} \, d\alpha^\lambda \, d\alpha^\mu. \tag{4.25}$$

Next, if we take an infinitesimal parallelepiped enclosed by six surfaces, α^λ = constant, $\alpha^\lambda + d\alpha^\lambda$ = constant ($\lambda = 1, 2, 3$), its volume is given by

$$dV = \sqrt{g}\, d\alpha^1 \, d\alpha^2 \, d\alpha^3, \tag{4.26}$$

where

$$\sqrt{g} = \mathbf{g}_1 \cdot (\mathbf{g}_2 \times \mathbf{g}_3) = \mathbf{g}_2 \cdot (\mathbf{g}_3 \times \mathbf{g}_1) = \mathbf{g}_3 \cdot (\mathbf{g}_1 \times \mathbf{g}_2), \tag{4.27}$$

and we have

$$g = |g_{\lambda\mu}| = \begin{vmatrix} g_{11} & g_{12} & g_{13} \\ g_{21} & g_{22} & g_{23} \\ g_{31} & g_{32} & g_{33} \end{vmatrix}. \tag{4.28}$$

With these geometrical preliminaries before deformation, we shall proceed to the analysis of stress and strain.

4.2. Analysis of Strain and Conditions of Compatibility

After deformation, the point $P^{(0)}$ moves to a new position P, the position vector to which will be denoted by

$$\mathbf{r} = \mathbf{r}(\alpha^1, \alpha^2, \alpha^3). \tag{4.29}$$

We shall define the covariant base vector after deformation by

$$\mathbf{G}_\lambda = \mathbf{r}_{,\lambda} \tag{4.30}$$

and the covariant and contravariant metric tensors after deformation by

$$G_{\lambda\mu} = \mathbf{G}_\lambda \cdot \mathbf{G}_\mu = G_{\mu\lambda} \tag{4.31}$$

and

$$G^{\lambda\varkappa} G_{\varkappa\mu} = \delta^\lambda_\mu, \tag{4.32}$$

respectively, where δ^λ_μ is the Kronecker symbol. Differentiation of $\mathbf{G}_\mu$ with respect to α^ν yields

$$\mathbf{G}_{\mu,\nu} = \left\{\begin{Bmatrix} \lambda \\ \mu\nu \end{Bmatrix}\right\} \mathbf{G}_\lambda, \tag{4.33}$$

in a manner similar to Eq. (4.11), where

$$\left\{\begin{Bmatrix} \lambda \\ \mu\nu \end{Bmatrix}\right\} = \frac{1}{2} G^{\lambda\varkappa}(G_{\varkappa\mu,\nu} + G_{\varkappa\nu,\mu} - G_{\mu\nu,\varkappa}) = \left\{\begin{Bmatrix} \lambda \\ \nu\mu \end{Bmatrix}\right\}. \tag{4.34}$$

The distance between the two points P and Q after deformation is given by

$$(ds)^2 = d\mathbf{r} \cdot d\mathbf{r} = (\mathbf{G}_\lambda \, d\alpha^\lambda) \cdot (\mathbf{G}_\mu \, d\alpha^\mu)$$
$$= G_{\lambda\mu} \, d\alpha^\lambda \, d\alpha^\mu . \tag{4.35}$$

By using the relations (4.25) and (4.35), we may define the components of the strain tensor in general curvilinear coordinates as follows:

$$f_{\lambda\mu} = \tfrac{1}{2}(G_{\lambda\mu} - g_{\lambda\mu}) = f_{\mu\lambda} . \tag{4.36}$$

Equation (4.36) is a natural extension to curvilinear coordinates of the definition (3.14). The quantities $f_{\lambda\mu}$ specify the state of strain of the infinitesimal parallelepiped which was bounded by the six surfaces, $\alpha^\lambda =$ constant and $\alpha^\lambda + d\alpha^\lambda =$ constant, before deformation.

Let us consider the strain–displacement relations. Defining the displacement vector $\mathbf{u}(\alpha^1, \alpha^2, \alpha^3)$ by

$$\mathbf{r} = \mathbf{r}^{(0)} + \mathbf{u} , \tag{4.37}$$

and its components by

$$\mathbf{u} = v^\lambda \mathbf{g}_\lambda , \tag{4.38†}$$

we have

$$\mathbf{G}_\lambda = (\delta_\lambda^\varkappa + v^\varkappa{}_{:\lambda}) \, \mathbf{g}_\varkappa . \tag{4.39}$$

Substituting Eq. (4.39) into Eq. (4.36), we obtain expressions of the strain in terms of the displacement components as follows:

$$f_{\lambda\mu} = \tfrac{1}{2}(g_{\lambda\varkappa} v^\varkappa{}_{:\mu} + g_{\varkappa\mu} v^\varkappa{}_{:\lambda} + g_{\varkappa\varrho} v^\varkappa{}_{:\lambda} v^\varrho{}_{:\mu}) . \tag{4.40}$$

By the use of Eqs. (4.19), (4.23) and (4.24), the above relations may be written alternatively as follows:

$$f_{\lambda\mu} = \tfrac{1}{2}(v_{\lambda:\mu} + v_{\mu:\lambda} + v_{\varkappa:\lambda} v^\varkappa{}_{:\mu}) . \tag{4.41}$$

Next, we shall consider the conditions of compatibility in the curvilinear coordinate system, namely, the necessary and sufficient conditions that strain components $f_{\lambda\mu}$ are derivable from a single-valued vector function $\mathbf{r}(\alpha^1, \alpha^2, \alpha^3)$. It is known in the theory of tensor calculus that the conditions of compatibility are given by

$$R^\lambda_{\cdot\mu\nu\omega} = 0 , \tag{4.42}$$

where $R^\lambda_{\cdot\mu\nu\omega}$ is the Riemann–Christoffel curvature tensor defined by

$$R^\lambda_{\cdot\mu\nu\omega} = \frac{\partial \left\{\!\!\left\{ \begin{matrix} \lambda \\ \mu\omega \end{matrix} \right\}\!\!\right\}}{\partial \alpha^\nu} - \frac{\partial \left\{\!\!\left\{ \begin{matrix} \lambda \\ \mu\nu \end{matrix} \right\}\!\!\right\}}{\partial \alpha^\omega} + \left\{\!\!\left\{ \begin{matrix} \varkappa \\ \mu\omega \end{matrix} \right\}\!\!\right\} \left\{\!\!\left\{ \begin{matrix} \lambda \\ \varkappa\nu \end{matrix} \right\}\!\!\right\} - \left\{\!\!\left\{ \begin{matrix} \varkappa \\ \mu\nu \end{matrix} \right\}\!\!\right\} \left\{\!\!\left\{ \begin{matrix} \lambda \\ \varkappa\omega \end{matrix} \right\}\!\!\right\} . \tag{4.43}$$

† We note here that the relation (4.38) is not the only way of defining the components of $\mathbf{u}$. For example, $\mathbf{u}$ may be resolved into the directions of $\mathbf{i}_\lambda$ as expressed by the equation (3.16), or in the directions of $\mathbf{g}^\lambda$ as expressed by the equation (4.18). Whatever the definitions of the components may be, the definition of the strain (4.36) remains unchanged.

The proof that the conditions (4.42) are necessary is easily given by using the relations

$$\frac{\partial}{\partial\alpha^{\omega}}\left(\frac{\partial \mathbf{G}_{\mu}}{\partial\alpha^{\nu}}\right)=\frac{\partial^{3}\mathbf{r}}{\partial\alpha^{\omega}\,\partial\alpha^{\nu}\,\partial\alpha^{\mu}}=\frac{\partial^{3}\mathbf{r}}{\partial\alpha^{\nu}\,\partial\alpha^{\omega}\,\partial\alpha^{\mu}}=\frac{\partial}{\partial\alpha^{\nu}}\left(\frac{\partial \mathbf{G}_{\mu}}{\partial\alpha^{\omega}}\right), \tag{4.44}$$

together with the relations (4.33). However, the proof that they are sufficient is rather lengthy, and will not be given here. The reader interested in the proof is advised to refer to books on tensor calculus or advanced elasticity. It is noted in this connection that the covariant curvature tensor $R_{\lambda\mu\nu\omega}$ defined by

$$R_{\lambda\mu\nu\omega}=G_{\lambda\varkappa}R^{\varkappa}_{\cdot\mu\nu\omega} \tag{4.45}$$

is frequently used instead of $R^{\lambda}_{\cdot\mu\nu\omega}$, and the conditions of compatibility (4.42) can be written alternatively as

$$R_{\lambda\mu\nu\omega}=0, \tag{4.46}$$

where it may be shown that

$$R_{\lambda\mu\nu\omega}=\frac{1}{2}\left(\frac{\partial^{2}G_{\lambda\omega}}{\partial\alpha^{\mu}\,\partial\alpha^{\nu}}+\frac{\partial^{2}G_{\mu\nu}}{\partial\alpha^{\lambda}\,\partial\alpha^{\omega}}-\frac{\partial^{2}G_{\mu\omega}}{\partial\alpha^{\lambda}\,\partial\alpha^{\nu}}-\frac{\partial G_{\lambda\nu}}{\partial\alpha^{\mu}\,\partial\alpha^{\omega}}\right)$$
$$+\,G_{\alpha\beta}\left(\left\{\!\!\left\{\begin{matrix}\alpha\\ \mu\nu\end{matrix}\right\}\!\!\right\}\left\{\!\!\left\{\begin{matrix}\beta\\ \lambda\omega\end{matrix}\right\}\!\!\right\}-\left\{\!\!\left\{\begin{matrix}\alpha\\ \mu\omega\end{matrix}\right\}\!\!\right\}\left\{\!\!\left\{\begin{matrix}\beta\\ \lambda\nu\end{matrix}\right\}\!\!\right\}\right). \tag{4.47}$$

The Greek letters in Eq. (4.42) are assigned in place of (1, 2, 3). Therefore, it appears that relations totaling $3^4 = 81$ in number are contained there. However, there exist relations among the components of $R^{\lambda}_{\cdot\mu\nu\omega}$ such that

$$R^{\lambda}_{\cdot\mu\nu\omega}=-R^{\lambda}_{\cdot\mu\omega\nu},$$

and it can be shown that the number of the independent conditions of compatibility for three-dimensional space reduces to 6. Moreover, we can prove the existence of Bianchi's identities

$$R^{\alpha}_{\cdot\lambda\mu\nu;\omega}+R^{\alpha}_{\cdot\lambda\nu\omega;\mu}+R^{\alpha}_{\cdot\lambda\omega\mu;\nu}=0 \tag{4.48}$$

among the Riemann–Christoffel curvature tensors.

When the above-mentioned relations are linearized and applied to the small displacement theory in rectangular Cartesian coordinates we find that

$$R_x=-R_{2323},\quad R_y=-R_{3131},\quad R_z=-R_{1212},$$
$$U_x=-R_{1231},\quad U_y=-R_{2312},\quad U_z=-R_{3123}, \tag{4.49}$$

and the conditions of compatibility (4.46) and Bianchi's identities (4.48) are reduced to Eqs. (1.15) and (1.17), respectively.

4.3. Analysis of Stress and Equations of Equilibrium

We shall consider the equilibrium of the infinitesimal parallelepiped (see Fig. 4.2) which, before deformation, was bounded by the six surfaces: α^λ = constant and $\alpha^\lambda + d\alpha^\lambda$ = constant. Let the internal forces acting on one of the surfaces, the sides of which are $\mathbf{G}_2\, d\alpha^2$ and $\mathbf{G}_3\, d\alpha^3$ after deforma-

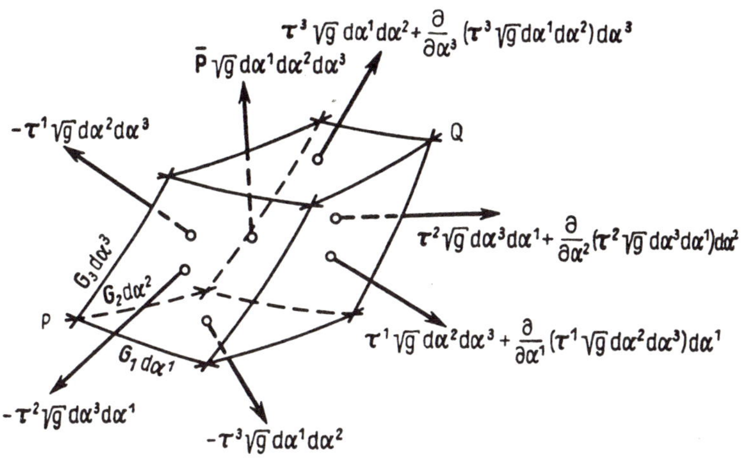

FIG. 4.2. Equilibrium of an infinitesimal parallelepiped.

tion, be denoted by $-\boldsymbol{\tau}^1\sqrt{g}\, d\alpha^2\, d\alpha^3$, where g is defined by Eq. (4.28). The quantities $\boldsymbol{\tau}^2$ and $\boldsymbol{\tau}^3$ are defined in a similar manner. By resolving $\boldsymbol{\tau}^\lambda$ such that

$$\boldsymbol{\tau}^\lambda = \tau^{\lambda\mu}\mathbf{G}_\mu, \tag{4.50}$$

we have the following equations of equilibrium for the infinitesimal parallelepiped after deformation:

$$(\sqrt{g}\,\boldsymbol{\tau}^\varkappa)_{,\varkappa} + \bar{\mathbf{P}}\sqrt{g} = 0, \tag{4.51}$$

$$\tau^{\lambda\mu} = \tau^{\mu\lambda}, \tag{4.52}$$

where $\bar{\mathbf{P}}$ is the body force vector defined per unit volume of the body before deformation (see Section 3.2 for the similar development in rectangular Cartesian coordinates).

Equation (4.51) is a vector equation; one way of expressing it in scalar form is to resolve it in the directions of $\mathbf{g}_\lambda$. By the use of Eqs. (4.11) and (4.39), we obtain

$$[\sqrt{g}(\delta^\lambda_\mu + v^\lambda{}_{:\mu})\,\tau^{\varkappa\mu}]_{,\varkappa} + \sqrt{g}(\delta^\varrho_\mu + v^\varrho{}_{:\mu})\,\tau^{\varkappa\mu}\begin{Bmatrix}\lambda\\ \varkappa\varrho\end{Bmatrix}$$

$$+ \sqrt{g}\,\bar{P}^\lambda_g = 0 \quad (\lambda = 1, 2, 3), \tag{4.53}$$

where

$$\bar{\mathbf{P}} = \bar{P}^{\lambda}_{g} \mathbf{g}_{\lambda}, \tag{4.54}$$

and the suffix $(\;)_g$ denotes that the components are taken in the directions of the base vectors $\mathbf{g}_\lambda$ of the generalized coordinate system.

4.4. Transformation of the Strain and Stress Tensors

Let us assume a local rectangular Cartesian coordinate system (y^1, y^2, y^3) issuing from the point $P^{(0)}$ before deformation as shown in Fig. 4.3, and denote the unit vector in the direction of the y^λ-axis by $\mathbf{j}_\lambda$. Since

$$\partial(\;)/\partial\alpha^\lambda = [\partial(\;)/\partial y^\mu]\,[\partial y^\mu/\partial\alpha^\lambda],$$
$$\partial(\;)/\partial y^\lambda = [\partial(\;)/\partial\alpha^\mu]\,[\partial\alpha^\mu/\partial y^\lambda],$$

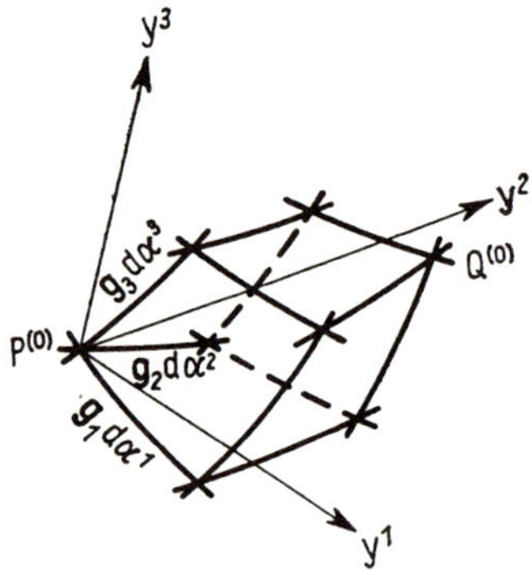

FIG. 4.3. A local rectangular Cartesian coordinate system (y^1, y^2, y^3).

we have

$$\mathbf{g}_\mu = \partial\mathbf{r}^{(0)}/\partial\alpha^\mu = (\partial y^\lambda/\partial\alpha^\mu)\,\mathbf{j}_\lambda, \tag{4.55}$$

and

$$\mathbf{j}_\lambda = \partial\mathbf{r}^{(0)}/\partial y^\lambda = (\partial\alpha^\mu/\partial y^\lambda)\,\mathbf{g}_\mu. \tag{4.56}$$

Scalar multiplication of Eq. (4.55) by $\mathbf{j}_\lambda$ yields

$$\frac{\partial y^\lambda}{\partial\alpha^\mu} = \mathbf{j}_\lambda \cdot \mathbf{g}_\mu. \tag{4.57}$$

Similarly, scalar multiplication of Eq. (4.56) by $\mathbf{g}_\varkappa$ yields

$$\frac{\partial\alpha^\lambda}{\partial y^\mu} = g^{\lambda\varkappa}(\mathbf{j}_\mu \cdot \mathbf{g}_\varkappa), \tag{4.58}$$

where the indices λ and μ have been interchanged.

We may now formulate the transformation law for strain. Let the strain tensor defined with respect to the local rectangular Cartesian coordinates by denoted by $e_{\lambda\mu}$, namely:

$$2e_{\lambda\mu} = \frac{\partial \mathbf{r}}{\partial y^\lambda} \cdot \frac{\partial \mathbf{r}}{\partial y^\mu} - \frac{\partial \mathbf{r}^{(0)}}{\partial y^\lambda} \cdot \frac{\partial \mathbf{r}^{(0)}}{\partial y^\mu}. \tag{4.59}$$

Then, we have from Eq. (4.36),

$$\begin{aligned} 2f_{\lambda\mu} &= \frac{\partial \mathbf{r}}{\partial \alpha^\lambda} \cdot \frac{\partial \mathbf{r}}{\partial \alpha^\mu} - \frac{\partial \mathbf{r}^{(0)}}{\partial \alpha^\lambda} \cdot \frac{\partial \mathbf{r}^{(0)}}{\partial \alpha^\mu} \\ &= \left[\frac{\partial \mathbf{r}}{\partial y^\varkappa} \cdot \frac{\partial \mathbf{r}}{\partial y^\varrho} - \frac{\partial \mathbf{r}^{(0)}}{\partial y^\varkappa} \cdot \frac{\partial \mathbf{r}^{(0)}}{\partial y^\varrho} \right] \frac{\partial y^\varkappa}{\partial \alpha^\lambda} \frac{\partial y^\varrho}{\partial \alpha^\mu} \\ &= 2e_{\varkappa\varrho} \frac{\partial y^\varkappa}{\partial \alpha^\lambda} \frac{\partial y^\varrho}{\partial \alpha^\mu}. \end{aligned}$$

Consequently, the transformation law between $f_{\lambda\mu}$ and $e_{\varkappa\varrho}$ may be written

$$f_{\lambda\mu} = \frac{\partial y^\varkappa}{\partial \alpha^\lambda} \frac{\partial y^\varrho}{\partial \alpha^\mu} e_{\varkappa\varrho}, \tag{4.60}$$

or conversely

$$e_{\lambda\mu} = \frac{\partial \alpha^\varkappa}{\partial y^\lambda} \frac{\partial \alpha^\varrho}{\partial y^\mu} f_{\varkappa\varrho}. \tag{4.61}$$

These relations show that $f_{\lambda\mu}$ and $e_{\lambda\mu}$ are covariant tensors of order two.

Next, we shall formulate the transformation law for stress. Let us isolate an infinitesimal tetrahedron $P^{(0)}R^{(0)}S^{(0)}T^{(0)}$ which is defined by the three sides $\mathbf{g}_1\, d\alpha^1$, $\mathbf{g}_2\, d\alpha^2$ and $\mathbf{g}_3\, d\alpha^3$ issuing from the point $P^{(0)}$ before deformation, and consider its equilibrium after deformation, as shown in Fig. 4.4. Let the internal forces acting on the oblique surface of the tetrahedron be denoted by $\mathbf{F}\, d\Sigma$, where $d\Sigma$ is the area of the oblique surface before deformation. The condition of equilibrium of the internal forces acting on the tetrahedron is

$$\mathbf{F}\, d\Sigma = \boldsymbol{\tau}^1 \sqrt{g}(d\alpha^2\, d\alpha^3/2) + \boldsymbol{\tau}^2 \sqrt{g}(d\alpha^3\, d\alpha^1/2) + \boldsymbol{\tau}^3 \sqrt{g}(d\alpha^1\, d\alpha^2/2). \tag{4.62}$$

From the geometry of the infinitesimal tetrahedron before deformation, we have

$$\begin{aligned} &\sqrt{g}\, d\alpha^2\, d\alpha^3 = 2(\mathbf{g}_1 \cdot \boldsymbol{\nu})\, d\Sigma, \quad \sqrt{g}\, d\alpha^3\, d\alpha^1 = 2(\mathbf{g}_2 \cdot \boldsymbol{\nu})\, d\Sigma, \\ &\sqrt{g}\, d\alpha^1\, d\alpha^2 = 2(\mathbf{g}_3 \cdot \boldsymbol{\nu})\, d\Sigma, \end{aligned} \tag{4.63†}$$

† In Fig. 4.4, we have

$$\begin{aligned} 2\boldsymbol{\nu}\, d\Sigma &= \overrightarrow{R^{(0)}S^{(0)}} \times \overrightarrow{R^{(0)}T^{(0)}} = (\mathbf{g}_2\, d\alpha^2 - \mathbf{g}_1\, d\alpha^1) \times (\mathbf{g}_3\, d\alpha^3 - \mathbf{g}_1\, d\alpha^1) \\ &= \mathbf{g}_1 \times \mathbf{g}_2\, d\alpha^1\, d\alpha^2 + \mathbf{g}_2 \times \mathbf{g}_3\, d\alpha^2\, d\alpha^3 + \mathbf{g}_3 \times \mathbf{g}_1\, d\alpha^3\, d\alpha^1. \end{aligned}$$

Combining this equation with Eqs. (4.27) and

$$\mathbf{g}_i \cdot (\mathbf{g}_i \times \mathbf{g}_j) = 0, \quad i \text{ not summed}$$

we obtain the relations (4.63).

where $\boldsymbol{\nu}$ is the unit normal vector drawn on the oblique surface before deformation. Substituting the relations (4.63) into Eq. (4.62), we obtain

$$\mathbf{F} = (\mathbf{g}_\varkappa \cdot \boldsymbol{\nu})\, \boldsymbol{\tau}^\varkappa. \tag{4.64}$$

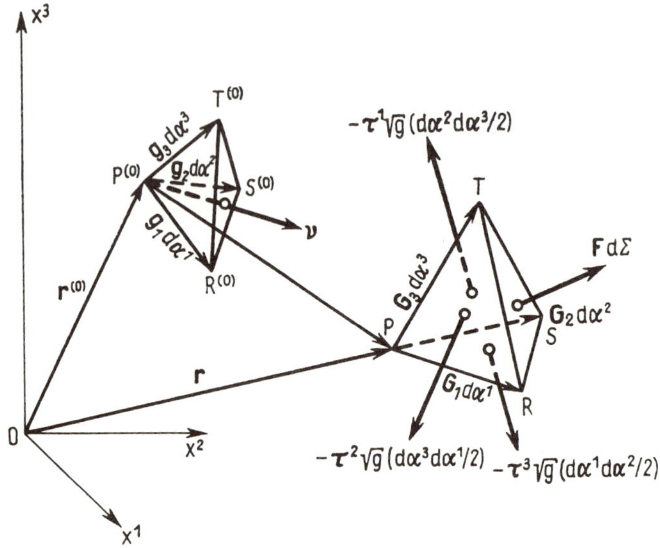

FIG. 4.4. Equilibrium of an infinitesimal tetrahedron.
(a) before deformation. (b) after deformation.

If we take an arbitrary set of local rectangular Cartesian coordinates (y^1, y^2, y^3) issuing from the point $P^{(0)}$ before deformation as a special case of the curvilinear coordinates and denote the stress vector defined with respect to the local Cartesian coordinates by $\boldsymbol{\sigma}^\lambda$, we have, instead of Eq. (4.64),

$$\mathbf{F} = (\mathbf{j}_\lambda \cdot \boldsymbol{\nu})\, \boldsymbol{\sigma}^\lambda. \tag{4.65}$$

Since $\mathbf{F}$ is a physical quantity, its magnitude and direction do not depend upon the choice of coordinate system. Therefore, combination of Eqs. (4.64) and (4.65) yields:

$$(\mathbf{j}_\lambda \cdot \boldsymbol{\nu})\, \boldsymbol{\sigma}^\lambda = (\mathbf{g}_\mu \cdot \boldsymbol{\nu})\, \boldsymbol{\tau}^\mu. \tag{4.66}$$

By taking the direction $\boldsymbol{\nu}$ coincident with the $y^\varkappa$-axis, we obtain:

$$\sigma^{\varkappa\varrho} \frac{\partial \mathbf{r}}{\partial y^\varrho} = \frac{\partial y^\varkappa}{\partial \alpha^\mu} \tau^{\mu\lambda} \frac{\partial \mathbf{r}}{\partial \alpha^\lambda},$$

which, after an interchange of indices, leads to

$$\sigma^{\lambda\mu} = \frac{\partial y^\lambda}{\partial \alpha^\varkappa} \frac{\partial y^\mu}{\partial \alpha^\varrho} \tau^{\varkappa\varrho}, \tag{4.67}$$

or conversely

$$\tau^{\lambda\mu} = \frac{\partial\alpha^\lambda}{\partial y^\varkappa}\frac{\partial\alpha^\mu}{\partial y^\varrho}\sigma^{\varkappa\varrho}. \tag{4.68}$$

These relations show that $\sigma^{\lambda\mu}$ and $\tau^{\lambda\mu}$ thus defined are contravariant tensors of order two.

For later convenience, we note that the resolution of **F** in the directions of the base vectors reads:

$$\mathbf{F} = \tau^{\varkappa\mu}\nu_\varkappa(\delta^\lambda_\mu + v^\lambda{}_{;\mu})\,\mathbf{g}_\lambda, \tag{4.69}$$

where $\nu_\varkappa$ is defined by

$$\boldsymbol{\nu} = \nu_\varkappa \mathbf{g}^\varkappa. \tag{4.70}$$

4.5. Stress–Strain Relations in Curvilinear Coordinates

Following the formulations in Section 3.4, we assume stress–strain relations in the local rectangular Cartesian coordinate system to be given by

$$\sigma^{\lambda\mu} = \sigma^{\lambda\mu}(e_{\alpha\beta}), \tag{4.71}$$

and, when the strain components are sufficiently small, by

$$\sigma^{\lambda\mu} = a^{\lambda\mu\alpha\beta}e_{\alpha\beta}. \tag{4.72}$$

When a problem of elasticity must be solved in curvilinear coordinates, the stress–strain relations must be expressed as

$$\tau^{\lambda\mu} = \tau^{\lambda\mu}(f_{\alpha\beta}), \tag{4.73}$$

or, in linearized form as

$$\tau^{\lambda\mu} = c^{\lambda\mu\alpha\beta}f_{\alpha\beta}. \tag{4.74}$$

Since the transformation laws for stress and strain have already been developed (Eqs. (4.60), (4.61), (4.67) and (4.68)), it is rather easy to derive the required relations. For example, Eqs. (4.72) can be written as,

$$\frac{\partial y^\lambda}{\partial\alpha^\varkappa}\frac{\partial y^\mu}{\partial\alpha^\varrho}\tau^{\varkappa\varrho} = a^{\lambda\mu\alpha\beta}\frac{\partial\alpha^\nu}{\partial y^\alpha}\frac{\partial\alpha^\omega}{\partial y^\beta}f_{\nu\omega},$$

or, after multiplication, summation and interchange of indices as,

$$\tau^{\lambda\mu} = a^{\varkappa\varrho\nu\omega}\frac{\partial\alpha^\lambda}{\partial y^\varkappa}\frac{\partial\alpha^\mu}{\partial y^\varrho}\frac{\partial\alpha^\alpha}{\partial y^\nu}\frac{\partial\alpha^\beta}{\partial y^\omega}f_{\alpha\beta}, \tag{4.75}$$

which shows that

$$c^{\lambda\mu\alpha\beta} = \frac{\partial\alpha^\lambda}{\partial y^\varkappa}\frac{\partial\alpha^\mu}{\partial y^\varrho}\frac{\partial\alpha^\alpha}{\partial y^\nu}\frac{\partial\alpha^\beta}{\partial y^\omega}a^{\varkappa\varrho\nu\omega}. \tag{4.76}$$

When the material is isotropic, Eq. (3.37) may be employed for the stress–strain relations in the rectangular Cartesian coordinate system and Eq. (4.76) takes the form

$$c^{\lambda\mu\alpha\beta} = \frac{E\nu}{(1+\nu)(1-2\nu)}g^{\lambda\mu}g^{\alpha\beta} + G(g^{\lambda\alpha}g^{\mu\beta} + g^{\lambda\beta}g^{\mu\alpha}). \tag{4.77}$$

4.6. Principle of Virtual Work

The problem which will be treated in the present section is the same as stated in Section 3.5, namely, to find the equilibrium configuration of a body subjected to prescribed mechanical boundary conditions $\mathbf{F} = \bar{\mathbf{F}}$ on S_1 and prescribed geometrical boundary conditions $\mathbf{u} = \bar{\mathbf{u}}$ on S_2, together with the imposition of body forces $\mathbf{P} = \bar{\mathbf{P}}$. The object of the present section is to derive for this problem the principle of virtual work expressed in general curvilinear coordinates.

From Eq. (4.51) and the mechanical boundary conditions, we have

$$-\iiint_V [(\sqrt{g}\,\boldsymbol{\tau}^\lambda)_{,\lambda} + \bar{\mathbf{P}}\sqrt{g}] \cdot \delta\mathbf{r}\, d\alpha^1\, d\alpha^2\, d\alpha^3 + \iint_{S_1} (\mathbf{F} - \bar{\mathbf{F}}) \cdot \delta\mathbf{r}\, dS = 0. \tag{4.78}$$

By the use of geometrical relations

$$\sqrt{g}\, d\alpha^2\, d\alpha^3 = \pm\nu_1\, dS, \qquad \sqrt{g}\, d\alpha^3\, d\alpha^1 = \pm\nu_2\, dS,$$
$$\sqrt{g}\, d\alpha^1\, d\alpha^2 = \pm\nu_3\, dS, \tag{4.79}†$$

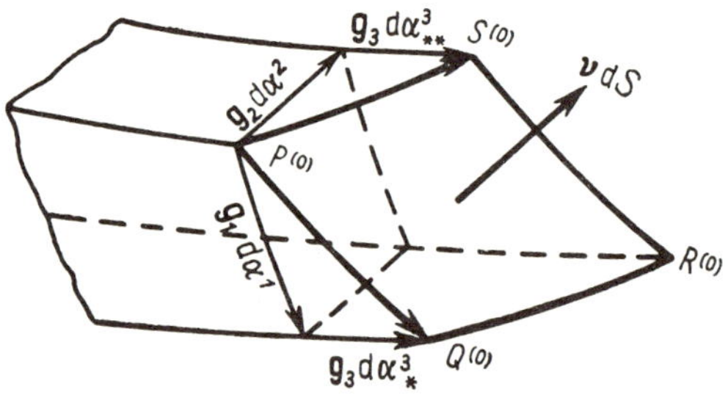

FIG. 4.5. An area element $P^{(0)}Q^{(0)}R^{(0)}S^{(0)}$ on the boundary.

which hold on the boundary, and following a development similar to Section 3.6, we may transform Eq. (4.78) into

$$\iiint_V (\tau^{\lambda\mu}\, \delta f_{\lambda\mu} - \bar{\mathbf{P}} \cdot \delta\mathbf{r})\, dV - \iint_{S_1} \bar{\mathbf{F}} \cdot \delta\mathbf{r}\, dS = 0 \tag{4.80}$$

† In Fig. 4.5, we have

$$\boldsymbol{\nu}\, dS = \overrightarrow{P^{(0)}Q^{(0)}} \times \overrightarrow{P^{(0)}S^{(0)}} = (\mathbf{g}_1\, d\alpha^1 + \mathbf{g}_3\, d\alpha^3_*) \times (\mathbf{g}_2\, d\alpha^2 + \mathbf{g}_3\, d\alpha^3_{**})$$
$$= \mathbf{g}_1 \times \mathbf{g}_2\, d\alpha^1\, d\alpha^2 - \mathbf{g}_2 \times \mathbf{g}_3\, d\alpha^2\, d\alpha^3_* - \mathbf{g}_3 \times \mathbf{g}_1\, d\alpha^3_{**}\, d\alpha^1,$$

and consequently

$$(\boldsymbol{\nu} \cdot \mathbf{g}_3)\, dS = \mathbf{g}_3 \cdot (\mathbf{g}_1 \times \mathbf{g}_2)\, d\alpha^1\, d\alpha^2 = \sqrt{g}\, d\alpha^1\, d\alpha^2,$$

where $\boldsymbol{\nu}$ is unit normal drawn on the oblique surface. Other relations of Eqs. (4.79) can be derived similarly.

where dV is given by Eq. (4.26). This is the principle of virtual work expressed in a general curvilinear coordinate system. By the use of Eqs. (4.18), (4.54) and

$$\bar{\mathbf{F}} = F_g^\lambda \mathbf{g}_\lambda, \tag{4.81}$$

the principle can be expressed alternatively as follows:

$$\iiint_V (\tau^{\lambda\mu}\delta f_{\lambda\mu} - \bar{P}_g^\lambda \, \delta v_\lambda) \, dV - \iint_{S_1} \bar{F}_g^\lambda \, \delta v_\lambda \, dS = 0. \tag{4.82}$$

From Eqs. (4.60) and (4.68), we have

$$\tau^{\lambda\mu} \, \delta f_{\lambda\mu} = \sigma^{\lambda\mu} \, \delta e_{\lambda\mu}. \tag{4.83}$$

Substituting Eq. (4.83) into Eq. (4.80), we obtain another expression for the principle of virtual work:

$$\iiint_V (\sigma^{\lambda\mu} \, \delta e_{\lambda\mu} - \bar{\mathbf{P}} \cdot \delta \mathbf{r}) \, dV - \iint_{S_1} \bar{\mathbf{F}} \cdot \delta \mathbf{r} \, dS = 0. \tag{4.84}$$

It may appear that the principle (4.84) is identical with the principle (3.48). However, the physical meaning is different, because the quantities $\sigma^{\lambda\mu}$ and $e_{\lambda\mu}$ appearing in Eq. (4.84) are defined with respect to a local rectangular Cartesian coordinate system, while $\sigma^{\lambda\mu}$ and $e_{\lambda\mu}$ appearing in Eq. (3.48) are defined with respect to the fixed Cartesian coordinates.

Thus far, the principle of virtual work has been written with respect to the curvilinear coordinate system. The approximate method of solution, mentioned in Section 1.5, and the technique of finding stress functions with knowledge of the conditions of compatibility, mentioned in Section 1.8, can be applied similarly to the present problem in curvilinear coordinates, once the principle of virtual work has been established. As will be shown in subsequent chapters, the principle plays a very important role in formulations of elasticity problems where curvilinear coordinates can be favorably employed.

4.7. Principle of Stationary Potential Energy and its Generalizations

We shall assume, as in Section 3.8, the existence of the functions A, Φ and Ψ. These are state functions and do not depend on the choice of coordinate system employed, i.e. the functional defined by Eq. (3.69) is invariant. Consequently, once the principle of stationary potential energy has been established in rectangular Cartesian coordinates, the principle in a general curvilinear coordinate system can be expressed in terms of v^λ by the use of the transformation law for strain (4.61), the strain–displacement relations (4.40), and the transformation law for displacement components

$$u^\lambda = (\mathbf{i}_\lambda \cdot \mathbf{g}_\mu) \, v^\mu, \tag{4.85}$$

where

$$\mathbf{u} = u^\lambda \mathbf{i}_\lambda = v^\mu \mathbf{g}_\mu. \tag{4.86}$$

It is obvious that the principle can be expressed in terms of v_λ by the use of Eq. (4.19).

The generalization of the principle of stationary potential energy is rather straightforward. We shall be satisfied with writing one of its generalizations which is given as follows:

$$\begin{aligned}\Pi_I = \iiint_V \{&A(f_{\lambda\mu}) + \Phi(v_\lambda) \\ &- \tau^{\lambda\mu}[f_{\lambda\mu} - \tfrac{1}{2}(v_{\lambda;\mu} + v_{\mu;\lambda} + v_{\varkappa;\lambda}v^{\varkappa}{}_{;\mu})]\}\, dV \\ &+ \iint_{S_1} \Psi(v_\lambda)\, dS - \iint_{S_2} p^\lambda(v_\lambda - \bar{v}_\lambda)\, dS,\end{aligned} \tag{4.87}$$

where $\bar{v}_\lambda$ is the prescribed displacement components defined by

$$\bar{\mathbf{u}} = \bar{v}_\lambda \mathbf{g}^\lambda \quad \text{on} \quad S_2. \tag{4.88}$$

In the functional (4.87), the independent quantities subject to variation are $f_{\lambda\mu}$, v_λ, $\tau^{\lambda\mu}$ and p^λ, while v^λ is dependent of v_μ by

$$v^\lambda = g^{\lambda\mu} v_\mu. \tag{4.89}$$

The stationary conditions are shown to be the governing equations of the problem, together with

$$p^\lambda = \tau^{\varkappa\mu}\nu_\varkappa(\delta^\lambda_\mu + v^\lambda{}_{;\mu}), \tag{4.90}$$

which determined the Lagrange multiplier p^λ on S_2.

4.8. Some Specializations to Small Displacement Theory in Orthogonal Curvilinear Coordinates

In the conclusion of this chapter, some of the results thus far obtained will be applied to an orthogonal curvilinear coordinate system, where the relation (4.25) reduces to

$$(ds^{(0)})^2 = g_{11}(d\alpha^1)^2 + g_{22}(d\alpha^2)^2 + g_{33}(d\alpha^3)^2. \tag{4.91}$$

The summation convention will not be employed in the present section, although Roman letters will be used in place of the numbers 1, 2 and 3. First, we note that Eq. (4.28) reduces to

$$g = g_{11}g_{22}g_{33}, \tag{4.92}$$

and the Christoffel symbols given by Eq. (4.13) may be reduced to[12]

$$\begin{aligned}\left\{\begin{matrix} i \\ i\,i \end{matrix}\right\} &= \frac{1}{\sqrt{g_{ii}}}\frac{\partial\sqrt{g_{ii}}}{\partial\alpha^i}, \\ \left\{\begin{matrix} i \\ i\,j \end{matrix}\right\} &= \left\{\begin{matrix} i \\ j\,i \end{matrix}\right\} = \frac{1}{\sqrt{g_{ii}}}\frac{\partial\sqrt{g_{ii}}}{\partial\alpha^j}, \\ \left\{\begin{matrix} i \\ j\,j \end{matrix}\right\} &= -\frac{\sqrt{g_{jj}}}{g_{ii}}\frac{\partial\sqrt{g_{jj}}}{\partial\alpha^i}, \\ \left\{\begin{matrix} i \\ j\,k \end{matrix}\right\} &= 0 \quad \text{for } i, j, k \text{ all different.}\end{aligned} \tag{4.93}$$

We shall confine our problem to the small displacement theory and derive the equations of equilibrium from the principle of virtual work. Since the magnitudes of the displacement components are assumed small, we may obtain the following linear strain–displacement relations from Eqs. (4.40):

$$2f_{ii} = 2g_{ii}\frac{\partial v^i}{\partial \alpha^i} + \sum_{j=1}^{3}\frac{\partial g_{ii}}{\partial \alpha^j}v^j,$$

$$2f_{ij} = g_{ii}\frac{\partial v^i}{\partial \alpha^j} + g_{jj}\frac{\partial v^j}{\partial \alpha^i} \quad \text{for} \quad i \neq j. \tag{4.94}$$

Next, we shall consider a set of local rectangular Cartesian coordinates y^i $(i = 1, 2, 3)$ coincident with the direction of $\mathbf{g}_i$ $(i = 1, 2, 3)$ at the point $P^{(0)}$, and denote the unit vector in the direction of $\mathbf{g}_i$ by $\mathbf{j}_i$. We then have

$$\mathbf{g}_i = \sqrt{g_{ii}}\,\mathbf{j}_i. \tag{4.95}$$

From Eqs. (4.57), (4.58) and (4.95), we obtain

$$\frac{\partial y^i}{\partial \alpha^j} = \sqrt{g_{ii}}\,\delta_{ij}, \quad \frac{\partial \alpha^i}{\partial y^j} = \frac{1}{\sqrt{g_{ii}}}\delta_{ij}, \tag{4.96}$$

where δ_{ij} is the Kronecker symbol. The components of the displacements and body forces may be alternatively defined with respect to the local Cartesian coordinates by

$$\mathbf{u} = \sum_{i=1}^{3} u^i\mathbf{j}_i, \tag{4.97}$$

$$\bar{\mathbf{P}} = \sum_{i=1}^{3} \bar{Y}^i\mathbf{j}_i. \tag{4.98}$$

Denoting the stresses and strains defined with respect to the local rectangular Cartesian coordinates by $\sigma^{11}, \ldots, \sigma^{33}$ and $\varepsilon_{11}, \ldots, \varepsilon_{33}$, respectively, we may use Eqs. (4.38), (4.61), (4.67) and (4.94) to obtain

$$u^i = \sqrt{g_{ii}}\,v^i, \tag{4.99}$$

$$\varepsilon_{ij} = \frac{1}{\sqrt{g_{ii}\,g_{jj}}}f_{ij}, \tag{4.100}$$

$$\sigma^{ij} = \sqrt{g_{ii}g_{jj}}\,\tau^{ij}, \tag{4.101}$$

and

$$\varepsilon_{ii} = \frac{1}{2g_{ii}}\left[2g_{ii}\frac{\partial}{\partial \alpha^i}\left(\frac{u^i}{\sqrt{g_{ii}}}\right) + \sum_{j=1}^{3}\frac{\partial g_{ii}}{\partial \alpha^j}\frac{u^j}{\sqrt{g_{jj}}}\right],$$

$$\varepsilon_{ij} = \frac{1}{2\sqrt{g_{ii}\,g_{jj}}}\left[g_{ii}\frac{\partial}{\partial \alpha^j}\left(\frac{u^i}{\sqrt{g_{ii}}}\right) + g_{jj}\frac{\partial}{\partial \alpha^i}\left(\frac{u^j}{\sqrt{g_{jj}}}\right)\right] \quad \text{for} \quad i \neq j. \tag{4.102}$$

When Eqs. (4.97), (4.98) and (4.102) are substituted into the principle of virtual work (4.84), we obtain the following equations of equilibrium:[13]

$$\frac{\partial(\sigma^{11}\sqrt{g_{22}g_{33}})}{\partial\alpha^1} - \sigma^{22}\sqrt{g_{33}}\frac{\partial\sqrt{g_{22}}}{\partial\alpha^1} - \sigma^{33}\sqrt{g_{22}}\frac{\partial\sqrt{g_{33}}}{\partial\alpha^1}$$
$$+\frac{1}{\sqrt{g_{11}}}\left[\frac{\partial(\sigma^{12}g_{11}\sqrt{g_{33}})}{\partial\alpha^2}+\frac{\partial(\sigma^{31}g_{11}\sqrt{g_{22}})}{\partial\alpha^3}\right]+\bar{Y}^1\sqrt{g_{11}g_{22}g_{33}}=0, \quad (4.103)$$

the other two equations being obtained by cylic permutations of the indices. It is obvious that Eqs. (4.103) are obtainable alternatively by specializing Eqs. (4.53) to small displacement theory in the orthogonal curvilinear coordinate system.

Bibliography

1. A. E. Green and W. Zerna, *Theoretical Elasticity*, Oxford University Press, 1959.
2. V. V. Novozhilov, *Theory of Elasticity*, Pergamon Press, 1961.
3. Y. C. Fung, *Foundations of Solid Mechanics*, Prentice-Hall, 1965.
4. E. Koppe, Methoden der nichtlinearen Elastizitätstheorie mit Anwendungen auf die dünne Platte endlicher Durchbiegung, *Zeitschrift für Angewandte Mathematik und Mechanik*, Vol. 36, No. 11/12, pp. 455–62, November/December 1956.
5. K. Kondo, Geometry of Elastic Deformation and Incompatibility, *and* A Theory of Stresses and Stress Densities, *Memoirs of the Unifying Study of the Basic Problems in Engineering Sciences by Means of Geometry*, Vol. 1, pp. 361–73 and 374–91, Gakujutsu Bunken Fukyu-kai, Tokyo, 1955.
6. Y. Yoshimura, *Meta-theory of Mechanics of Continua Subject to Deformation of Arbitrary Magnitudes*, Aeronautical Research Institute, University of Tokyo, Report No. 343, May 1959.
7. L. Brand, *Vector and Tensor Analysis*, John Wiley, 1947.
8. S. L. Synge and A. Schild, *Tensor Calculus*, University of Toronto Press, 1949.
9. D. J. Struik, *Lectures on Classical Differential Geometry*, Addison-Wesley, 1950.
10. I. S. Sokolnikoff, *Tensor Analysis*, John Wiley, 1951.
11. H. D. Block, *Introduction to Tensor Analysis*, Charles E. Merrill, 1962.
12. P. Morse and H. Feshbach, *Methods of Theoretical Physics*, Parts I and II, McGraw-Hill, 1953.
13. C. Biezeno and R. Grammel, *Technische Dynamik*, Springer Verlag, 1939.
14. L. S. D. Morley, *Skew Plates and Structures*, Pergamon Press, 1963.

CHAPTER 5

EXTENSIONS OF THE PRINCIPLE OF VIRTUAL WORK AND RELATED VARIATIONAL PRINCIPLES

5.1. Initial Stress Problems

We have derived the principle of virtual work and related variational principles of the boundary value problem in Chapter 3. We shall extend these principles to other problems of elasticity in the present chapter.† We shall formulate each problem in the finite displacement theory, specializing to the small displacement theory whenever necessary. The rectangular Cartesian coordinate system will be employed for describing the behavior of the elastic body. However, due to the invariant character observed in Chapter 4, expressions of the principles in the general curvilinear coordinate system are obtainable through coordinate transformation.

The Lagrangian approach will be employed throughout the present chapter. The set of values (x^1, x^2, x^3) which locate an arbitrary point of the body in a reference state will be used for specification during the subsequent behavior. Determination of the reference state depends upon the specific problem under consideration. Unless otherwise stated, the displacement is measured from the reference state.

We shall first consider an initial stress problem.[1, 2] By initial stress, we mean those stresses which have existed in a body in the initial state, that is, before the start of a deformation of interest. We choose the initial state as the reference state of an initial stress problem.

Let a rectangular Cartesian coordinate system (x^1, x^2, x^3) be fixed in space. Form an infinitesimal rectangular parallelepiped enclosed by the six surfaces: x^λ = constant and $x^\lambda + dx^\lambda$ = constant ($\lambda = 1, 2, 3$). Denoting the initial internal forces per unit area acting on the surface x^λ = constant by $-\boldsymbol{\sigma}^{(0)\lambda}$, we define components of the initial stress as

$$\boldsymbol{\sigma}^{(0)\lambda} = \sigma^{(0)\lambda\mu}\mathbf{i}_\mu, \tag{5.1}$$

† It is noted that some of the principles which will be derived in this chapter may have fields of application outside of elasticity problems.

where $\mathbf{i}_\mu$ is the unit vector in the direction of the x^μ-axis. Initial body forces and surface tractions are denoted by $\bar{\mathbf{P}}^{(0)}$ and $\bar{\mathbf{F}}^{(0)}$, respectively, and their components by

$$\bar{\mathbf{P}}^{(0)} = \bar{P}^{(0)\lambda}\mathbf{i}_\lambda, \quad \bar{\mathbf{F}}^{(0)} = F^{(0)\lambda}\mathbf{i}_\lambda. \tag{5.2}$$

For the sake of simplicity, we assume that these initial stresses and forces form a self-equilibrating system, i.e.

$$\sigma^{(0)\lambda\mu}{}_{,\mu} + \bar{P}^{(0)\lambda} = 0 \tag{5.3}$$

in the interior of the body and

$$\sigma^{(0)\lambda\mu}n_\mu = F^{(0)\lambda} \tag{5.4}$$

on the surface of the body, where $(\ )_{,\mu} = \partial(\ )/\partial x^\mu$.

We define a boundary value problem of the body with initial stresses by prescribing additional body forces $\bar{P}^\lambda$, additional surface forces F^λ on S_1 and surface displacements $\bar{u}^\lambda$ on S_2, where the displacements are measured from the initial state.

Consider the equilibrium of the infinitesimal parallelepiped after deformation in a manner similar to the development in Section 3.2, and denote the internal forces acting on the surface with the sides $\mathbf{E}_2\, dx^2$ and $\mathbf{E}_3\, dx^3$ by $-(\sigma^{(0)1\mu} + \sigma^{1\mu})\,\mathbf{E}_\mu\, dx^2\, dx^3$. Those acting on the other surfaces are defined in a similar manner. The quantities $\sigma^{\lambda\mu}$ thus defined will be called incremental stresses. Then, we find that the equations of equilibrium and mechanical boundary condition equations for the initial stress problem are derivable from Eqs. (3.27) and (3.42) by replacing $\sigma^{\lambda\mu}$, $\bar{P}^\lambda$ and F^λ with $\sigma^{(0)\lambda\mu} + \sigma^{\lambda\mu}$, $\bar{P}^{(0)\lambda} + \bar{P}^\lambda$ and $F^{(0)\lambda} + F^\lambda$, respectively. Consequently following the development in Section 3.6, we have the principle of virtual work for the initial stress problem as follows:

$$\iiint_V [(\sigma^{(0)\lambda\mu} + \sigma^{\lambda\mu})\,\delta e_{\lambda\mu} - (\bar{P}^{(0)\lambda} + \bar{P}^\lambda)\,\delta u^\lambda]\, dV$$

$$- \iint_{S_1} (F^{(0)\lambda} + F^\lambda)\,\delta u^\lambda\, dS = 0, \tag{5.5}$$

where

$$e_{\lambda\mu} = \tfrac{1}{2}(u^\lambda_{,\mu} + u^\mu_{,\lambda} + u^\varkappa_{,\lambda}\, u^\varkappa_{,\mu}), \tag{5.6}$$

and δu^λ is required to vanish on S_2. When the initial stresses are in self-equilibrium, we may employ Eqs. (5.3) and (5.4) to transform Eq. (5.5) into:

$$\iiint_V [\sigma^{\lambda\mu}\,\delta e_{\lambda\mu} + \sigma^{(0)\lambda\mu}u^\varkappa_{,\lambda}\,\delta u^\varkappa_{,\mu} - \bar{P}^\lambda\,\delta u^\lambda]\, dV$$

$$- \iint_{S_1} F^\lambda\,\delta u^\lambda\, dS = 0. \tag{5.7}$$

Next, a formulation of the principle of stationary potential energy and related variational principles will be considered. First, relations between the incremental stresses and strains are assumed to exist such that

$$\sigma^{\lambda\mu} = \sigma^{\lambda\mu}(e_{\alpha\beta};\ \sigma^{(0)\alpha\beta}), \tag{5.8†}$$

or conversely

$$e_{\lambda\mu} = e_{\lambda\mu}(\sigma^{\alpha\beta};\ \sigma^{(0)\alpha\beta}), \tag{5.9†}$$

where the initial stresses may be contained as parameters. Second Eqs. (5.8) are assumed to satisfy Eqs. (3.53). Then, the existence of the strain energy function $A(e_{\lambda\mu};\ \sigma^{(0)\lambda\mu})$ defined by

$$dA = \sigma^{\lambda\mu}\, de_{\lambda\mu} \tag{5.10}$$

is assured. Then Eq. (5.7) may be transformed into

$$\delta \iiint_V [A(u^\lambda;\ \sigma^{(0)\lambda\mu}) + \tfrac{1}{2}\sigma^{(0)\lambda\mu} u^{\varkappa}_{.\lambda}\, u^{\varkappa}_{.\mu}]\, dV$$
$$- \iiint_V \bar{P}^\lambda\, \delta u^\lambda\, dV - \iint_{S_1} \bar{F}^\lambda\, \delta u^\lambda\, dS = 0, \tag{5.11}$$

where $A(u^\lambda;\ \sigma^{(0)\lambda\mu})$ is obtained from $A(e_{\lambda\mu};\ \sigma^{(0)\lambda\mu})$ by writing $e_{\lambda\mu}$ in terms of u^λ by the use of Eqs. (5.6), and the variation is taken with respect to u^λ.

If the existence of the two potential functions defined by Eqs. (3.66) is also assured, we have a functional for the principle of stationary potential energy, which is then generalized by the use of Lagrange multipliers. Here, we record only the expression for Π_I:

$$\Pi_I = \iiint_V \{A(e_{\lambda\mu};\ \sigma^{(0)\lambda\mu}) + \tfrac{1}{2}\sigma^{(0)\lambda\mu} u^{\varkappa}_{.\lambda} u^{\varkappa}_{.\mu} + \Phi(u^\lambda)$$
$$- \sigma^{\lambda\mu}[e_{\lambda\mu} - \tfrac{1}{2}(u^\lambda_{.\mu} + u^\mu_{.\lambda} + u^{\varkappa}_{.\lambda} u^{\varkappa}_{.\mu})]\}\, dV$$
$$+ \iint_{S_1} \Psi(u^\lambda)\, dS - \iint_{S_2} p^\lambda(u^\lambda - \bar{u}^\lambda)\, dS, \tag{5.12}$$

where the independent quantities subject to variation are $e_{\lambda\mu}$, u^λ, $\sigma^{\lambda\mu}$ and p^λ with no subsidiary conditions. The stationary conditions are shown to be the governing equations of the initial stress problem, together with

$$p^\lambda = \sigma^{(0)\varkappa\mu} n_\varkappa u^\lambda_{.\mu} + \sigma^{\varkappa\mu}\, n_\varkappa(\delta^\lambda_\mu + u^\lambda_{.\mu}), \tag{5.13}$$

which determines the Lagrange multiplier p^λ on S_2. The expression (5.12) suggests that the complementary energy function $B(\sigma^{\lambda\mu};\ \sigma^{(0)\lambda\mu})$ corresponding to $A(e_{\lambda\mu};\ \sigma^{(0)\lambda\mu})$ is given by

$$B = \sigma^{\lambda\mu} e_{\lambda\mu} - A, \tag{5.14}$$

where Eq. (5.9) is used to express strains in terms of stresses.

We shall derive in this connection a linearized formulation of the initial stress problem, assuming that the displacements are of infinitesimal magnitude, i.e. $u^\lambda = O(\varepsilon)$‡ and the initial stresses are of finite magnitude, i.e.

† We assume throughout the present chapter that stress–strain relations have unique inverse relation, unless otherwise stated.

‡ The notation $O(\varepsilon)$ stands for "order of ε", where ε denotes an infinitesimal magnitude.

$\sigma^{(0)\lambda\mu} = O(1)$. This assumption leads to the reduction of the principle of virtual work (5.7), and to the linearization of the strain–displacement relations (5.6) and the stress–strain relations (5.8) as follows:

$$\iiint_V [\sigma^{\lambda\mu}\,\delta\varepsilon_{\lambda\mu} + \sigma^{(0)\lambda\mu}u^{\varkappa}_{.\lambda}\,\delta u^{\varkappa}_{.\mu} - \bar{P}^{\lambda}\delta u^{\lambda}]\,dV$$

$$-\iint_{S_1} \bar{F}^{\lambda}\,\delta u^{\lambda}\,dS = 0, \tag{5.15}$$

$$\varepsilon_{\lambda\mu} = \tfrac{1}{2}(u^{\lambda}_{.\mu} + u^{\mu}_{.\lambda}), \tag{5.16}$$

$$\sigma^{\lambda\mu} = a^{\lambda\mu\alpha\beta}\varepsilon_{\alpha\beta}. \tag{5.17}$$

The principle (5.15) yields the equation of equilibrium

$$\sigma^{\lambda\mu}{}_{,\mu} + (\sigma^{(0)\varkappa\mu}u^{\lambda}_{.\mu})_{,\varkappa} + \bar{P}^{\lambda} = 0, \tag{5.18}$$

and the mechanical boundary conditions

$$\sigma^{\lambda\varkappa}n_{\varkappa} + \sigma^{(0)\varkappa\mu}n_{\varkappa}u^{\lambda}_{.\mu} = \bar{F}^{\lambda} \quad \text{on} \quad S_2. \tag{5.19}$$

By combining Eqs. (5.16), (5.17), (5.18), (5.19) and the geometrical boundary conditions

$$u^{\lambda} = \bar{u}^{\lambda} \quad \text{on} \quad S_2, \tag{5.20}$$

we may obtain the governing equations for the desired linearization. Related variational principles for the linearized problem can be derived in the usual manner.

5.2. Stability Problems of a Body with Initial Stresses

Consider a body with initial stresses $k\sigma^{(0)\lambda\mu}$, under body forces $k\bar{P}^{(0)\lambda}$ in V and surface tractions $k\bar{F}^{(0)\lambda}$ on S, where k is a monotonically increasing factor of proportionality. The quantities $\sigma^{(0)\lambda\mu}$, $\bar{P}^{(0)\lambda}$ and $\bar{F}^{(0)\lambda}$ are assumed to be prescribed. When k is sufficiently small, the equilibrating configuration will be stable. However, with increase of k, a critical condition may be reached, beyond which the body ceases to be stable. The present section will be concerned with finding the critical initial stress distribution under the assumption that, in spite of the increase of k, changes in the geometrical configuration of the body remain negligible until the instability occurs.[2,3,4] We shall confine our problem by assuming that the prescribed body forces as well as the surface forces on S_1 vary neither their magnitudes nor their directions, while the body is rigidly fixed on S_2, during the period of instability. It is observed that this instability problem may be considered as a special case of the formulation developed in Section 3.11.

We shall employ as a criterion of instability the existence of an adjacent equilibrium configuration introduced in Section 3.11. It is then obvious that the linearized formulation in the preceding section yields the governing equations of this instability problem. By replacing $\sigma^{(0)\lambda\mu}$ by $k\sigma^{(0)\lambda\mu}$ and requiring that the incremental body forces $\bar{P}^\lambda$, surface forces $\bar{F}^\lambda$ and displacements $\bar{u}^\lambda$ vanish in Eqs. (5.15), (5.18), (5.19) and (5.20), we obtain

$$\iiint_V [\sigma^{\lambda\mu}\,\delta\varepsilon_{\lambda\mu} + k\sigma^{(0)\lambda\mu}\,u^{\varkappa}_{.\lambda}\,\delta u^{\varkappa}_{.\mu}]\,dV = 0, \tag{5.21}$$

$$\sigma^{\lambda\mu}{}_{,\mu} + (k\sigma^{(0)\varkappa\mu}u^{\lambda}_{.\mu})_{,\varkappa} = 0, \tag{5.22}$$

$$\sigma^{\lambda\varkappa}n_\varkappa + k\sigma^{(0)\varkappa\mu}n_\varkappa u^{\lambda}_{.\mu} = 0 \quad \text{on} \quad S_1, \tag{5.23}$$

$$u^\lambda = 0 \quad \text{on} \quad S_2. \tag{5.24}$$

The equations (5.16), (5.17), (5.22), (5.23) and (5.24) describe completely the instability problem under consideration. The solution of these equations reduces to an eigenvalue problem, where the critical value of the parameter k is determined as an eigenvalue and the adjacent equilibrium configuration as the corresponding eigenfunction.

When the elastic constants in Eq. (5.17) satisfy the symmetry relation $a^{\lambda\mu\alpha\beta} = a^{\alpha\beta\lambda\mu}$, the principle (5.21) is transformed into the principle of stationary potential energy, of which the functional is given by

$$\Pi = \iiint_V [A(u^\lambda;\sigma^{(0)\lambda\mu}) + \tfrac{1}{2}k\sigma^{(0)\lambda\mu}u^{\varkappa}_{.\lambda}u^{\varkappa}_{.\mu}]\,dV, \tag{5.25}$$

where $A(u^\lambda;\sigma^{(0)\lambda\mu})$ is obtained from

$$A(\varepsilon_{\lambda\mu};\sigma^{(0)\lambda\mu}) = \tfrac{1}{2}a^{\lambda\mu\alpha\beta}\varepsilon_{\lambda\mu}\varepsilon_{\alpha\beta}, \tag{5.26}$$

by writing $\varepsilon_{\lambda\mu}$ in terms of u^λ by the use of Eq. (5.16). In the functional (5.25), the variation is taken with respect to u^λ under the subsidiary conditions (5.24), while k is treated as a parameter not subject to variation.

Once the principle of stationary potential energy is thus derived, it can be generalized through the use of Lagrange multipliers. Only the expression of Π_I is shown below:

$$\Pi_I = \iiint_V \{A(\varepsilon_{\lambda\mu};\sigma^{(0)\lambda\mu}) + \tfrac{1}{2}k\sigma^{(0)\lambda\mu}u^{\varkappa}_{.\lambda}u^{\varkappa}_{.\mu} - \sigma^{\lambda\mu}[\varepsilon_{\lambda\mu} - \tfrac{1}{2}(u^{\lambda}_{.\mu} + u^{\mu}_{.\lambda})]\}\,dV - \iint_{S_2} p^\lambda u^\lambda dS, \tag{5.27}$$

where the quantities subject to variation are $\sigma^{\lambda\mu}$, $\varepsilon_{\lambda\mu}$, u^λ and p^λ with no subsidiary conditions. The stationary conditions are shown to be the governing equations of the instability problem, together with

$$p^\lambda = k\sigma^{(0)\varkappa\mu}n_\varkappa u^{\lambda}_{.\mu} + \sigma^{\lambda\varkappa}n_\varkappa, \tag{5.28}$$

which determines the Lagrange multipliers p^λ on S_2.

Retracing the development in Section 2.7, it is observed that the principle (5.25) is equivalent to finding, among admissible displacements u^λ, those which make the quotient

$$k = -\frac{\iiint_V A(u^\lambda)\,dV}{\tfrac{1}{2}\iiint_V \sigma^{(0)\lambda\mu}u^{\varkappa}_{.\lambda}u^{\varkappa}_{.\mu}\,dV} \tag{5.29}$$

stationary. When the smallest positive and largest negative values of k are found to be K^+ and K^-, respectively, the body is stable as long as the value of k lies within the region bounded by the two extremities, i.e. $K^- < k < K^+$. The expression (5.29) is closely related to the Rayleigh quotient in eigenvalue problems.[5, 6]

5.3. Initial Strain Problems

Assume that there exists a body in a reference state with neither stresses nor strains. Let this body be cut into a number of infinitesimal rectangular parallelepipeds and let each piece be given strains of arbitrary magnitude. These strains are called initial strains and will be denoted by $e^{(0)}_{\lambda\mu}$ in the following formulation. Then, let the pieces be reassembled and brought again into a continuous body.

A set of strains must be added to the initial strains to reproduce continuous body, because of the incompatibility of the arbitrary initial strains. These incremental strains cause internal stresses, even if neither external forces nor displacements are applied. We shall generalize the present problem further by prescribing, together with the initial strains, body forces $\bar{P}^\lambda$, surface forces $\bar{F}^\lambda$ on S_1, and surface displacements $\bar{u}^\lambda$ on S_2, where the displacements u^λ are measured from the reference state.

The set of incremental strains required to produce the final configuration is denoted by $\epsilon_{\lambda\mu}$. In general, neither the initial strains nor the incremental strains satisfy the conditions of compatibility. But their sums,

$$e_{\lambda\mu} = e^{(0)}_{\lambda\mu} + \epsilon_{\lambda\mu}, \tag{5.30}$$

must satisfy the conditions of compatibility, namely, they should be derivable from u^λ, measured from the reference state, such that

$$e_{\lambda\mu} = \tfrac{1}{2}(u^\lambda_{.\mu} + u^\mu_{.\lambda} + u^\varkappa_{.\lambda} u^\varkappa_{.\mu}). \tag{5.31}$$

Then, retracing the development in Section 3.6, we find that the principle of virtual work for the initial strain problem is also given by Eq. (3.48) through the use of the strain–displacement relations (5.31).

Next, we shall derive the principle of stationary potential energy and related variational principles. First, we assume that stress–strain relations are given by

$$\sigma^{\lambda\mu} = \sigma^{\lambda\mu}(e_{\alpha\beta};\ e^{(0)}_{\alpha\beta}), \tag{5.32}$$

or conversely

$$e_{\lambda\mu} = e_{\lambda\mu}(\sigma^{\alpha\beta};\ e^{(0)}_{\alpha\beta}), \tag{5.33}$$

where the initial strains appear as parameters and $\sigma^{\lambda\mu}(e^{(0)}_{\alpha\beta};\ e^{(0)}_{\alpha\beta}) = 0$. Second, Eqs. (5.32) are assumed to satisfy Eqs. (3.53), thus assuring the existence of a strain energy function $A(e_{\lambda\mu};\ e^{(0)}_{\lambda\mu})$ defined by

$$dA = \sigma^{\lambda\mu}\, de_{\lambda\mu}. \tag{5.34}$$

Third, the existence of the two potential functions $\Phi(u^\lambda)$ and $\Psi(u^\lambda)$ is assumed. We can now derive the principle of stationary potential energy for the initial strain problem. The principle can be generalized by following familiar procedures. For example Π_I may be shown to be

$$\begin{aligned}\Pi_I = \iiint_V \{&A(e_{\lambda\mu};\ e_{\lambda\mu}^{(0)}) + \Phi(u^\lambda)\\ &- \sigma^{\lambda\mu}[e_{\lambda\mu} - \tfrac{1}{2}(u^\lambda_{.\mu} + u^\mu_{.\lambda} + u^\varkappa_{.\lambda}u^\varkappa_{.\mu})]\}\, dV\\ &+ \iint_{S_1} \Psi(u^\lambda)\, dS - \iint_{S_1} p^\lambda(u^\lambda - \bar{u}^\lambda)\, dS, \end{aligned} \tag{5.35}$$

where the independent quantities subject to variation are $\sigma^{\lambda\mu}$, $e_{\lambda\mu}$, u^λ and p^λ with no subsidiary conditions. The expression (5.35) suggests that the complementary energy function $B(\sigma^{\lambda\mu};\ e_{\lambda\mu}^{(0)})$ corresponding to $A(e_{\lambda\mu};\ e_{\lambda\mu}^{(0)})$ is

$$B = \sigma^{\lambda\mu}e_{\lambda\mu} - A, \tag{5.36}$$

where Eqs. (5.33) are used to express $e_{\lambda\mu}$ in terms of $\sigma^{\alpha\beta}$.

We may then conclude that the principle of virtual work and related variational principles are derived in the same forms as those in Chapter 3, except for the difference in the expressions for A and B. Similar statements can be made for initial strain problems of the small displacement theory.

5.4. Thermal Stress Problems

Consider an elastic body in a reference state with neither stresses nor strains and of uniform absolute temperature T_0. Then, let a temperature distribution $T(x^1, x^2, x^3)$ be given to the body, while body forces and surface boundary conditions are applied as prescribed in Section 3.5. Our problem is to find the stress distribution thus created in the body.†

Since we know that the coupling between the elastic deformation and the heat transfer is very weak and can usually be neglected, we shall assume that the temperature distribution is prescribed and the stress–strain relations are given as

$$\sigma^{\lambda\mu} = \sigma^{\lambda\mu}(e_{\alpha\beta};\ \theta), \tag{5.37}$$

where

$$\sigma^{\lambda\mu}(0;\ 0) = 0 \quad \text{and} \quad \theta = T - T_0.$$

Once the above assumption is employed, the equations which govern the thermal stress problem are found to be the same as those of the problem in Chapter 3, except that Eqs. (3.33) are now replaced by Eqs. (5.37), in which the temperature θ appears as a parameter. Thus, the principle of virtual work for the thermal stress problem is also given by Eq. (3.48).

Retracing the development of Section 3.7 and keeping in mind that the temperature distribution is prescribed, we find that the strain energy func-

† Refs. 7 through 14.

tion exists for each element of the elastic body in the thermoelastic problem and is equal to the Helmholtz free energy given by Eq. (3.63). Consequently, we need assume only the existence of the two state functions Φ and Ψ for the establishment of the principle of stationary potential energy, of which the functional is shown to be

$$\Pi = \iiint_V [A(u^\lambda;\ \theta) + \Phi(u^\lambda)]\, dV + \iint_{S_1} \Psi(u^\lambda)\, dS, \tag{5.38}$$

where the quantities subject to variation are u^λ under the geometrical boundary conditions on S_2, while the temperature θ is treated as prescribed and not subject to variation. Since the strain energy function appearing in the functional (5.38) is equal to the free energy function, the principle of stationary potential energy for the thermoelastic problem is frequently called the principle of stationary free energy.[9] Once the variational principle is thus established, it can be generalized by the use of Lagrange multipliers in a manner similar to the development in Section 3.9.

We may then conclude that the principle of virtual work and related variational principles for the thermoelastic problem are expressed in the same forms as those in Chapter 3, except for the difference in the expressions for A and B. Similar statements can be made for thermoelastic problems of the small displacement theory.

Mention is made of the stress–strain relations and the expressions for A and B. From the free energy function defined in Eq. (3.63), we may derive the relation:

$$dF_0 = dU_0 - T\, dS - S\, dT. \tag{5.39}$$

Combining the above with Eq. (3.59), we obtain

$$dF_0 = \sigma^{\lambda\mu}\, de_{\lambda\mu} - S\, dT, \tag{5.40}$$

which yields:

$$\sigma^{\lambda\mu} = \frac{\partial F_0}{\partial e_{\lambda\mu}}, \quad S = -\frac{\partial F_0}{\partial T}. \tag{5.41 a, b}$$

This means that once the free energy function is given explicitly in terms of $e_{\lambda\mu}$ and T, the stress–strain relations (5.37) are derived functions.

We shall look for linear stress–strain relations for the thermoelastic problem by assuming

$$F_0 = a_0 + a_I^{\lambda\mu} e_{\lambda\mu} + \tfrac{1}{2} a^{\lambda\mu\alpha\beta} e_{\lambda\mu} e_{\alpha\beta}, \tag{5.42}$$

where a_0, $a_I^{\lambda\mu}$ and $a^{\lambda\mu\alpha\beta}$ are functions of T and $a^{\lambda\mu\alpha\beta} = a^{\alpha\beta\lambda\mu}$. Then, from Eqs. (5.41 a) and (5.42), we have

$$\sigma^{\lambda\mu} = a^{\lambda\mu\alpha\beta} e_{\alpha\beta} + a_I^{\lambda\mu}. \tag{5.43}$$

If we denote the thermal strain by $e^\theta_{\lambda\mu}$, we have

$$a_I^{\lambda\mu} = -a^{\lambda\mu\alpha\beta} e^\theta_{\alpha\beta}, \tag{5.44}$$

because it is required that $\sigma^{\lambda\mu} = 0$ for $e_{\lambda\mu} = e^\theta_{\lambda\mu}$. The inverse relations of Eq. (5.43) may be obtained as

$$e_{\lambda\mu} = b_{\lambda\mu\alpha\beta}\sigma^{\alpha\beta} + e^\theta_{\lambda\mu}. \tag{5.45}$$

Equations (5.45) show that for a linear relationship, the thermal strains may be treated as initial strains. The expressions of the two functions defined by

$$dA = \sigma^{\lambda\mu}\, de_{\lambda\mu}, \quad dB = e_{\lambda\mu}\, d\sigma^{\lambda\mu} \tag{5.46}$$

are derived from Eqs. (5.43) and (5.45):

$$A = \tfrac{1}{2}a^{\lambda\mu\alpha\beta}e_{\lambda\mu}e_{\alpha\beta} - a^{\lambda\mu\alpha\beta}e^{\theta}_{\alpha\beta}e_{\lambda\mu}, \tag{5.47}$$

$$B = \tfrac{1}{2}b_{\lambda\mu\alpha\beta}\sigma^{\lambda\mu}\sigma^{\alpha\beta} + e^{\theta}_{\lambda\mu}\sigma^{\lambda\mu}. \tag{5.48}$$

When the material is isotropic both elastically and thermally, we may choose

$$e^{\theta}_{\lambda\mu} = e^{\theta}\,\delta_{\lambda\mu}, \tag{5.49}$$

and derive the following relations:

$$\sigma^{\lambda\mu} = \frac{E}{(1-2\nu)}\, e\, \delta_{\lambda\mu} + 2Ge'_{\lambda\mu} - \frac{E}{(1-2\nu)}\, e^{\theta}\delta_{\lambda\mu}, \tag{5.50}$$

$$e_{\lambda\mu} = \frac{(1-2\nu)}{E}\,\sigma\delta^{\lambda\mu} + \frac{1}{2G}\,\sigma^{\lambda\mu'} + e^{\theta}\delta^{\lambda\mu}, \tag{5.51}$$

$$A = \frac{3E}{2(1-2\nu)}\, e^2 + Ge'_{\lambda\mu}e'_{\lambda\mu} - \frac{3E}{(1-2\nu)}\, e^{\theta}e, \tag{5.52}$$

$$B = \frac{3(1-2\nu)}{2E}\,\sigma^2 + \frac{1}{4G}\,\sigma^{\lambda\mu'}\sigma^{\lambda\mu'} + 3e^{\theta}\sigma, \tag{5.53}$$

where $\delta_{\lambda\mu}$ and $\delta^{\lambda\mu}$ are the Kronecker symbols. If a linear relation is postulated between the thermal strain e^{θ} and the temperature difference θ, we may write

$$e^{\theta} = \alpha\theta, \tag{5.54}$$

where α is the coefficient of thermal expansion.

Thermal stress is closely associated with high speed flight and has been one of the primary problems in the design of flight vehicles in recent years. A great number of papers have been written on the subject, some of which are listed in the bibliography for the reader's reference.

5.5. Quasi-static Problems

Consider a body in a reference state with neither stresses nor strains. Let this body be subject to time-dependent body forces $\bar{P}^{\lambda}(x^1, x^2, x^3, t)$, surface forces $\bar{F}^{\lambda}(x^1, x^2, x^3, t)$ on S_1 and surface displacements $\bar{u}^{\lambda}$ (x^1, x^2, x^3, t) on S_2, where t and $\bar{u}^{\lambda}$ are measured from the reference state. Our problem is to find the deformation and stress distribution of the body due to the motion.

In the present section, we shall consider a quasi-static formulation of the dynamical problem presented above. By quasi-static, we mean that the time rate of change of the prescribed body forces, surface forces and displacements is so gradual that inertial terms can be neglected in the equations of motion. It is then obvious that the principle of virtual work and related variational principles can be formulated in the same manner as in Chapter 3, except that the time t now appears as a parameter. Consequently, we shall be rather interested in the quasi-static problem expressed in terms of rate as follows: given the distribution of stresses $\sigma^{\lambda\mu}$ and displacements u^λ in the body at the generic time, find the time rates of change of the stresses, $\dot{\sigma}^{\lambda\mu}$, and displacements, $\dot{u}^\lambda$ induced in the body, where a dot denotes differentiation with respect to time.

Since the equations of equilibrium and boundary conditions are written in terms of rate as

$$\frac{d}{dt}\{[(\delta^\lambda_\mu + u^\lambda{}_{,\mu})\,\sigma^{\varkappa\mu}]_{,\varkappa} + \bar{P}^\lambda\} = 0 \quad \text{in} \quad V, \tag{5.55}$$

$$\dot{F}^\lambda = \dot{\bar{F}}^\lambda \quad \text{on} \quad S_1, \tag{5.56}$$

$$\dot{u}^\lambda = \dot{\bar{u}}^\lambda \quad \text{on} \quad S_2, \tag{5.57}$$

where

$$\dot{F}^\lambda = \dot{\sigma}^{\varkappa\mu} n_\varkappa (\delta^\lambda_\mu + u^\lambda{}_{,\mu}) + \sigma^{\varkappa\mu} n_\varkappa \dot{u}^\lambda{}_{,\mu}, \tag{5.58}$$

we have

$$-\iiint_V \frac{d}{dt}\{[(\delta^\lambda_\mu + u^\lambda{}_{,\mu})\,\sigma^{\varkappa\mu}]_{,\varkappa} + \bar{P}^\lambda\}\,\delta\dot{u}^\lambda\,dV$$

$$+\iint_{S_1} (\dot{F}^\lambda - \dot{\bar{F}}^\lambda)\,\delta\dot{u}^\lambda\,dS = 0. \tag{5.59}$$

After some calculation, we may reduce Eq. (5.59) to

$$\iiint_V [\dot{\sigma}^{\lambda\mu}\,\delta\dot{e}_{\lambda\mu} + \sigma^{\lambda\mu}\dot{u}^\varkappa{}_{,\lambda}\,\delta\dot{u}^\varkappa{}_{,\mu} - \dot{\bar{P}}^\lambda\,\delta\dot{u}^\lambda]\,dV$$

$$-\iint_{S_1} \dot{\bar{F}}^\lambda\,\delta\dot{u}^\lambda\,dS = 0, \tag{5.60}$$

where $\delta\dot{e}_{\lambda\mu}$ denotes the variation of $\dot{e}_{\lambda\mu}$ with respect to $\dot{u}^\varkappa$ only, namely,

$$2\dot{e}_{\lambda\mu} = (\delta^\varkappa_\lambda + u^\varkappa{}_{,\lambda})\,\dot{u}^\varkappa{}_{,\mu} + (\delta^\varkappa_\mu + u^\varkappa{}_{,\mu})\,\dot{u}^\varkappa{}_{,\lambda}, \tag{5.61}$$

$$2\delta\dot{e}_{\lambda\mu} = (\delta^\varkappa_\lambda + u^\varkappa{}_{,\lambda})\,\delta\dot{u}^\varkappa{}_{,\mu} + (\delta^\varkappa_\mu + u^\varkappa{}_{,\mu})\,\delta\dot{u}^\varkappa{}_{,\lambda}. \tag{5.62}$$

Equation (5.60) is the principle of virtual work for the quasi-static problem.

Next, we shall consider variational principles of the quasi-static problem. First, we assume relations between the stress-rate and strain-rate to be given by,

$$\dot{\sigma}^{\lambda\mu} = \dot{\sigma}^{\lambda\mu}(\dot{e}_{\alpha\beta};\ \sigma^{\alpha\beta}, e_{\alpha\beta}), \tag{5.63}$$

or, conversely,

$$\dot{e}_{\lambda\mu} = \dot{e}_{\lambda\mu}(\dot{\sigma}^{\alpha\beta};\ \sigma^{\alpha\beta}, e_{\alpha\beta}), \tag{5.64}$$

where $\sigma^{\alpha\beta}$ and $e_{\alpha\beta}$ may be contained as parameters. Second, the relations (5.63) are assumed to satisfy the equations

$$\frac{\partial \dot{\sigma}^{\lambda\mu}}{\partial \dot{e}_{\alpha\beta}} = \frac{\partial \dot{\sigma}^{\alpha\beta}}{\partial \dot{e}_{\lambda\mu}}, \tag{5.65}$$

which assure the existence of a state function $A^*(\dot{e}_{\lambda\mu}; \sigma^{\lambda\mu}, e_{\lambda\mu})$ defined by

$$dA^* = \dot{\sigma}^{\lambda\mu}\, d\dot{e}_{\lambda\mu}. \tag{5.66}$$

Third, two state functions $\Phi^*(\dot{u}^\lambda)$ and $\Psi^*(\dot{u}^\lambda)$ defined by

$$\delta\Phi^* = -\dot{\bar{P}}^\lambda\, \delta\dot{u}^\lambda, \quad \delta\Psi^* = -\dot{\bar{F}}^\lambda\, \delta\dot{u}^\lambda \tag{5.67}$$

are assumed to exist. Then, we may obtain the principle of stationary potential energy for the quasi-static problem from Eq. (5.60). The principle thus obtained can be generalized by the use of Lagrange multipliers; the generalized expression Π_I may be shown to be

$$\begin{aligned} \Pi_I = {} & \iiint_V [A^*(\dot{e}_{\lambda\mu}) + \tfrac{1}{2}\sigma^{\lambda\mu}\dot{u}^{\varkappa}_{,\lambda}\dot{u}^{\varkappa}_{,\mu} + \Phi^*(\dot{u}^\lambda)]\, dV \\ & - \iiint_V \dot{\sigma}^{\lambda\mu}\{\dot{e}_{\lambda\mu} - \tfrac{1}{2}[(\delta^{\varkappa}_{\lambda} + u^{\varkappa}_{,\lambda})\, \dot{u}^{\varkappa}_{,\mu} + (\delta^{\varkappa}_{\mu} + u^{\varkappa}_{,\mu})\, \dot{u}^{\varkappa}_{,\lambda})\}\, dV \\ & + \iint_{S_1} \Psi^*(\dot{u}^\lambda)\, dS - \iint_{S_2} \dot{p}^\lambda(\dot{u}^\lambda - \dot{\bar{u}}^\lambda)\, dS, \end{aligned} \tag{5.68}$$

where the independent quantities subject to variation are $\dot{\sigma}^{\lambda\mu}$, $\dot{e}_{\lambda\mu}$, $\dot{u}^\lambda$ and $\dot{p}^\lambda$ with no subsidiary conditions, while the quantities $\sigma^{\lambda\mu}$, $e_{\lambda\mu}$ and u^λ are treated as parameters not subject to variation. The stationary conditions are shown to be the governing equations of the quasi-static problem, together with

$$\dot{p}^\lambda = \dot{F}^\lambda, \tag{5.69}$$

which determine the Lagrange multipliers $\dot{p}^\lambda$ on S_2, The functional (5.68) is equivalent to that formulated by Sanders, McComb and Schlechte.[15]

The expression (5.68) suggests the function $B^*(\dot{\sigma}^{\lambda\mu}; \sigma^{\lambda\mu}, e_{\lambda\mu})$ corresponding to A^* to be

$$B^* = \dot{\sigma}^{\lambda\mu}\dot{e}_{\lambda\mu} - A^*, \tag{5.70}$$

where $\dot{e}_{\lambda\mu}$ are expressed in terms of $\dot{\sigma}^{\alpha\beta}$ by the use of Eqs. (5.64).

When the quasi-static problem is confined to the small displacement theory, the governing equations corresponding to Eqs. (5.55), (5.56), (5.57) and (5.61) are

$$\dot{\sigma}_{ij,j} + \dot{\bar{X}}_i = 0 \quad \text{in} \quad V, \tag{5.71}$$

$$\dot{\sigma}_{ij}n_j = \dot{\bar{F}}_i \quad \text{on} \quad S_1, \tag{5.72}$$

$$\dot{u}_i = \dot{\bar{u}}_i \quad \text{on} \quad S_2, \tag{5.73}$$

$$2\dot{\varepsilon}_{ij} = \dot{u}_{i,j} + \dot{u}_{j,i} \quad \text{in} \quad V, \tag{5.74}$$

respectively. Here, components of the stress, strain and displacement are denoted respectively by σ_{ij}, ε_{ij} and u_i, while the quantities $\bar{X}_i$, $\bar{F}_i$ and $\bar{u}_i$ are respectively components of the prescribed body forces, surface forces and displacements. From the above relations, we obtain the principle of virtual work for the quasi-static problem of the small displacement theory as,

$$\iiint_V (\dot{\sigma}_{ij}\,\delta\dot{\varepsilon}_{ij} - \dot{\bar{X}}_i\,\delta\dot{u}_i)\,dV - \iint_{S_1} \dot{\bar{F}}_i\,\delta\dot{u}_i\,dS = 0, \tag{5.75}$$

where Eqs. (5.74) have been substituted. Since Eqs. (5.71) through (5.74) hold without the dot notation, we have the principle in mixed forms as

$$\iiint_V (\dot{\sigma}_{ij}\,\delta\varepsilon_{ij} - \dot{\bar{X}}_i\,\delta u_i)\,dV - \iint_{S_1} \dot{\bar{F}}_i\,\delta u_i dS = 0, \tag{5.76}$$

$$\iiint_V (\sigma_{ij}\,\delta\dot{\varepsilon}_{ij} - \bar{X}_i\,\delta\dot{u}_i)\,dV - \iint_{S_1} \bar{F}_i\,\delta\dot{u}_i\,dS = 0. \tag{5.77}$$

It is obvious that we may also obtain the principle of complementary virtual work corresponding to Eq. (5.75) as

$$\iiint_V \dot{\varepsilon}_{ij}\,\delta\dot{\sigma}_{ij}\,dV - \iint_{S_2} \delta\dot{\sigma}_{ij} n_j \dot{\bar{u}}_i\,dS = 0. \tag{5.78}$$

The expressions which correspond to Eqs. (5.76) and (5.77) may be derived in a similar manner. When the relations between stress or stress-rate and strain or strain-rate assure quantities such as $\dot{\sigma}_{ij}\,d\dot{\varepsilon}_{ij}$ to be perfect differentials, the above principles may lead to variational formulations.

5.6. Dynamical Problems

We shall now consider the dynamical problem defined in Section 5.5 without requiring the motion of the body to be quasi-static. The equations of motion for the dynamical problem are obtained from Eqs. (3.22) and (3.25) by replacing $\bar{\mathbf{P}}$ with $\bar{\mathbf{P}} - \varrho(d^2\mathbf{r}/dt^2)$:

$$\sigma^{\lambda}_{\cdot\lambda} + \bar{\mathbf{P}} - \varrho\frac{d^2\mathbf{r}}{dt^2} = 0, \tag{5.79}$$

$$\sigma^{\lambda\mu} = \sigma^{\mu\lambda}, \tag{5.80}$$

where ϱ is the density of the body per unit volume in the reference state. Consequently, Eq. (3.48) holds for the dynamical problem if the above replacement is made. By integrating the equation thus replaced with respect to time between $t = t_1$ and $t = t_2$, and employing the convention that values of $\mathbf{r}$ at $t = t_1$ and $t = t_2$ are prescribed such that $\delta\mathbf{r}(x^1, x^2, x^3, t_1) = \delta\mathbf{r}(x^1, x^2, x^3, t_2) = 0$, we finally obtain the principle of virtual work for the dynamical problem as follows:

$$\int_{t_1}^{t_2} \left[\iiint_V \sigma^{\lambda\mu}\,\delta e_{\lambda\mu} dV - \delta T - \iiint_V \bar{\mathbf{P}}\cdot\delta\mathbf{r}\,dV - \iint_{S_1} \bar{\mathbf{F}}\cdot\delta\mathbf{r}\,dS\right] dt = 0, \tag{5.81}$$

Trevithick Libraries

Items on loan 10/09/2012 15:52

XXXXX06896

tem Details	Due Date
Finite elements of nonlinear	30/11/2012
Mathematical foundations of	15/11/2012
Functional analysis / Kôsaku	19/11/2012
Analysis of the finite elemen	03/10/2012
ntroductory functional analy	19/11/2012
Mathematical theory of finite	30/11/2012
NURBS book / Les Piegl, W	30/11/2012
Introduction to the mathema	30/11/2012
* Variational methods in elas	10/12/2012
Green's functions and bounc	14/09/2012
Introduction to the mechanic	04/12/2012

* Indicates items borrowed today

Overdue items will incur charges.

Renew items at: http://library.cf.ac.uk

where

$$e_{\lambda\mu} = \tfrac{1}{2}(u^{\lambda}_{.\mu} + u^{\mu}_{.\lambda} + u^{\varkappa}_{.\lambda}u^{\varkappa}_{.\mu}), \tag{5.82}$$

and T is the kinetic energy of the body defined by

$$T = \frac{1}{2}\iiint\limits_V \left(\frac{d\mathbf{r}}{dt}\right)^2 \varrho\, dV = \frac{1}{2}\iiint\limits_V \dot{u}^{\lambda}\dot{u}^{\lambda}\varrho\, dV. \tag{5.83}$$

If the existence of the strain energy function defined by Eq. (3.51) is assured, we find that the principle (5.81) becomes

$$\int\limits_{t_1}^{t_2} \left[\delta T - \delta U + \iiint\limits_V \bar{\mathbf{P}}\cdot\delta\mathbf{r}\, dV + \int\limits_{S_1} \bar{\mathbf{F}}\cdot\delta\mathbf{r}\, dS\right] dt = 0, \tag{5.84}$$

where U is the strain energy of the elastic body:

$$U = \iiint\limits_V A(u^{\lambda})\, dV. \tag{5.85}$$

The principle (5.84) is useful in application to dynamical problems of elastic bodies in which external forces are not derivable from potential functions.

If the existence of the two potentials Φ and Ψ defined by Eqs. (3.66) is also assured, the above principle reduces to

$$\delta\int\limits_{t_1}^{t_2} \left[T - U - \iiint\limits_V \Phi\, dV - \iint\limits_{S_1} \Psi\, dS\right] dt = 0, \tag{5.86}$$

where the variation is taken with respect to u^{λ}. Equation (5.86) is Hamilton's principle applied to the dynamical problem of the elastic body. It states that *among all admissible displacements which satisfy the prescribed geometrical boundary conditions on S_2 and the prescribed conditions at the limits $t = t_1$ and $t = t_2$, the actual solution makes the functional*

$$\int\limits_{t_1}^{t_2} \left[T - U - \iiint\limits_V \Phi\, dV - \iint\limits_{S_1} \Psi\, dS\right] dt$$

stationary.

It is an extension of the principle of stationary potential energy (3.68) to the dynamical problem. Its generalization can be formulated in a manner similar to the development in Section 3.9.

We shall employ Eq. (5.84) in subsequent formulations, in order to account for forces not derivable from a potential function, and we shall consider an approximate solution of the problem under the assumption that the displaced components of the body can be expressed in terms of a discrete number of generalized coordinates q_r ($r = 1, 2, \ldots, n$) as follows:

$$u^{\lambda} = u^{\lambda}(x^1, x^2, x^3;\ q_1, q_2, \ldots, q_n, t), \tag{5.87}$$

where the generalized coordinates are functions of time. The expressions (5.87) are so chosen as to satisfy the prescribed geometrical boundary conditions on S_2, irrespectively of the values of the generalized coordinates. From Eqs. (5.87), we obtain

$$\dot{u}^\lambda = \sum_{r=1}^{n} \frac{\partial u^\lambda}{\partial q_r} \dot{q}_r + \frac{\partial u^\lambda}{\partial t}, \tag{5.88}$$

$$\delta u^\lambda = \sum_{r=1}^{n} \frac{\partial u^\lambda}{\partial q_r} \delta q_r. \tag{5.89}$$

Introducing Eqs. (5.87) and (5.88) into Eqs. (5.83) and (5.85), we can express the Lagrangian function

$$L = T - U \tag{5.90}$$

in terms of q_r and $\dot{q}_r$. With these preliminaries, the first two terms on the left hand side of Eq. (5.84) are transformed into

$$\begin{aligned}\int_{t_1}^{t_2} \delta L\, dt &= \int_{t_1}^{t_2} \left[\sum_{r=1}^{n} \left(\frac{\partial L}{\partial q_r} \delta q_r + \frac{\partial L}{\partial \dot{q}_r} \delta \dot{q}_r\right)\right] dt \\ &= \sum_{r=1}^{n} \frac{\partial L}{\partial \dot{q}_r} \delta q_r \Big|_{t_1}^{t_2} - \int_{t_1}^{t_2} \sum_{r=1}^{n} \left[\frac{d}{dt}\left(\frac{\partial L}{\partial \dot{q}_r}\right) - \frac{\partial L}{\partial q_r}\right] \delta q_r\, dt \\ &= -\int_{t_1}^{t_2} \sum_{r=1}^{n} \left[\frac{d}{dt}\left(\frac{\partial L}{\partial \dot{q}_r}\right) - \frac{\partial L}{\partial q_r}\right] \delta q_r\, dt,\end{aligned} \tag{5.91}$$

where the convention $\delta q_r(t_1) = \delta q_r(t_2) = 0$ $(r = 1, 2, \ldots, n)$ is employed.† Introducing Eq. (5.89) and remembering that $\delta \mathbf{r} = \delta \mathbf{u}$, we find that the remaining terms of Eq. (5.84) become

$$\iiint_V \bar{\mathbf{P}} \cdot \delta \mathbf{r}\, dV + \iint_{S_1} \bar{\mathbf{F}} \cdot \delta \mathbf{r}\, dS = \sum_{r=1}^{n} Q_r\, \delta q_r, \tag{5.92}$$

where

$$Q_r = \iiint_V \bar{P}^\lambda \cdot \frac{\partial u^\lambda}{\partial q_r} dV + \iint_{S_1} \bar{F}^\lambda \cdot \frac{\partial u^\lambda}{\partial q_r} dS, \quad (r = 1, 2, \ldots, n) \tag{5.93}$$

is called the generalized force. Introducing Eqs. (5.91) and (5.92) into Eq. (5.84), we have

$$\int_{t_1}^{t_2} \sum_{r=1}^{n} \left[\frac{d}{dt}\left(\frac{\partial L}{\partial \dot{q}_r}\right) - \frac{\partial L}{\partial q_r} - Q_r\right] \delta q_r\, dt = 0. \tag{5.94}$$

Since the δq_r are independent, the above equation leads to n simultaneous equations:

$$\frac{d}{dt}\left(\frac{\partial L}{\partial \dot{q}_r}\right) - \frac{\partial L}{\partial q_r} = Q_r \quad (r = 1, 2, \ldots, n). \tag{5.95}$$

† This corresponds to our earlier assumption that

$$\delta \mathbf{r}(x^1, x^2, x^3, t_1) = \delta \mathbf{r}(x^1, x^2, x^3, t_2) = 0.$$

These are Lagrange's equations of motion for the elastic body. For applications of these equations to motions of elastic bodies, the reader is directed to Refs. 16, 17 and 18. The above formulations have been made in the finite displacement theory. However, they may be specialized to the small displacement theory through the familiar procedure, i.e. linearization of the strain–displacement relations (5.82).

5.7. Dynamical Problems of an Unrestrained Body

In the last section of this chapter we shall consider a dynamical problem of an unrestrained body.[18, 19, 20] Let a rectangular Cartesian coordinate system (X^1, X^2, X^3) be fixed in space and let the unit vector in the direction of the X^λ-axis be denoted by $\mathbf{i}_\lambda$, as shown in Fig. 5.1. Let another rectangular Cartesian coordinate system (x^1, x^2, x^3), called the body axes, be fixed to the body in a reference state with neither stresses nor strains. The Lagrangian approach will be employed, i.e. the set of the values (x^1, x^2, x^3) which locate an arbitrary point of the body with respect to the body axes in the reference state will be used for specification during the motion. The position vector of the point of the body at the generic time t is given by

$$\mathbf{r} = \mathbf{r}_G + \mathbf{r}^{(0)} + \mathbf{u}. \tag{5.96}$$

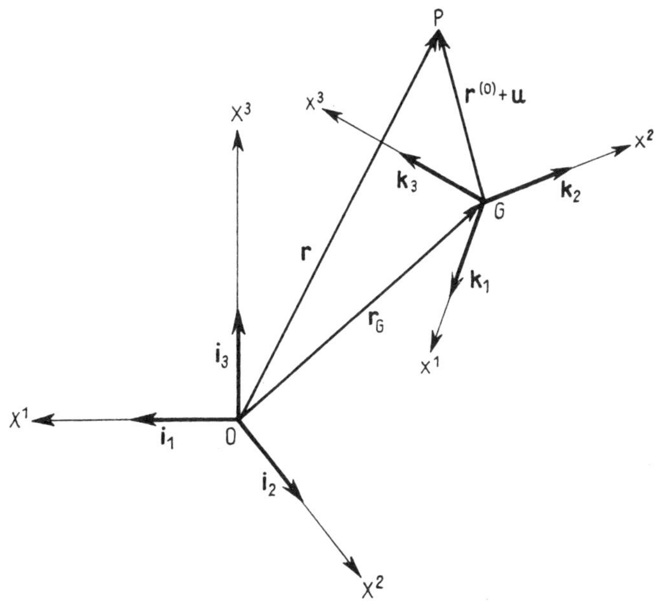

FIG. 5.1. Fixed and moving coordinate systems.

Here, $\mathbf{r}_G$ is the position vector from the origin of the space-fixed coordinates to the origin of the body axes, $\mathbf{r}^{(0)}$ is the position vector from the origin of the body axes to a point of the body in the reference state and $\mathbf{u}$ is the deformation vector. Components of the last two vectors are defined with respect to the body axes as follows:

$$\mathbf{r}^{(0)} = x^\lambda \mathbf{k}_\lambda, \quad \mathbf{u} = u^\lambda \mathbf{k}_\lambda, \tag{5.97}$$

where $\mathbf{k}_\lambda$ is the unit vector in the direction of the x^λ-axis. Since

$$\frac{d\mathbf{k}_\lambda}{dt} = \boldsymbol{\omega} \times \mathbf{k}_\lambda, \tag{5.98}$$

where

$$\boldsymbol{\omega} = p\mathbf{k}_1 + q\mathbf{k}_2 + r\mathbf{k}_3 \tag{5.99}$$

is the angular velocity vector of the moving coordinate system, we have

$$\frac{d\mathbf{r}^{(0)}}{dt} = \boldsymbol{\omega} \times \mathbf{r}^{(0)}, \tag{5.100}$$

$$\frac{d\mathbf{u}}{dt} = \frac{d^*\mathbf{u}}{dt} + \boldsymbol{\omega} \times \mathbf{u}, \tag{5.101}$$

where $d^*(\ \ )/dt$ denotes a partial differentiation, the unit vectors $\mathbf{k}_\lambda$ ($\lambda = 1, 2, 3$) being held constant. For example, we have $d^*\mathbf{u}/\mathrm{dt} = \dot{u}^\lambda \mathbf{k}_\lambda$.

We shall define the orientation of the body axes with respect to the space-fixed coordinate system by the Eulerian angles ϕ, θ and ψ, as shown in Fig. 5.2 to obtain the following geometrical and kinematical relations:(21, 22, 23)

$$\begin{bmatrix} \mathbf{k}_1 \\ \mathbf{k}_2 \\ \mathbf{k}_3 \end{bmatrix} = \begin{bmatrix} \cos\theta\cos\psi, & \cos\theta\sin\psi, & -\sin\theta \\ -\cos\phi\sin\psi + \sin\phi\sin\theta\cos\psi, & \cos\phi\cos\psi + \sin\phi\sin\theta\sin\psi, & \sin\phi\cos\theta \\ \sin\phi\sin\psi + \cos\phi\sin\theta\cos\psi, & -\sin\phi\cos\psi + \cos\phi\sin\theta\sin\psi, & \cos\phi\cos\theta \end{bmatrix} \begin{bmatrix} \mathbf{i}_1 \\ \mathbf{i}_2 \\ \mathbf{i}_3 \end{bmatrix}, \tag{5.102}$$

$$\begin{bmatrix} p \\ q \\ r \end{bmatrix} = \begin{bmatrix} 1 & 0 & -\sin\theta \\ 0 & \cos\phi & \sin\phi\cos\theta \\ 0 & -\sin\phi & \cos\phi\cos\theta \end{bmatrix} \begin{bmatrix} \dot\phi \\ \dot\theta \\ \dot\psi \end{bmatrix}. \tag{5.103}$$

Combining Eqs. (5.96) and (5.100), we may express the kinetic energy of the body as follows:

$$\begin{aligned} T = {} & \frac{1}{2}\left(\frac{d\mathbf{r}_G}{dt}\right)^2 \iiint \varrho\, dV + \frac{1}{2}\iiint (\boldsymbol{\omega} \times \mathbf{r}^{(0)})^2 \varrho\, dV \\ & + \frac{1}{2}\iiint \left(\frac{d\mathbf{u}}{dt}\right)^2 \varrho\, dV + \left(\frac{d\mathbf{r}_G}{dt} \times \boldsymbol{\omega}\right)\cdot \iiint \mathbf{r}^{(0)} \varrho\, dV \\ & + \frac{d\mathbf{r}_G}{dt}\cdot \iiint \frac{d\mathbf{u}}{dt} \varrho\, dV + \boldsymbol{\omega}\cdot \iiint \left(\mathbf{r}^{(0)} \times \frac{d\mathbf{u}}{dt}\right) \varrho\, dV. \end{aligned} \tag{5.104}$$

The vector $\mathbf{r}_G$ is a function of time only. Consequently, we have $\partial\mathbf{r}/\partial x^\lambda = \partial(\mathbf{r}^{(0)} + \mathbf{u})/\partial x^\lambda$ and we find that the strain–displacement relations are given by

$$e_{\lambda\mu} = \tfrac{1}{2}(u^\lambda_{,\mu} + u^\mu_{,\lambda} + u^\varkappa_{,\lambda} u^\varkappa_{,\mu}). \tag{5.105}$$

Combination of Eqs. (5.96), (5.100) and (5.101) yields the following result for the virtual displacement vector $\delta\mathbf{r}$:

$$\delta\mathbf{r} = \delta\mathbf{r}_G + \delta u^\lambda \mathbf{k}_\lambda + \delta'\boldsymbol{\Theta} \times (\mathbf{r}^{(0)} + \mathbf{u}), \tag{5.106}$$

where

$$\delta'\boldsymbol{\Theta} = (\delta\phi - \delta\psi \sin\theta)\,\mathbf{k}_1 + (\delta\theta\cos\phi + \delta\psi\sin\phi\cos\theta)\,\mathbf{k}_2 + (-\delta\theta\sin\phi + \delta\psi\cos\phi\cos\theta)\,\mathbf{k}_3. \tag{5.107}$$

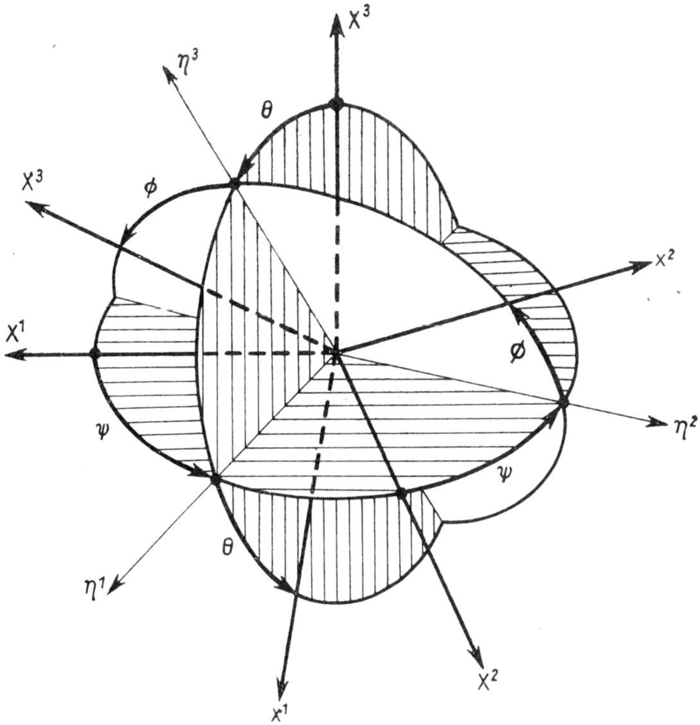

FIG. 5.2. The Eulerian angles.†

† The transformation from the space fixed axes (X^1, X^2, X^3) to the moving axes (x^1, x^2, x^3) is defined by three successive angles of rotation. First, the (X^1, X^2, X^3) axes are rotated around the X^3-axis by the angle ψ to obtain (η^1, η^2, X^3) axes. Second, the (η^1, η^2, X^3) axes are rotated around the η^2-axis by the angle θ to obtain (x^1, η^2, η^3) axes. Finally, the (x^1, η^2, η^3) axes are rotated around the x^1-axis by the angle ϕ to obtain (x^1, x^2, x^3) axes. The three angles ϕ, θ and ψ, called the Eulerian angles, specify the orientation of the (x^1, x^2, x^3) axes uniquely.

We shall define a dynamical problem of the unrestrained body by prescribing, in addition to the body forces $\bar{\mathbf{P}}(x^1, x^2, x^3, t)$, the external forces $\bar{\mathbf{F}}(x^1, x^2, x^3, t)$ applied on the surface S of the body. We then find that the principle of virtual work for the dynamical problem of the unrestrained body is given by

$$\int_{t_1}^{t_2} \left[\iiint_V \sigma^{\lambda\mu}\, \delta e_{\lambda\mu}\, dV - \delta T - \iiint_V \bar{\mathbf{P}} \cdot \delta\mathbf{r}\, dV - \iint_S \bar{\mathbf{F}} \cdot \delta\mathbf{r}\, ds \right] dt = 0, \quad (5.108)$$

which, when the strain energy function exists, becomes

$$\int_{t_1}^{t_2} \left[\delta T - \delta U + \iiint_V \bar{\mathbf{P}} \cdot \delta\mathbf{r}\, dV + \iint_S \bar{\mathbf{F}} \cdot \delta\mathbf{r}\, ds \right] dt = 0. \quad (5.109)$$

Since dynamical problems are beyond the intended scope of this book, we shall mention briefly only two special cases. The first example is the motion of a rigid body, where the deformation vector $\mathbf{u}$ vanishes throughout the body. We shall take the origin of the body coordinates in coincidence with the center of gravity of the body. Equations (5.104), (5.106) and (5.109) then reduce to

$$T = \frac{1}{2}\left(\frac{d\mathbf{r}_G}{dt}\right)^2 \iiint_V \varrho\, dV + \frac{1}{2} \iiint_V (\boldsymbol{\omega} \times \mathbf{r}^{(0)})^2\, \varrho\, dV, \quad (5.110)$$

$$\delta\mathbf{r} = \delta\mathbf{r}_G + \delta'\boldsymbol{\Theta} \times \mathbf{r}^{(0)}, \quad (5.111)$$

and

$$\int_{t_1}^{t_2} \left[\delta T + \iiint_V \bar{\mathbf{P}} \cdot \delta\mathbf{r}\, dV + \iint_S \bar{\mathbf{F}} \cdot \delta\mathbf{r}\, ds \right] dt = 0, \quad (5.112)$$

respectively. Since $\delta\mathbf{r}_G$, $\delta\phi$, $\delta\theta$ and $\delta\psi$ are arbitrary, we obtain from the principle (5.112) six equations of motion of the rigid body which are written in vector forms as follows:

$$M \frac{d^2\mathbf{r}_G}{dt^2} = \iiint_V \bar{\mathbf{P}}\, dV + \iint_S \bar{\mathbf{F}}\, dS, \quad (5.113)$$

and

$$\frac{d\mathbf{H}}{dt} = \iiint_V (\mathbf{r}^{(0)} \times \bar{\mathbf{P}}) dV + \iint_S (\mathbf{r}^{(0)} \times \bar{\mathbf{F}}) dS, \quad (5.114)\dagger$$

† Since $\delta \iiint_V \frac{1}{2}(\boldsymbol{\omega} \times \mathbf{r}^{(0)})^2 \varrho\, dV = H_x\, \delta p + H_y\, \delta q + H_z\, \delta r$ and $\delta p = \delta\dot{\phi} - \sin\theta\, \delta\dot{\psi} - \dot{\psi}\, \delta\theta \cos\theta, \ldots$, we have through integrations by parts $\delta \int_{t_1}^{t_2} \left[\iiint_V \frac{1}{2}(\boldsymbol{\omega} \times \mathbf{r}^{(0)})^2 \varrho\, dV \right] dt = (\quad)\Big|_{t_2}^{t_1} - \int_{t_1}^{t_2} \delta'\boldsymbol{\Theta} \cdot \mathbf{N}\, dt$, where $\mathbf{N} = (\dot{H}_x + qH_z - rH_y)\, \mathbf{k}_1 + (\dot{H}_y + rH_x - pH_z)\, \mathbf{k}_2 + (\dot{H}_z + pH_y - qH_x)\, \mathbf{k}_3 = d\mathbf{H}/dt$.[24]

In the above equations, M and $\mathbf{H}$ are the total mass and the total angular momentum around the centre of mass of the rigid body, respectively, defined by the following relations:

$$M = \iiint \varrho \, dV, \tag{5.115}$$

$$\mathbf{H} = H_x \mathbf{k}_1 + H_y \mathbf{k}_2 + H_z \mathbf{k}_3, \tag{5.116}$$

$$\begin{bmatrix} H_x \\ H_y \\ H_z \end{bmatrix} = \begin{bmatrix} I_x & J_{xy} & J_{xz} \\ J_{xy} & I_y & J_{yz} \\ J_{xz} & J_{yz} & I_z \end{bmatrix} \begin{bmatrix} p \\ q \\ r \end{bmatrix}, \tag{5.117}$$

$$I_x = \iiint_V (y^2 + z^2) \varrho \, dV, \quad I_y = \iiint_V (z^2 + x^2) \varrho \, dV, \quad I_z = \iiint_V (x^2 + y^2) \varrho \, dV,$$

$$J_{xy} = -\iiint_V xy\varrho \, dV, \quad J_{yz} = -\iiint_V yz\varrho \, dV, \quad J_{xz} = -\iiint_V xz\varrho \, dV, \tag{5.118}$$

where notations x, y, z are used in places of x^1, x^2, x^3.

The second example is a small disturbed motion of an elastic body. We assume that the body is in rectilinear flight with a constant velocity before it is subjected to small external disturbances. The steady flight state is taken as a reference state, in which the origin of the body axes is located at the center of gravity of the body. The x^1-axis is taken in coincidence with the direction of the constant velocity. The Eulerian angles are measured from a system of reference axes, the directions of which are coincident with those of the body axes in rectilinear flight.

Now, the body is assumed subject to small external disturbances. Following Ref. 5.18, we may express the elastic deformation in terms of the normal modes of the unrestrained elastic body as

$$\mathbf{u} = \sum_{i=1}^{\infty} \boldsymbol{\Phi}_i(x^1, x^2, x^3) \, \xi_i(t), \tag{5.119}$$

where $\boldsymbol{\Phi}_i$ is the i-th natural mode. Then, we obtain the equations of motion of the unrestrained body from the principle of virtual work (5.109), where the independent quantities are $\delta \mathbf{r}_G$, $\delta\phi$, $\delta\theta$, $\delta\psi$ and $\delta\xi_i$ $(i = 1, 2, \ldots)$. After neglecting higher order terms, we may finally reduce these equations to linearized form. The reader is directed to Ref. 18 for further detail.

Thus far, we have derived the principle of virtual work and related variational principles for several elasticity problems. In the five chapters which follow, these principles will be applied to special problems such as torsion of bars, beams, plates, shells and structures. In these applications, the material is assumed isotropic and homogeneous, and problems are treated in the small displacement theory, unless otherwise stated. Moreover, we shall employ conventional notations in these problems. For example, U, V and W will be used instead of u^λ in Chapters 7, 8 and 9, while the symbols u, v and w will be reserved for expressing displacement components of the centroid locus of the beam or of the middle surface of plates and shells. As a second example, we note that σ_x, σ_y, ... and τ_{xy} will be employed even in the finite displacement theory with an understanding that these symbols now represent $\sigma^{\lambda\mu}$ defined in Chapters 3 and 5.

Bibliography

1. E. TREFFTZ, Zur Theorie der Stabilität des elastischen Gleichgewichts, *Zeitschrift für Angewandte Mathematik und Mechanik*, Vol. 13, No. 2, pp. 160–5, April 1933.
2. V. V. NOVOZHILOV, *Foundations of the Nonlinear Theory of Elasticity*, Graylock, 1953.
3. W. PRAGER, The General Variational Principle of the Theory of Structural Stability, *Quarterly of Applied Mathematics*, Vol. 4, No. 4, pp. 378–84, January 1947.
4. J. N. GOODIER and H. J. PLASS, Energy Theorems and Critical Load Approximations in the General Theory of Elastic Stability, *Quarterly of Applied Mathematics*, Vol. IX, No. 4, pp. 371–80, 1952.
5. L. COLLATZ, *Eigenwertaufgaben mit technischen Anwendungen*, Akademische Verlagsgesellschaft, Leipzig, 1949.
6. G. TEMPLE and W. G. BICKLEY, *Rayleigh's Principle and its Application to Engineering*, Oxford University Press, London, 1933.
7. S. TIMOSHENKO and J. N. GOODIER, *Theory of Elasticity*, McGraw-Hill, 1951.
8. W. S. HEMP, Fundamental Principles and Methods of Thermal Elasticity, *Aircraft Engineering*, Vol. 26, No. 302, pp. 126–7, April 1954.
9. W. S. HEMP, Fundamental Principles and Theorems of Thermoelasticity, *Aeronautical Quarterly*, Vol. 7 Part 3, pp. 184–92, August 1956.
10. W. S. HEMP, *Methods for the Theoretical Analysis of Aircraft Structures*, AGARD Lecture Course, April 1957.
11. B. E. GATEWOOD, *Thermal Stresses*, McGraw-Hill, 1957.
12. B. A. BOLEY and J. H. WEINER, *Theory of Thermal Stresses*, John Wiley, 1960.
13. R. L. BISPLINGHOFF, Some Structural and Aeroelastic Considerations of High Speed Flight, *Journal of the Aeronautical Sciences*, Vol. 23, No. 4, pp. 289–327, April 1956.
14. *High Temperature Effects in Aircraft Structures*, edited by N. J. Hoff, AGARDograph 28, Pergamon Press, 1958.
15. J. L. SANDERS, JR., H. G. MCCOMB, JR., and F. R. SCHLECHTE, *A Variational Theorem for Creep with Applications to Plates and Columns*, NACA TN 4003.
16. Y. C. FUNG, *Introduction to the Theory of Aeroelasticity*, John Wiley, 1955.
17. R. L. BISPLINGHOFF, H. ASHLEY and R. L. HALFMAN, *Aeroelasticity*, Addison-Wesley, 1955.
18. R. L. BISPLINGHOFF and H. ASHLEY, *Principles of Aeroelasticity*, John Wiley, 1962.
19. H. GOLDSTEIN, *Classical Mechanics*, Addison-Wesley, 1959.
20. J. L. SYNGE and B. A. GRIFFITH, *Principles of Mechanics*, McGraw-Hill, 1959.
21. R. A. FRAZER, W. J. DUNCAN and A. R. COLLAR, *Elementary Matrices*, Cambridge University Press, 1938.
22. M. J. ABZUG, Applications of Matrix Operators to the Kinematics of Airplane Motion, *Journal of Aeronautical Sciences*, Vol. 23, No. 7, pp. 679–84, July 1956.
23. B. ETKIN, *Dynamics of Flight*, John Wiley, 1959.
24. B. L. BYRUM and E. R. GRADY, General Air-frame Dynamics of a Guided Missile, *Journal of Aeronautical Sciences*, Vol. 22, No. 8, pp. 534–40, August 1955.

CHAPTER 6

TORSION OF BARS

6.1. Saint-Venant Theory of Torsion

In the present section, the Saint-Venant theory of torsion of a cylindrical bar is treated. Unless otherwise stated, the cross section of the bar, denoted by the area S, is assumed to be simply connected. Let the z-axis be taken in the direction of the generating line of the cylinder, and the x- and y-axes in the sectional plane, as shown in Fig. 6.1. Torsion of a bar is defined as the application of twisting moments at both ends of the bar, while keeping the side surface of the bar traction free. Consequently, the mechanical boundary conditions at the ends, $z = 0$ and $z = l$, are given as

$$X_\nu = -\bar{X}_z, \quad Y_\nu = -\bar{Y}_z, \quad Z_\nu = 0, \tag{6.1}$$

and

$$X_\nu = \bar{X}_z, \quad Y_\nu = \bar{Y}_z, \quad Z_\nu = 0, \tag{6.2}$$

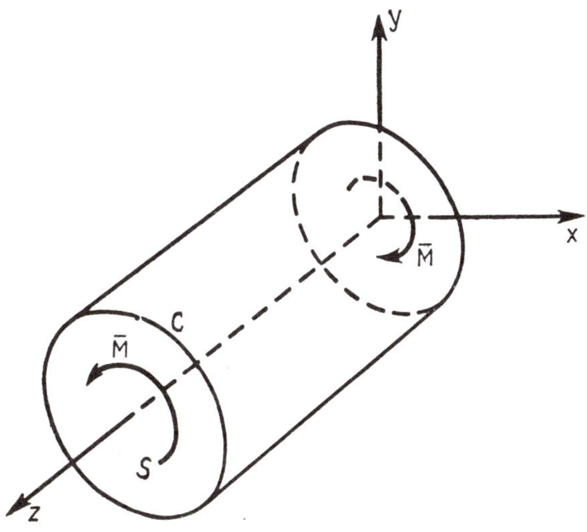

FIG. 6.1. Torsion of a bar.

respectively, to produce the twisting couple

$$\bar{M} = \iint_S (\bar{Y}_z x - \bar{X}_z y)\, dx\, dy. \tag{6.3}$$

The displacement components u, v and w for a cylindrical bar in torsion are assumed to be[1, 2]

$$u = -\vartheta y, \quad v = \vartheta x, \quad w = w(x, y, z), \tag{6.4}$$

where $\vartheta = \vartheta(z)$ is the twist angle of the cross section around the z-axis, and is a function of z. The relation (6.4) assures us that the only non-vanishing strain components are ε_z, γ_{xz} and γ_{yz}, which are given by

$$\varepsilon_z = \frac{\partial w}{\partial z}, \quad \gamma_{xz} = \frac{\partial w}{\partial x} - \frac{d\vartheta}{dz} y, \quad \gamma_{yz} = \frac{\partial w}{\partial y} + \frac{\partial \vartheta}{\partial z} x. \tag{6.5}†$$

It is assumed in the Saint-Venant theory of torsion that the deformation of the bar takes place independently of z. This means that $w(x, y, z)$ and $d\vartheta/dz$ are independent of z. Therefore, we may write

$$u = -\theta yz, \quad v = \theta xz, \quad w = w(x, y), \tag{6.6}$$

and

$$\varepsilon_z = 0, \quad \gamma_{xz} = \frac{\partial w}{\partial x} - y\theta, \quad \gamma_{yz} = \frac{\partial w}{\partial y} + x\theta, \tag{6.7}$$

where $\theta = d\vartheta/dz$ is the rate of twist. Accordingly, the only non-vanishing stress components are τ_{xz} and τ_{yz}, which are related to γ_{xz} and γ_{yz} by

$$\tau_{xz} = G\gamma_{xz}, \quad \tau_{yz} = G\gamma_{yz}. \tag{6.8}†$$

With the above preliminaries the principle of virtual work for the Saint-Venant theory of torsion can be expressed as follows:

$$\iint_S \left[\tau_{xz}\left(\frac{\partial\, \delta w}{\partial x} - y\, \delta\theta\right) + \tau_{yz}\left(\frac{\partial\, \delta w}{\partial y} + x\, \delta\theta\right)\right] dx\, dy - \bar{M}\, \delta\theta = 0, \tag{6.9}‡$$

where the length of the cylinder is taken as unity, due to the uniformity of the deformation in the direction of the z-axis. After some calculation the principle (6.9) is transformed into:

$$-\iint_S \left(\frac{\partial \tau_{xz}}{\partial x} + \frac{\partial \tau_{yz}}{\partial y}\right) \delta w\, dx\, dy + \int_C (\tau_{xz} l + \tau_{yz} m)\, \delta w\, ds$$
$$+ \left\{\iint_S (\tau_{yz} x - \tau_{xz} y)\, dx\, dy - \bar{M}\right\} \delta\theta = 0, \tag{6.10}$$

where l and m are the direction cosines of the normal ν drawn outward on the boundary C. If the contour of the boundary C is given by $x = x(s)$

† Notations τ_{xz} and γ_{xz} will be preferably employed instead of τ_{zx} and γ_{zx} in Chapters 6, 7 and 8.

‡ See Eq. (1.32).

and $y = y(s)$, where s is measured along the contour as shown in Fig. 6.2, we have

$$l = dy/ds, \quad m = -dx/ds. \tag{6.11}$$

Since δw and $\delta\theta$ are arbitrary, we have, from Eq. (6.10), the equation of equilibrium and boundary conditions as follows:

$$\frac{\partial \tau_{xz}}{\partial x} + \frac{\partial \tau_{yz}}{\partial y} = 0 \quad \text{in} \quad S, \tag{6.12}$$

$$\tau_{xz} l + \tau_{yz} m = 0 \quad \text{on} \quad C, \tag{6.13}$$

and

$$\bar{M} = \iint_S (\tau_{yz} x - \tau_{xz} y)\, dx\, dy. \tag{6.14}$$

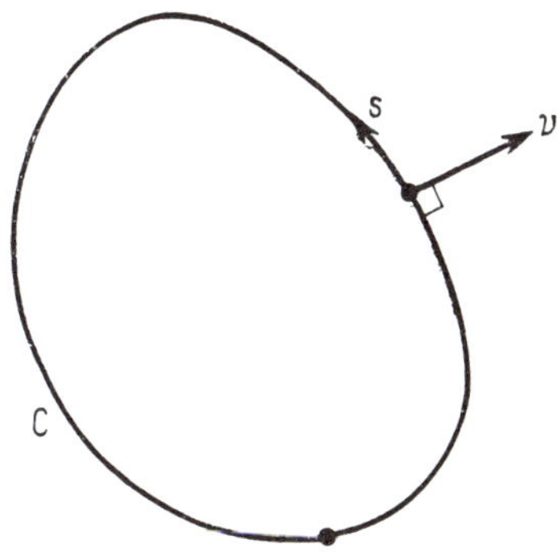

FIG. 6.2. Directions of s and ν.

One way of obtaining the solution is to eliminate τ_{xz}, τ_{yz}, γ_{xz} and γ_{yz} from Eqs. (6.7), (6.8), (6.12) and (6.13). Using Eqs. (6.11), the elimination finally leads to

$$\frac{\partial^2 \varphi}{\partial x^2} + \frac{\partial^2 \varphi}{\partial y^2} = 0 \quad \text{in} \quad S, \tag{6.15}$$

$$\frac{\partial \varphi}{\partial \nu} = \frac{1}{2} \frac{d}{ds}(x^2 + y^2) \quad \text{on} \quad C, \tag{6.16}$$

where

$$w = \theta \varphi(x, y), \tag{6.17}$$

and

$$\frac{\partial \varphi}{\partial \nu} = l \frac{\partial \varphi}{\partial x} + m \frac{\partial \varphi}{\partial y}. \tag{6.18}$$

Thus, the function φ, called the warping function of the cross section, is a plane harmonic function which satisfies the boundary condition (6.16). Once the solution has been obtained, we have from Eq. (6.14),

$$\bar{M} = G\theta \iint_S \left[\frac{\partial \varphi}{\partial y} x - \frac{\partial \varphi}{\partial x} y + x^2 + y^2 \right] dx\, dy, \tag{6.19}$$

which means that the torsional rigidity of the cross section is given by GJ, where

$$J = \iint\limits_S \left[\frac{\partial \varphi}{\partial y} x - \frac{\partial \varphi}{\partial x} y + x^2 + y^2\right] dx\, dy. \tag{6.20}$$

Many studies of the Saint-Venant torsion problem have been made. The problem has been solved for various cross sectional forms such as the circle, ellipse, square, rectangle and so forth. The interested reader is advised to read, for example, Refs. 1, 2, and 3.

6.2. The Principle of Minimum Potential Energy and its Transformation

It is observed that the functional (2.12), when combined with Eqs. (6.1), (6.2), (6.6) and (6.7), provides the total potential energy for the Saint-Venant torsion problem:

$$\Pi = \frac{1}{2} \iint\limits_S G \left[\left(\frac{\partial w}{\partial x} - \theta y\right)^2 + \left(\frac{\partial w}{\partial y} + \theta x\right)^2\right] dx\, dy - \theta \bar{M}, \tag{6.21}†$$

where the length of the cylinder is taken as unity, and the absolute minimum property of the total potential energy for the actual solution can be proved. Following a development similar to that in Chapter 2, we may generalize the expression (6.21) as follows:

$$\begin{aligned} \Pi_I = \iint\limits_S \Big\{ & \frac{G}{2}(\gamma_{xz}^2 + \gamma_{yz}^2) - \left(\gamma_{xz} - \frac{\partial w}{\partial x} + \theta y\right)\tau_{xz} \\ & - \left(\gamma_{yz} - \frac{\partial w}{\partial y} - \theta x\right)\tau_{yz}\Big\}\, dx\, dy - \theta \bar{M}, \end{aligned} \tag{6.22}$$

where the independent functions and scalar quantity subject to variations are γ_{xz}, γ_{yz}, τ_{xz}, τ_{yz}, w and θ. Since the first variation of Π_I is given by

$$\begin{aligned} \delta \Pi_I = \iint\limits_S & [(G\gamma_{xz} - \tau_{xz})\,\delta\gamma_{xz} + (G\gamma_{yz} - \tau_{yz})\,\delta\gamma_{yz} \\ & - \left(\gamma_{xz} - \frac{\partial w}{\partial x} + \theta y\right)\delta\tau_{xz} - \left(\gamma_{yz} - \frac{\partial w}{\partial y} - \theta x\right)\delta\tau_{yz} \\ & - \left(\frac{\partial \tau_{xz}}{\partial x} + \frac{\partial \tau_{yz}}{\partial y}\right)\delta w\Big]\, dx\, dy \\ & + \int\limits_C (\tau_{xz} l + \tau_{yz} m)\,\delta w\, ds - \left\{\bar{M} - \iint\limits_S (\tau_{yz} x - \tau_{xz} y)\, dx\, dy\right\}\delta\theta, \end{aligned} \tag{6.23}$$

† This functional is also obtainable from the principle of virtual work (6.9) combined with Eqs. (6.7) and (6.8).

it is easily shown that all the conditions which define the torsion problem under consideration can be obtained from the requirement that Π_{I} be stationary.

Now, let us employ the following stationary conditions as subsidiary conditions:

$$G\gamma_{xz} = \tau_{xz}, \quad G\gamma_{yz} = \tau_{yz}, \tag{6.24}$$

$$\frac{\partial \tau_{xz}}{\partial x} + \frac{\partial \tau_{yz}}{\partial y} = 0 \quad \text{in} \quad S, \tag{6.25}$$

$$\tau_{xz} l + \tau_{yz} m = 0 \quad \text{on} \quad C, \tag{6.26}$$

which mean elimination of γ_{xz}, γ_{yz} and w from Π_{I}. Since Eq. (6.25) is automatically satisfied through the introduction of a stress function $\phi(x, y)$ defined by

$$\tau_{xz} = \frac{\partial \phi}{\partial y}, \quad \tau_{yz} = -\frac{\partial \phi}{\partial x}, \tag{6.27}$$

we shall use ϕ in place of Eq. (6.25) in subsequent formulations. Then, by using the relations (6.11), we may write Eq. (6.26) as

$$\frac{\partial \phi}{\partial y}\frac{dy}{ds} + \frac{\partial \phi}{\partial x}\frac{dx}{ds} = \frac{d\phi}{ds} = 0,$$

or equivalently

$$\phi = c_0, \tag{6.28}$$

on the boundary C, where c_0 is an integration constant. Since the cross section of the bar is assumed to be simply connected, we may put

$$\phi = 0 \quad \text{on} \quad C, \tag{6.29}$$

without loss of generality. Thus, elimination of γ_{xz} and γ_{yz} through Eqs. (6.24) and introduction of the stress function ϕ defined by Eq. (6.27) transform Π_{I} into Π_{II} defined by

$$\Pi_{II} = -\iint_S \left\{\frac{1}{2G}\left[\left(\frac{\partial \phi}{\partial x}\right)^2 + \left(\frac{\partial \phi}{\partial y}\right)^2\right] - 2\theta\phi\right\} dx\, dy + \int_C \frac{\partial \phi}{\partial s} w\, ds - \theta \int_C (xl + ym)\, \phi\, ds - \theta \bar{M}. \tag{6.30}$$

Imposition of the condition (6.29) further simplifies Eq. (6.30) to

$$\Pi_{III} = -\iint_S \left\{\frac{1}{2G}\left[\left(\frac{\partial \phi}{\partial x}\right)^2 + \left(\frac{\partial \phi}{\partial y}\right)^2\right] - 2\theta\phi\right\} dx\, dy - \theta \bar{M}, \tag{6.31}$$

which is the final result obtained from Π_{I} through the elimination of γ_{xz}, γ_{yz} and w by the use of the relations (6.24), (6.25) and (6.26).

In Π_{III} derived above, the independent function and scalar subject to variation are ϕ and θ, respectively, where ϕ satisfies the condition (629).

Further reduction of Eq. (6.31) leads to the following stationary conditions:

$$\frac{\partial^2 \phi}{\partial x^2} + \frac{\partial^2 \phi}{\partial y^2} + 2G\theta = 0 \quad \text{in} \quad S, \tag{6.32}$$

$$\bar{M} = 2 \iint_S \phi \, dx \, dy, \tag{6.33}$$

where Eq. (6.29) is taken as the subsidiary condition. The equation (6.32) thus obtained is the condition of compatibility for the torsion problem. A direct proof will be given as follows: By the use of Eqs. (6.7), (6.8) and (6.27), we have

$$\begin{aligned} dw &= \frac{\partial w}{\partial x} dx + \frac{\partial w}{\partial y} dy \\ &= (\gamma_{xz} + \theta y)\, dx + (\gamma_{yz} - \theta x)\, dy \\ &= \left(\frac{1}{G} \frac{\partial \phi}{\partial y} + \theta y \right) dx - \left(\frac{1}{G} \frac{\partial \phi}{\partial x} + \theta x \right) dy. \end{aligned} \tag{6.34}$$

By integrating the above relation along an arbitrary closed path within the region S, we have

$$\begin{aligned} \underset{\rightarrow}{\overset{\leftarrow}{[\, w \,]}} &= \oint \left[\left(\frac{1}{G} \frac{\partial \phi}{\partial y} + \theta y \right) dx - \left(\frac{1}{G} \frac{\partial \phi}{\partial x} + \theta x \right) dy \right] \\ &= -\frac{1}{G} \iint \left(\frac{\partial^2 \phi}{\partial x^2} + \frac{\partial^2 \phi}{\partial y^2} + 2G\theta \right) dx \, dy, \end{aligned} \tag{6.35}$$

where the bracket notation $\underset{\rightarrow}{\overset{\leftarrow}{[\quad]}}$ indicates the increase of the value of the enclosed function with respect to one complete circuit of the closed path. The notation $\oint$ is the integral around the closed path, and the area integral is defined with respect to the region enclosed by the closed path. Thus, Eq. (6.32) assures that no dislocation is allowed for the displacement w.

Furthermore, if the condition (6.33) is taken as one more subsidiary condition, Π_{III} can be transformed into $\tilde{\Pi}_c$ written as follows:

$$\tilde{\Pi}_c = -\iint \frac{1}{2G} \left[\left(\frac{\partial \phi}{\partial x} \right)^2 + \left(\frac{\partial \phi}{\partial y} \right)^2 \right] dx \, dy, \tag{6.36}$$

where the independent function subject to variation is ϕ under the subsidiary conditions (6.29) and (6.33). It is easily verified that among admissible functions ϕ, the actual solution makes the functional $\tilde{\Pi}_c$ a maximum and that the variational principle thus obtained is equivalent to the principle of minimum complementary energy. If we trace backward from $\tilde{\Pi}_c$ to Π_{III}, it is easily seen that the scalar quantity θ appearing in Eq. (6.31) plays the role of a Lagrange multiplier via which the subsidiary condition (6.33) is introduced into the framework of the variational expression.

Finally, it is noted that in the Saint-Venant theory of torsion of a bar, the strain energy and complementary energy stored per unit length of the bar are given by

$$\frac{1}{2}G\theta^2\iint_S\left[\left(\frac{\partial\varphi}{\partial x}-y\right)^2+\left(\frac{\partial\varphi}{\partial y}+x\right)^2\right]dx\,dy=\frac{1}{2}GJ\left(\frac{d\vartheta}{dz}\right)^2, \quad (6.37)$$

and

$$\frac{1}{2G}\iint_S\left[\left(\frac{\partial\phi}{\partial x}\right)^2+\left(\frac{\partial\phi}{\partial y}\right)^2\right]dx\,dy=\frac{1}{2GJ}M^2, \quad (6.38)$$

respectively, where M is the twisting stress couple at the cross section.

6.3. Torsion of a Bar with a Hole

In the present section, we shall derive a variational formulation for the torsion problem of a cylindrical bar with a hole, as shown in Fig. 6.3. Let the outer and inner boundaries of the cross section be denoted by C_0 and C_i,

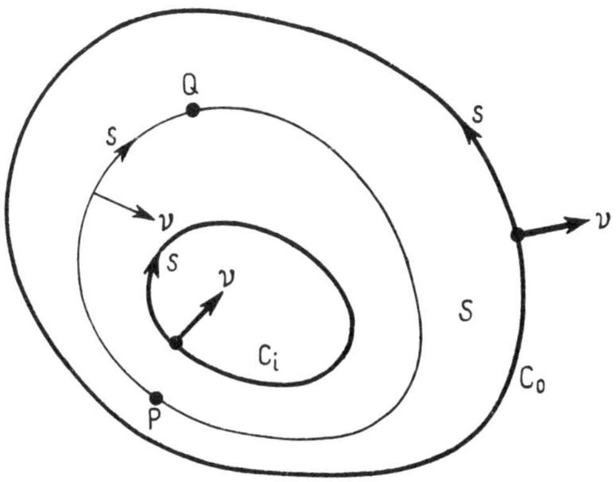

FIG. 6.3. A bar with a hole.

respectively. The assumption of Saint-Venant torsion theory asserts that the equations defining the problem are the same as stated in Section 6.1, except for one additional condition on the boundary C_i:

$$\tau_{xz}l+\tau_{yz}m=0, \quad (6.39)$$

where l and m are direction cosines of the normal drawn on the inner boundary from the interior of the bar S into the hole. By the use of the stress function ϕ defined by Eq. (6.27), the condition (6.39) can be written as

$$\phi=c_i \quad \text{on the boundary} \quad C_i, \quad (6.40)$$

where c_i is an arbitrary integration constant. We have put the integration constant c_0 given in Eq. (6.28) equal to zero before. However, the same disposition cannot be applied to the integration constant c_i. A formula for determining the value of c_i will be given in the following.

With the above preliminaries, it is easily shown that expression of the total potential energy Π is the same as (6.21), if the integration is taken over the region between the two curves C_0 and C_i. However, care should be taken in generalizing the principle. The generalization Π_{I} is the same as (6.22), except for the region of integration. But in the avenue leading from Π_{I} to Π_{II}, we should notice that we now have two boundaries. Thus, we have,

$$\Pi_{II} = -\iint_S \left\{\frac{1}{2G}\left[\left(\frac{\partial\phi}{\partial x}\right)^2 + \left(\frac{\partial\phi}{\partial y}\right)^2\right] - 2\theta\phi\right\} dx\,dy - \theta\bar{M}$$
$$+ \int_{C_0} \frac{\partial\phi}{\partial s} w\,ds - \theta\int_{C_0} (xl + ym)\,\phi\,ds$$
$$+ \int_{C_i} \frac{\partial\phi}{\partial s} w\,ds - \theta\int_{C_i} (xl + ym)\,\phi\,ds, \tag{6.41}$$

where the directions of s and ν on the C_i are shown in Fig. 6.3. By use of the relations (6.29) and (6.40), Π_{II} can be reduced to

$$\Pi_{III} = -\iint_S \left\{\frac{1}{2G}\left[\left(\frac{\partial\phi}{\partial x}\right)^2 + \left(\frac{\partial\phi}{\partial y}\right)^2\right] - 2\theta\phi\right\} dx\,dy - \theta\bar{M}$$
$$- c_i\theta\int_{C_i} (xl + ym)\,ds. \tag{6.42}$$

The quantities subject to independent variation in the above are ϕ, c_i and θ, and the stationary conditions thereof are the same as obtained in Section 6.2, except for one additional condition that

$$\int_{C_i} \frac{\partial\phi}{\partial\nu} ds + G\theta\int_{C_i} (xl + ym)\,ds = 0, \tag{6.43}$$

which is obtained by requiring the coefficient of δc_i in $\delta\Pi_{\text{III}}$ to vanish. Noting that the direction s is defined clockwise on the boundary C_i, we have

$$\int_{C_i} (xl + ym)\,ds = \int_{C_i} (x\,dy - y\,dx) = -2A_i, \tag{6.44}$$

where A_i is the area enclosed by the curve C_i. Consequently, Eq. (6.43) is reduced to[2, 3]

$$\int_{C_i} \frac{\partial\phi}{\partial\nu} ds = 2G\theta A_i. \tag{6.45}$$

We have shown that, for a bar consisting of a simply connected region, Eq. (6.32) gives the condition of compatibility. For a bar with a hole, Eq.

(6.45) gives an additional condition of compatibility, which is used to determine the value of c_i. For the proof, we shall again use the relation (6.34), which, when integrated along an arbitrary path between two arbitrary points P and Q, provides

$$w(Q) - w(P) = -\frac{1}{G}\int_P^Q \frac{\partial \phi}{\partial \nu}\, ds + \theta \int_P^Q (y\, dx - x\, dy), \tag{6.46}$$

where the directions of s and ν on the path PQ are shown in Fig. 6.3. Since no dislocation is allowed for the displacement w, it is required that

$$\oint \frac{\partial \phi}{\partial \nu}\, ds + G\theta \oint (x\, dy - y\, dx) = 0 \tag{6.47}$$

for any arbitrary closed path of integration within the region enclosed by C_0 and C_i. If the closed path of integration is taken in coincidence with the inner boundary C_i, Eq. (6.47) may be shown to reduce to Eq. (6.45). Thus, Eq. (6.45) is an additional condition of compatibility defined along the inner boundary. By using the relations derived above, we can prove that Eq. (6.47) holds for any arbitrary closed path on the cross section, provided that the relations (6.32) and (6.45) are assumed to hold. Equation (6.45) is called *the condition of compatibility in the large* for the torsion of a bar with a hole.

So far, only Saint-Venant torsion problems have been treated. Namely, the strain of the bar is assumed to bc independent of z. It is obvious that for the complete realization of Saint-Venant torsion, the mechanical boundary conditions at the two ends, Eqs. (6.1) and (6.2), must be prescribed in a manner such that they are exactly coincident with the stress distribution given by the Saint-Venant solution. When a bar of finite length is subjected to twisting moments applied on both ends in an arbitrary manner, stress distribution induced in the bar may be different from that obtained from the Saint-Venant torsion theory. However, it is assured from the Saint-Venant principle mentioned in the Introduction of this book that the end constraints disturb the stress distributions derived from Saint-Venant theory only locally. The spread of the disturbed regions in the z-direction is of the order of magnitude of the lateral dimensions of the bar, and the Saint-Venant theory of torsion can apply quite well to regions away from the ends of the bar. Approximate solutions have been obtained by other authors through variational methods for end constraint problems in the torsion of a bar.[2, 4]

6.4. Torsion of a Bar with Initial Stresses

Next, we shall consider the problem of torsion of a bar with initial stresses. For the sake of simplicity, the initial stresses are assumed to consist of $\sigma_z^{(0)}$ only, which is a function of (x, y) and independent of z. The governing equations for this torsion problem will be derived through the principle of virtual work for the initial stress problem, Eq. (5.5).

Following Ref. 5, we assume the displacement components to be given by

$$\begin{aligned} u &= -x(1-\cos\vartheta) - y\sin\vartheta, \\ v &= x\sin\vartheta - y(1-\cos\vartheta), \\ w &= \theta\Phi(x, y) + \varepsilon_0 z, \end{aligned} \tag{6.48†}$$

where $\vartheta(z)$ is the twist angle and $\theta = d\vartheta/dz$ is the rate of twist. The deformation is assumed to take place independently of z. This assumption makes θ and ε_0 constant and allows us to take the length of the cylinder as unity. The quantities to be determined are $\Phi(x, y)$ and the two constants θ and ε_0. From Eqs. (6.48), we obtain

$$\begin{aligned} &\frac{\partial u}{\partial x} = -1 + \cos\vartheta, \quad \frac{\partial u}{\partial y} = -\sin\vartheta, \quad \frac{\partial u}{\partial z} = -(x\sin\vartheta + y\cos\vartheta)\,\theta, \\ &\frac{\partial v}{\partial x} = \sin\vartheta, \quad \frac{\partial v}{\partial y} = -1 + \cos\vartheta, \quad \frac{\partial v}{\partial z} = (x\cos\vartheta - y\sin\vartheta)\,\theta, \\ &\frac{\partial w}{\partial x} = \frac{\partial \Phi}{\partial x}\theta, \quad \frac{\partial w}{\partial y} = \frac{\partial \Phi}{\partial y}\theta, \quad \frac{\partial w}{\partial z} = \varepsilon_0. \end{aligned} \tag{6.49}$$

Consequently, Eqs. (3.19) provide:

$$\begin{aligned} &e_{xx} = \frac{1}{2}\left(\frac{\partial \Phi}{\partial x}\right)^2 \theta^2, \quad 2e_{yz} = \left(\frac{\partial \Phi}{\partial y} + x\right)\theta + \frac{\partial \Phi}{\partial y}\theta\varepsilon_0, \\ &e_{yy} = \frac{1}{2}\left(\frac{\partial \Phi}{\partial y}\right)^2 \theta^2, \quad 2e_{xz} = \left(\frac{\partial \Phi}{\partial x} - y\right)\theta + \frac{\partial \Phi}{\partial x}\theta\varepsilon_0, \\ &e_{zz} = \varepsilon_0 + \frac{1}{2}(x^2 + y^2)\,\theta^2 + \frac{1}{2}\varepsilon_0^2, \quad 2e_{xy} = \frac{\partial \Phi}{\partial x}\frac{\partial \Phi}{\partial y}\theta^2. \end{aligned} \tag{6.50}$$

These strain components are assumed to be related to incremental stresses, denoted by σ_x, σ_y, ... and τ_{xy}, by Eqs. (3.38).

We shall be interested in formulating a linearized theory of the problem of pure torsion and we shall follow the development in the latter half of Section 5.1. Since the constant ε_0 proves to make no contribution to final results of interest as far as the linearized theory is concerned, we allow ε_0 to vanish in the following formulation. To begin with, the principle of virtual work, Eq. (5.5), is written for the present problem. After neglecting higher order terms, we find that the contribution from the volume integral of the principle becomes:

$$\iint_S (\tau_{xz}\,\delta\gamma_{xz} + \tau_{yz}\delta\gamma_{yz} + \sigma_z^{(0)}\,\delta e_{zz})\,dx\,dy, \tag{6.51}$$

† Compare with Eqs. (6.4) and (6.17).

where

$$\gamma_{xz} = \left(\frac{\partial \Phi}{\partial x} - y\right)\theta, \quad \gamma_{yz} = \left(\frac{\partial \Phi}{\partial y} + x\right)\theta, \tag{6.52}$$

and

$$e_{zz} = \tfrac{1}{2}(x^2 + y^2)\,\theta^2. \tag{6.53}$$

We define components of the external force $\bar{\mathbf{F}}$ by

$$\bar{\mathbf{F}} = \bar{F}_x \mathbf{i}_1 + \bar{F}_y \mathbf{i}_2 + \bar{F}_z \mathbf{i}_3, \tag{6.54}$$

and prescribe them as follows:

$$\bar{F}_x = -\bar{X}_z, \quad \bar{F}_y = -\bar{Y}_z, \quad \bar{F}_z = -\sigma_z^{(0)} \tag{6.55}$$

on the end at $z = 0$, and

$$\begin{aligned} \bar{F}_x &= \bar{X}_z \cos\theta - \bar{Y}_z \sin\theta, \\ \bar{F}_y &= \bar{X}_z \sin\theta + \bar{Y}_z \cos\theta, \\ \bar{F}_z &= \sigma_z^{(0)} \end{aligned} \tag{6.56}$$

on the other end at $z = 1$. The side boundary is traction-free. Here, $\bar{X}_z$ and $\bar{Y}_z$ are the external forces applied at both ends to produce the twisting moment $\bar{M}$ given by Eq. (6.3). Since the virtual displacements are obtained from Eqs. (6.48) as

$$\delta u = 0, \quad \delta v = 0, \quad \delta w = \Phi\,\delta\theta + \theta\,\delta\Phi \tag{6.57}$$

on the end at $z = 0$, and

$$\begin{aligned} \delta u &= -x\,\delta\theta \sin\theta - y\,\delta\theta\cos\theta, \\ \delta v &= x\,\delta\theta\cos\theta - y\,\delta\theta\sin\theta, \\ \delta w &= \Phi\,\delta\theta + \theta\,\delta\Phi \end{aligned} \tag{6.58}$$

on the other end at $z = 1$, the contribution from the surface integral of the principle reduces to

$$-\bar{M}\,\delta\theta. \tag{6.59}$$

By the use of the relations (6.51) and (6.59), we obtain the principle of virtual work for the present problem. Through partial integration, the principle becomes

$$\begin{aligned} &-\theta \iint_S \left(\frac{\partial \tau_{xz}}{\partial x} + \frac{\partial \tau_{yz}}{\partial y}\right)\delta\Phi\,dx\,dy + \theta\int_C (\tau_{xz} l + \tau_{yz} m)\,\delta\Phi\,ds \\ &+ \left\{\iint_S \left[\tau_{xz}\left(\frac{\partial \Phi}{\partial x} - y\right) + \tau_{yz}\left(\frac{\partial \Phi}{\partial y} + x\right)\right] dx\,dy \right. \\ &\left. + \theta \iint_S (x^2 + y^2)\,\sigma_z^{(0)}\,dx\,dy - \bar{M}\right\}\delta\theta = 0, \end{aligned} \tag{6.60}$$

which yields following equations:

$$\frac{\partial \tau_{xz}}{\partial x} + \frac{\partial \tau_{yz}}{\partial y} = 0 \quad \text{in} \quad S, \tag{6.61}$$

$$\tau_{xz} l + \tau_{yz} m = 0 \quad \text{on} \quad C, \tag{6.62}$$

$$\bar{M} = \iint_S (\tau_{yz} x - \tau_{xz} y)\, dx\, dy + \theta \iint_S (x^2 + y^2)\, \sigma_z^{(0)}\, dx\, dy. \tag{6.63}$$

To complete the formulation the stress–strain relations (3.38) are linearized, and we obtain

$$\tau_{xz} = G\gamma_{xz}, \quad \tau_{yz} = G\gamma_{yz}. \tag{6.64}$$

Substituting Eqs. (6.52) and (6.64) into Eqs. (6.61) and (6.62), we find that the function Φ thus determined is equivalent to the warping function φ defined in Section 6.1. Consequently, Eq. (6.63) may be written as

$$\bar{M} = \left[GJ + \iint_S (x^2 + y^2)\, \sigma_z^{(0)}\, dx\, dy \right] \theta, \tag{6.65}$$

to yield the effective torsional rigidity,

$$GJ_{\text{eff}} = GJ + \iint_S (x^2 + y^2)\, \sigma_z^{(0)}\, dx\, dy, \tag{6.66}$$

where J is defined by Eq. (6.20).

The last term in Eq. (6.66) shows the effect of the initial normal stress $\sigma_z^{(0)}$ on the torsional rigidity. The effect may be explained as follows: Since the position vector of an arbitrary point of the bar after deformation is

$$\mathbf{r} = (x + u)\, \mathbf{i}_1 + (y + v)\, \mathbf{i}_2 + (z + w)\, \mathbf{i}_3, \tag{6.67}$$

we have

$$\begin{aligned} \frac{\partial \mathbf{r}}{\partial z} = &-(x \sin\vartheta + y \cos\vartheta)\, \theta\, \mathbf{i}_1 \\ &+ (x \cos\vartheta - y \sin\vartheta)\, \theta\, \mathbf{i}_2 + (1 + \varepsilon_0)\, \mathbf{i}_3, \end{aligned} \tag{6.68}$$

with the aid of Eq. (6.49). Therefore, the stress $\sigma_z^{(0)}(\partial \mathbf{r}/\partial z)$ produces a twisting moment

$$\begin{aligned} &[(x \cos\vartheta - y \sin\vartheta)(x + u) + (x \sin\vartheta + y \cos\vartheta)(y + v)]\, \sigma_z^{(0)}\, \theta \\ &= (x^2 + y^2)\, \sigma_z^{(0)}\, \theta \end{aligned} \tag{6.69}$$

around the z-axis, as shown in Fig. 6.4.

It is seen that the resultant of the stress $\sigma_z^{(0)}(\partial \mathbf{r}/\partial z)$ in the sectional plane may not vanish, but instead, may produce bending of the bar, unless the axis of rotation is chosen properly. The effective torsional rigidity depends on the location of the axis of rotation through the term $(x^2 + y^2)$ involved in the surface integral in Eq. (6.66). However, if the initial stress $\sigma_z^{(0)}$ is given such that

$$\iint_S \sigma_z^{(0)}\, dx\, dy = \iint_S \sigma_z^{(0)} x\, dx\, dy = \iint \sigma_z^{(0)} y\, dx\, dy = 0, \tag{6.70}$$

the torsion does not produce bending and any location of the axis of rotation may be used for the computation of the effective torsional rigidity.[6]

We note that the governing equations for a torsion problem of a cylinder with initial stresses $\sigma_x^{(0)}$, $\sigma_y^{(0)}$, $\sigma_z^{(0)}$, $\tau_{yz}^{(0)}$, $\tau_{xz}^{(0)}$ and $\tau_{xy}^{(0)}$ can be derived in a manner similar to the above development. When these initial stresses are functions of (x, y) only and in self-equilibrium in the cylinder with the side boundary surface traction-free, we obtain, in place of Eq. (6.61), the following equation:

$$\frac{\partial \tau_{xz}}{\partial x} + \frac{\partial \tau_{yz}}{\partial y} + \left[\sigma_x^{(0)} \frac{\partial^2 \Phi}{\partial x^2} + \sigma_y^{(0)} \frac{\partial^2 \Phi}{\partial y^2} + 2\tau_{xy}^{(0)} \frac{\partial^2 \Phi}{\partial x\, \partial y}\right] \theta = 0, \qquad (6.71)$$

while Eqs. (6.52), (6.62), (6.63) and (6.64) remain unchanged.

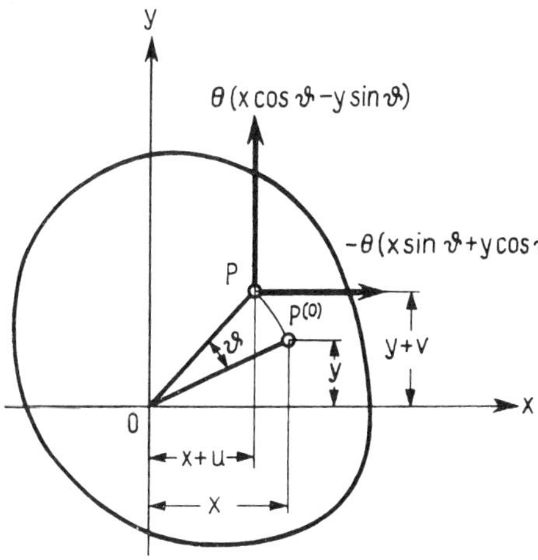

FIG. 6.4. Components of $\partial \mathbf{r}/\partial z$.

It is well known that the presence of axial tensile or compressive stress can cause an increase or decrease in the torsional rigidity of a bar.[7, 8] In recent years, thermal stresses induced in structural members of high speed flight vehicles due to aerodynamic heating have been one of the greatest engineering concerns. Among the difficulties caused by the thermal stresses is the loss of torsional rigidity of the lifting surfaces of flight vehicles.[6, 9] This loss is responsible for the reduction of safety margins for static and dynamic aeroelastic phenomena in high speed flight.

6.5. Upper and Lower Bounds of Torsional Rigidity†

The topic of the last section of this chapter is to show that formulae providing upper and lower bounds for torsional rigidity are derivable by the simultaneous use of the principles of minimum potential and comple-

† Refs. 10 through 15.

mentary energy. For the sake of simplicity, the bar is assumed to be simply connected and the torsion problem is defined in the same manner as in Section 6.1.

First, let us derive a lower bound formula employing the principle of minimum complementary energy. Let ϕ and Π_c represent the stress function and the total complementary energy corresponding to the exact solution, and let ϕ^* and Π_c^* represent the corresponding quantities of an admissible function which satisfies the subsidiary conditions:

$$\bar{M} = 2\iint_S \phi^* \, dx \, dy, \tag{6.72}$$

and

$$\phi^* = 0 \quad \text{on} \quad C. \tag{6.73}$$

Then, the principle of minimum complementary energy assures that:

$$\Pi_c \leqq \Pi_c^*, \tag{6.74}$$

where

$$\Pi_c = \frac{1}{2G}\iint_S \left[\left(\frac{\partial \phi}{\partial x}\right)^2 + \left(\frac{\partial \phi}{\partial y}\right)^2\right] dx \, dy = \frac{\bar{M}^2}{2GJ}, \tag{6.75}$$

and

$$\Pi_c^* = \frac{1}{2G}\iint \left[\left(\frac{\partial \phi^*}{\partial x}\right)^2 + \left(\frac{\partial \phi^*}{\partial y}\right)^2\right] dx \, dy. \tag{6.76}$$

Now, we shall assume ϕ^* as a linear combination of $\phi_i(x, y)$ $(i = 1, 2, \ldots, m)$ such that

$$\phi^* = \sum_{i=1}^{m} a_i \phi_i(x, y) \tag{6.77}$$

and consider the minimum value of Π_c^*. The functions $\phi_i(x, y)$ are chosen to satisfy the required continuity and differentiability conditions in the domain S and the boundary conditions $\phi_i(x, y) = 0$ on C, and a_i are arbitrary constants. The function ϕ^* thus expressed must satisfy the subsidiary condition (6.72). Since the subsidiary condition can be introduced into the variational expression through the use of a Lagrange multiplier λ, the minimum value is given by the extreme value of

$$\iint_S \left\{\frac{1}{2G}\left[\left(\frac{\partial \phi^*}{\partial x}\right)^2 + \left(\frac{\partial \phi^*}{\partial y}\right)^2\right] - 2\lambda\phi^*\right\} dx \, dy + \lambda\bar{M} \tag{6.78}$$

where the quantities subject to variation are λ and a_i $(i = 1, 2, \ldots, m)$. After some calculation, we obtain the stationary conditions of the expression (6.78) with respect to these quantities as follows:

$$\sum_{j=1}^{m}\left[\iint_S \left(\frac{\partial \phi_i}{\partial x}\frac{\partial \phi_j}{\partial x} + \frac{\partial \phi_i}{\partial y}\frac{\partial \phi_j}{\partial y}\right) dx \, dy\right] a_j = 2G\lambda \iint_S \phi_i \, dx \, dy,$$
$$(i = 1, 2, \ldots, m), \tag{6.79}$$

and

$$\bar{M} = 2\sum_{i=1}^{m} a_i \iint_S \phi_i \, dx\, dy. \tag{6.80}$$

By solving these equations, we may express a_i and λ in terms of $\bar{M}$. Substitution of a_i and λ thus obtained into Eq. (6.76) yields:

$$\Pi_c^* = \frac{\bar{M}^2}{2GJ^*}, \tag{6.81}$$

where

$$GJ^* = \bar{M}/\lambda. \tag{6.82}$$

Combining Eqs. (6.74), (6.75) and (6.81), we obtain:

$$GJ^* \leqq GJ, \tag{6.83}$$

which shows that GJ^* thus obtained provides a lower bound to the torsional rigidity.

Second, we may derive an upper bound formula via the principle of minimum potential energy. Let w, θ and Π represent displacement, twist angle per unit length and the total potential energy corresponding to the exact solution, and let w^{**}, θ^{**} and Π^{**} represent the corresponding quantities of an admissible function. The principle of minimum potential energy then assures that:

$$\Pi \leqq \Pi^{**}, \tag{6.84}$$

where

$$\Pi = \frac{1}{2} G \iint_S \left[\left(\frac{\partial w}{\partial x} - \theta y\right)^2 + \left(\frac{\partial w}{\partial y} + \theta x\right)^2\right] dx\, dy - \theta\bar{M} = -\frac{\bar{M}^2}{2GJ}, \tag{6.85}$$

and

$$\Pi^{**} = \frac{1}{2} G \iint_S \left[\left(\frac{\partial w^{**}}{\partial x} - \theta^{**} y\right)^2 + \left(\frac{\partial w^{**}}{\partial y} + \theta^{**} x\right)^2\right] dx\, dy - \theta^{**}\bar{M}. \tag{6.86}$$

Now, let us assume w^{**} as a linear combination of $w_i(x, y)$ $(i = 1, 2, \ldots, n)$, such that

$$w^{**} = \sum_{i=1}^{n} b_i w_i(x, y), \tag{6.87}$$

and consider the minimum value of Π^{**}. The functions $w_i(x, y)$ may be chosen arbitrarily, except that they must satisfy the required differentiability and continuity conditions in the domain S, and b_i are arbitrary constants. Substituting Eq. (6.87) into Eq. (6.86), and taking variations with respect to b_i and θ^{**}, we obtain:

$$\sum_{j=1}^{n} \left[\iint_S \left(\frac{\partial w_i}{\partial x}\frac{\partial w_j}{\partial x} + \frac{\partial w_i}{\partial y}\frac{\partial w_j}{\partial y}\right) dx\, dy\right] b_j = \theta^{**} \iint_S \left(y\frac{\partial w_i}{\partial x} - x\frac{\partial w_i}{\partial y}\right) dx\, dy,$$

$$(i, = 1, 2, \ldots, n) \tag{6.88}$$

and

$$\bar{M} = G\left\{\sum_{i=1}^{n}\left[\iint\limits_{S}\left(\frac{\partial w_i}{\partial y}x - \frac{\partial w_i}{\partial x}y\right)dx\,dy\right]b_i + \theta^{**}\iint\limits_{S}(x^2+y^2)\,dx\,dy\right\}. \tag{6.89}$$

By solving these equations, we may express the quantities b_i and θ^{**} in terms of $\bar{M}$. Substitution of b_i and θ^{**} thus obtained into Eq. (6.86) then yields:

$$\Pi^{**} = -\frac{\bar{M}^2}{2GJ^{**}}, \tag{6.90}$$

where

$$GJ^{**} = \bar{M}/\theta^{**}. \tag{6.91}$$

Combining Eqs. (6.84), (6.85) and (6.90), we obtain the following formula providing an upper bound to the torsional rigidity of the bar:

$$GJ \leqq GJ^{**}. \tag{6.92}$$

So far, no conditions have been prescribed for the admissible functions w_i. However, since the exact solution w should satisfy Laplace's equation, it will be more convenient to choose w_i so that they satisfy Laplace's equation

$$\frac{\partial^2 w_i}{\partial x^2} + \frac{\partial^2 w_i}{\partial y^2} = 0 \quad (i = 1, 2, \ldots, n). \tag{6.93}$$

Then, the surface integrals on the left hand side of Eq. (6.88) can be replaced by line integrals as follows:

$$\iint\limits_{S}\left(\frac{\partial w_i}{\partial x}\frac{\partial w_j}{\partial x} + \frac{\partial w_i}{\partial y}\frac{\partial w_j}{\partial y}\right)dx\,dy = \int w_i\frac{\partial w_j}{\partial \nu}\,ds = \int w_j\frac{\partial w_i}{\partial \nu}\,ds, \tag{6.94}$$

where the line integrals are taken along the contour of the bar, and ν is in the direction of the outward normal to the contour.

Thus, combining Eqs. (6.83) and (6.92), we finally obtain the upper and lower bound formulae for the torsional rigidity of the bar:

$$GJ^* \leqq GJ \leqq GJ^{**}. \tag{6.95}$$

The accuracy of the bounds obtained in this way can be improved by increasing the number of admissible functions.

As an example of the procedure outlined above, bounds for the torsional rigidity of the prismatic bar with a square cross section, as shown in Fig. 6.5, will be calculated following Ref. 10. To begin with, we shall obtain a lower bound by choosing

$$\phi_1(x, y) = a^2(x^2 - a^2)(y^2 - a^2),$$

$$\phi_2(x, y) = (x^2 + y^2)(x^2 - a^2)(y^2 - a^2), \tag{6.96}$$

thus assuming $\phi^* = a_1\phi_1 + a_2\phi_2$. Then Eqs. (6.79) and (6.80) may be written, after several integrations and cancellation of a common factor, as:

$$\begin{bmatrix} 26880 & 9216 \\ 9216 & 11264 \end{bmatrix} \begin{bmatrix} a_1 \\ a_2 \end{bmatrix} = \left(\frac{G\lambda}{a^4}\right) \begin{bmatrix} 16800 \\ 6720 \end{bmatrix}, \tag{6.97}$$

and

$$\bar{M} = \frac{32}{45} a^8(5a_1 + 2a_2), \tag{6.98}$$

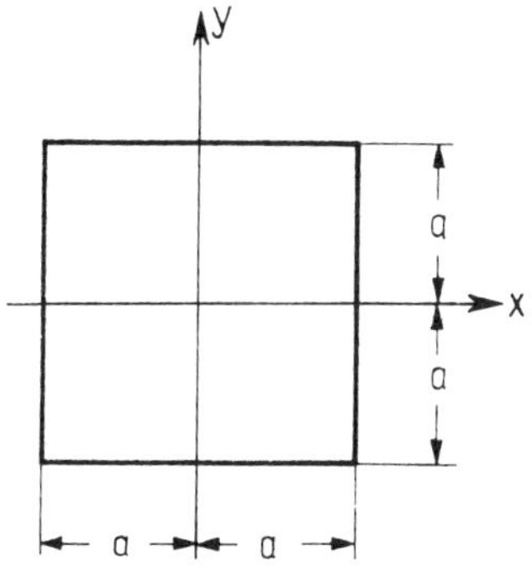

FIG. 6.5. A square section.

which yield:

$$\begin{aligned} a_1 &= (3885/\ 6648)\,(G\lambda/a^4), \\ a_2 &= (1575/13296)\,(G\lambda/a^4), \\ \bar{M} &= (5600/\ 2493)\, Ga^4\lambda. \end{aligned} \tag{6.99}$$

Thus, we obtain for the lower bound:

$$(5600/2493)\, Ga^4 \leqq GJ. \tag{6.100}$$

Next, we proceed to obtain an upper bound by choosing

$$w_1 = x^3y - xy^3, \tag{6.101}$$

and expressing w^{**} in terms of w_1 only, namely, $w^{**} = b_1w_1$. It is easily seen that w_1 is the imaginary part of $(x + iy)^4$ where $i = \sqrt{-1}$, which ensures that w_1 is a plane harmonic function. After further calculation, Eqs. (6.88) and (6.89) can be written as follows:

$$(96/35)\, a^8 \cdot b_1 = (16/15)\, a^6\theta^{**}, \tag{6.102}$$

$$\bar{M} = G[-(16/15)\, a^6b_1 + (8/3)\, a^4\theta^{**}], \tag{6.103}$$

which yield:

$$\begin{aligned} b_1 &= (7/18)\,(\theta^{**}/a^2), \\ \bar{M} &= (304/135)\, Ga^4\theta^{**}. \end{aligned} \tag{6.104}$$

Thus, we obtain for the upper bound:

$$GJ \leqq (304/135)\, Ga^4. \tag{6.105}$$

Combining Eqs. (6.100) and (6.105), we have the following upper and lower bounds for the torsional rigidity:

$$(5600/2493)\, Ga^4 \leqq GJ \leqq (304/135)\, Ga^4, \tag{6.106}$$

or

$$2.24628\, Ga^4 \leqq GJ \leqq 2.25186\, Ga^4. \tag{6.107}$$

The exact value of the torsional rigidity is:

$$GJ = 2.2496\, Ga^4, \tag{6.108}$$

as shown in books on the theory of elasticity. It is seen that the accuracy of the bounds is excellent. It should be noted, however, that they do not guarantee the same accuracy for displacements or stresses thus determined approximately. More complex techniques are necessary for obtaining pointwise bound formulae for displacements or stresses at an arbitrary point of the bar.[16-18]

Bibliography

1. A. E. H. Love, *Mathematical Theory of Elasticity*, Cambridge University Press, 4th edition, 1927.
2. S. Timoshenko and J. N. Goodier, *Theory of Elasticity*, McGraw-Hill, 1951.
3. I. S. Sokolnikoff, *Mathematical Theory of Elasticity*, McGraw-Hill, 1956.
4. E. Reissner, On Non-uniform Torsion of Cylindrical Rods, *Journal of Mathematics and Physics*, Vol. 31, No. 2, pp. 214–21, July 1952. Note on Torsion with Variable Twist, *Journal of Applied Mechanics*, Vol. 23, No. 2, p. 315, June 1956. On Torsion with Variable Twist, *Österreichisches Ingenieur-Archiv*, Vol. 9, No. 2–3, pp. 218–24, 1955.
5. R. Kappus, Zur Elastizitätstheorie endlicher Verschiebungen, *Zeitschrift für Angewandte Mathematik und Mechanik*, Vol. 19, No. 5, pp. 344–61, December 939.
6. R. L. Bisplinghoff *et al.*, *Aerodynamic Heating of Aircraft Structures in High Speed Flight*, Notes for a Special Summer Program, Department of Aeronautical Engineering, Massachusetts Institute of Technology, June 25–July 6, 1956.
7. H. Wagner, *Verdrehung und Knickung von offenen Profilen*, 25th Anniversary Publication, Technische Hochschule Danzig, 1904–29, pp. 329–44, Druck und Verlag von A. W. Kafemann GmbH, Danzig, 1929. Translated in NACA TM 807, October 1936.
8. F. Bleich and H. Bleich, *Buckling Strength of Metal Structures*, McGraw-Hill, 1952.
9. B. Budiansky and J. Mayers, Influence of Aerodynamic Heating on the Effective Torsional Stiffness of Thin Wings, *Journal of Aeronautical Sciences*, Vol. 23, No. 12, pp. 1081–93, December 1956.
10. E. Trefftz, Ein Gegenstück zum Ritzschen Verfahren, *Proceedings of the 2nd International Congress for Applied Mechanics*, Zürich, pp. 131–7, 1926.
11. N. M. Basu, On an Application of the New Methods of Calculus of Variations to Some Problems in the Theory of Elasticity, *Philosophical Magazine*, Vol. 10, No. 66, pp. 886–904, November 1930.
12. J. B. Diaz and A. Weinstein, The Torsional Rigidity and Variational Methods, *American Journal of Mathematics*, Vol. 70, No. 1, pp. 107–16, January 1948.

13. A. Weinstein, New Methods for the Estimation of Torsional Rigidity, *Proceedings of Symposia in Applied Mathematics*, Vol. 3, pp. 141–61, McGraw-Hill, 1950.
14. Lin Hung-Sun, On Variational Methods in the Problem of Torsion for Multiply-connected Cross Sections, *Acta Sci-Sinica*, Vol. 3, pp. 171–86, 1954.
15. S. G. Mikhlin, *Variational Methods in Mathematical Physics*, Pergamon Press, 1964.
16. H. J. Greenberg, The Determination of Upper and Lower Bounds for Solution of the Dirichlet Problem, *Journal of Mathematics and Physics*, Vol. 27, No. 3, pp. 161–82, October 1948.
17. J. L. Synge, The Dirichlet Problem: Bound at a Point for the Solution and its Derivatives, *Quarterly of Applied Mathematics*, Vol. 8, No. 3, pp. 213–28, October 1950.
18. K. Washizu, Bounds for Solution of Boundary Value Problems in Elasticity, *Journal of Mathematics and Physics*, Vol. 32, No. 2–3, pp. 117–28, July–October 1953.
19. S. Timoshenko, Theory of Bending, Torsion and Buckling of Thin-walled Members of Open Cross Section, *Journal of the Franklin Institute*, Vol. 239, No. 3, pp. 201–19, March 1945.

CHAPTER 7

BEAMS

7.1. Elementary Theory of a Beam

We shall treat slender beams in the present chapter. It will be assumed that the locus of the centroid of the cross section of the beam is a straight line and that the envelopes of the principal axes through the centroid are two flat planes perpendicular to each other. We shall take the x-axis in the direction of the centroid locus and the y- and z-axes parallel to the principal directions. Thus, the x-, y- and z-axes form a right-handed rectangular Cartesian coordinate system (see Fig. 7.1).

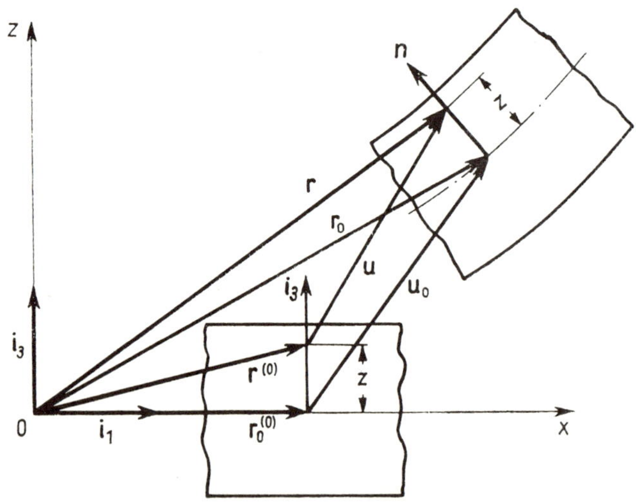

FIG. 7.1. Geometrical relations.†

Saint-Venant has formulated a method of solution for bending of a cylindrical cantilever beam of constant cross section by a terminal load.[1, 2] Solutions of the problem have been obtained for beams having circular, elliptic, rectangular and several other cross sections. These results show that both bending and torsional deformations occur due to the terminal load. Consequently, it is considered convenient to define the center of shear of

† Only the projections onto the (x, z) plane are shown in Fig. 7.1.

the cross section as a point through which a shear force can be applied without producing torsion, thus realizing torsion-free bending. It is known from the above definition that once the shearing stress distribution over the cross section due to the torsion-free bending has been obtained, the center of shear is determined as the point of application of the resultant of the shearing force.† When the cross section has an axis of symmetry, the center of shear lies on that axis. When the beam has a doubly symmetric cross section, the center of shear coincides with the centroid of the cross section. Exact general solutions for bending of a beam with arbitrary cross section along the span under arbitrary external loads have not been obtained.

The present chapter will treat, unless otherwise stated, the elementary theory of a beam under the tacit assumptions that variation of the geometry of the cross section along the x-axis is gradual and that torsion-free bending has been realized in the (x, z)-plane by proper application of external loads. Since the longitudinal dimension of a slender beam is much larger than its lateral dimensions, it is a common practice in the elementary theory to employ the following two assumptions. First, we assume that the stress components σ_y, σ_z and τ_{yz} may be neglected in comparison with the other stress components and may be set

$$\sigma_y = \sigma_z = \tau_{yz} = 0. \tag{7.1}$$

Then Eqs. (1.10) and Eqs. (3.38) reduce to

$$\sigma_x = E\varepsilon_x, \quad \tau_{xz} = G\gamma_{xz}, \quad \tau_{xy} = G\gamma_{xy}, \qquad \text{(7.2a, b, c)}$$

and

$$\sigma_x = Ee_{xx}, \quad \tau_{xz} = 2Ge_{xz}, \quad \tau_{xy} = 2Ge_{xy}, \qquad \text{(7.3a, b, c)}$$

respectively. Second, we employ the Bernoulli–Euler hypothesis that the cross sections which are perpendicular to the centroid locus before bending remain plane and perpendicular to the deformed locus and suffer no strains in their planes.

We shall show that expressions for the displacements are greatly simplified by the introduction of the hypothesis. We consider an arbitrary point of a beam having the coordinates (x, y, z) before deformation, and denote its position vectors before and after deformation by $\mathbf{r}^{(0)}$ and $\mathbf{r}$, respectively, which are related to the displacement vector $\mathbf{u}$ by

$$\mathbf{r} = \mathbf{r}^{(0)} + \mathbf{u}, \qquad \text{(7.4)‡}$$

where $\mathbf{r}^{(0)} = x\mathbf{i}_1 + y\mathbf{i}_2 + z\mathbf{i}_3$, and $\mathbf{i}_1$, $\mathbf{i}_2$, $\mathbf{i}_3$ are the unit vectors in the directions of the x-, y-, z-axes, respectively. Similarly, we denote position

† The analytical determination of the center of shear depends upon the definition of "torsion-free bending", and there exist several different definitions (see Refs. 2 through 6). A detailed and lucid discussion on the center of shear, center of twist, elastic axis, etc., is given in Ref. 7.

‡ The superscript $^{(0)}$ and subscript $_0$ used in Chapters 7, 8 and 9 mean that the quantity is referred to the state before deformation and to the centroid locus or the middle surface, respectively.

vectors of a point $(x, 0, 0)$ of the centroid locus before and after deformation by $\mathbf{r}_0^{(0)}$ and $\mathbf{r}_0$, respectively, which are related to the displacement vector $\mathbf{u}_0$ by

$$\mathbf{r}_0 = \mathbf{r}_0^{(0)} + \mathbf{u}_0, \tag{7.5}‡$$

where $\mathbf{r}_0^{(0)} = x\mathbf{i}_1$. We define components of $\mathbf{u}$ and $\mathbf{u}_0$ as follows:

$$\mathbf{u} = U\mathbf{i}_1 + V\mathbf{i}_2 + W\mathbf{i}_3, \tag{7.6}$$

$$\mathbf{u}_0 = u\mathbf{i}_1 + w\mathbf{i}_3, \tag{7.7}$$

where u and w are functions of x only. It is seen that the hypothesis allows us to express $\mathbf{r}$ as

$$\mathbf{r} = \mathbf{r}_0 + z\mathbf{n} + y\mathbf{i}_2, \tag{7.8}$$

where $\mathbf{n}$ is the unit normal to the deformed locus and is given by

$$\mathbf{n} = \mathbf{r}_0' \times \mathbf{i}_2/|\mathbf{r}_0'|. \tag{7.9}$$

In Eq. (7.9) and throughout the present chapter, the prime denotes differentiation with respect to x, namely, $(\ )' = d(\ )/dx$. Since

$$\mathbf{r}_0 = (x + u)\,\mathbf{i}_1 + w\mathbf{i}_3, \tag{7.10}$$

we may express $\mathbf{n}$ in terms of u and w:

$$\mathbf{n} = \frac{-w'\mathbf{i}_1 + (1 + u')\,\mathbf{i}_3}{\sqrt{(1 + u')^2 + (w')^2}}. \tag{7.11}$$

From Eqs. (7.4), (7.5) and (7.8), we obtain

$$\mathbf{u} = \mathbf{u}_0 + z(\mathbf{n} - \mathbf{i}_3). \tag{7.12}$$

This is the expression for the displacements of a beam under the Bernoulli–Euler hypothesis. It is observed that the degree of freedom of the beam deformation implied by Eq. (7.12) is two, namely $u(x)$ and $w(x)$.

When a beam problem is confined to small displacement theory, Eq. (7.12) may be linearized with respect to the displacement components to yield

$$U = u - zw', \quad V = 0, \quad W = w. \tag{7.13}$$

Then, we find that in the elementary theory of bending, the only non-vanishing strain component is

$$\varepsilon_x = \frac{\partial U}{\partial x} = u' - zw'', \tag{7.14}$$

which is related to σ_x by Eq. (7.2a).

7.2. Bending of a Beam

As a simple example of bending of a beam, let us consider a problem shown in Fig. 7.2: a beam of span l is clamped at one end $x = 0$ and is subjected to a distributed lateral load $\bar{p}(x)$ per unit span acting in the direc-

‡ See footnote to p. 133.

tion of the z-axis. At $x = l$, it is subjected to end forces $\bar{P}_x$ and $\bar{P}_z$ in the directions of the x- and z-axes, respectively and to an external end moment $\bar{M}$. We may write the principle of virtual work for the present problem as follows:

$$\iiint_V \sigma_x \, \delta\varepsilon_x \, dx \, dy \, dz - \int_0^l \bar{p} \, \delta w \, dx$$
$$- \bar{P}_x \, \delta u(l) - \bar{P}_z \, \delta w(l) + \bar{M} \, \delta w'(l) = 0, \qquad (7.15)\dagger$$

where Eq. (7.14) has been substituted, and δu and δw should satisfy the geometrical boundary conditions:

$$u(0) = 0, \qquad (7.16)$$

and

$$w(0) = w'(0) = 0, \qquad (7.17)$$

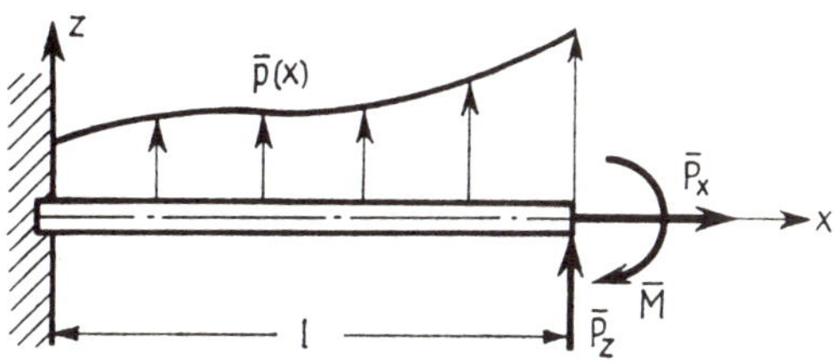

FIG. 7.2. A cantilever beam.

respectively. Integration of the first term in Eq. (7.15) with respect to y and z leads to

$$\int_0^l (N \, \delta u' - M \, \delta w'') \, dx - \int_0^l \bar{p} \, \delta w \, dx$$
$$- \bar{P}_x \, \delta u(l) - \bar{P}_z \, \delta w(l) + \bar{M} \, \delta w'(l) = 0, \qquad (7.18)$$

where we define

$$N = \iint_S \sigma_x \, dy \, dz, \qquad (7.19)$$

$$M = \iint_S \sigma_x z \, dy \, dz, \qquad (7.20)$$

integrations being taken over the cross section S of the beam. The quantities N and M are axial force and bending moment of the cross section as shown in Fig. 7.3.

Now, we may proceed to derive approximate equations of equilibrium implied by Eq. (7.18). By the use of the conditions (7.16) and (7.17) and

† See Eq. (1.32).

partial integrations, we have

$$-\int_0^l [N'\,\delta u + (M'' + \bar{p})\,\delta w]\,dx + (N - \bar{P}_x)\,\delta u(l)$$
$$+ (M' - \bar{P}_z)\,\delta w(l) - (M - \bar{M})\,\delta w'(l) = 0, \tag{7.21}$$

from which, we obtain

$$N' = 0, \quad 0 \leqq x \leqq l, \tag{7.22}†$$
$$N = \bar{P}_x \quad \text{at} \quad x = l, \tag{7.23}$$

and

$$M'' + \bar{p} = 0, \quad 0 \leqq x \leqq l, \tag{7.24}†$$
$$M' = \bar{P}_z, \quad M = \bar{M} \quad \text{at} \quad x = l. \tag{7.25}$$

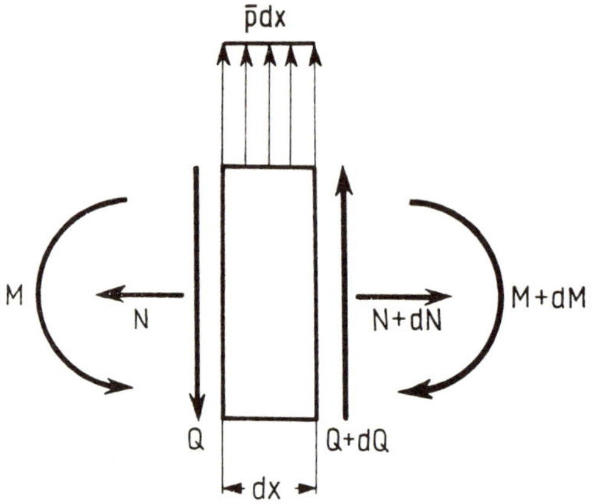

FIG. 7.3. Positive directions of N, Q and M.

In order to solve the problem, we must use the stress–strain relation (7.2a), which, when combined with Eqs. (7.14), (7.19) and (7.20), provide N and M in terms of u and w as follows:

$$N = EA_0 u', \tag{7.26}$$
$$M = -EIw'', \tag{7.27}$$

where

$$A_0 = \iint_S dy\,dz, \quad I = \iint_S z^2\,dy\,dz \tag{7.28}$$

are the area and the moment of inertia of the cross section, respectively.

† As is well known, these equations can be obtained alternatively by writing the equilibrium conditions with respect to forces and moments of the beam element shown in Fig. 7.3 as

$$N' = 0, \quad Q' + \bar{p} = 0, \quad M' - Q = 0$$

and then eliminating Q, where Q is the shearing force of the cross section.

Using the relations above obtained, we have the governing differential equations and the boundary conditions for the beam problem. Combining Eqs. (7.16), (7.22), (7.23) and (7.26), we obtain a differential equation and boundary conditions which determine the stretching of the beam. Alternatively, combination of Eqs. (7.17), (7.24), (7.25) and (7.27) yields a differential equation and boundary conditions which determine the bending of the beam. Thus, in the small displacement theory of a beam where the displacement components are assumed to be of the form (7.13), the stretching and bending do not couple with each other and can be treated separately. It is observed from the above relations that in the elementary theory of bending of a beam, the stress σ_x and the strain energy U are given by

$$\sigma_x = \frac{N}{A_0} + \frac{M}{I} z, \tag{7.29}$$

and

$$U = \tfrac{1}{2} \iiint_V E\varepsilon_x^2 \, dx \, dy \, dz$$

$$= \tfrac{1}{2} \int_0^l [EA_0(u')^2 + EI(w'')^2] \, dx, \tag{7.30}$$

respectively.

Before leaving the present section, we note that if the distributed load $\bar{p}(x)$ is discontinuous at some point along the span, care should be taken in deriving Eq. (7.21). For example, if the beam is subjected to a concentrated load $\bar{P}$ acting at $x = \xi$ in the direction of the z-axis, Eq. (7.15) is appended with a term $-\bar{P}\,\delta w(\xi)$ and we have

$$\begin{aligned}\int_0^l M\,\delta w''\,dx &= \int_0^{\xi-0} M\,\delta w''\,dx + \int_{\xi+0}^{l} M\,\delta w''\,dx \\ &= \int_0^{\xi-0} M''\,\delta w\,dx + \int_{\xi+0}^{l} M''\,\delta w\,dx + M\,\delta w'\Big|_0^l - M'\,\delta w\Big|_0^l \\ &\quad + [M(\xi-0) - M(\xi+0)]\,\delta w'(\xi) \\ &\quad - [M'(\xi-0) - M'(\xi+0)]\,\delta w(\xi).\end{aligned} \tag{7.31}$$

Consequently, the principle of virtual work yields the connection conditions at $x = \xi$ as follows:

$$\begin{aligned} &w\Big|_{\xi-0}^{\xi+0} = 0, \quad w'\Big|_{\xi-0}^{\xi+0} = 0, \\ &M\Big|_{\xi-0}^{\xi+0} = 0, \quad M'\Big|_{\xi-0}^{\xi+0} + \bar{P} = 0. \end{aligned} \tag{7.32}$$

7.3. Principle of Minimum Potential Energy and its Transformation

We shall consider in the present section variational principles of a beam problem shown in Fig. 7.4: a beam is clamped at one end and is subjected to a distributed lateral load $\bar{p}(x)$, while at the other end, it is supported and is subjected to an end moment $\bar{M}$. Since no external forces are applied in the x-direction, we may take

$$U = -zw', \quad V = 0, \quad W = w, \tag{7.33}$$

$$\varepsilon_x = -zw'', \tag{7.34}$$

$$\sigma_x = \frac{M}{I} z. \tag{7.35}$$

The functional for the principle of minimum potential energy of this problem is given, as suggested by the relations derived in the preceding section, by

$$\Pi = \tfrac{1}{2}\int_0^l EI(w'')^2\,dx - \int_0^l \bar{p}w\,dx + \bar{M}w'(l), \tag{7.36}$$

where w must satisfy the geometrical boundary conditions.

$$w(0) = w'(0) = w(l) = 0. \tag{7.37}$$

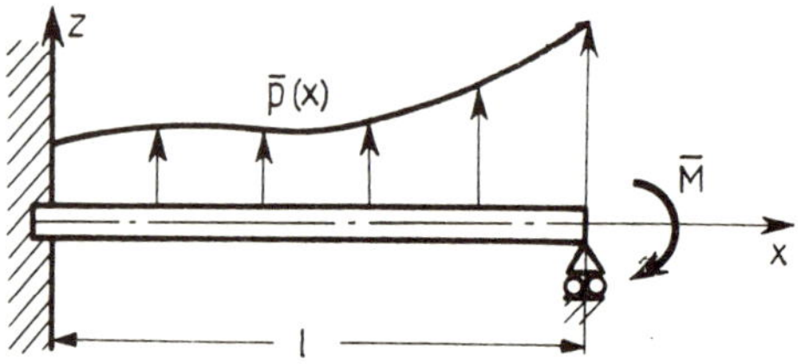

FIG. 7.4. A beam with clamped and supported ends.

Next, let us consider transformations of the principle of minimum potential energy. By the introduction of an auxiliary function defined by

$$\varkappa = w'' \tag{7.38}$$

and with the use of Lagrange multipliers $M(x)$, P^*, Q^* and R^*, the functional (7.36) is generalized as follows:

$$\Pi_I = \tfrac{1}{2}\int_0^l EI\varkappa^2\,dx - \int_0^l \bar{p}w\,dx + \bar{M}w'(l) + \int_0^l (\varkappa - w'')\,M\,dx + P^*w(o) + Q^*w'(o) + R^*w(l), \tag{7.39}$$

where the quantities subject to variation are $\varkappa$, w, M, P^*, Q^* and R^* with no subsidiary conditions. After some calculation including partial integrations, the first variation is shown to be

$$\begin{aligned}\delta\Pi_I = &\int_0^l [(M + EI\varkappa)\,\delta\varkappa - (M'' + \bar{p})\,\delta w + (\varkappa - w'')\,\delta M]\,dx \\ &+ [P^* - M'(o)]\,\delta w(o) + [Q^* + M(o)]\,\delta w'(o) \\ &+ [R^* + M'(l)]\,\delta w(l) - [M(l) - \bar{M}]\,\delta w'(l) \\ &+ w(o)\,\delta P^* + w'(o)\,\delta Q^* + w(l)\,\delta R^*.\end{aligned} \tag{7.40}$$

Consequently, the stationary conditions are shown to be the governing equations which define the problem, together with

$$P^* = M'(o), \quad Q^* = -M(o), \quad R^* = -M'(l) \tag{7.41}$$

which determine the Lagrange multipliers P^*, Q^* and R^*.

Familiar techniques lead to specializations of the generalized expression (7.39). For example, by requiring that the coefficients of $\delta\varkappa$, δw and $\delta w'$ in Eq. (7.40) vanish, thus eliminating $\varkappa$ and w, we obtain the functional for the principle of minimum complementary energy as follows:

$$\Pi_c = \frac{1}{2}\int_0^l \frac{M^2}{EI}\, dx, \tag{7.42}$$

where the function subject to variation is $M(x)$ under the subsidiary conditions

$$M'' + \bar{p} = 0, \tag{7.43}$$

and

$$M(l) = \bar{M}. \tag{7.44}$$

We note that the principle of minimum complementary energy for the beam problem is directly obtainable from the principle (2.23), assuming that the stress component σ_x is given by Eq. (7.35) and all the other stress components make negligible contributions in establishing the complementary energy function (see also Appendix C).

7.4. Free Lateral Vibration of a Beam†

Let us consider a free lateral vibration of a beam which is clamped at $x = 0$ and simply supported at $x = l$, as shown in Fig. 7.5. Following the development in Section 2.7, we may express the total potential energy for the free lateral vibration problem as

$$\Pi = \tfrac{1}{2}\int_0^l EI(w'')^2\, dx - \tfrac{1}{2}\lambda\int_0^l mw^2\, dx, \tag{7.45‡}$$

where $\lambda = \omega^2$ and

$$m(x) = \iint_S \varrho\, dy\, dz \tag{7.46}$$

† Refs. 8 through 11.

‡ In the derivation of the last term of Eq. (7.45) by the use of the functional (2.69) and Eq. (7.33), the term $\iint_S \varrho z^2(w')^2\, dy\, dz$ is neglected in comparison with the term $\iint_S \varrho w^2\, dy\, dz$ due to the slenderness of the beam—a common practice in the elementary theory of the lateral vibration of a beam.

is the mass per unit span. In the functional (7.45), the function subject to variation is w under the subsidiary boundary conditions

$$w(o) = w'(o) = w(l) = 0, \tag{7.47}$$

while λ is treated as a parameter and not subject to variation. The stationary conditions of the functional (7.45) are shown to be the equations of motion,

$$(EIw'')'' - \lambda mw = 0, \tag{7.48}$$

and the boundary condition

$$EIw'' = 0 \quad \text{at} \quad x = l. \tag{7.49}$$

Therefore, the problem reduces to an eigenvalue problem, where the required natural modes and frequencies are determined as eigenfunctions and eigenvalues of the differential equation (7.48) under the boundary conditions

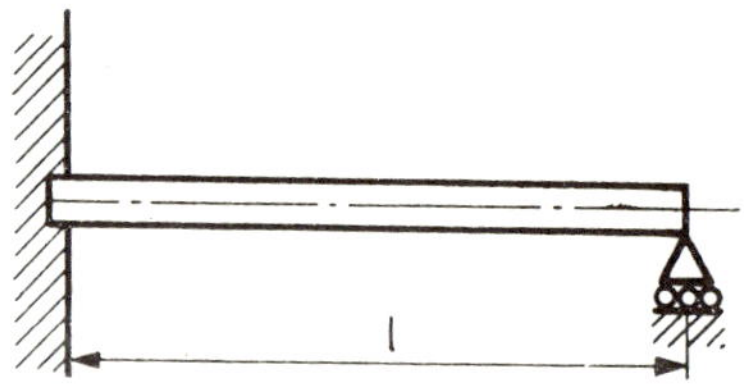

FIG. 7.5. A beam with clamped and simply supported ends.

(7.47) and (7.49). It is observed that this eigenvalue problem is equivalent to finding, among admissible functions w, those which make the quotient

$$\lambda = \frac{\frac{1}{2}\int_0^l EI(w'')^2\, dx}{\frac{1}{2}\int_0^l mw^2\, dx} \tag{7.50}$$

stationary.[12]

Next, let us consider a generalization of the principle of stationary potential energy.[13] Through the usual procedure, the functional (7.45) may be generalized as follows:

$$\Pi_I = \tfrac{1}{2}\int_0^l EI\varkappa^2\, dx - \tfrac{1}{2}\lambda \int_0^l mw^2\, dx$$
$$+ \int_0^l (\varkappa - w'')\, M\, dx + P^*w(o) + Q^*w'(o) + R^*w(l), \tag{7.51}$$

where the quantities subject to variation are $\varkappa$, w ,M, P^*, Q^* and R^* with no subsidiary conditions. The stationary conditions with respect to $\varkappa$ and

w are shown to be

$$EI\varkappa + M = 0, \tag{7.52}$$

$$M'' + \lambda m w = 0, \tag{7.53}$$

$$P^* - M'(o) = 0, \quad Q^* + M(o) = 0, \quad R^* + M'(l) = 0, \tag{7.54}$$

$$M(l) = 0. \tag{7.55}$$

We eliminate $\varkappa$, P^*, Q^* and R^* by the use of Eqs. (7.52), (7.54) and (7.55), and with the aid of Eq. (7.53), to transform the functional (7.51) into:

$$\Pi_c = \frac{1}{2}\int_0^l \frac{M^2}{EI}\,dx - \frac{1}{2}\lambda\int_0^l mw^2\,dx, \tag{7.56}$$

where it is assumed that $\lambda \neq 0$. In the functional (7.56), the quantities subject to variation are M and w under the subsidiary conditions (7.53) and (7.55). The expression (7.56) is a functional for the principle of stationary complementary energy of the free vibration problem.

As mentioned in Section 2.8, the Rayleigh–Ritz method can be applied for obtaining approximate eigenvalues of the free vibration problem, once the variational expressions have been established. When the method is applied to the principle of stationary potential energy (7.45), we may assume

$$w = c_1 w_1 + c_2 w_2 \tag{7.57}$$

where

$$w_1 = x^2(x - l), \quad w_2 = x^3(x - l) \tag{7.58}$$

are coordinate functions which satisfy Eqs. (7.47). Substituting Eq. (7.57) into the functional and requiring that

$$\frac{\partial \Pi}{\partial c_i} = 0, \quad i = 1, 2 \tag{7.59}$$

we obtain approximate eigenvalues.

Numerical results have been obtained for a beam with constant EI and m, as shown in Table 7.1.

TABLE 7.1. EXACT AND APPROXIMATE EIGENVALUES

$\omega_i = k_i\sqrt{EI/ml^4}$

	Exact eigenvalues	Approximate eigenvalues	
		Rayleigh–Ritz method applied to the functional (7.45)	Modified Rayleigh–Ritz method applied to the functional (7.56)
k_1	15.42	15.45	15.42
k_2	49.96	75.33	51.93

Next, we shall apply the modified Rayleigh–Ritz method to the principle of stationary complementary energy (7.56). We choose w as given by Eq. (7.57). As the derivation of the functional (7.56) shows, it is not necessary for the coordinate functions $w_1(x)$ and $w_2(x)$ to satisfy Eqs. (7.47) for the establishment of the principle. However, this imposition is desirable for improving the accuracy of approximate eigenvalues and is essential for obtaining the inequality relations (2.93). We substitute Eq. (7.57) into Eq. (7.53) and perform integrations with the boundary condition (7.55) to obtain

$$(1/\lambda)\, M = c(x - l) - \sum_{i=1}^{2} c_i \int_x^l \left[\int_\eta^l m(\xi)\, w_i(\xi)\, d\xi \right] d\eta, \tag{7.60}$$

where c is an integration constant. Substituting Eqs. (7.57) and (7.60) into the principle (7.56) and requiring that

$$\partial \Pi_c / \partial c = 0 \tag{7.61}$$

and

$$\partial \Pi_c / \partial c_i = 0, \quad i = 1, 2, \tag{7.62}$$

we obtain approximate eigenvalues. Numerical results have been obtained for a beam with constant EI and m, and are shown in Table 7.1. It is observed that the inequality relations (2.93) hold between them. See Refs. 11, 12, 14 and 15 for other numerical examples of the Rayleigh–Ritz method and modified Rayleigh–Ritz method applied to free vibration problems.

7.5. Large Deflection of a Beam

We shall consider large deflection of an elastic beam in this section and take as an example the beam problem treated in Section 7.2. It is obvious that since the displacements are given by Eq. (7.12) and the strains $e_{\lambda\mu}$ can be calculated in terms of u and w by the use of Eqs. (3.19), a finite displacement theory of the beam under the Bernoulli–Euler hypothesis may be formulated by the principle of virtual work (3.49). However, we shall be satisfied with confining the problem by assuming that although the deflection of the beam is no longer small in comparison with the height of the beam, it is still small in comparison with the longitudinal dimension of the beam and employ the following expressions for displacements and strain–displacement relations:

$$U = u - zw', \quad V = 0, \quad W = w, \tag{7.63}^\dagger$$

$$e_{xx} = u' + \tfrac{1}{2}(w')^2 - zw''. \tag{7.64}^\dagger$$

† These equations may be derived from Eqs. (7.12) and (3.19) by assuming that $u' \sim (w')^2 \ll 1$, and terms containing z^2 may be neglected in view of the hypothesis and slenderness of the beam, where the notation $\sim$ stands for "same order of magnitude". The first assumption states that the square of the slope and the strain of the centroid locus are very small compared to unity.

Then, the principle of virtual work for the present problem may be written as

$$\iiint_V \sigma_x \,\delta e_{xx}\, dx\, dy\, dz - \int_0^l \bar{p}\,\delta w\, dx$$
$$- \bar{P}_x\,\delta u(l) - \bar{P}_z\,\delta w(l) + \bar{M}\,\delta w'(l) = 0, \qquad (7.65)\dagger$$

where Eq. (7.64) has been substituted. By introducing the stress resultants defined by Eqs. (7.19) and (7.20), we may transform the principle (7.65) into

$$\int_0^l [N(\delta u' + w'\,\delta w') - M\,\delta w'' - \bar{p}\,\delta w]\, dx$$
$$- \bar{P}_x\,\delta u(l) - \bar{P}_z\,\delta w(l) + \bar{M}\,\delta w'(l) = 0, \qquad (7.66)$$

where the independent variables are δu and δw under the subsidiary conditions (7.16) and (7.17). After some calculation, we obtain from Eq. (7.66), the governing differential equations

$$N' = 0, \quad M'' + (Nw')' + \bar{p} = 0 \qquad (7.67, 7.68)$$

and the mechanical boundary conditions at $x = l$:

$$N = \bar{P}_x, \quad M = \bar{M}, \quad Nw' + M' = \bar{P}_z. \qquad (7.69, 7.70, 7.71)$$

Comparing Eqs. (7.68), (7.70) and (7.71) with Eqs. (7.24) and (7.25), we find that when the deflection of the beam becomes large, the axial force N_x has a contribution to the equations of equilibrium in the direction of the z-axis due to the inclination of the centroid locus. From Eqs. (7.67) and (7.69), we have

$$N(x) = \bar{P}_x = \text{constant}. \qquad (7.72)$$

Combining Eqs. (7.3a), (7.19), (7.20) and (7.64), we obtain the stress resultant–displacement relations as follows:

$$N = EA_0[u' + \tfrac{1}{2}(w')^2], \qquad (7.73)$$
$$M = -EIw''. \qquad (7.74)$$

The equations (7.68), (7.72), (7.73) and (7.74), together with the boundary conditions (7.16), (7.17), (7.70) and (7.71), constitute the governing equations for the large deflection problem. It is observed that in the large deflection theory of a beam, the stretching and bending couple with each other and must be treated simultaneously.

We note that, in the large deflection theory of a beam, the stress σ_x is given by Eq. (7.29) and the strain energy U by

$$U = \tfrac{1}{2}\iiint_V Ee_{xx}^2\, dx\, dy\, dz$$
$$= \tfrac{1}{2}\int_0^l \{EA_0[u' + \tfrac{1}{2}(w')^2]^2 + EI(w'')^2\}\, dx. \qquad (7.75)$$

† See Eq. (3.49).

7.6. Buckling of a Beam†

Next, let us consider the buckling problem of a beam, as shown in Fig. 7.6. The beam is fixed at one end, while at the other end it is simply supported and under a compressive load $\bar{P}$ applied in the negative direction of the x-axis. When the load reaches a critical value, denoted by P_{cr}, the column may buckle. We shall treat the column as a body with initial stress $\sigma_x^{(0)}$, the magnitude of which is given from the equilibrium conditions as

$$N^{(0)\prime} = 0, \quad N^{(0)}(l) = -P_{cr}, \tag{7.76}$$

where $N^{(0)} = A_0 \sigma_x^{(0)}$. It is assumed that the force P_{cr} changes neither its magnitude nor its direction during the buckling. Following the derivation in Section 5.1, we may write the principle of virtual work for the present problem as

$$\iiint_V (\sigma_x^{(0)} + \sigma_x)\, \delta e_{xx}\, dx\, dy\, dz + P_{cr}\, \delta u(l) = 0, \qquad (7.77)‡$$

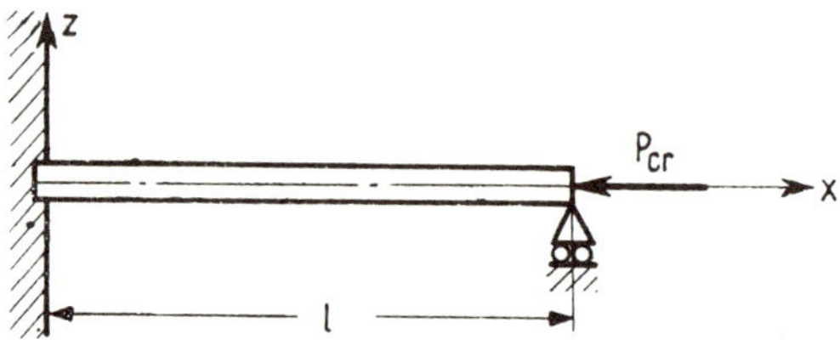

FIG. 7.6. A beam under a critical axial load.

where σ_x is incremental stress and e_{xx} is given by Eq. (7.64). The subsidiary boundary conditions of the displacements are

$$u(o) = 0 \tag{7.78}$$

and

$$w(o) = w'(o) = w(l) = 0. \tag{7.79}$$

Since we are interested in the determination of the critical load, we assume that $\sigma_x^{(0)} = O(1)$ and σ_x, u, $w = O(\varepsilon)$ to neglect terms higher than $O(\varepsilon^2)$ in the principle of virtual work (see Sections 5.1 and 5.2 for a similar development). Thus, by introducing stress resultants defined by Eqs. (7.19) and (7.20), we have

$$\int_0^l (N^{(0)}\, \delta u' + N\, \delta u' + N^{(0)} w'\, \delta w' - M\, \delta w'')\, dx + P_{cr}\, \delta u(l) = 0. \tag{7.80}$$

By the use of Eqs. (7.76) and (7.78), we find that the terms relating to δu in Eq. (7.80) reduce to

$$-\int_0^l N'\, \delta u\, dx + N(l)\, \delta u(l). \tag{7.81}$$

† Refs. 16 and 17.
‡ See Eq. (5.5).

Consequently, we have $N'(x) = 0$ and $N(l) = 0$, and we conclude that $N(x) = 0$ throughout the beam. Thus, the principle (7.80) is simplified into

$$\int_0^l (M\,\delta w'' + P_{cr} w'\,\delta w')\,dx = 0, \tag{7.82}$$

or after some calculation, to

$$\int_0^l [M'' - (P_{cr} w')']\,\delta w\,dx + M\,\delta w'\Big|_0^l - (M' - P_{cr} w')\,\delta w\Big|_0^l = 0. \tag{7.83}$$

Consequently, taking account of Eqs. (7.79), we have from Eq. (7.83) the equation of equilibrium

$$M'' - P_{cr} w'' = 0 \tag{7.84}$$

and a boundary condition

$$M(l) = 0. \tag{7.85}$$

Combining Eqs. (7.3a), (7.20) and (7.64), we obtain the stress resultant–displacement relation as follows:

$$M = -EIw''. \tag{7.86}$$

The equations (7.84) and (7.86), together with the boundary conditions (7.79) and (7.85), constitute the governing equations for the buckling problem.

When combined with Eq. (7.86), the principle of virtual work (7.82) may be transformed into the principle of stationary potential energy, of which the functional is given by

$$\Pi = \tfrac{1}{2}\int_0^l EI(w'')^2\,dx - \tfrac{1}{2}P_{cr}\int_0^l (w')^2\,dx, \tag{7.87}$$

where the function subject to variation is w under the subsidiary conditions (7.79). It is observed that the principle (7.87) is equivalent to finding, among admissible functions w, those which make the quotient

$$P_{cr} = \frac{\tfrac{1}{2}\int_0^l EI(w'')^2\,dx}{\tfrac{1}{2}\int_0^l (w')^2 dx} \tag{7.88}$$

stationary.[12]

Next, let us consider a generalization of the principle of stationary potential energy.[18] Through the usual procedure, the functional (7.87) may be generalized as

$$\Pi_I = \tfrac{1}{2}\int_0^l EI\varkappa^2\,dx - \tfrac{1}{2}P_{cr}\int_0^l (w')^2\,dx + \int_0^l (\varkappa - w'')\,M\,dx + P^* w(o) + Q^* w'(o) + R^* w(l), \tag{7.89}$$

where the quantities subject to variation are $\varkappa$, w, M, P^*, Q^* and R^* with no subsidiary conditions. The stationary conditions with respect to $\varkappa$ and w are shown to be

$$EI\varkappa + M = 0, \tag{7.90}$$

$$M - P_{cr}w'' = 0, \tag{7.91}$$

$$P^* + P_{cr}w'(o) - M'(o) = 0, \quad Q^* + M(o) = 0,$$
$$R^* - P_{cr}w'(l) + M'(l) = 0, \tag{7.92}$$

$$M(l) = 0. \tag{7.93}$$

We eliminate $\varkappa$, P^*, Q^* and R^* by the use of Eqs. (7.90), (7.92) and (7.93), and with the aid of Eq. (7.91), to transform the functional (7.89) into:

$$\Pi_c = \frac{1}{2}\int_0^l \frac{M^2}{EI}\,dx - \frac{1}{2}P_{cr}\int_0^l (w')^2\,dx, \tag{7.94}$$

where it is assumed that $P_{cr} \neq 0$. In the functional (7.94), the quantities subject to variation are M and w under the subsidiary conditions (7.91) and (7.93). The expression (7.94) is a functional for the principle of stationary complementary energy for the buckling problem.

TABLE 7.2. EXACT AND APPROXIMATE EIGENVALUES

$P_{cr}^{(i)} = k^{(i)}(EI/l^2)$

	Exact eigenvalues	Approximate eigenvalues: Rayleigh–Ritz method applied to the functional (7.87)	Approximate eigenvalues: Modified Rayleigh–Ritz method applied to the functional (7.94)
$k^{(1)}$	20.19	20.92	20.30
$k^{(2)}$	59.69	107.1	67.70

Once the variational principles have thus been established, we can apply the Rayleigh–Ritz method and the modified Rayleigh–Ritz method for obtaining approximate eigenvalues. A numerical example is shown by taking w as given by Eqs. (7.57) for a beam with constant EI. Numerical results are listed in Table 7.2 and compared with the exact eigenvalues. See Refs. 16 and 17 for other numerical examples of the Rayleigh–Ritz method and modified Rayleigh–Ritz method applied to buckling problems. We note that nonconservative problems of the stability of elastic beams have been extensively treated in Ref. 19.

7.7. A Beam Theory Including the Effect of Transverse Shear Deformation

The elementary beam theory which has been considered in the preceeding sections is based on the Bernoulli–Euler hypothesis, in which no transverse shear deformation is allowed to occur. We shall consider in the present section an approximate formulation for a dynamical beam problem taking account of the effect of the transverse shear deformation. A dynamical problem defined in a manner similar to the presentation in Section 7.2 will be taken as an example, except that the external forces are now time-dependent. The principle of virtual work is an avenue which leads to an approximate formulation.

Since the displacement vector $\mathbf{u}$ is a function of (x, y, z), we may expand it into a Taylor series about $z = 0$:

$$\mathbf{u}(x, y, z) = \mathbf{u}(x, y, o) + \left(\frac{\partial \mathbf{u}}{\partial z}\right)_{z=0} z + \frac{1}{2!}\left(\frac{\partial^2 \mathbf{u}}{\partial z^2}\right)_{z=0} z^2 + \cdots. \tag{7.95}$$

Therefore, one of the simplest expressions for displacements to include the effect of transverse shear deformation may be given by retaining the first two terms only:

$$\mathbf{u} = \mathbf{u}_0 + z\mathbf{u}_1, \tag{7.96}$$

where components of $\mathbf{u}_1$ are defined by

$$\mathbf{u}_1 = u_1\mathbf{i}_1 + w_1\mathbf{i}_3, \tag{7.97}$$

and u_1 and w_1 are functions of x only. The degree of freedom implied by Eq. (7.96) is four, namely, u, w, u_1 and w_1. However, if we continue to use the assumption (7.1), and employ Eqs. (7.3) as the stress–strain relations, we may take

$$2e_{zz} = u_1^2 + (1 + w_1)^2 - 1 = 0 \tag{7.98}$$

as an additional geometrical constraint to reduce the degree of freedom to three. Equations (7.96) and (7.98) state that the cross sections perpendicular to the undeformed locus remain plane and suffer no strains in their planes although they are no longer perpendicular to the deformed locus.

We shall confine our problem to small displacement theory. Then, Eq. (7.98) is linearized with respect to the displacements to yield

$$w_1 = 0. \tag{7.99}$$

Consequently, the simplest expression for displacements to include the transverse shear deformation is to assume that

$$U = u + zu_1, \quad V = 0, \quad W = w, \tag{7.100}$$

which provide the following nonvanishing strain components

$$\varepsilon_x = u' + zu_1', \quad \gamma_{xz} = w' + u_1. \tag{7.101†}$$

† It is seen that the Bernoulli–Euler hypothesis imposes the constraint condition $u_1 = -w'$.

Then, the principle of virtual work for the present dynamical problem is written as

$$\int_{t_1}^{t_2} \left\{ \iiint_V (\sigma_x \, \delta\varepsilon_x + \tau_{xz} \, \delta\gamma_{xz}) \, dx \, dy \, dz - \delta \iiint_V \tfrac{1}{2}[\dot{u}^2 + \dot{w}^2] \varrho \, dx \, dy \, dz - \int_0^l \bar{p} \, \delta w \, dx - \bar{P}_x \, \delta u(l) - \bar{P}_z \, \delta w(l) - \bar{M} \, \delta u_1 \right\} dt = 0, \quad (7.102)\dagger$$

where Eqs. (7.100) and (7.101) have been substituted. Here, we introduce new quantities defined by

$$Q = \iint_S \tau_{xz} \, dy \, dz, \quad (7.103)$$

$$I_m = \iint_S \varrho z^2 \, dy \, dz, \quad (7.104)$$

in addition to the stress resultants defined by Eqs. (7.19) and (7.20). The quantity defined by Eq. (7.103) is the shearing force of the cross section as shown in Fig. 7.3, while the quantity defined by Eq. (7.104) is the mass moment of inertia of the cross section. By the use of these quantities, the first two terms of Eq. (7.102) may be written as follows:

$$\int_{t_1}^{t_2} \left\{ \int_0^l [N \, \delta u' + M \, \delta u_1' + Q(\delta w' + \delta u_1)] \, dx - \delta \int_0^l [\tfrac{1}{2} m(\dot{u}^2 + \dot{w}^2) + \tfrac{1}{2} I_m \dot{u}_1^2] \, dx \right\} dt. \quad (7.105)$$

Consequently, after some calculation including integrations by parts, the principle (7.102) is transformed into

$$\int_{t_1}^{t_2} \left\{ \int_0^l \{(m\ddot{u} - N') \, \delta u + (m\ddot{w} - Q' - \bar{p}) \, \delta w + (I_m \ddot{u}_1 - M' + Q) \, \delta u_1\} \, dx + [(N - \bar{P}_x) \, \delta u + (Q - \bar{P}_z) \, \delta w + (M - \bar{M}) \, \delta u_1]_{x=l} - [N \, \delta u + Q \, \delta w + M \, \delta u_1]_{x=0} \right\} dt = 0, \quad (7.106)$$

from which we obtain the equations of motion

$$m\ddot{u} = N', \quad (7.107)\ddagger$$

$$m\ddot{w} = Q' + \bar{p}, \quad (7.108)\ddagger$$

$$I_m \ddot{u}_1 = M' - Q, \quad (7.109)\ddagger$$

† See Eq. (5.81).

‡ Compare with Eqs. (7.22) and (7.24).

and the mechanical boundary conditions

$$N = \bar{P}_x, \quad Q = \bar{P}_z, \quad M = \bar{M} \quad \text{at} \quad x = l, \tag{7.110}$$

while it is suggested that the geometrical boundary conditions may be specified approximately as

$$u = 0, \quad w = 0, \quad u_1 = 0 \quad \text{at} \quad x = 0. \tag{7.111}$$

Combining Eqs. (7.2a, b), (7.19), (7.20), (7.101) and (7.103), we have the following stress resultant–displacement relations:

$$N = EA_0 u' \tag{7.112}$$

$$M = EIu_1', \tag{7.113}$$

$$Q = GkA_0(w' + u_1), \tag{7.114}$$

where $k = 1$. The factor k in Eq. (7.114) is appended to take account of the nonuniformity of γ_{xz} over the cross section and the effect of γ_{yz}. An approximate method of determining the value of k for a beam in static equilibrium is shown in Appendix C, where the minimum complementary energy method is employed. Another method may be to determine the value of k so that some results obtained from the above approximate equations may be coincident with those obtained from the exact theory of vibrations or wave propagations (see Refs. 20 and 21). Substituting Eqs. (7.112), (7.113) and (7.114) into Eqs. (7.107), (7.108) and (7.109), we obtain

$$m\ddot{u} = (EA_0 u')', \tag{7.115}$$

$$m\ddot{w} = [GkA_0(w' + u_1)]' + \bar{p}, \tag{7.116}$$

$$I_m \ddot{u}_1 = (EIu_1')' - GkA_0(w' + u_1). \tag{7.117}$$

These equations constitute the governing equations for the dynamical beam problem including the effect of transverse shear deformations, the so-called Timoshenko beam theory.[10] From the above relations, it is observed that the strain energy of the Timoshenko beam is given by

$$U = \tfrac{1}{2} \int_0^l [EA_0(u')^2 + EI(u_1')^2 + GkA_0(w' + u_1)^2] \, dx. \tag{7.118}$$

Effects of shear flexibility and rotary inertia play a very important role in theories of beam vibration and dynamical behavior under impulsive loading (see Refs. 21 through 25).

7.8. Some Remarks

The elementary theory of the beam formulated in Section 7.1 is based on the assumption (7.1) and the Bernoulli–Euler hypothesis. However, we have $\varepsilon_y = \varepsilon_z = 0$ from Eqs. (7.13), and we find that the simultaneous use of the assumption (7.1) and the hypothesis does not satisfy the stress–strain relations (1.10), and will not lead to correct results. The same kind of contradiction exists in the formulations of Sections 7.5 and 7.7. We have tried to remove this difficulty approximately by setting $\sigma_y = \sigma_z = \tau_{yz} = 0$ in the three-dimensional stress–strain relations and then eliminating ε_y and ε_z. For a complete removal of the inconsistency and an improvement of the accuracy of the beam theory, we may assume

$$U = \sum_{m=0,\, n=0} u_{mn}(x)\, y^m z^n, \quad V = \sum_{m=0,\, n=0} v_{mn}(x)\, y^m z^n,$$

$$W = \sum_{m=0,\, n=0} w_{mn}(x)\, y^m z^n, \tag{7.119}$$

where the numbers of terms should be chosen properly. Equations governing u_{mn}, v_{mn} and w_{mn} are obtainable by the use of the principle of virtual work. It is noted here that a theory has been developed in Ref. 26 for expressions of displacements and strain–displacement relations applied to rod problems.

A naturally curved and twisted slender beam presents a classical problem in elasticity.[1] The principle of virtual work may provide an avenue to an approximate formulation of the problem, where a curvilinear coordinate system may be conveniently employed for describing the curved centroid locus and the two curved surfaces constituted by the envelopes of the principal axes through the centroid.[27, 28] Variational formulations have been proposed for the problem, and Ref. 29 is among recent contributions in this field.

Bibliography

1. A. E. H. LOVE, *Mathematical Theory of Elasticity*, Cambridge University Press, 4th edition, 1927.
2. S. TIMOSHENKO and J. N. GOODIER, *Theory of Elasticity*, McGraw-Hill, 1951.
3. E. TREFFTZ, Über den Schubmittelpunkt in einem durch eine Einzellast gebogenen Balken. *Zeitschrift für Angewandte Mathematik und Mechanik*, Vol. 15, No. 4, pp. 220–5, July 1935.
4. A. WEINSTEIN, The Center of Shear and the Center of Twist, *Quarterly of Applied Mathematics*, Vol. 5, No. 1, pp. 97–9, 1947.
5. A. G. STEVENSON, Flexure with Shear and Associated Torsion in Prisms of Uni-axial and Asymmetric Cross Section, *Philosophical Transaction of Royal Society*, Vol. A237, No. 2, pp. 161–229, 1938.
6. J. N. GOODIER, A Theorem on the Shearing Stress in Beams with Applications to Multicellular Sections, *Journal of the Aeronautical Sciences*, Vol. 11, No. 3, pp. 272–80, July 1944.
7. Y. C. FUNG, *An Introduction to the Theory of Aeroelasticity*, John Wiley, 1955.
8. LORD RAYLEIGH, *Theory of Sound*, Macmillan, 1877.
9. J. P. DEN HARTOG, *Mechanical Vibrations*, McGraw-Hill, 1934.
10. S. TIMOSHENKO, *Vibration Problems in Engineering*, D. van Nostrand, 1928.
11. R. L. BISPLINGHOFF, H. ASHLEY and R. L. HALFMAN, *Aeroelasticity*, Addison-Wesley, 1955.
12. L. COLLATZ, *Eigenwertaufgaben mit technischen Anwendungen*, Academische Verlagsgesellschaft, 1949.

13. K. Washizu, Note on the Principle of Stationary Complementary Energy Applied to Free Lateral Vibration of An Elastic Body, *International Journal of Solids and Structures*, Vol. 2, No. 1, pp. 27–35, January 1966.
14. P. A. Libby and R. C. Sauer, Comparison of the Rayleigh–Ritz and Complementary Energy Methods in Vibration Analysis, Reader's Forum, *Journal of Aeronautical Sciences*, Vol. 16, No. 11, pp. 700–2, November 1949.
15. S. H. Crandall, *Engineering Analysis*, McGraw-Hill, 1956.
16. S. Timoshenko, *Theory of Elastic Stability*, McGraw-Hill, 1936.
17. N. J. Hoff, *The Analysis of Structures*, John Wiley, 1956.
18. K. Washizu, Note on the Principle of Stationary Complementary Energy Applied to Buckling of a Column, *Transactions of Japan Society for Aeronautical and Space Sciences*, Vol. 7, No. 12, pp. 18–22, 1965.
19. V. V. Bolotin, *Nonconservative Problems of the Theory of Elastic Stability*, Translated by T. K. Lusher and edited by G. Herrmann, Pergamon Press, 1963.
20. R. D. Mindlin and G. A. Herrmann, A One-dimensional Theory of Compressional Waves in an Elastic Rod, *Proceedings of the 1st National Congress for Applied Mechanics*, Chicago, pp. 187–91, 1951.
21. Y. C. Fung, *Foundations of Solid Mechanics*, Prentice-Hall, 1965.
22. R. W. Trail-Nash and A. R. Collar, Effects of Shear Flexibility and Rotary Inertia on the Bending Vibrations of Beams, *Quarterly Journal of Mechanics and Applied Mathematics*, Vol. 6, No. 2, pp. 186–222, June 1953.
23. H. N. Abramson, H. J. Plass and E. A. Ripperge, Stress Wave Propagation in Rods and Beams, *Advances in Applied Mechanics*, Vol. 5, pp. 111–94, Academic Press, 1958.
24. R. W. Leonard and B. Budiansky, *On Travelling Waves in Beams*, NACA Report 1173, 1954.
25. R. W. Leonard, *On Solutions for the Transient Response of Beams*, NASA, Technical Report R-21, 1959.
26. V. V. Novozhilov, *Foundations of the Nonlinear Theory of Elasticity*, Graylock Press, 1953.
27. K. Washizu, Some Considerations on a Naturally Curved and Twisted Slender Beam, *Journal of Mathematics and Physics*, Vol. 43, No. 2, pp. 111–16, June 1964.
28. K. Washizu, Some Considerations on the Center of Shear, *Transactions of Japan Society for Aeronautical and Space Sciences*, Vol. 9, No. 15, pp. 77–83, 1966.
29. E. Reissner, Variational Considerations for Elastic Beams and Shells, *Journal of the Engineering Mechanics Division, Proceedings of the American Society of Civil Engineers*, Vol. 88, No. EMI, pp. 23–57, February 1962.
30. R. Kappus, Drillknicken zentrisch gedrückter Stäbe mit offenem Profil im elastischen Bereich, *Luftfahrtforschung*, Vol. 14, pp. 444–57, 1937.
31. J. N. Goodier, Torsion and Flexural Buckling of a Bar of Thin-walled Open Section under Compression and Bending Loads, *Journal of Applied Mechanics*, Vol. 9, No. 3, pp. A-103–A-107, September 1942.
32. S. Timoshenko, Theory of Bending, Torsion and Buckling of Thin-walled Members of Open Cross Section, *Journal of the Franklin Institute*, Vol. 239, No. 3, pp. 201–19, March 1945; Vol. 239, No. 4, pp. 249–68, April 1945; Vol. 239, No. 5, pp. 343–61, May 1945.
33. F. Bleich and H. Bleich, *Buckling Strength of Metal Structures*, McGraw-Hill, 1952.
34. K. Marguerre, Die Durchschlagskraft eines schwach gekrümmten Balkens, *Sitzungsberichte der Berliner Mathematischen Gesellschaft*, pp. 22–40, 1938.

CHAPTER 8

PLATES

8.1. Stretching and Bending of a Plate

Let us consider in the present chapter the stretching and bending of a thin plate, the middle surface of which is assumed to be flat. Concerning the coordinate system employed, the x- and y-axes are taken in coincidence with the middle surface and the z-axis in the direction of the normal to the middle surface, so that the x-, y- and z-axes constitute a right handed rectangular Cartesian coordinate system. The plate is assumed to be simply-connected, and its side boundary surfaces to be cylindrical, i.e. parallel

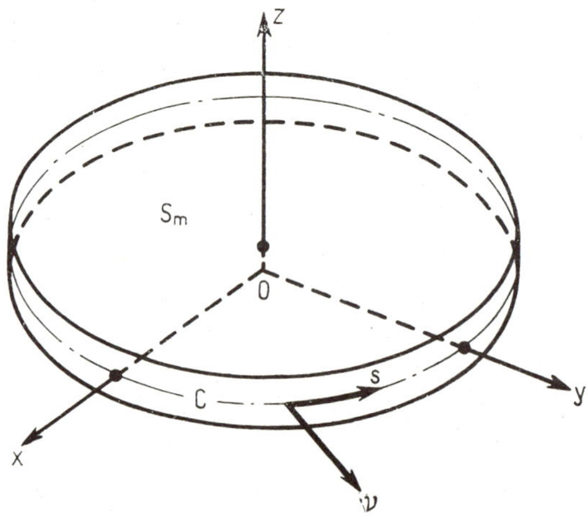

FIG. 8.1. Coordinate system for a plate.

to the z-axis, as shown in Fig. 8.1. We shall denote the region and periphery which constitute the middle surface of the plate by S_m and C, respectively. The direction cosines of the normal ν, drawn outwardly on the boundary C, are denoted by (l, m, o) namely, $l = \cos(x, \nu)$ and $m = \cos(y, \nu)$. A coordinate s is taken along the boundary C, such that ν, s and z form a right handed system.

In formulating an approximate theory of the thin plate in stretching and bending, we shall employ the following assumptions based on the thinness of the plate. First, we assume that the transverse normal stress may be neglected in comparison with the other stress components and may be set

$$\sigma_z = 0. \tag{8.1}$$

Then, as shown in Appendix B, we have the following stress–strain relations for linear theories of the thin plate:

$$\sigma_x = \frac{E}{(1-\nu^2)}(\varepsilon_x + \nu\varepsilon_y), \quad \sigma_y = \frac{E}{(1-\nu^2)}(\nu\varepsilon_x + \varepsilon_y),$$
$$\tau_{xy} = G\gamma_{xy}, \quad \tau_{xz} = G\gamma_{xz}, \quad \tau_{yz} = G\gamma_{yz}. \tag{8.2}$$

For nonlinear theories, we may have

$$\sigma_x = \frac{E}{(1-\nu^2)}(e_{xx} + \nu e_{yy}), \quad \sigma_y = \frac{E}{(1-\nu^2)}(\nu e_{xx} + e_{yy}),$$
$$\tau_{xy} = 2Ge_{xy}, \quad \tau_{xz} = 2Ge_{xz}, \quad \tau_{yz} = 2Ge_{yz}. \tag{8.3}$$

Second, we shall employ the Kirchhoff hypothesis that the linear filaments of the plate initially perpendicular to the middle surface remain straight and perpendicular to the deformed middle surface and suffer no extensions.[1, 2]†

We shall derive expresions for the displacements under this hypothesis. We consider an arbitrary point of a plate having the coordinates (x, y, z) before deformation, and denote its position vectors before and after deformation by $\mathbf{r}^{(0)}$ and $\mathbf{r}$, respectively, which are related to the displacement vector $\mathbf{u}$ by

$$\mathbf{r} = \mathbf{r}^{(0)} + \mathbf{u}, \tag{8.4}$$

where $\mathbf{r}^{(0)} = x\mathbf{i}_1 + y\mathbf{i}_2 + z\mathbf{i}_3$, and $\mathbf{i}_1, \mathbf{i}_2, \mathbf{i}_3$ are the unit vectors in the directions of the x-, y-, z-axes, respectively. Similarly, we denote position vectors of a point $(x, y, 0)$ of the middle surface before and after deformation by $\mathbf{r}_0^{(0)}$ and $\mathbf{r}_0$, respectively, which are related to the displacement vector $\mathbf{u}_0$ by

$$\mathbf{r}_0 = \mathbf{r}_0^{(0)} + \mathbf{u}_0, \tag{8.5}$$

where $\mathbf{r}_0^{(0)} = x\mathbf{i}_1 + y\mathbf{i}_2$. We define components of $\mathbf{u}$ and $\mathbf{u}_0$ as follows:

$$\mathbf{u} = U\mathbf{i}_1 + V\mathbf{i}_2 + W\mathbf{i}_3, \tag{8.6}$$
$$\mathbf{u}_0 = u\mathbf{i}_1 + v\mathbf{i}_2 + w\mathbf{i}_3, \tag{8.7}$$

where u, v and w are functions of (x, y) only. It is seen that the hypothesis allows us to express $\mathbf{r}$ as

$$\mathbf{r} = \mathbf{r}_0 + z\mathbf{n}, \tag{8.8}$$

† The Kirchhoff hypothesis is usually understood to include the first assumption $\sigma_z = 0$ as well as the second assumption. However, only the second assumption will be called the Kirchhoff hypothesis in this book.

where $\mathbf{n}$ is a unit normal to the deformed middle surface and is given by

$$\mathbf{n} = \frac{\partial \mathbf{r}_0}{\partial x} \times \frac{\partial \mathbf{r}_0}{\partial y} \bigg/ \left| \frac{\partial \mathbf{r}_0}{\partial x} \times \frac{\partial \mathbf{r}_0}{\partial y} \right|. \tag{8.9}$$

Since

$$\mathbf{r}_0 = (x + u)\,\mathbf{i}_1 + (y + v)\,\mathbf{i}_2 + w\,\mathbf{i}_3, \tag{8.10}$$

we may express $\mathbf{n}$ in terms of u, v and w as follows:

$$\mathbf{n} = \frac{L\mathbf{i}_1 + M\mathbf{i}_2 + N\mathbf{i}_3}{\sqrt{L^2 + M^2 + N^2}}, \tag{8.11}$$

where

$$\begin{aligned} L &= -\frac{\partial w}{\partial x} + \frac{\partial v}{\partial x}\frac{\partial w}{\partial y} - \frac{\partial v}{\partial y}\frac{\partial w}{\partial x}, \\ M &= -\frac{\partial w}{\partial y} + \frac{\partial u}{\partial y}\frac{\partial w}{\partial x} - \frac{\partial u}{\partial x}\frac{\partial w}{\partial y}, \\ N &= 1 + \frac{\partial u}{\partial x} + \frac{\partial v}{\partial y} + \frac{\partial u}{\partial x}\frac{\partial v}{\partial y} - \frac{\partial u}{\partial y}\frac{\partial v}{\partial x}. \end{aligned} \tag{8.12}$$

From Eqs. (8.4), (8.5) and (8.8), we obtain

$$\mathbf{u} = \mathbf{u}_0 + z(\mathbf{n} - \mathbf{i}_3). \tag{8.13}$$

This is the expression for the displacements of a plate under the Kirchhoff hypothesis. It is observed that the degree of freedom of the plate deformation implied by Eq. (8.13) is three, namely $u(x, y)$, $v(x, y)$ and $w(x, y)$. When a plate problem is confined to small displacement theory, Eq. (8.13) may be linearized with respect to the displacements to obtain

$$U = u - z\frac{\partial w}{\partial x}, \quad V = v - z\frac{\partial w}{\partial y}; \quad W = w. \tag{8.14}$$

Consequently, the strain components are given by

$$\begin{aligned} \varepsilon_x &= \frac{\partial u}{\partial x} - z\frac{\partial^2 w}{\partial x^2}, \quad \varepsilon_y = \frac{\partial v}{\partial y} - z\frac{\partial^2 w}{\partial y^2}, \\ \gamma_{xy} &= \frac{\partial u}{\partial y} + \frac{\partial v}{\partial x} - 2z\frac{\partial^2 w}{\partial x\,\partial y}, \\ \varepsilon_z &= \gamma_{xz} = \gamma_{yz} = 0, \end{aligned} \tag{8.15}$$

which are related to the stress components by Eqs. (8.2).

8.2. A Problem of Stretching and Bending of a Plate

We consider a problem of a plate stated as follows: Let the plate be subject to a distributed lateral load $\bar{p}(x, y)$ per unit area of the middle surface in the direction of the z-axis. The lateral load may consist of body forces as well as external forces on the upper and lower surfaces of the plate. On part of the side boundary, denoted by S_1, external forces are prescribed. They are defined per unit area of the side boundary, and their components in the direction of the x-, y- und z-axes are denoted by $\bar{F}_x$, $\bar{F}_y$ and $\bar{F}_z$, respectively. On the remaining part of the side boundary, denoted by S_2, geometrical boundary conditions are prescribed.

The principle of virtual work for the present problem can be written as follows:

$$\iiint_V (\sigma_x \,\delta\varepsilon_x + \sigma_y \,\delta\varepsilon_y + \tau_{xy} \,\delta\gamma_{xy})\, dx\, dy\, dz - \iint_{S_m} \bar{p}\, \delta w\, dx\, dy$$
$$- \iint_{S_1} (\bar{F}_x \,\delta u + \bar{F}_y \,\delta v + \bar{F}_z \,\delta w)\, ds\, dz = 0, \tag{8.16†}$$

where Eqs. (8.14) and (8.15) have been substituted. Here, we define the following stress resultants:

$$N_x = \int_{-h/2}^{h/2} \sigma_x\, dz, \quad N_y = \int_{-h/2}^{h/2} \sigma_y\, dz, \quad N_{xy} = \int_{-h/2}^{h/2} \tau_{xy}\, dz,$$
$$M_x = \int_{-h/2}^{h/2} \sigma_x z\, dz, \quad M_y = \int_{-h/2}^{h/2} \sigma_y z\, dz, \quad M_{xy} = \int_{-h/2}^{h/2} \tau_{xy} z\, dz, \tag{8.17}$$

and

$$\bar{N}_{x\nu} = \int_{-h/2}^{h/2} \bar{F}_x\, dz, \quad \bar{N}_{y\nu} = \int_{-h/2}^{h/2} \bar{F}_y\, dz, \quad \bar{V}_z = \int_{-h/2}^{h/2} \bar{F}_z\, dz,$$
$$\bar{M}_{x\nu} = \int_{-h/2}^{h/2} \bar{F}_x z\, dz, \quad \bar{M}_{y\nu} = \int_{-h/2}^{h/2} \bar{F}_y z\, dz, \tag{8.18}$$

and perform integrations with respect to z in Eq. (8.16), where $h(x, y)$ is the thickness of the plate. Then, through the use of an integration by parts,

$$\iint_{S_m} [M_x \,\delta w_{,xx} + M_y \,\delta w_{,yy} + 2M_{xy} \,\delta w_{,xy}]\, dx\, dy$$
$$= \int_{C_1+C_2} [M_{x\nu} \,\delta w_{,x} + M_{y\nu} \,\delta w_{,y}]\, ds - \iint_{S_m} [\tilde{Q}_x \,\delta w_{,x} + \tilde{Q}_y \,\delta w_{,y}]\, dx\, dy, \tag{8.19‡}$$

and the geometrical conditions

$$\frac{\partial}{\partial x} = l\frac{\partial}{\partial \nu} - m\frac{\partial}{\partial s}, \quad \frac{\partial}{\partial y} = m\frac{\partial}{\partial \nu} + l\frac{\partial}{\partial s}, \tag{8.20}$$

which hold on the boundary C, we may transform Eq. (8.16) into

$$- \iint_{S_m} [(N_{x,x} + N_{xy,y})\,\delta u + (N_{xy,x} + N_{y,y})\,\delta v + (\tilde{Q}_{x,x} + \tilde{Q}_{y,y} + \bar{p})\,\delta w]\, dx\, dy$$
$$+ \int_{C_1} [(N_{x\nu} - \bar{N}_{x\nu})\,\delta u + (N_{y\nu} - \bar{N}_{y\nu})\,\delta v + (V_z - \bar{V}_z)\,\delta w$$
$$- (M_\nu - \bar{M}_\nu)\,\delta w_{,\nu} - (M_{\nu s} - \bar{M}_{\nu s})\,\delta w_{,s}]\, ds$$
$$+ \int_{C_2} [N_{x\nu}\,\delta u + N_{y\nu}\,\delta v + V_z\,\delta w - M_\nu\,\delta w_{,\nu} - M_{\nu s}\,\delta w_{,s}]\, ds = 0, \tag{8.21}$$

† See Eq. (1.32).

‡ Notations $(\ )_{,x} = \partial(\ )/\partial x$, $(\ )_{,y} = \partial(\ )/\partial y$, $(\ )_{,s} = \partial(\ )/\partial s$ and $(\ )_{,\nu} = \partial(\ )/\partial \nu$ will be used for the sake of brevity whenever convenient.

where C_1 and C_2 are parts of the boundary C which correspond to S_1 and S_2, respectively, and it is defined that

$$\tilde{Q}_x = \frac{\partial M_x}{\partial x} + \frac{\partial M_{xy}}{\partial y}, \quad \tilde{Q}_y = \frac{\partial M_{xy}}{\partial x} + \frac{\partial M_y}{\partial y}, \tag{8.22}$$

$$N_{x\nu} = N_x l + N_{xy} m, \qquad N_{y\nu} = N_{xy} l + N_y m, \tag{8.23}$$

$$\begin{aligned} M_{x\nu} &= M_x l + M_{xy} m, \quad M_{y\nu} = M_{xy} l + M_y m, \\ M_\nu &= M_{x\nu} l + M_{y\nu} m = M_x l^2 + 2M_{xy} lm + M_y m^2, \\ M_{\nu s} &= -M_{x\nu} m + M_{y\nu} l = -(M_x - M_y)\, lm + M_{xy}(l^2 - m^2), \end{aligned} \tag{8.24}$$

$$V_z = \tilde{Q}_x l + \tilde{Q}_y m, \tag{8.25}$$

$$\bar{M}_\nu = \bar{M}_{x\nu} l + \bar{M}_{y\nu} m, \quad \bar{M}_{\nu s} = -\bar{M}_{x\nu} m + \bar{M}_{y\nu} l. \tag{8.26}$$

The quantities defined by Eqs. (8.17) are stress resultants and moments per unit length of the lines x and y of the middle surface as shown in Fig. 8.2.

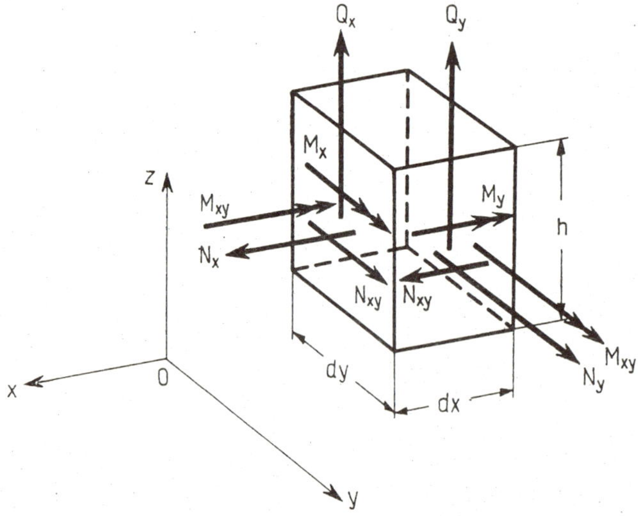

FIG. 8.2. Stress resultants and moments.

The quantities N_x, N_y and N_{xy} are in-plane stress resultants, while M_x, M_y and M_{xy} are bending and twisting moments. The quantities $\tilde{Q}_x$ and $\tilde{Q}_y$ defined by Eqs. (8.22) are proved equal to shearing forces Q_x and Q_y per unit length of the lines x and y of the middle surface by considering the equilibrium conditions of the infinitesimal rectangular parallelepiped in the figure with respect to moments around the axes parallel to the y- and x-axes, respectively.† The quantities defined by Eqs. (8.18) and (8.26) are prescribed external

† See the footnote of Section 7.2 for a similar development.

forces and moments per unit length along the boundary. It is seen that $\bar{V}_z$ is the shearing force acting in the direction of the z-axis, while $\bar{M}_\nu$ and $\bar{M}_{\nu s}$ are bending and twisting moments as shown in Fig. 8.3.

Returning back to Eq. (8.21), we find that some of the line integral terms must be transformed through integrations by parts. For example, we have

$$\int_{C_1} [(V_z - \bar{V}_z)\,\delta w - (M_{\nu s} - \bar{M}_{\nu s})\,\delta w_{,s}]\,ds$$
$$= -(M_{\nu s} - \bar{M}_{\nu s})\,\delta w\Big|_{C_1} + \int_{C_1} [(V_z + M_{\nu s,s}) - (\bar{V}_z + \bar{M}_{\nu s,s})]\,\delta w\,ds \quad (8.27)$$

where the notation $(\quad)|_{C_1}$ indicates that the difference in the values at the ends of C_1 is taken. The above equation shows that under the Kirchhoff hypothesis, the action of the twisting moments $M_{\nu s}$ and $\bar{M}_{\nu s}$ distributed

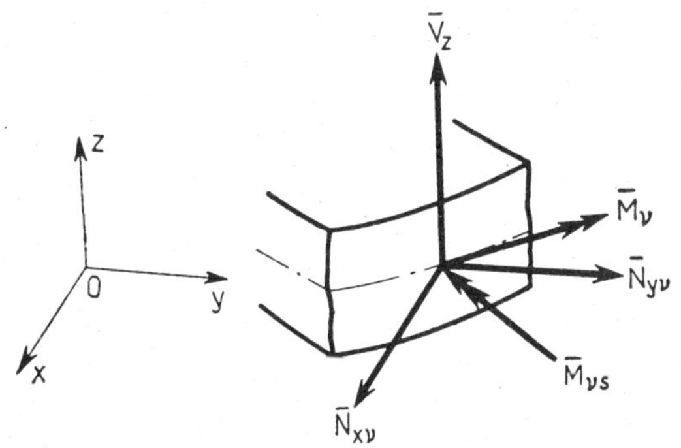

FIG. 8.3. Resultant forces and couples on the boundary.

along the boundary is replaced by that of the shearing forces V_z and $\bar{V}_z$, respectively, while $M_{\nu s}$ and $\bar{M}_{\nu s}$ at the ends of C_1 remain as concentrated forces in the $\pm z$-directions, respectively.[1, 2] A similar transformation is applied to the line integral on C_2, and it is suggested that the geometrical boundary conditions on S_2 can be specified approximately as,

$$u = \bar{u}, \quad v = \bar{v}, \quad w = \bar{w}, \quad \frac{\partial w}{\partial \nu} = \overline{\frac{\partial w}{\partial \nu}} \quad \text{on} \quad C_2. \quad (8.28\text{a, b, c, d})$$

Consequently, we may put $\delta u = \delta v = \delta w = \partial \delta w/\partial \nu = 0$ on C_2.

In view of the above development, the principle of virtual work, (8.21), is finally reduced to

$$-\iint_{S_m} [(N_{x,x} + N_{xy,y})\,\delta u + (N_{xy,x} + N_{y,y})\,\delta v$$
$$+ (\tilde{Q}_{x,x} + \tilde{Q}_{y,y} + \bar{p})\,\delta w]\,dx\,dy + \int_{C_1} \{(N_{x\nu} - \bar{N}_{x\nu})\,\delta u + (N_{y\nu} - \bar{N}_{y\nu})\,\delta v$$
$$+ [(V_z + M_{\nu s,s}) - (\bar{V}_z + \bar{M}_{\nu s,s})]\,\delta w - (M_\nu - \bar{M}_\nu)\,\delta w_{,\nu}\}\,ds = 0. \quad (8.29)$$

Since δu, δv, δw and $\delta w_{,\nu}$ are arbitrary in S_m and on C_1, we obtain the equations of equilibrium,

$$\frac{\partial N_x}{\partial x} + \frac{\partial N_{xy}}{\partial y} = 0, \quad \frac{\partial N_{xy}}{\partial x} + \frac{\partial N_y}{\partial y} = 0,$$
$$\frac{\partial^2 M_x}{\partial x^2} + 2\frac{\partial^2 M_{xy}}{\partial x\, \partial y} + \frac{\partial^2 M_y}{\partial y^2} + \bar{p} = 0, \qquad (8.30\text{a, b, c})$$

and the mechanical boundary conditions,

$$N_{x\nu} = \bar{N}_{x\nu}, \quad N_{y\nu} = \bar{N}_{y\nu}, \quad V_z + \frac{\partial M_{\nu s}}{\partial s} = \bar{V}_z + \frac{\partial \bar{M}_{\nu s}}{\partial s}, \quad M_\nu = \bar{M}_\nu \quad \text{on } C_1. \qquad (8.31\text{a, b, c, d})$$

We now seek the stress resultant–displacement relations. Combining Eqs. (8.2), (8.15) and (8.17), we find

$$N_x = \frac{Eh}{(1-\nu^2)}\left(\frac{\partial u}{\partial x} + \nu\frac{\partial v}{\partial y}\right), \quad N_y = \frac{Eh}{(1-\nu^2)}\left(\nu\frac{\partial u}{\partial x} + \frac{\partial v}{\partial y}\right)$$
$$N_{xy} = Gh\left(\frac{\partial u}{\partial y} + \frac{\partial v}{\partial x}\right), \qquad (8.32)$$

and

$$M_x = -D\left(\frac{\partial^2 w}{\partial x^2} + \nu\frac{\partial^2 w}{\partial y^2}\right), \quad M_y = -D\left(\nu\frac{\partial^2 w}{\partial x^2} + \frac{\partial^2 w}{\partial y^2}\right),$$
$$M_{xy} = -D(1-\nu)\frac{\partial^2 w}{\partial x\, \partial y}, \qquad (8.33)$$

where $D = Eh^3/12(1-\nu^2)$ is the bending rigidity of the plate.

Combining Eqs. (8.28a, b), (8.30a, b), (8.31a, b) and (8.32), we obtain two simultaneous differential equations and boundary conditions in terms of u and v. By solving this boundary value problem, we can determine the stretching of the plate. Alternatively, combination of Eqs. (8.28c, d) (8.30c), (8.31c, d) and (8.33) yields a differential equation and boundary conditions in terms of w, which determine the bending of the plate. When the plate is of uniform bending rigidity, they take the form[1]

$$D\Delta\Delta w = \bar{p}, \qquad (8.34)$$

$$-D\left\{\frac{\partial}{\partial \nu}(\Delta w) + (1-\nu)\frac{\partial}{\partial s}\left[\frac{\partial}{\partial \nu}\left(\frac{\partial w}{\partial s}\right)\right]\right\} = \bar{V}_z + \frac{\partial \bar{M}_{\nu s}}{\partial s},$$
$$-D\left[\frac{\partial^2 w}{\partial \nu^2} + \nu\left(\frac{\partial^2 w}{\partial s^2} + \frac{1}{\varrho}\frac{\partial w}{\partial \nu}\right)\right] = \bar{M}_\nu \quad \text{on } C_1, \qquad (8.35)$$

$$w = \bar{w}, \quad \frac{\partial w}{\partial \nu} = \overline{\frac{\partial w}{\partial \nu}} \quad \text{on } C_2, \qquad (8.36)$$

where $\Delta(\ ) = \partial^2(\ )/\partial x^2 + \partial^2(\ )/\partial y^2$ is the two-dimensional Laplace operator. The quantity ϱ in Eq. (8.35) is the local radius of curvature of the periphery

C_1 defined by $1/\varrho = d\vartheta/ds$, where ϑ is the angle between the tangent to the periphery and the x-axis as shown in Fig. 8.4. Thus, in small displacement theory of a plate, where the displacement components are assumed to be of the form (8.14), the stretching and bending do not couple with each other and can be treated separately.

The stress–strain relations (8.2) ensure the existence of the strain energy function as shown in Appendix B. Consequently, with the aid of Eq. (8.15), we have the expression for the strain energy of the plate as follows:

$$U = \frac{1}{2}\iint_{S_m} \left\{\frac{Eh}{(1-\nu^2)}\left(\frac{\partial u}{\partial x} + \frac{\partial v}{\partial y}\right)^2 + Gh\left[\left(\frac{\partial u}{\partial y} + \frac{\partial v}{\partial x}\right)^2 - 4\frac{\partial u}{\partial x}\frac{\partial v}{\partial y}\right]\right\} dx\,dy$$
$$+ \frac{1}{2}\iint_{S_m} D\left\{\left(\frac{\partial^2 w}{\partial x^2} + \frac{\partial^2 w}{\partial y^2}\right)^2 + 2(1-\nu)\left[\left(\frac{\partial^2 w}{\partial x\,\partial y}\right)^2 - \frac{\partial^2 w}{\partial x^2}\frac{\partial^2 w}{\partial y^2}\right]\right\} dx\,dy. \tag{8.37}$$

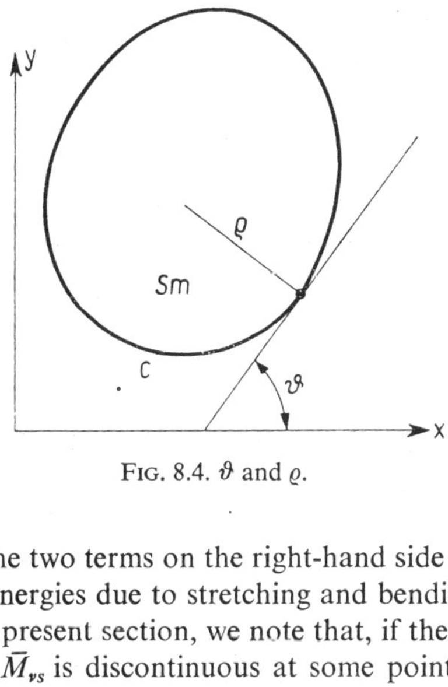

FIG. 8.4. ϑ and ϱ.

It is observed that the two terms on the right-hand side of Eq. (8.37) correspond to the strain energies due to stretching and bending, respectively.

Before leaving the present section, we note that, if the slope ϑ of the contour or the quantity $\bar{M}_{\nu s}$ is discontinuous at some points on the boundary C_1, care should be taken in deriving Eq. (8.27). For example, if $\bar{M}_{\nu s}$ is discontinuous at a point $s = s^*$, we should have

$$\int_{C_1} (-\bar{V}_z\,\delta w + \bar{M}_{\nu s}\,\delta w_{,s})\,ds = \bar{M}_{\nu s}\,\delta w\Big|_{C_1}$$
$$- [\bar{M}_{\nu s}(s^* + 0) - \bar{M}_{\nu s}(s^* - 0)]\,\delta w(s^*) - \int_{C_1} (\bar{V}_z + \bar{M}_{\nu s,s})\,\delta w\,ds. \tag{8.38}$$

Similar care should be taken in the transformation of the line integral on C_2. However, in subsequent sections, we shall assume that such singular points do not exist on the boundary.

8.3. Principle of Minimum Potential Energy and its Transformation for the Stretching of a Plate

The formulation in the preceding section suggests that the expression for the total potential energy of the plate in stretching is given by

$$\Pi = \frac{1}{2}\iint_{S_m}\left\{\frac{Eh}{(1-\nu^2)}\left(\frac{\partial u}{\partial x}+\frac{\partial v}{\partial y}\right)^2 + Gh\left[\left(\frac{\partial u}{\partial y}+\frac{\partial v}{\partial x}\right)^2 - 4\frac{\partial u}{\partial x}\frac{\partial v}{\partial y}\right]\right\} dx\, dy$$
$$- \int_{C_1} (\bar{N}_{x\nu} u + \bar{N}_{y\nu} v)\, ds, \tag{8.39}$$

where the independent quantities subject to variation are u and v under the subsidiary conditions (8.28a, b). By the introduction of three auxiliary functions defined by

$$\varepsilon_{x0} = \frac{\partial u}{\partial x}, \quad \varepsilon_{y0} = \frac{\partial v}{\partial y}, \quad \gamma_{xy0} = \frac{\partial u}{\partial y} + \frac{\partial v}{\partial x}, \tag{8.40}$$

the functional (8.39) is generalized into

$$\Pi_I = \iint_{S_m}\left[\frac{Eh}{2(1-\nu^2)}(\varepsilon_{x0}+\varepsilon_{y0})^2 + \frac{Gh}{2}(\gamma_{xy0}^2 - 4\varepsilon_{x0}\varepsilon_{y0})\right.$$
$$- \left(\varepsilon_{x0} - \frac{\partial u}{\partial x}\right) N_x - \left(\varepsilon_{y0} - \frac{\partial v}{\partial y}\right) N_y$$
$$\left. - \left(\gamma_{xy0} - \frac{\partial u}{\partial y} - \frac{\partial v}{\partial x}\right) N_{xy}\right] dx\, dy - \int_{C_1} (\bar{N}_{x\nu} u + \bar{N}_{y\nu} v)\, ds$$
$$- \int_{C_2} [(u - \bar{u}) N_{x\nu} + (v - \bar{v}) N_{y\nu}]\, ds. \tag{8.41}$$

If we eliminate ε_{x0}, ε_{y0}, γ_{xy0}, u and v through the use of the stationary conditions:

$$N_x = \frac{Eh}{(1-\nu^2)}(\varepsilon_{x0} + \nu\varepsilon_{y0}), \quad N_y = \frac{Eh}{(1-\nu^2)}(\nu\varepsilon_{x0} + \varepsilon_{y0}), \quad N_{xy} = Gh\gamma_{xy0}, \tag{8.42}$$

$$N_{x,x} + N_{xy,y} = 0, \quad N_{xy,x} + N_{y,y} = 0 \quad \text{in} \quad S_m, \tag{8.43}$$

$$N_{x\nu} = \bar{N}_{x\nu}, \quad N_{y\nu} = \bar{N}_{y\nu} \quad \text{on} \quad C_1, \tag{8.44}$$

the functional (8.41) reduces to

$$\Pi_c = \frac{1}{2}\iint_{S_m}\left[\frac{1}{Eh}(N_x + N_y)^2 + \frac{1}{Gh}(N_{xy}^2 - N_x N_y)\right] dx\, dy$$
$$- \int_{C_2} (\bar{u} N_{x\nu} + \bar{v} N_{y\nu})\, ds, \tag{8.45}$$

where the functions subject to variation are N_x, N_y and N_{xy} under the subsidiary conditions (8.43) and (8.44). If the Airy stress function $F(x, y)$ defined by

$$N_x = \frac{\partial^2 F}{\partial y^2}, \quad N_y = \frac{\partial^2 F}{\partial x^2}, \quad N_{xy} = -\frac{\partial^2 F}{\partial x\, \partial y} \tag{8.46}$$

is employed, the functional (8.45) may be written as

$$\Pi_C^* = \frac{1}{2}\iint\limits_{S_m} \left\{ \frac{1}{Eh}\left(\frac{\partial^2 F}{\partial x^2} + \frac{\partial^2 F}{\partial y^2}\right)^2 + \frac{1}{Gh}\left[\left(\frac{\partial^2 F}{\partial x\,\partial y}\right)^2 - \frac{\partial^2 F}{\partial x^2}\frac{\partial^2 F}{\partial y^2}\right]\right\} dx\,dy$$

$$- \int\limits_{C_2} \left[\bar{u}\frac{d}{ds}\left(\frac{\partial F}{\partial y}\right) - \bar{v}\frac{d}{ds}\left(\frac{\partial F}{\partial x}\right)\right] ds, \tag{8.47}$$

where the independent function subject to variation is F under the subsidiary boundary condition (8.44), namely,

$$\frac{d}{ds}\left(\frac{\partial F}{\partial y}\right) = \bar{N}_{x\nu}, \quad -\frac{d}{ds}\left(\frac{\partial F}{\partial x}\right) = \bar{N}_{y\nu} \quad \text{on} \quad C_1. \tag{8.48}$$

The stationary conditions of the functional (8.47) are an equation in S_m and boundary conditions on C_2. The equations in S_m comprises the condition of compatibility between the strain components ε_{x0}, ε_{y0}, and γ_{xy0}. When the plate is of uniform thickness, the equation becomes

$$\Delta\Delta F = 0 \quad \text{in} \quad S_m, \tag{8.49}$$

where Δ is the two-dimensional Laplace operator. It is obvious that the boundary conditions on C_2 are equivalent to Eqs. (8.28a, b). It is noted here that the functional (8.45) can be obtained directly from the functional (2.23) by assuming that

$$\sigma_x = \frac{N_x}{h}, \quad \sigma_y = \frac{N_y}{h}, \quad \tau_{xy} = \frac{N_{xy}}{h}, \tag{8.50}$$

and all the other stress components vanish.

The problem of a plate in stretching has been extensively investigated, and a great number of papers have been written on the subject (see Refs. 3 through 6, for example). Variational principles combined with the Rayleigh–Ritz method have been employed to obtain approximate solutions for the analysis of plates in stretching (see Refs. 3, 7 and 8, for example).

8.4. Principle of Minimum Potential Energy and its Transformation for the Bending of a Plate

As is observed in Section 8.2, the total potential energy for the plate in bending is given by

$$\Pi = \frac{1}{2}\iint\limits_{S_m} D\left\{\left(\frac{\partial^2 w}{\partial x^2} + \frac{\partial^2 w}{\partial y^2}\right)^2 + 2(1-\nu)\left[\left(\frac{\partial^2 w}{\partial x\,\partial y}\right)^2 - \frac{\partial^2 w}{\partial x^2}\frac{\partial^2 w}{\partial y^2}\right]\right\} dx\,dy$$

$$- \iint\limits_{S_m} \bar{p}w\,dx\,dy + \int\limits_{C_1}\left[\bar{M}_\nu\frac{\partial w}{\partial \nu} - \left(\bar{V}_z + \frac{\partial \bar{M}_{\nu s}}{\partial s}\right) w\right] ds, \tag{8.51}$$

where the independent function subject to variation is w under the subsidiary condition (8.28c, d). By the introduction of three auxiliary functions defined by

$$\varkappa_x = \frac{\partial^2 w}{\partial x^2}, \quad \varkappa_y = \frac{\partial^2 w}{\partial y^2}, \quad \varkappa_{xy} = \frac{\partial^2 w}{\partial x \, \partial y}, \tag{8.52}$$

the functional (8.51) can be generalized as follows:

$$\begin{aligned} \Pi_I = \iint_{S_m} \Big\{ & \frac{D}{2} [(\varkappa_x + \varkappa_y)^2 + 2(1 - \nu)(\varkappa_{xy}^2 - \varkappa_x \varkappa_y)] \\ & + \left(\varkappa_x - \frac{\partial^2 w}{\partial x^2}\right) M_x + \left(\varkappa_y - \frac{\partial^2 w}{\partial y^2}\right) M_y \\ & + 2\left(\varkappa_{xy} - \frac{\partial^2 w}{\partial x \partial y}\right) M_{xy} - \bar{p} w \Big\} \, dx \, dy \\ & + \int_{C_1} \left[\bar{M}_\nu \frac{\partial w}{\partial \nu} - \left(\bar{V}_z + \frac{\partial \bar{M}_{\nu s}}{\partial s} \right) w \right] ds \\ & + \int_{C_2} \left[\left(\frac{\partial w}{\partial \nu} - \overline{\frac{\partial w}{\partial \nu}} \right) P^* - (w - \bar{w}) Q^* \right] ds, \end{aligned} \tag{8.53}$$

where M_x, M_y, M_{xy}, P^* and Q^* are Lagrange multipliers. Eliminating $\varkappa_x$, $\varkappa_y$, $\varkappa_{xy}$ and w through the use of following stationary conditions:

$$M_x = -D(\varkappa_x + \nu\varkappa_y), \quad M_y = -D(\nu\varkappa_x + \varkappa_y),$$
$$M_{xy} = -D(1 - \nu)\varkappa_{xy}, \tag{8.54}$$

$$M_{x,xx} + 2M_{xy,xy} + M_{y,yy} + \bar{p} = 0, \tag{8.55}$$

$$M_\nu = \bar{M}_\nu, \quad V_z + M_{\nu s,s} = \bar{V}_z + \bar{M}_{\nu s,s} \quad \text{on} \quad C_1 \tag{8.56}$$

and

$$P^* = M_\nu, \quad Q^* = V_z + M_{\nu s,s} \quad \text{on} \quad C_2, \tag{8.57}$$

the generalized functional yields the following functional for the principle of minimum complementary energy:

$$\begin{aligned} \Pi_c = & \frac{1}{2} \iint_{S_m} \left(\frac{12}{Eh^3}\right) [(M_x + M_y)^2 + 2(1 + \nu)(M_{xy}^2 - M_x M_y)] \, dx \, dy \\ & + \int_{C_2} \left[M_\nu \overline{\frac{\partial w}{\partial \nu}} - \left(V_z + \frac{\partial M_{\nu s}}{\partial s} \right) \bar{w} \right] ds, \end{aligned} \tag{8.58}$$

where the quantities subject to variation are M_x, M_y and M_{xy} under the subsidiary conditions (8.55) and (8.56). The stationary conditions of the functional (8.58) are shown to be the conditions of compatibility, which are equivalent to Eqs. (8.52), and the geometrical boundary conditions (8.28c, d). It is noted that the first term of the functional (8.58) can be obtained from the first term of the functional (2.23) by assuming that

$$\sigma_x = \frac{M_x}{(h^2/6)} \frac{z}{(h/2)}, \quad \sigma_y = \frac{M_y}{(h^2/6)} \frac{z}{(h/2)}, \quad \tau_{xy} = \frac{M_{xy}}{(h^2/6)} \frac{z}{(h/2)}, \tag{8.59}$$

and all the other stress components vanish (see also Appendix D).

The variational principles derived above can be applied to the solution of problems of plates in bending. The principle of minimum potential energy (8.51), combined with the Rayleigh–Ritz method, has been successfully employed for obtaining approximate solutions for the deflection of plates in bending (see Refs. 2, 9 and 10, for example).

8.5. Large Deflection of a Plate in Stretching and Bending

We shall consider a large deflection theory of a plate proposed by T. v. Kármán,† prescribing the plate problem in the same manner as in Section 8.2. It is assumed that, although the deflection of the plate is no longer small in comparison with the thickness of the plate, it is still small in comparison with the lateral dimensions of the plate and the following expressions may be employed for the displacements U, V and W and for the strain–displacement relations:

$$U = u - z\frac{\partial w}{\partial x}, \quad V = v - z\frac{\partial w}{\partial y}, \quad W = w \tag{8.60‡}$$

$$\begin{aligned} e_{xx} &= \frac{\partial u}{\partial x} + \frac{1}{2}\left(\frac{\partial w}{\partial x}\right)^2 - z\frac{\partial^2 w}{\partial x^2}, \\ e_{yy} &= \frac{\partial v}{\partial y} + \frac{1}{2}\left(\frac{\partial w}{\partial y}\right)^2 - z\frac{\partial^2 w}{\partial y^2}, \\ 2e_{xy} &= \frac{\partial u}{\partial y} + \frac{\partial v}{\partial x} + \frac{\partial w}{\partial x}\frac{\partial w}{\partial y} - 2z\frac{\partial^2 w}{\partial x\,\partial y}, \end{aligned} \tag{8.61‡}$$

higher order terms being neglected.

Since we are dealing with the large deflection theory, we must employ Eq. (3.49) for establishing the principle of virtual work for the present problem, and we have

$$\begin{aligned} \iiint_V (\sigma_x\,\delta e_{xx} + \sigma_y\,\delta e_{yy} + 2\tau_{xy}\,\delta e_{xy})\,dx\,dy\,dz - \iint_{S_m} \bar{p}\,\delta w\,dx\,dy \\ - \iint_{S_1} [\bar{F}_x\,\delta U + \bar{F}_y\,\delta V + \bar{F}_z\,\delta W]\,dz\,ds = 0, \end{aligned} \tag{8.62}$$

where Eqs. (8.60) and (8.61) have been substituted. After some calculation and introduction of the quantities defined in Section 8.2, we find that Eq. (8.62) reduces to an equation which may also be obtained from Eq. (8.29) by making the following replacements:

$$\begin{aligned} &\tilde{Q}_x \quad \text{by} \quad \tilde{Q}_x + N_x w_{,x} + N_{xy} w_{,y}, \\ &\tilde{Q}_y \quad \text{by} \quad \tilde{Q}_y + N_{xy} w_{,x} + N_y w_{,y}. \end{aligned} \tag{8.63}$$

These replacements mean that when the deflection of the plate becomes large, the in-plane stress resultants N_x, N_y and N_{xy} have contributions to

† See Refs. 2, 11 and 12.

‡ These equations may be derived from Eqs. (8.13) and (3.19) by assuming that $u_{,x} \sim u_{,y} \sim v_{,x} \sim v_{,y} \sim (w_{,x})^2 \sim (w_{,y})^2 \ll 1$ and terms containing z^2 may be neglected. The first assumption states that the quantities $(w_{,x})^2$, $(w_{,y})^2$, the strain of the middle surface as well as the rotation of the plate around the z-axis are very small compared to unity.[12]

the equation of equilibrium in the direction of the z-axis due to the inclination of the middle surface. Thus we obtain the equations of equilibrium

$$\frac{\partial N_x}{\partial x} + \frac{\partial N_{xy}}{\partial y} = 0, \quad \frac{\partial N_{xy}}{\partial x} + \frac{\partial N_y}{\partial y} = 0,$$

$$\frac{\partial^2 M_x}{\partial x^2} + 2\frac{\partial^2 M_{xy}}{\partial x\,\partial y} + \frac{\partial^2 M_y}{\partial y^2} + \frac{\partial}{\partial x}\left(N_x \frac{\partial w}{\partial x} + N_{xy}\frac{\partial w}{\partial y}\right)$$
$$+ \frac{\partial}{\partial y}\left(N_{xy}\frac{\partial w}{\partial x} + N_y \frac{\partial w}{\partial y}\right) + \bar{p} = 0 \tag{8.64}$$

and the mechanical boundary conditions on C_1,

$$N_{xv} = \bar{N}_{xv}, \quad N_{yv} = \bar{N}_{yv},$$
$$\tilde{Q}_x l + \tilde{Q}_y m + N_{xv}\frac{\partial w}{\partial x} + N_{yv}\frac{\partial w}{\partial y} + \frac{\partial M_{vs}}{\partial s} = \bar{V}_z + \frac{\partial \bar{M}_{vs}}{\partial s},$$
$$M_v = \bar{M}_v. \tag{8.65}$$

Combining Eqs. (8.3), (8.17) and (8.61), we obtain the in-plane stress resultant–displacement relations as follows:

$$N_x = \frac{Eh}{(1-\nu^2)}(e_{xx0} + \nu e_{yy0}), \quad N_y = \frac{Eh}{(1-\nu^2)}(\nu e_{xx0} + e_{yy0}),$$
$$N_{xy} = 2Ghe_{xy0}, \tag{8.66}$$

where

$$e_{xx0} = \frac{\partial u}{\partial x} + \frac{1}{2}\left(\frac{\partial w}{\partial x}\right)^2, \quad e_{yy0} = \frac{\partial v}{\partial y} + \frac{1}{2}\left(\frac{\partial w}{\partial y}\right)^2,$$
$$2e_{xy0} = \frac{\partial u}{\partial y} + \frac{\partial v}{\partial x} + \frac{\partial w}{\partial x}\frac{\partial w}{\partial y}, \tag{8.67}$$

while the bending moment–curvature relations are still given by Eqs. (8.33). The equations thus obtained, together with the geometrical boundary conditions (8.28a, b, c, d), formulate the problem of the flat plate in large deflection. It is observed that the stretching and bending couple with each other in the large deflection theory and cannot be treated independently.

Next, let us consider variational formulations of the problem. Following the development similar to that in small displacement theory, we can write the principle of stationary potential energy, from which we obtain the following generalized form, Π_I:

$$\Pi_I = \iint\left\{\frac{Eh}{2(1-\nu^2)}[(e_{xx0}+e_{yy0})^2 + 2(1-\nu)(e_{xy0}^2 - e_{xx0}e_{yy0})]\right.$$
$$+ \frac{D}{2}[(\varkappa_x + \varkappa_y)^2 + 2(1-\nu)(\varkappa_{xy}^2 - \varkappa_x\varkappa_y)] - \bar{p}w$$
$$- \left[e_{xx0} - \left(\frac{\partial u}{\partial x} + \frac{1}{2}\left(\frac{\partial w}{\partial x}\right)^2\right)\right]N_x - \left[e_{yy0} - \left(\frac{\partial v}{\partial y} + \frac{1}{2}\left(\frac{\partial w}{\partial y}\right)^2\right)\right]N_y$$
$$- \left[2e_{xy0} - \left(\frac{\partial u}{\partial y} + \frac{\partial v}{\partial x} + \frac{\partial w}{\partial x}\frac{\partial w}{\partial y}\right)\right]N_{xy} + \left(\varkappa_x - \frac{\partial^2 w}{\partial x^2}\right)M_x$$
$$\left. + \left(\varkappa_y - \frac{\partial^2 w}{\partial y^2}\right)M_y + 2\left(\varkappa_{xy} - \frac{\partial^2 w}{\partial x \partial y}\right)M_{xy}\right\}dx\,dy$$
$$+ (\text{terms on } C_1 \text{ and } C_2). \tag{8.68}$$

We shall eliminate the strain components e_{xx0}, e_{yy0} and e_{xy0} by the use of the stationary conditions with respect to these quantities, while Eqs. (8.52) will be substituted to eliminate $\varkappa_x$, $\varkappa_y$, $\varkappa_{xy}$, M_x, M_y and M_{xy}. These having been eliminated, introduction of the Airy stress function defined by Eqs. (8.46) then allows us to transform the expression (8.68) into,

$$\Pi^* = \iint_{S_m} \Big\{ -\frac{1}{2Eh}\left[\left(\frac{\partial^2 F}{\partial x^2}+\frac{\partial^2 F}{\partial y^2}\right)^2 + 2(1+\nu)\left(\left(\frac{\partial^2 F}{\partial x\,\partial y}\right)^2 - \frac{\partial^2 F}{\partial x^2}\frac{\partial^2 F}{\partial y^2}\right)\right]$$
$$+\frac{D}{2}\left[\left(\frac{\partial^2 w}{\partial x^2}+\frac{\partial^2 w}{\partial y^2}\right)^2 + 2(1-\nu)\left(\left(\frac{\partial^2 w}{\partial x\,\partial y}\right)^2 - \frac{\partial^2 w}{\partial x^2}\frac{\partial^2 w}{\partial y^2}\right)\right]$$
$$+\frac{1}{2}\left[\frac{\partial^2 F}{\partial y^2}\left(\frac{\partial w}{\partial x}\right)^2 + \frac{\partial^2 F}{\partial x^2}\left(\frac{\partial w}{\partial y}\right)^2 - 2\frac{\partial^2 F}{\partial x\,\partial y}\frac{\partial w}{\partial x}\frac{\partial w}{\partial y}\right] - \bar{p}w\Big\}\,dx\,dy$$
$$+ \text{(integrals on } C_1 \text{ and } C_2\text{)}, \tag{8.69}$$

where the functions subject to variation are F and w. Assuming, for the sake of simplicity, that the thickness h is constant, we obtain as the stationary conditions of the functional Π^* the following two equations in S_m:

$$D\Delta\Delta w = \bar{p} + \frac{\partial^2 F}{\partial y^2}\frac{\partial^2 w}{\partial x^2} + \frac{\partial^2 F}{\partial x^2}\frac{\partial^2 w}{\partial y^2} - 2\frac{\partial^2 F}{\partial x\,\partial y}\frac{\partial^2 w}{\partial x\,\partial y}, \tag{8.70}$$

and

$$\Delta\Delta F = Eh\left[\left(\frac{\partial^2 w}{\partial x\,\partial y}\right)^2 - \frac{\partial^2 w}{\partial x^2}\frac{\partial^2 w}{\partial y^2}\right], \tag{8.71}$$

where Δ is the two-dimensional Laplace operator. It is observed that Eq. (8.70) is the equation of equilibrium in the direction of the z-axis, while Eq. (8.71) comprises the condition of compatibility between the strain components e_{xx0}, e_{yy0} and e_{xy0}. Some of papers related to the large deflection theory of flat plates are listed in the bibliography (Refs. 13 through 18).

8.6. Buckling of a Plate

We shall now formulate a buckling problem for the flat plate.(19) It is assumed that before buckling occurs, the plate is subject to a system of two-dimensional stresses $k\sigma_x^{(0)}$, $k\sigma_y^{(0)}$ and $k\tau_{xy}^{(0)}$, where k is a monotonically increasing factor of proportionality and the distribution of the stresses $\sigma_x^{(0)}$, $\sigma_y^{(0)}$ and $\tau_{xy}^{(0)}$ is prescribed. The stress system will be treated as initial stresses which satisfy the following equations of equilibrium and mechanical boundary conditions:

$$\frac{\partial N_x^{(0)}}{\partial x} + \frac{\partial N_{xy}^{(0)}}{\partial y} = 0, \quad \frac{\partial N_{xy}^{(0)}}{\partial x} + \frac{\partial N_y^{(0)}}{\partial y} = 0 \quad \text{in} \quad S_m, \tag{8.72a}$$

$$N_{x\nu}^{(0)} = \bar{N}_{x\nu}^{(0)}, \quad N_{y\nu}^{(0)} = \bar{N}_{y\nu}^{(0)} \quad \text{on} \quad C, \tag{8.72b}$$

where

$$N_x^{(0)} = h\sigma_x^{(0)}, \quad N_y^{(0)} = h\sigma_y^{(0)}, \quad N_{xy}^{(0)} = h\tau_{xy}^{(0)},$$
$$N_{xv}^{(0)} = N_x^{(0)}l + N_{xy}^{(0)}m, \quad N_{yv}^{(0)} = N_{xy}^{(0)}l + N_y^{(0)}m.$$

We shall measure the displacement components u, v and w from the state just prior to the occurrence of buckling, and assume that they are given by Eqs. (8.60). We also assume that the external forces $kN_{xv}^{(0)}$ and $kN_{yv}^{(0)}$ on C_1 vary neither in magnitude nor in direction during buckling, and that the plate is fixed on C_2, requiring that

$$u = 0, \quad v = 0, \quad w = 0, \quad \frac{\partial w}{\partial v} = 0 \quad \text{on} \quad C_2. \qquad (8.73\text{a, b, c, d})$$

Since we are dealing with an initial stress problem, we must employ the nonlinear expressions (8.61) for the strain components in the principle of virtual work, which is written for the present problem as follows:

$$\iiint_V [(k\sigma_x^{(0)} + \sigma_x)\,\delta e_{xx} + (k\sigma_y^{(0)} + \sigma_y)\,\delta e_{yy} + 2(k\tau_{xy}^{(0)} + \tau_{xy})\,\delta e_{xy}]\,dx\,dy\,dz$$
$$- \int_{C_1} (k\bar{N}_{xv}^{(0)}\,\delta u + k\bar{N}_{yv}^{(0)}\,\delta v)\,ds = 0, \qquad (8.74)†$$

where σ_x, σ_y and τ_{xy} are incremental stresses. Since we are interested only in the configuration and critical load for the buckling, higher order terms must be neglected in the principle (8.74). The strains are linearized with respect to the displacement components in the incremental stress–strain relations for the same reason. In the present problem we assume that the incremental stress–strain relations are given by Eqs. (8.2). Then, employing the stress resultants defined by Eqs. (8.17), we find that the incremental stress resultant–displacement relations are given by Eqs. (8.32) and (8.33).

Returning to the principle (8.74) and employing Eqs. (8.72a, b) and (8.73a, b), we find that contributions from the δu and δv terms in Eq. (8.74) provide

$$N_{x,x} + N_{xy,y} = 0, \quad N_{xy,x} + N_{y,y} = 0 \quad \text{in} \quad S_m, \qquad (8.75\text{a})$$

and

$$N_{xv} = 0, \quad N_{yv} = 0 \quad \text{on} \quad C_1. \qquad (8.75\text{b})$$

Combining these equations with Eqs. (8.32) and (8.73a, b), we conclude that $N_x = N_y = N_{xy} = 0$ throughout the plate. Consequently, we may reduce the principle to the following form:

$$-\iint_{S_m} \left[M_x \frac{\partial^2 \delta w}{\partial x^2} + M_y \frac{\partial^2 \delta w}{\partial y^2} + 2M_{xy} \frac{\partial^2 \delta w}{\partial x\,\partial y}\right] dx\,dy$$
$$+ k\iint_{S_m} \left[\left(N_x^{(0)} \frac{\partial w}{\partial x} + N_{xy}^{(0)} \frac{\partial w}{\partial y}\right) \frac{\partial\,\delta w}{\partial x} + \left(N_{xy}^{(0)} \frac{\partial w}{\partial x} + N_y^{(0)} \frac{\partial w}{\partial y}\right) \frac{\partial \delta w}{\partial y}\right] dx\,dy = 0, \qquad (8.76)$$

† See Eq. (5.5).

which, through a familiar process, provides the equations of equilibrium in S_m,

$$\frac{\partial^2 M_x}{\partial x^2} + 2\frac{\partial^2 M_{xy}}{\partial x\,\partial y} + \frac{\partial^2 M_y}{\partial y^2}$$
$$+ k\left[\frac{\partial}{\partial x}\left(N_x^{(0)}\frac{\partial w}{\partial x} + N_{xy}^{(0)}\frac{\partial w}{\partial y}\right) + \frac{\partial}{\partial y}\left(N_{xy}^{(0)}\frac{\partial w}{\partial x} + N_y^{(0)}\frac{\partial w}{\partial y}\right)\right] = 0, \quad (8.77)$$

and the mechanical boundary conditions on C_1,

$$\tilde{Q}_x l + \tilde{Q}_y m + k\left[\bar{N}_{xv}^{(0)}\frac{\partial w}{\partial x} + \bar{N}_{yv}^{(0)}\frac{\partial w}{\partial y}\right] + \frac{\partial M_{vs}}{\partial s} = 0, \quad M_v = 0. \quad (8.78)$$

Equations (8.77) and (8.78), together with Eqs. (8.33) and (8.73c, d), formulate the buckling problem under consideration. In the case of uniform bending rigidity these equations may be written as follows:

$$D\Delta\Delta w = k\left[\frac{\partial}{\partial x}\left(N_x^{(0)}\frac{\partial w}{\partial x} + N_{xy}^{(0)}\frac{\partial w}{\partial y}\right) + \frac{\partial}{\partial y}\left(N_{xy}^{(0)}\frac{\partial w}{\partial x} + N_y^{(0)}\frac{\partial w}{\partial y}\right)\right] \quad \text{in } S_m, \quad (8.79)$$

$$-D\left\{\frac{\partial}{\partial v}(\Delta w) + (1-\nu)\frac{\partial}{\partial s}\left[\frac{\partial}{\partial v}\left(\frac{\partial w}{\partial s}\right)\right]\right\} + k\left(\bar{N}_{xv}^{(0)}\frac{\partial w}{\partial x} + \bar{N}_{yv}^{(0)}\frac{\partial w}{\partial y}\right) = 0,$$
$$-D\left[\frac{\partial^2 w}{\partial v^2} + \nu\left(\frac{\partial^2 w}{\partial s^2} + \frac{1}{\varrho}\frac{\partial w}{\partial v}\right)\right] = 0 \quad \text{on } C_1, \quad (8.80)$$

and

$$w = 0, \quad \frac{\partial w}{\partial v} = 0 \quad \text{on } C_2. \quad (8.81)$$

Consequently, the buckling configurations and critical loads can be determined by solving the differential equation (8.79) under the boundary conditions (8.80) and (8.81).

When the principle (8.76) is combined with Eqs. (8.33), we obtain the principle of stationary potential energy for the buckling problem as follows:

$$\delta\Pi = 0, \quad (8.82)$$

where

$$\Pi = \frac{1}{2}\iint_{S_m} D\left\{\left(\frac{\partial^2 w}{\partial x^2} + \frac{\partial^2 w}{\partial y^2}\right)^2 + 2(1-\nu)\left[\left(\frac{\partial^2 w}{\partial x\,\partial y}\right)^2 - \frac{\partial^2 w}{\partial x^2}\frac{\partial^2 w}{\partial y^2}\right]\right\} dx\,dy$$
$$+ \frac{1}{2}k\iint_{S_m}\left[N_x^{(0)}\left(\frac{\partial w}{\partial x}\right)^2 + N_y^{(0)}\left(\frac{\partial w}{\partial y}\right)^2 + 2N_{xy}^{(0)}\frac{\partial w}{\partial x}\frac{\partial w}{\partial y}\right] dx\,dy, \quad (8.83)$$

and the independent function subject to variation is w under the subsidiary conditions (8.73c, d). By employing the auxiliary functions defined by Eqs. (8.52), we may generalize the principle (8.82) in a manner similar to the usual development, obtaining the principle of stationary complementary energy. Due to the limited space available, the derivation will not be shown here.

It is noted here that the principle (8.82) is equivalent to finding, among admissible w, that function which makes the quotient defined by

$$k = -\frac{\frac{1}{2}\iint_{S_m} D\left\{\left(\frac{\partial^2 w}{\partial x^2} + \frac{\partial^2 w}{\partial y^2}\right)^2 + 2(1-\nu)\left[\left(\frac{\partial^2 w}{\partial x\, \partial y}\right)^2 - \frac{\partial^2 w}{\partial x^2}\frac{\partial^2 w}{\partial y^2}\right]\right\} dx\, dy}{\frac{1}{2}\iint_{S_m}\left[N_x^{(0)}\left(\frac{\partial w}{\partial x}\right)^2 + N_y^{(0)}\left(\frac{\partial w}{\partial y}\right)^2 + 2N_{xy}^{(0)}\frac{\partial w}{\partial x}\frac{\partial w}{\partial y}\right] dx\, dy} \tag{8.84}$$

stationary.

8.7. Thermal Stresses in a Plate†

We shall now consider a problem of thermal stresses in a flat plate subject to a temperature distribution $\theta(x, y, z)$. The temperature θ is measured from a reference state of uniform temperature distribution, in which the plate has neither stresses nor strains. Confining the problem to the small displacement theory of elasticity and introducing results from Appendix B. we may employ the following stress–strain relations:

$$\begin{aligned} \sigma_x &= \frac{E}{(1-\nu^2)}(\varepsilon_x + \nu\varepsilon_y) - \frac{Ee^\theta}{(1-\nu)}, \\ \sigma_y &= \frac{E}{(1-\nu^2)}(\nu\varepsilon_x + \varepsilon_y) - \frac{Ee^\theta}{(1-\nu)}, \\ \tau_{xy} &= G\gamma_{xy}, \quad \tau_{xz} = G\gamma_{xz}, \quad \tau_{yz} = G\gamma_{yz}, \end{aligned} \tag{8.85}$$

where e^θ denotes the thermal strain. We assume that the displacement components measured from the reference state may be expressed by Eqs. (8.14). For the sake of simplicity, the boundary S is prescribed to be fixed, namely

$$u = 0, \quad v = 0, \quad w = 0, \quad \frac{\partial w}{\partial \nu} = 0 \quad \text{on} \quad C, \qquad \text{(8.86a, b, c, d)}$$

while the surfaces $z = \pm h/2$ are assumed to be traction free.

The derivation of governing equations for the thermal stress problem proceeds in a manner similar to the development in Section 8.2, the effect of thermal expansion having been accounted for by including e^θ in the stress–strain relations. Combining Eqs. (8.85) with Eqs. (8.17), we have

$$\begin{aligned} N_x &= \frac{Eh}{(1-\nu^2)}\left(\frac{\partial u}{\partial x} + \nu\frac{\partial v}{\partial y}\right) - \frac{N_T}{(1-\nu)}, \\ N_y &= \frac{Eh}{(1-\nu^2)}\left(\nu\frac{\partial u}{\partial x} + \frac{\partial v}{\partial y}\right) - \frac{N_T}{(1-\nu)}, \\ N_{xy} &= Gh\left(\frac{\partial u}{\partial y} + \frac{\partial v}{\partial x}\right), \end{aligned} \tag{8.87}$$

† Ref. 20.

and

$$M_x = -D\left(\frac{\partial^2 w}{\partial x^2} + \nu \frac{\partial^2 w}{\partial y^2}\right) - \frac{M_T}{(1-\nu)},$$

$$M_y = -D\left(\nu \frac{\partial^2 w}{\partial x^2} + \frac{\partial^2 w}{\partial y^2}\right) - \frac{M_T}{(1-\nu)}, \qquad (8.88)$$

$$M_{xy} = -D(1-\nu)\frac{\partial^2 w}{\partial x\, \partial y},$$

where

$$N_T = \int_{-h/2}^{h/2} E e^{\theta}\, dz, \quad M_T = \int_{-h/2}^{h/2} E e^{\theta} z\, dz. \qquad (8.89)$$

These relations show that thermal stretching and bending are decoupled in small displacement theory in which the displacement components are given by Eqs. (8.14).

Let us now consider variational formulations of the problem. With the aid of Eqs. (8.15) and some results from Appendix B, we may express the strain energy of the plate as follows:

$$U = \frac{1}{2}\iint_{S_m}\left\{\frac{Eh}{(1-\nu^2)}\left(\frac{\partial u}{\partial x} + \frac{\partial v}{\partial y}\right)^2 + Gh\left[\left(\frac{\partial u}{\partial y} + \frac{\partial v}{\partial x}\right)^2 - 4\frac{\partial u}{\partial x}\frac{\partial v}{\partial y}\right]\right.$$
$$\left. - \frac{2N_T}{(1-\nu)}\left(\frac{\partial u}{\partial x} + \frac{\partial v}{\partial y}\right)\right\} dx\, dy + \frac{1}{2}\iint_{S_m}\left\{D\left[\left(\frac{\partial^2 w}{\partial x^2} + \frac{\partial^2 w}{\partial y^2}\right)^2\right.\right.$$
$$\left.\left. + 2(1-\nu)\left(\left(\frac{\partial^2 w}{\partial x\, \partial y}\right)^2 - \frac{\partial^2 w}{\partial x^2}\frac{\partial^2 w}{\partial y^2}\right)\right] + \frac{2M_T}{(1-\nu)}\left(\frac{\partial^2 w}{\partial x^2} + \frac{\partial^2 w}{\partial y^2}\right)\right\} dx\, dy. \qquad (8.90)$$

Consequently, the total potential energy for thermal stretching of the plate is given by

$$\Pi = \iint_{S_m}\left\{\frac{Eh}{2(1-\nu^2)}\left(\frac{\partial u}{\partial x} + \frac{\partial v}{\partial y}\right)^2 + \frac{Gh}{2}\left[\left(\frac{\partial u}{\partial y} + \frac{\partial v}{\partial x}\right)^2 - 4\frac{\partial u}{\partial x}\frac{\partial v}{\partial y}\right]\right.$$
$$\left. - \frac{N_T}{(1-\nu)}\left(\frac{\partial u}{\partial x} + \frac{\partial v}{\partial y}\right)\right\} dx\, dy, \qquad (8.91)$$

where the functions subject to variation are u and v under the subsidiary conditions (8.86a, b).

The functional (8.91) is generalized in a manner similar to the development in Section 8.3, and we obtain the following functional for the principle of stationary complementary energy:

$$\Pi_C = \iint_{S_m} \frac{1}{2Eh}\{[(N_x + N_y)^2 + 2(1+\nu)(N_{xy}^2 - N_x N_y)]$$
$$+ 2N_T(N_x + N_y)\}\, dx\, dy, \qquad (8.92)$$

where the functions subject to variation are N_x, N_y and N_{xy} under the subsidiary conditions (8.43). Introduction of the Airy stress function defined by Eqs. (8.46) reduces the functional further to

$$\Pi_C^* = \iint_{S_m} \frac{1}{2Eh} \left\{ \left(\frac{\partial^2 F}{\partial x^2} + \frac{\partial^2 F}{\partial y^2} \right)^2 + 2(1+\nu) \left[\left(\frac{\partial^2 F}{\partial x \, \partial y} \right)^2 - \frac{\partial^2 F}{\partial x^2} \frac{\partial^2 F}{\partial y^2} \right] \right.$$
$$\left. + 2N_T \left(\frac{\partial^2 F}{\partial x^2} + \frac{\partial^2 F}{\partial y^2} \right) \right\} dx \, dy, \tag{8.93}$$

where the only function subject to variation is $F(x, y)$, upon which no subsidiary conditions are imposed.

Thus, far, the variational formulations have been made for the thermal stretching. Formulations can also be made for the thermal bending in a manner similar to Section 8.4, by employing the latter half of the right hand side of Eq. (8.90). Extensions of the above formulations to thermal stress problems of plates in large deflection may be made in a manner similar to the developments in Section 8.5. These variational principles have been used for obtaining approximate solutions in combination with the Rayleigh–Ritz method.[21, 22] Thermal stresses in a plate are responsible for phenomena such as thermal buckling or variation of stiffeness and vibration frequencies of the plate.[23, 24]

8.8. A Thin Plate Theory Including the Effect of Transverse Shear Deformation

So far, theories of a thin plate have been established on the Kirchhoff hypothesis. In this section, we shall consider a small displacement theory of a thin plate including the effect of transverse shear deformation. In making this extension, we are forced to abandon the hypothesis; its alternative must be chosen judiciously.

Since the displacement vector $\mathbf{u}$ is a function of (x, y, z), we may expand it in power series of z:

$$\mathbf{u}(x, y, z) = \mathbf{u}(x, y, 0) + \left(\frac{\partial \mathbf{u}}{\partial z} \right)_{z=0} z + \frac{1}{2!} \left(\frac{\partial^2 \mathbf{u}}{\partial z^2} \right)_{z=0} z^2 + \cdots. \tag{8.94}$$

Therefore, one of the simplest expressions for displacements to include the effect of transverse shear deformation may be given by retaining the first two terms only:

$$\mathbf{u} = \mathbf{u}_0 + z\mathbf{u}_1, \tag{8.95}$$

where components of $\mathbf{u}_1$ are defined by

$$\mathbf{u}_1 = u_1 \mathbf{i}_1 + v_1 \mathbf{i}_2 + w_1 \mathbf{i}_3, \tag{8.96}$$

and u_1, v_1, w_1 are functions of (x, y) only. The degree of freedom implied by Eq. (8.95) is six, namely, u, v, w, u_1, v_1 and w_1. However, if we continue to use the assumption (8.1), and employ Eqs. (8.3) as the stress–strain relations, we may take

$$2e_{zz} = u_1^2 + v_1^2 + (1 + w_1)^2 - 1 = 0 \tag{8.97}$$

as an additional geometrical constraint to reduce the degree of freedom to five. Equations (8.95) and (8.97) state that the linear filaments perpendicular to the undeformed middle surface remain straight and suffer no strains although they are no longer perpendicular to the deformed middle surface.

Since we are interested in a small displacement theory,† Eq. (8.97) is linearized with respect to the displacements to yield

$$w_1 = 0. \tag{8.98}$$

Consequently, we observe that the most natural and simplest expression to include the effect of transverse shear deformation is to assume that

$$U = u + zu_1, \quad V = v + zv_1, \quad W = w. \tag{8.99}$$

In a manner similar to the development in Section 8.2, it can be shown that the functions u and v are related to the stretching of the plate, while the functions u_1, v_1 and w are related to the bending of the plate, and these two problems can be treated separately. Therefore, we confine our subsequent interest to bending only by assuming that

$$U = zu_1, \quad V = zv_1, \quad W = w, \tag{8.100}$$

and we obtain

$$\begin{aligned} \varepsilon_x &= z\frac{\partial u_1}{\partial x}, \quad \varepsilon_y = z\frac{\partial v_1}{\partial y}, \quad \varepsilon_z = 0, \\ \gamma_{xy} &= z\left(\frac{\partial v_1}{\partial x} + \frac{\partial u_1}{\partial y}\right), \\ \gamma_{xz} &= \frac{\partial w}{\partial x} + u_1, \quad \gamma_{yz} = \frac{\partial w}{\partial y} + v_1. \end{aligned} \tag{8.101}$$

It is seen from Eqs. (8.15) and (8.101) that the Kirchhoff hypothesis imposes the constraint conditions,

$$u_1 = -\frac{\partial w}{\partial x}, \quad v_1 = -\frac{\partial w}{\partial y}. \tag{8.102}$$

We shall consider a dynamical problem defined in a manner similar to the presentation in Section 8.2, except that the external forces and geometrical boundary conditions are now time-dependent. The form of the principle of virtual work for this dynamical problem is suggested by Eq. (5.81) to be

$$\begin{aligned} \int_{t_1}^{t_2} \Big\{ &\iiint_V (\sigma_x\,\delta\varepsilon_x + \sigma_y\,\delta\varepsilon_y + \tau_{xy}\,\delta\gamma_{xy} + \tau_{xz}\,\delta\gamma_{xz} + \tau_{yz}\,\delta\gamma_{yz})\,dx\,dy\,dz \\ &- \delta\iiint_V \tfrac{1}{2}(\dot{U}^2 + \dot{V}^2 + \dot{W}^2)\,\varrho\,dx\,dy\,dz - \iint_{S_m} \bar{p}\,\delta w\,dx\,dy \\ &- \iint_S [\bar{F}_x\,\delta U + \bar{F}_y\,\delta V + \bar{F}_z\,\delta W]\,ds\,dz \Big\}\,dt = 0, \end{aligned} \tag{8.103}$$

† See Ref. 12 for a finite displacement theory in which Eq. (8.95) is employed as an expression of displacements.

where Eqs. (8.100) and (8.101) have been substituted. Here we shall introduce new resultants defined as follows:

$$Q_x = \int_{-h/2}^{h/2} \tau_{xz}\, dz, \quad Q_y = \int_{-h/2}^{h/2} \tau_{yz}\, dz, \tag{8.104}$$

$$m = \int_{-h/2}^{h/2} \varrho\, dz, \quad I_m = \int_{-h/2}^{h/2} \varrho\, z^2\, dz. \tag{8.105}$$

The quantities defined by Eqs. (8.104) are shearing forces per unit length acting in the direction of the z-axis.† The quantities defined by Eqs. (8.105) are the mass and mass moment of inertia per unit area of the middle surface. With these preliminaries and some calculation including integrations by parts, Eq. (8.103) is finally reduced to

$$\begin{aligned}\int_{t_1}^{t_2} \Big\{ \iint_{S_m} & [(I_m \ddot{u}_1 - M_{x,x} - M_{xy,y} + Q_x)\,\delta u_1 \\ & + (I_m \ddot{v}_1 - M_{xy,x} - M_{y,y} + Q_y)\,\delta v_1 \\ & + (m\ddot{w} - Q_{x,x} - Q_{y,y} - \bar{p})\,\delta w]\, dx\, dy \\ & + \int_{C_1} [(M_{x\nu} - \bar{M}_{x\nu})\,\delta u_1 + (M_{y\nu} - \bar{M}_{y\nu})\,\delta v_1 + (Q_x l + Q_y m - \bar{V}_z)\,\delta w]\, ds \\ & + \int_{C_2} [M_{x\nu}\,\delta u_1 + M_{y\nu}\,\delta v_1 + (Q_x l + Q_y m)\,\delta w]\, ds \Big\}\, dt = 0. \end{aligned} \tag{8.106}$$

Thus, the principle provides the equations of motion,

$$I_m \ddot{u}_1 = \frac{\partial M_x}{\partial x} + \frac{\partial M_{xy}}{\partial y} - Q_x, \tag{8.107}$$

$$I_m \ddot{v}_1 = \frac{\partial M_{xy}}{\partial x} + \frac{\partial M_y}{\partial y} - Q_y, \tag{8.108}$$

$$m\ddot{w} = \frac{\partial Q_x}{\partial x} + \frac{\partial Q_y}{\partial y} + \bar{p}, \tag{8.109}$$

and the mechanical boundary conditions on C_1,

$$M_{x\nu} = \bar{M}_{x\nu}, \quad M_{y\nu} = \bar{M}_{y\nu}, \quad Q_x l + Q_y m = \bar{V}_z, \tag{8.110}$$

while it suggests that the geometrical boundary conditions on C_2 can be specified approximately as

$$u_1 = \bar{u}_1, \quad v_1 = \bar{v}_1, \quad w = \bar{w}. \tag{8.111}$$

† Thus, the shearing forces Q_x and Q_y appear as independent quantities in the thin plate theory including the effect of transverse shear deformation. Compare Eqs. (8.104), (8.107) and (8.108) with Eqs. (8.22).

The stress resultant-displacement relations are obtained from Eqs. (8.2), (8.17), (8.101) and (8.104) as

$$M_x = D\left(\frac{\partial u_1}{\partial x} + \nu\frac{\partial v_1}{\partial y}\right), \quad M_y = D\left(\nu\frac{\partial u_1}{\partial x} + \frac{\partial v_1}{\partial y}\right),$$

$$M_{xy} = \frac{1}{2}D(1-\nu)\left(\frac{\partial v_1}{\partial x} + \frac{\partial u_1}{\partial y}\right), \tag{8.112}$$

and

$$Q_x = Gkh\left(\frac{\partial w}{\partial x} + u_1\right), \quad Q_y = Gkh\left(\frac{\partial w}{\partial y} + v_1\right), \tag{8.113}$$

where $k = 1$. The factor k in Eqs. (8.113) has been included to account for the nonuniformity of the shearing strains over the cross section. In Appendix D, a theory of a thin plate based on the principle of minimum complementary energy is introduced following E. Reissner's papers† and the value of k is found to be 5/6 for the isotropic plate. On the other hand, from the result obtained in a vibrational problem of a thin plate, Mindlin(29) suggests that $k = \pi^2/12$, which is very close to 5/6 obtained from the formulation based on the complementary energy principle. Introducing Eqs. (8.112) and (8.113) into Eqs. (8.107) through (8.109), we obtain three simultaneous differential equations in terms of u_1, v_1 and w. Consequently, the dynamical problem is reduced to solving these differential equations under the boundary conditions (8.110) and (8.111).

It is seen from the above formulation that we have three mechanical boundary conditions (8.110) on C_1 and three geometrical boundary conditions (8.111) on C_2 in the thin plate theory including the effect of transverse shear deformation. We have replaced through integrations by parts the action of $\bar{M}_{\nu s}$ and $M_{\nu s}$ by that of $\bar{V}_z$ and V_z respectively in the thin plate theory under the Kirchhoff hypothesis. However, such replacements are no longer necessary in the thin plate theory including the effect of transverse shear deformation.

8.9. Thin Shallow Shell

In the present section we shall consider a nonlinear theory of thin, shallow shells proposed by K. Marguerre.(30) Let the rectangular Cartesian coordinates fixed in space be (x, y, z) and let the middle surface of the thin, shallow shell be represented by

$$z = z(x, y) \tag{8.114}$$

as shown in Fig. 8.5. The position vector of an arbitrary point $P_0^{(0)}$ in the undeformed middle surface is given by

$$\mathbf{r}_0^{(0)} = x\mathbf{i}_1 + y\mathbf{i}_2 + z(x, y)\,\mathbf{i}_3, \tag{8.115}$$

† Refs. 25 through 28.

where $\mathbf{i}_1$, $\mathbf{i}_2$ and $\mathbf{i}_3$ are unit vectors in the directions of the x-, y- and z-axes, respectively. Then, the position vector of an arbitrary point $P^{(0)}$ outside the middle surface before deformation may be given by

$$\mathbf{r}^{(0)} = \mathbf{r}_0^{(0)} + \zeta\mathbf{n}^{(0)}, \tag{8.116}$$

where $\mathbf{n}^{(0)}$ is a unit vector drawn perpendicular to the undeformed middle surface and is calculated by

$$\mathbf{n}^{(0)} = \frac{\partial \mathbf{r}_0^{(0)}}{\partial x} \times \frac{\partial \mathbf{r}_0^{(0)}}{\partial y} \Bigg/ \left| \frac{\partial \mathbf{r}_0^{(0)}}{\partial x} \times \frac{\partial \mathbf{r}_0^{(0)}}{\partial y} \right|, \tag{8.117}$$

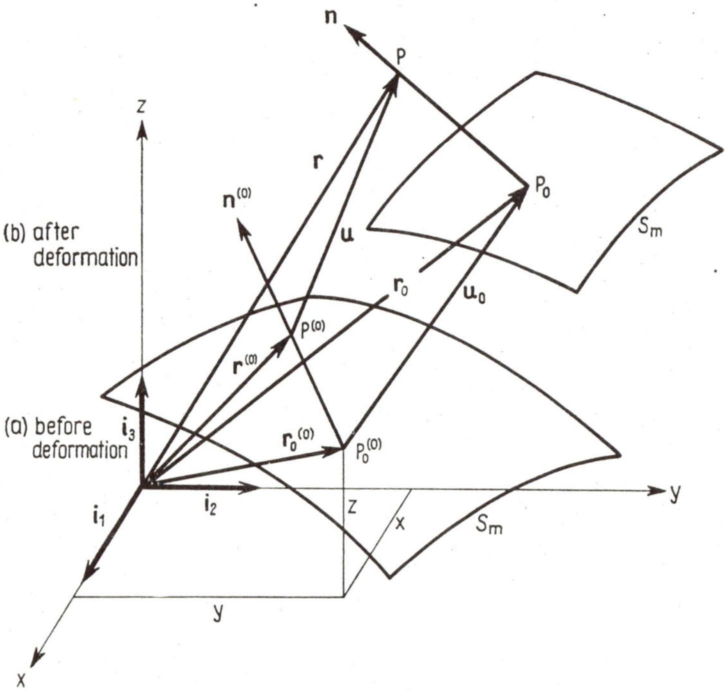

FIG. 8.5. A plate with small initial deflection.

and where ζ is the distance from the middle surface to the point. Equation (8.116) suggests that an arbitrary point in the shell can be specified by the coordinates (x, y, ζ), which form a curvilinear coordinate system. Consequently, by taking $\alpha^1 = x$, $\alpha^2 = y$, $\alpha^3 = \zeta$, we may apply the formulations developed in Chapter 4.

The shell is now assumed to be subject to deformation and the position vectors of the two points P_0 and P after deformation are represented by

$$\mathbf{r}_0 = \mathbf{r}_0^{(0)} + \mathbf{u}_0 \tag{8.118}$$

and

$$\mathbf{r} = \mathbf{r}^{(0)} + \mathbf{u}, \tag{8.119}$$

respectively, where $\mathbf{u}_0$ and $\mathbf{u}$ are displacement vectors, their components being defined by

$$\mathbf{u}_0 = u\mathbf{i}_1 + v\mathbf{i}_2 + w\mathbf{i}_3 \tag{8.120}$$

and

$$\mathbf{u} = U\mathbf{i}_1 + V\mathbf{i}_2 + W\mathbf{i}_3 \tag{8.121}$$

respectively, where u, v and w are functions of x and y only.

In subsequent formulations, we shall employ the Kirchhoff hypothesis, under which the position vector $\mathbf{r}$ is related to $\mathbf{r}_0$ by

$$\mathbf{r} = \mathbf{r}_0 + \zeta\mathbf{n}, \tag{8.122}$$

where $\mathbf{n}$ is the unit normal to the deformed middle surface and is given by

$$\mathbf{n} = \frac{\partial \mathbf{r}_0}{\partial x} \times \frac{\partial \mathbf{r}_0}{\partial y} \Big/ \left| \frac{\partial \mathbf{r}_0}{\partial x} \times \frac{\partial \mathbf{r}_0}{\partial y} \right|. \tag{8.123}$$

Combining Eqs. (8.116), (8.118), (8.119) and (8.122), we obtain

$$\mathbf{u} = \mathbf{u}_0 + \zeta(\mathbf{n} - \mathbf{n}^{(0)}). \tag{8.124}$$

Then, the strain tensors $f_{\lambda\mu}$ can be obtained in terms of the displacements u, v and w by the use of Eqs. (4.36), (8.116) and (8.122).

It is assumed hereafter that the shell is shallow and thin to the extent that

$$\left(\frac{\partial z}{\partial x}\right)^2, \quad \left(\frac{\partial z}{\partial y}\right)^2, \quad \left|\frac{\partial z}{\partial x}\frac{\partial z}{\partial y}\right| \ll 1, \tag{8.125}$$

$$\left|\zeta \frac{\partial^2 z}{\partial x^2}\right|, \quad \left|\zeta \frac{\partial^2 z}{\partial y^2}\right|, \quad \left|\zeta \frac{\partial^2 z}{\partial x \, \partial y}\right| \ll 1, \tag{8.126}$$

and terms containing ζ^2 can be neglected. Then, we have

$$\mathbf{n}^{(0)} = -\frac{\partial z}{\partial x}\mathbf{i}_1 - \frac{\partial z}{\partial y}\mathbf{i}_2 + \mathbf{i}_3, \tag{8.127}$$

and we observe that the (x, y, ζ) coordinate system can be taken approximately to be locally rectangular Cartesian. In addition to the above assumption, restrictions on the orders of magnitude of the displacements of the middle surface are introduced. It is assumed that the initial deflection $z(x, y)$ and the displacement $w(x, y)$ are of the same order of magnitude. Then, we have the following approximate expressions:†

$$\mathbf{n} = -\left(\frac{\partial z}{\partial x} + \frac{\partial w}{\partial x}\right)\mathbf{i}_1 - \left(\frac{\partial z}{\partial y} + \frac{\partial w}{\partial x}\right)\mathbf{i}_2 + \mathbf{i}_3, \tag{8.128}$$

$$\mathbf{u} = \left(u - \frac{\partial w}{\partial x}\zeta\right)\mathbf{i}_1 + \left(v - \frac{\partial w}{\partial y}\zeta\right)\mathbf{i}_2 + w\mathbf{i}_3, \tag{8.129}$$

$$f_{11} = \hat{e}_{xx0} - \zeta\varkappa_x, \quad f_{22} = \hat{e}_{yy0} - \zeta\varkappa_y,$$
$$f_{12} = \hat{e}_{xy0} - \zeta\varkappa_{xy}, \tag{8.130}$$

† These equations may be derived by assuming that $u_{,x} \sim u_{,y} \sim v_{,x} \sim v_{,y} \sim (w_{,x})^2 \sim (w_{,y})^2 \ll 1$ and terms containing ζ^2 may be neglected.[31]

where

$$\hat{e}_{xx0} = \frac{\partial u}{\partial x} + \frac{\partial z}{\partial x}\frac{\partial w}{\partial x} + \frac{1}{2}\left(\frac{\partial w}{\partial x}\right)^2,$$

$$\hat{e}_{yy0} = \frac{\partial v}{\partial y} + \frac{\partial z}{\partial y}\frac{\partial w}{\partial y} + \frac{1}{2}\left(\frac{\partial w}{\partial y}\right)^2, \tag{8.131}$$

$$2\hat{e}_{xy0} = \frac{\partial u}{\partial y} + \frac{\partial v}{\partial x} + \frac{\partial z}{\partial x}\frac{\partial w}{\partial y} + \frac{\partial z}{\partial y}\frac{\partial w}{\partial x} + \frac{\partial w}{\partial x}\frac{\partial w}{\partial y},$$

and

$$\varkappa_x = \frac{\partial^2 w}{\partial x^2}, \quad \varkappa_y = \frac{\partial^2 w}{\partial y^2}, \quad \varkappa_{xy} = \frac{\partial^2 w}{\partial x\,\partial y}, \tag{8.132}$$

higher order terms being neglected.

We may define the stress tensor $\tau^{\lambda\mu}$ with respect to the (x, y, ζ) coordinates in a manner similar to the definition in Chapter 4, and employ Eqs. (4.74) and (4.77) as the stress–strain relations. However, since the shell is thin and shallow, the transverse normal stress σ_ζ may be neglected and the (x, y, ζ) coordinate system can be considered approximately locally rectangular Cartesian. Consequently, we may have

$$\tau^{11} = \frac{E}{(1-\nu^2)}(f_{11} + \nu f_{22}), \quad \tau^{22} = \frac{E}{(1-\nu^2)}(\nu f_{11} + f_{22}),$$

$$\tau^{12} = 2Gf_{12}. \tag{8.133}$$

A problem of a thin shallow shell is stated as follows. External forces are prescribed per unit area of the (x, y) plane and their components in the directions of the x-, y- and z-axes are denoted by $\bar{X}$, $\bar{Y}$ and $\bar{Z}$, respectively. The side boundary generated by the envelope of normals drawn perpendicular to the middle surface divides into two parts, S_1 and S_2. External forces per unit area are prescribed on S_1, with components $\bar{F}_x$, $\bar{F}_y$ and $\bar{F}_z$, and geometrical boundary conditions are prescribed on S_2.

With the above preliminaries, we have the following expression for the principle of virtual work:

$$\iiint_V [\tau^{11}\,\delta f_{11} + \tau^{22}\,\delta f_{22} + 2\tau^{12}\,\delta f_{12}]\,dx\,dy\,d\zeta$$

$$- \iint_{S_m} [\bar{X}\,\delta u + \bar{Y}\,\delta v + \bar{Z}\,\delta w]\,dx\,dy$$

$$- \iint_{S_1} \left[\bar{F}_x\left(\delta u - \zeta\frac{\partial \delta w}{\partial x}\right) + \bar{F}_y\left(\delta v - \zeta\frac{\partial \delta w}{\partial y}\right) + \bar{F}_z\,\delta w\right] ds\,d\zeta = 0. \quad (8.134)†$$

† See Eqs. (4.80).

Here we introduce stress resultants defined as follows:

$$N_x = \int \tau^{11}\,\delta\zeta, \quad N_y = \int \tau^{22}\,d\zeta, \quad N_{xy} = \int \tau^{12}\,d\zeta,$$
$$M_x = \int \tau^{11}\zeta\,d\zeta, \quad M_y = \int \tau^{22}\zeta\,d\zeta, \quad M_{xy} = \int \tau^{12}\zeta\,d\zeta, \tag{8.135}$$
$$N_{x\nu} = N_x l + N_{xy} m, \quad N_{y\nu} = N_{xy} l + N_y m, \tag{8.136}$$
$$M_\nu = M_x l^2 + 2M_{xy} lm + M_y m^2,$$
$$M_{\nu s} = -(M_x - M_y)\,lm + M_{xy}(l^2 - m^2), \tag{8.137}$$
$$\bar{N}_{x\nu} = \int \bar{F}_x\,d\zeta, \quad \bar{N}_{y\nu} = \int \bar{F}_y\,d\zeta, \quad \bar{V}_z = \int \bar{F}_z\,d\zeta,$$
$$\bar{M}_{x\nu} = \int \bar{F}_x\zeta\,d\zeta, \quad \bar{M}_{y\nu} = \int \bar{F}_y\zeta\,d\zeta, \tag{8.138}$$
$$\bar{M}_\nu = \bar{M}_{x\nu} l + \bar{M}_{y\nu} m, \quad \bar{M}_{\nu s} = -\bar{M}_{x\nu} m + \bar{M}_{y\nu} l, \tag{8.139}$$

where integrations extend through the thickness of the shell. Following the usual procedure, we find that the principle (8.134) provides the equations of equilibrium,

$$\frac{\partial N_x}{\partial x} + \frac{\partial N_{xy}}{\partial y} + \bar{X} = 0, \quad \frac{\partial N_{xy}}{\partial x} + \frac{\partial N_y}{\partial y} + \bar{Y} = 0,$$
$$\frac{\partial^2 M_x}{\partial x^2} + 2\frac{\partial^2 M_{xy}}{\partial x\,\partial y} + \frac{\partial^2 M_y}{\partial y^2} + \frac{\partial}{\partial x}\left[\left(\frac{\partial z}{\partial x} + \frac{\partial w}{\partial x}\right)N_x + \left(\frac{\partial z}{\partial y} + \frac{\partial w}{\partial y}\right)N_{xy}\right]$$
$$+ \frac{\partial}{\partial y}\left[\left(\frac{\partial z}{\partial x} + \frac{\partial w}{\partial x}\right)N_{xy} + \left(\frac{\partial z}{\partial y} + \frac{\partial w}{\partial y}\right)N_y\right] + \bar{Z} = 0, \tag{8.140}$$

and the mechanical boundary conditions of C_1,

$$N_{x\nu} = \bar{N}_{x\nu}, \quad N_{y\nu} = \bar{N}_{y\nu}, \quad \left(\frac{\partial M_x}{\partial x} + \frac{\partial M_{xy}}{\partial y}\right)l + \left(\frac{\partial M_{xy}}{\partial x} + \frac{\partial M_y}{\partial y}\right)m$$
$$+ \left(\frac{\partial z}{\partial x} + \frac{\partial w}{\partial x}\right)N_{x\nu} + \left(\frac{\partial z}{\partial y} + \frac{\partial w}{\partial y}\right)N_{y\nu} + \frac{\partial M_{\nu s}}{\partial s} = \bar{V}_z + \frac{\partial \bar{M}_{\nu s}}{\partial s}, \quad M_\nu = \bar{M}_\nu, \tag{8.141}$$

while it suggests that the geometrical boundary conditions on C_2 are given by

$$u = \bar{u}, \quad v = \bar{v}, \quad w = \bar{w}, \quad \frac{\partial w}{\partial \nu} = \overline{\frac{\partial w}{\partial \nu}}. \tag{8.142}$$

Combining Eqs. (8.130), (8.133) and (8.135), we obtain the following relations between the stress resultants and strains:

$$N_x = \frac{Eh}{(1-\nu^2)}(\hat{e}_{xx0} + \nu\hat{e}_{yy0}), \quad N_y = \frac{Eh}{(1-\nu^2)}(\nu\hat{e}_{xx0} + \hat{e}_{yy0}),$$
$$N_{xy} = 2Gh\hat{e}_{xy0}, \tag{8.143}$$

and

$$M_x = -D(\varkappa_x + \nu\varkappa_y), \quad M_y = -D(\nu\varkappa_x + \varkappa_y),$$
$$M_{xy} = -D(1-\nu)\,\varkappa_{xy}. \tag{8.144}$$

These equations formulate the nonlinear theory of the thin, shallow shell. It is observed that the total potential energy for the present problem is given by

$$\begin{aligned}\Pi = \frac{1}{2}\iint_{S_m}\Big\{&\frac{Eh}{(1-\nu^2)}[(\hat{e}_{xx0}+\hat{e}_{yy0})^2 + 2(1-\nu)(\hat{e}_{xy0}^2 - \hat{e}_{xx0}\hat{e}_{yy0})]\\ &+ D[(\varkappa_x+\varkappa_y)^2 + 2(1-\nu)(\varkappa_{xy}^2 - \varkappa_x\varkappa_y)]\Big\}\,dx\,dy\\ &- \iint_{S_m}(\bar{X}u + \bar{Y}v + \bar{Z}w)\,dx\,dy\\ &- \int_{C_1}\left[\bar{N}_{x\nu}u + \bar{N}_{y\nu}v + \left(\bar{V}_z + \frac{\partial \bar{M}_s}{\partial s}\right)w - \bar{M}_\nu\frac{\partial w}{\partial \nu}\right]ds, \qquad (8.145)\end{aligned}$$

where Eqs. (8.131) and (8.132) have been substituted. Generalizations and transformations of the functional (8.145) can be formulated in the usual manner.

So far, the nonlinear theory of the thin shallow shell has been derived. It is noted in this connection that a linear theory can be obtained by linearizing the strain–displacement expressions (8.130) as

$$\begin{aligned} f_{11} &= \frac{\partial u}{\partial x} + \frac{\partial z}{\partial x}\frac{\partial w}{\partial x} - \zeta\frac{\partial^2 w}{\partial x^2},\\ f_{22} &= \frac{\partial v}{\partial y} + \frac{\partial z}{\partial y}\frac{\partial w}{\partial y} - \zeta\frac{\partial^2 w}{\partial y^2}, \qquad (8.146)\\ 2f_{12} &= \frac{\partial u}{\partial y} + \frac{\partial v}{\partial x} + \frac{\partial z}{\partial x}\frac{\partial w}{\partial y} + \frac{\partial z}{\partial y}\frac{\partial w}{\partial x} - 2\zeta\frac{\partial^2 w}{\partial x\,\partial y}, \end{aligned}$$

and deriving equations in a manner similar to the development for the nonlinear theory. Some related papers are listed in the bibliography.†

8.10. Some Remarks

The theories of thin plates developed in this chapter are based on the assumption that the transverse normal stress may be neglected in the stress–strain relations. Rigorously speaking, the transverse stress σ_z is induced in the plate. However, a consideration such as given in Appendix D shows that unless the surface forces are highly concentrated, the stress σ_z is in general of smaller order of magnitude than σ_x and σ_y. Consequently, the terms containing σ_z are usually neglected in the stress–strain relations. On the other hand, it is seen from Eqs. (8.15) that we have $\varepsilon_z = 0$ under the Kirchhoff hypothesis. The three dimensional stress–strain relations, Eqs. (1.10), show that a theory which includes $\sigma_z = 0$ as well as $\varepsilon_z = 0$ would

† See Refs. 32 through 36.

fail to produce correct results. The same kind of contradiction exists in the formulations of Sections 8.5 and 8.8. We have tried to avoid this difficulty by putting $\sigma_z = 0$ in the three-dimensional stress–strain relations and then eliminating ε_z. To remove the inconsistency completely, it would be necessary to employ

$$W(x, y, z) = w(x, y) + w_1(x, y)\, z + w_2(x, y)\, z^2, \tag{8.147}$$

instead of the last equation in Eqs. (8.14) or (8.99). However, these additional linear and quadratic terms are usually found small in comparison with the leading term as far as the small displacement theory of thin plates is concerned and may be omitted in a first approximate theory. The theories of thin plates developed in this chapter have been based on the above considerations, which are due mainly to Ref. 37. The accuracy of these plate theories may be improved by assuming the displacement components as

$$U = \sum_{m=0}^{n} u_m(x, y)\, z^m, \quad V = \sum_{m=0}^{n} v_m(x, y)\, z^m, \quad W = \sum_{m=0}^{n} w_m(x, y)\, z^m, \tag{8.148}$$

thus adding terms of high powers with respect to z. We note here that a theory of thin plates including the effect of transverse shear deformation has been derived by Yu with the use of generalized Hamilton's principle in which the variation is taken with respect to displacements, strains and stresses.[38]

Variational formulations can, of course, be made for free vibrational problems of elastic plates, although no mention has been made of this topic in this chapter.[39, 40, 41] An application of the variational method has been made to the problem of free vibration of non-isotropic, rectangular, AT-cut quartz plates.[42] We also note that the self-excited or forced vibration of plates due to aerodynamic forces has been one of the central problems in the theory of aeroelasticity.[43, 44]

Bibliography

1. A. E. H. Love, *Mathematical Theory of Elasticity*, Cambridge University Press, 4th edition, 1927.
2. S. Timoshenko and S. Woinowsky-Krieger, *Theory of Plates and Shells*, McGraw-Hill, 1959.
3. S. Timoshenko and J. N. Goodier, *Theory of Elasticity*, McGraw-Hill, 1951.
4. I. S. Sokolnikoff, *Mathematical Theory of Elasticity*, McGraw-Hill, 1956.
5. N. I. Muschelisvili, Practische Lösung der fundamentalen Randwertaufgaben der Elastizitäts-Theorie in der Ebene für einige Berandungsformen, *Zeitschrift für Angewandte Mathematik und Mechanik*, Vol. 13, No. 4, pp. 264–82, August 1933.
6. S. Moriguti, *Theory of Two-Dimensional Elasticity* (in Japanese), Series on Modern Applied Mathematics, Iwanami Book Publishing Co., 1957.
7. E. Reissner, Least Work Solutions of Shear Lag Problems, *Journal of the Aeronautical Sciences*, Vol. 8, No. 7, pp. 284–91, May 1941.
8. E. Reissner, Analysis of Shear Lag in Box Beams by the Principle of Minimum Potential Energy, *Quarterly of Applied Mathematics*, Vol. 4, No. 3, pp. 268–78, October 1946.

9. E. REISSNER and M. STEIN, *Torsion and Transverse Bending of Cantilever Plates*, NACA TN 2369, June 1951.
10. R. L. BISPLINGHOFF, H. ASHLEY and R. L. HALFMAN, *Aeroelasticity*, Addison-Wesley, 1955.
11. T.v. KÁRMÁN, Festigkeitsprobleme im Maschinenbau, *Encyklopädie der Mathematischen Wissenschaften*, Vol. IV, pp. 314–85, 1910.
12. V. V. NOVOZHILOV, *Foundation of the Nonlinear Theory of Elasticity*, Graylock, 1953.
13. K. MARGUERRE, Die über die Ausbeulgrenze belastete Platte, *Zeitschrift für Angewandte Mathematik und Mechanik*, Vol. 16, No. 6, pp. 353–5, December 1936.
14. K. MARGUERRE and E. TREFFTZ, Über die Tragfähigkeit eines längsbelasteten Plattenstreifens nach Überschreiten der Beullast, *Zeitschrift für Angewandte Mathematik und Mechanik*, Vol. 17, No. 2, pp. 85–100, April 1937.
15. A. FROMM and K. MARGUERRE, Verhalten eines von Schub- und Druckkräften beanspruchten Plattenstreifens oberhalb der Beulgrenze, *Luftfahrtforschung*, Vol. 14, No. 12, pp. 627–39, December 1937.
16. C. T. WANG, *Principle and Application of Complementary Energy Method for Thin Homogeneous and Sandwich Plates and Shells with Finite Deflections*. NACA TN 2620, 1952.
17. E. REISSNER, Finite Twisting and Bending of Thin Rectangular Elastic Plates, *Journal of Applied Mechanics*, Vol. 24, No. 3, pp. 391–6, September 1957.
18. R. L. BISPLINGHOFF, *The Finite Twisting and Bending of Heated Elastic Lifting Surfaces*, Mitteilung Nr. 4 aus dem Institut für Flugzeugstatik und Leichtbau, E. T. H., Zürich, 1957.
19. S. TIMOSHENKO, *Theory of Elastic Stability*, McGraw-Hill, 1936.
20. B. A. BOLEY and J. H. WEINER, *Theory of Thermal Stresses*, John Wiley, 1960.
21. R. R. HELDENFELS and W. M. ROBERTS, *Experimental and Theoretical Determination of Thermal Stresses in a Flat Plate*, NACA TN 2769, 1952.
22. M. L. GOSSARD, P. SEIDE and W. M. ROBERTS, *Thermal Buckling of Plates*, NACA TN 2771, 1952.
23. R. L. BISPLINGHOFF *et al.*, *Aerodynamic Heating of Aircraft Structures in High-speed Flight*, Notes for a Special Summer Program, Department of Aeronautical Engineering, Massachusetts Institute of Technology, June 25–July 6, 1956.
24. N. J. HOFF, Editor, *High Temperature Effects in Aircraft Structures*, AGARDograph 28, Pergamon Press, 1958.
25. E. REISSNER, On the Theory of Bending of Elastic Plates, *Journal of Mathematics and Physics*, Vol. 23, No. 4, pp. 184–91, November 1944.
26. E. REISSNER, The Effect of Transverse-Shear Deformation on the Bending of Elastic Plates, *Journal of Applied Mechanics*, Vol. 12, No. 2, pp. 69–77, June 1945.
27. E. REISSNER, On Bending of Elastic Plates, *Quarterly of Applied Mathematics*, Vol. 5, No. 1, pp. 55–68, April 1947.
28. E. REISSNER, On a Variational Theorem in Elasticity, *Journal of Mathematics and Physics*, Vol. XXIX, No. 2, pp. 90–5, July 1950.
29. R. D. MINDLIN, Thickness-Shear and Flexural Vibrations of Crystal Plates, *Journal of Applied Physics*, Vol. 23, No. 3, pp. 316–23, March 1951.
30. K. MARGUERRE, Zur Theorie der gekrümmten Platte großer Formänderung, *Proceedings of the 5th International Congress for Applied Mechanics*, pp. 93–101, 1938.
31. E. REISSNER, On Some Aspects of the Theory of Thin Elastic Shells, *Journal of the Boston Society for Civil Engineers*. Vol. XIII, No. 2, pp. 100–33, April 1955.
32. E. REISSNER, On Transverse Vibrations of Thin Shallow Elastic Shells, *Quarterly of Applied Mathematics*, Vol. 13, No. 2, pp. 169–76, July 1955.
33. R. R. HELDENFELS and L. F. VOSTEEN, *Approximate Analysis of Effects of Large Deflections and Initial Twist on Torsional Stiffness of a Cantilever Plate Subjected to Thermal Stresses*, NACA TN 4067, 1959.

34. E. L. Reiss, H. J. Greenberg and H. B. Keller, Nonlinear Deflections of Shallow Spherical Shells, *Journal of the Aeronautical Sciences*, Vol. 24, No. 7, pp. 533–43, July 1957.
35. E. L. Reiss, Axially Symmetric Buckling of Shallow Spherical Shells under External Pressure, *Journal of Applied Mechanics*, Vol. 25, No. 4, pp. 556–60, December 1958.
36. H. B. Keller and E. L. Reiss, Spherical Cap Snapping, *Journal of the Aero/Space Sciences*, Vol. 26, No. 10, pp. 643–52, October 1959.
37. F. B. Hildebrand, E. Reissner and G. B. Thomas, *Notes on the Foundations of the Theory of Small Displacements of Orthotropic Shells*, NACA TN 1833, 1949.
38. Y. Y. Yu, Generalized Hamilton's Principle and Variational Equation of Motion in Nonlinear Elasticity Theory, with Application to Plate Theory, *Journal of the Acoustical Society of America*, Vol. 36, No. 1, pp. 111–19, January 1964.
39. R. Weinstock, *Calculus of Variations with Applications to Physics and Engineering*, McGraw-Hill, 1952.
40. M. V. Barton, Vibration of Rectangular and Skew Cantilever Plates, *Journal of Applied Mechanics*, Vol. 18, No. 2, pp. 129–34, June 1951.
41. H. J. Plass Jr., J. H. Gaines, and C. D. Newsom, Application of Reissner's Variational Principle to Cantilever Plate Deflection and Vibration Problems, *Journal of Applied Mechanics*, Vol. 29, No. 1, pp. 127–35, March 1962.
42. I. Koga, Radio-Frequency Vibrations of Rectangular AT-Cut Quartz Plates, *Journal of Applied Physics*, Vol. 34, No. 8, pp. 2357–65, August 1963.
43. R. L. Bisplinghoff and H. Ashley, *Principles of Aeroelasticity*, John Wiley, 1962.
44. V. V. Bolotin, *Nonconservative Problems of the Theory of Elastic Stability*, Translated by T. K. Lusher and edited by G. Herrmann, Pergamon Press, 1963.
45. S. G. Mikhlin, *Variational Methods in Mathematical Physics*, Pergamon Press, 1964.
46. K. Washizu, Variational Methods Applied to Free Lateral Vibrations of a Plate with Initial Stresses, *Transactions of Japan Society for Aeronautical and Space Sciences*, Vol. 6, No. 9, pp. 36–42, 1963.
47. L. S. D. Morley, *Skew Plates and Structures*, Pergamon Press, 1963.

CHAPTER 9

SHELLS

9.1. Geometry before Deformation

We shall consider theories of thin shells in the present chapter. Let the middle surface of the shell, denoted by S_m, be taken as a reference curved surface which is defined by two curvilinear coordinates α and β in such a way that the position vector of an arbitrary point $P_0^{(0)}$ in S_m is represented by

$$\mathbf{r}_0^{(0)} = \mathbf{r}_0^{(0)}(\alpha, \beta), \tag{9.1}$$

as shown in Fig. 9.1. The coordinates α and β are chosen so as to coincide with the lines of curvature of the middle surface, and the unit vectors in the

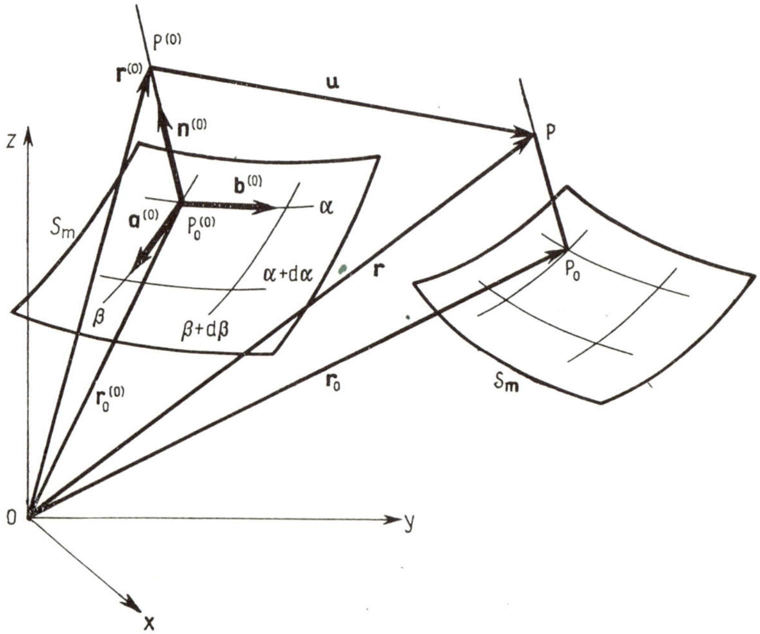

FIG. 9.1. Geometry of the shell before and after deformation.
(a) before deformation. (b) after deformation.

directions of α and β are denoted by $\mathbf{a}^{(0)}$ and $\mathbf{b}^{(0)}$, respectively:

$$\mathbf{a}^{(0)} = \frac{1}{A}\frac{\partial \mathbf{r}_0^{(0)}}{\partial \alpha}, \quad \mathbf{b}^{(0)} = \frac{1}{B}\frac{\partial \mathbf{r}_0^{(0)}}{\partial \beta}, \tag{9.2}$$

where

$$A^2 = \frac{\partial \mathbf{r}_0^{(0)}}{\partial \alpha}\cdot\frac{\partial \mathbf{r}_0^{(0)}}{\partial \alpha}, \quad B^2 = \frac{\partial \mathbf{r}_0^{(0)}}{\partial \beta}\cdot\frac{\partial \mathbf{r}_0^{(0)}}{\partial \beta}. \tag{9.3}$$

The length of a line element between two neighboring points in S_m, the coordinates of which are (α, β) and $(\alpha + d\alpha, \beta + d\beta)$, is given by

$$(ds_0^{(0)})^2 = d\mathbf{r}_0^{(0)}\cdot d\mathbf{r}_0^{(0)} = A^2(d\alpha)^2 + B^2(d\beta)^2. \tag{9.4}†$$

The unit vector perpendicular to S_m is denoted by $\mathbf{n}^{(0)}$, which is chosen so that $\mathbf{a}^{(0)}$, $\mathbf{b}^{(0)}$ and $\mathbf{n}^{(0)}$ form a right-handed orthogonal system:

$$\mathbf{n}^{(0)} = \mathbf{a}^{(0)} \times \mathbf{b}^{(0)}. \tag{9.5}$$

The radii of curvature in the directions of α and β will be denoted by R_α and R_β, and are taken positive when the centers of curvature lie in the positive direction of $\mathbf{n}^{(0)}$. The geometry of the middle surface gives rise to the following matrix relations:‡

$$\frac{\partial}{\partial \alpha}\begin{bmatrix}\mathbf{a}^{(0)}\\ \mathbf{b}^{(0)}\\ \mathbf{n}^{(0)}\end{bmatrix} = \begin{bmatrix}0 & -\dfrac{1}{B}\dfrac{\partial A}{\partial \beta} & \dfrac{A}{R_\alpha}\\ \dfrac{1}{B}\dfrac{\partial A}{\partial \beta} & 0 & 0\\ -\dfrac{A}{R_\alpha} & 0 & 0\end{bmatrix}\begin{bmatrix}\mathbf{a}^{(0)}\\ \mathbf{b}^{(0)}\\ \mathbf{n}^{(0)}\end{bmatrix}, \tag{9.6a}$$

$$\frac{\partial}{\partial \beta}\begin{bmatrix}\mathbf{a}^{(0)}\\ \mathbf{b}^{(0)}\\ \mathbf{n}^{(0)}\end{bmatrix} = \begin{bmatrix}0 & \dfrac{1}{A}\dfrac{\partial B}{\partial \alpha} & 0\\ -\dfrac{1}{A}\dfrac{\partial B}{\partial \alpha} & 0 & \dfrac{B}{R_\beta}\\ 0 & -\dfrac{B}{R_\beta} & 0\end{bmatrix}\begin{bmatrix}\mathbf{a}^{(0)}\\ \mathbf{b}^{(0)}\\ \mathbf{n}^{(0)}\end{bmatrix}. \tag{9.6b}$$

Using the above relations and the following identities:

$$\frac{\partial^2 \mathbf{a}^{(0)}}{\partial \alpha\, \partial \beta} = \frac{\partial^2 \mathbf{a}^{(0)}}{\partial \beta\, \partial \alpha}, \quad \frac{\partial^2 \mathbf{b}^{(0)}}{\partial \alpha\, \partial \beta} = \frac{\partial^2 \mathbf{b}^{(0)}}{\partial \beta\, \partial \alpha}, \quad \frac{\partial^2 \mathbf{n}^{(0)}}{\partial \alpha\, \partial \beta} = \frac{\partial^2 \mathbf{n}^{(0)}}{\partial \beta\, \partial \alpha},$$

we have,

$$\frac{\partial}{\partial \alpha}\left(\frac{B}{R_\beta}\right) = \frac{1}{R_\alpha}\frac{\partial B}{\partial \alpha}, \quad \frac{\partial}{\partial \beta}\left(\frac{A}{R_\alpha}\right) = \frac{1}{R_\beta}\frac{\partial A}{\partial \beta}, \tag{9.7}$$

† Specializations to several kinds of shells are given in Appendix E.

‡ Refs. 1 and 2. See also Problem 5 of Chapter 4 in Appendix H.

and

$$\frac{\partial}{\partial \alpha}\left(\frac{1}{A}\frac{\partial B}{\partial \alpha}\right) + \frac{\partial}{\partial \beta}\left(\frac{1}{B}\frac{\partial A}{\partial \beta}\right) + \frac{AB}{R_\alpha R_\beta} = 0. \tag{9.8}$$

These relations are known as the conditions of Codazzi and of Gauss, respectively.

Next, we shall consider an arbitrary point $P^{(0)}$ outside the middle surface of the shell. We represent its position vector by:

$$\mathbf{r}^{(0)} = \mathbf{r}_0^{(0)}(\alpha, \beta) + \zeta \mathbf{n}^{(0)}(\alpha, \beta), \tag{9.9}$$

where ζ is the distance of the point from the middle surface. The relation (9.9) shows that an arbitrary point in the shell can be specified by the set (α, β, ζ), which can be employed as a set of orthogonal curvilinear coordinates. Consequently, the formulations derived in Chapter 4 are applicable and the notation $\alpha^1 = \alpha$, $\alpha^2 = \beta$, $\alpha^3 = \zeta$ will be used whenever convenient. From Eq. (9.9), we have the local base vectors as

$$\mathbf{g}_1 = \frac{\partial \mathbf{r}^{(0)}}{\partial \alpha} = A\left(1 - \frac{\zeta}{R_\alpha}\right)\mathbf{a}^{(0)}, \quad \mathbf{g}_2 = \frac{\partial \mathbf{r}^{(0)}}{\partial \beta} = B\left(1 - \frac{\zeta}{R_\beta}\right)\mathbf{b}^{(0)},$$
$$\mathbf{g}_3 = \frac{\partial \mathbf{r}^{(0)}}{\partial \zeta} = \mathbf{n}^{(0)}. \tag{9.10}$$

The position vector connecting two neighboring points $P^{(0)}(\alpha, \beta, \zeta)$ and $Q^{(0)}(\alpha + d\alpha, \beta + d\beta, \zeta + d\zeta)$ is

$$\begin{aligned} d\mathbf{r}^{(0)} &= \mathbf{r}^{(0)}_{,\alpha}\, d\alpha + \mathbf{r}^{(0)}_{,\beta}\, d\beta + \mathbf{r}^{(0)}_{,\zeta}\, d\zeta \\ &= A\left(1 - \frac{\zeta}{R_\alpha}\right)\mathbf{a}^{(0)}\, d\alpha + B\left(1 - \frac{\zeta}{R_\beta}\right)\mathbf{b}^{(0)}\, d\beta + \mathbf{n}^{(0)}\, d\zeta, \end{aligned} \tag{9.11}$$

and its length, denoted by $ds^{(0)}$, is given by

$$(ds^{(0)})^2 = d\mathbf{r}^{(0)} \cdot d\mathbf{r}^{(0)} = \sum_{\lambda,\mu=1}^{3} g_{\lambda\mu}\, d\alpha^\lambda\, d\alpha^\mu, \tag{9.12}$$

where

$$g_{11} = A^2\left(1 - \frac{\zeta}{R_\alpha}\right)^2, \quad g_{22} = B^2\left(1 - \frac{\zeta}{R_\beta}\right)^2,$$
$$g_{33} = 1, \quad g_{23} = g_{31} = g_{12} = 0. \tag{9.13}$$

The volume of an infinitesimal parallelepiped, enclosed by the six surfaces: $\alpha = \text{const}$, $\beta = \text{const}$, $\zeta = \text{const}$, $\alpha + d\alpha = \text{const}$, $\beta + d\beta = \text{const}$, $\zeta + d\zeta = \text{const}$, is given by

$$dV = AB\left(1 - \frac{\zeta}{R_\alpha}\right)\left(1 - \frac{\zeta}{R_\beta}\right) d\alpha\, d\beta\, d\zeta. \tag{9.14}$$

For later convenience, we shall locate a system of local rectangular Cartesian coordinates (y^1, y^2, y^3) at the point $P^{(0)}$, where the directions of the coordinate axes are taken in coincidence with the unit vectors $\mathbf{a}^{(0)}$, $\mathbf{b}^{(0)}$ and $\mathbf{n}^{(0)}$ at the point $P_0^{(0)}$, respectively, as shown in Fig. 9.2. Then, from Eqs. (4.57) and (9.10), we obtain the following geometrical relations:

$$dy^1 = A\left(1 - \frac{\zeta}{R_\alpha}\right)d\alpha, \quad dy^2 = B\left(1 - \frac{\zeta}{R_\beta}\right)d\beta, \quad dy^3 = d\zeta. \tag{9.15}$$

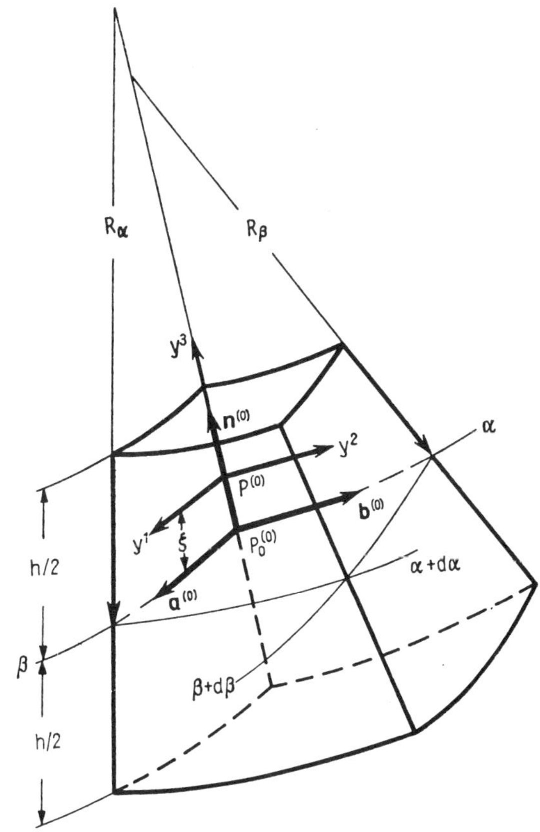

FIG. 9.2. A shell element.

Let us now consider the side surface of the shell. It is assumed that the middle surface is simply connected and that the side surface, denoted by S, is generated by the envelope of normals drawn perpendicular to the middle surface S_m. Let the intersection curve between S_m and S be denoted by C, and let the unit vector drawn outwards on C and perpendicular to S be denoted by $\boldsymbol{\nu}$, as shown in Fig. 9.3. Then, we have the area of an infinitesimal rectangle on S as

$$dS = \sqrt{\left[m\left(1 - \frac{\zeta}{R_\alpha}\right)\right]^2 + \left[l\left(1 - \frac{\zeta}{R_\beta}\right)\right]^2}\, ds\, d\zeta, \tag{9.16}$$

where $l = \mathbf{a}^{(0)} \cdot \boldsymbol{\nu}$ and $m = \mathbf{b}^{(0)} \cdot \boldsymbol{\nu}$, and s is measured along the curve C. We have the following relations on the boundary C:

$$A\,d\alpha = \pm m\,ds, \quad B\,d\beta = \pm l\,ds, \tag{9.17}$$

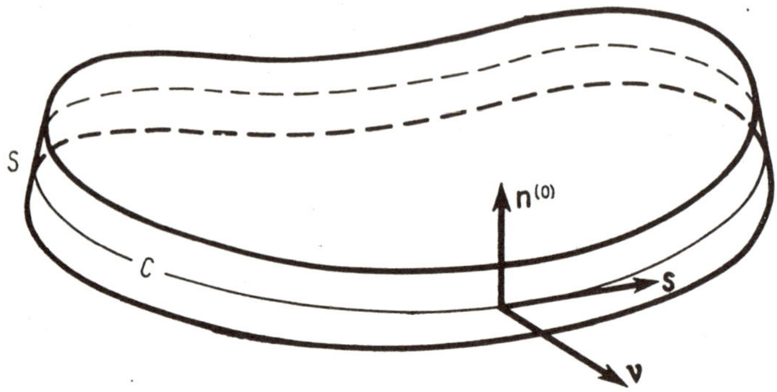

FIG. 9.3. Directions of $\boldsymbol{\nu}$, s and $\mathbf{n}^{(0)}$.

as shown in Fig. 9.4. If the direction of increasing s is chosen in such a way that the local vectors $\boldsymbol{\nu}$, s and $\mathbf{n}^{(0)}$ form a right-hand system, as shown in Fig. 9.3, we have

$$\frac{1}{A}\frac{\partial}{\partial\alpha} = l\frac{\partial}{\partial\nu} - m\frac{\partial}{\partial s}, \quad \frac{1}{B}\frac{\partial}{\partial\beta} = m\frac{\partial}{\partial\nu} + l\frac{\partial}{\partial s}, \tag{9.18}$$

where s is the unit tangent vector in the direction of increasing s.

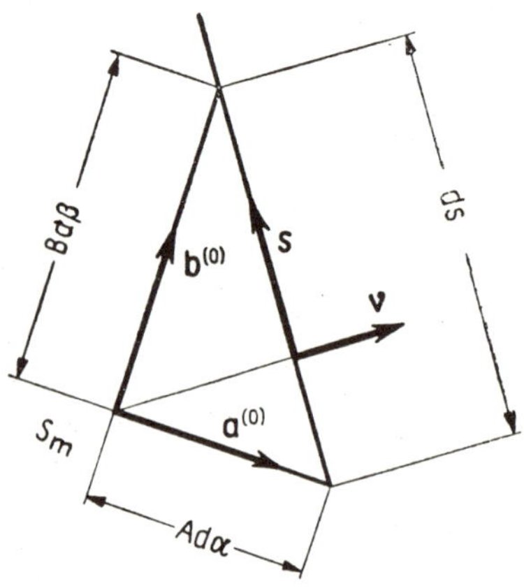

FIG. 9.4. Geometrical relations on the boundary C.

9.2. Analysis of Strain

The shell is now assumed to be subject to deformation. The point $P^{(0)}$ is displaced to a new position P, whose position vector is given by

$$\mathbf{r} = \mathbf{r}^{(0)} + \mathbf{u}, \tag{9.19}$$

where the displacement vector $\mathbf{u}$ is a function of (α, β, ζ), and its components, denoted by U, V and W, are defined in the directions of $\mathbf{a}^{(0)}$, $\mathbf{b}^{(0)}$ and $\mathbf{n}^{(0)}$:

$$\mathbf{u} = U\mathbf{a}^{(0)} + V\mathbf{b}^{(0)} + W\mathbf{n}^{(0)}. \tag{9.20}$$

By the use of Eqs. (4.36), (9.9), (9.19) and (9.20), we can calculate the strains $f_{\lambda\mu}$ defined with respect to the (α, β, ζ) coordinate system in terms of U, V and W. Then, the strains $e_{\lambda\mu}$ defined with respect to the (y^1, y^2, y^3) coordinate system may be obtained using the transformation law (4.61) and the geometrical relations (9.15):

$$\begin{aligned}
e_{\alpha\alpha} &= \frac{f_{\alpha\alpha}}{A^2(1-\zeta/R_\alpha)^2}, \quad e_{\beta\beta} = \frac{f_{\beta\beta}}{B^2(1-\zeta/R_\beta)^2},\\
e_{\zeta\zeta} &= f_{\zeta\zeta}, \quad e_{\alpha\beta} = \frac{f_{\alpha\beta}}{AB(1-\zeta/R_\alpha)(1-\zeta/R_\beta)},\\
e_{\alpha\zeta} &= \frac{f_{\alpha\zeta}}{A(1-\zeta/R_\alpha)}, \quad e_{\beta\zeta} = \frac{f_{\beta\zeta}}{B(1-\zeta/R_\beta)}.
\end{aligned} \tag{9.21}$$

It is obvious that the linearized strains $(\varepsilon_\alpha, \varepsilon_\beta, \varepsilon_\zeta, \ldots)$ defined with respect to the (y^1, y^2, y^3) coordinate system are obtainable from Eqs. (9.21) by linearizing the strains $f_{\lambda\mu}$ with respect to the displacements.

We shall now introduce two assumptions for thin shell theories which will be developed in the present chapter. First, we assume that transverse normal stress σ_ζ is small compared to other stress components and may be set

$$\sigma_\zeta = 0 \tag{9.22}$$

in the stress–strain relations, thus obtaining

$$\begin{aligned}
\sigma_\alpha &= \frac{E}{(1-\nu^2)}(\varepsilon_\alpha + \nu\varepsilon_\beta), \quad \sigma_\beta = \frac{E}{(1-\nu^2)}(\nu\varepsilon_\alpha + \varepsilon_\beta),\\
\tau_{\alpha\beta} &= G\gamma_{\alpha\beta}, \quad \tau_{\alpha\zeta} = G\gamma_{\alpha\zeta}, \quad \tau_{\beta\zeta} = G\gamma_{\beta\zeta},
\end{aligned} \tag{9.23}$$

for linearized theories and

$$\begin{aligned}
\sigma_\alpha &= \frac{E}{(1-\nu^2)}(e_{\alpha\alpha} + \nu e_{\beta\beta}), \quad \sigma_\beta = \frac{E}{(1-\nu^2)}(\nu e_{\alpha\alpha} + e_{\beta\beta}),\\
\tau_{\alpha\beta} &= 2Ge_{\alpha\beta}, \quad \tau_{\alpha\zeta} = 2Ge_{\alpha\zeta}, \quad \tau_{\beta\zeta} = 2Ge_{\beta\zeta},
\end{aligned} \tag{9.24}$$

for nonlinear theories, where the stresses as well as the strains are defined with respect to the local rectangular Cartesian coordinates. Second, we assume that the displacement vector $\mathbf{u}$ is approximated by

$$\mathbf{u} = \mathbf{u}_0 + \zeta\mathbf{u}_1, \tag{9.25}$$

where $\mathbf{u}_0$ and $\mathbf{u}_1$ are functions of (α, β) only, their components being defined by

$$\mathbf{u}_0 = u\mathbf{a}^{(0)} + v\mathbf{b}^{(0)} + w\mathbf{n}^{(0)}, \tag{9.26}$$

$$\mathbf{u}_1 = u_1\mathbf{a}^{(0)} + v_1\mathbf{b}^{(0)} + w_1\mathbf{n}^{(0)}, \tag{9.27}$$

and where a geometrical constraint is introduced by requiring that

$$e_{\zeta\zeta} = u_1^2 + v_1^2 + (1 + w_1)^2 - 1 = 0. \tag{9.28}$$

It is seen that Eq. (9.25) under the constraint condition (9.28) provides one of the simplest expressions for displacements including the effect of transverse shear deformation (see Section 8.8 for a similar development).

We may obtain strain–displacement relations for finite displacement theory by substituting Eq. (9.25) into Eq. (9.19) and following familiar procedures. However, we shall be satisfied with deriving those for small displacement theory in the remainder of this section. To begin with, we find that due to the assumption of small displacement, Eq. (9.28) is linearized with respect to the displacements to yield

$$w_1 = 0, \tag{9.29}$$

and consequently, Eqs. (9.25) reduce to

$$U = u + \zeta u_1, \quad V = v + \zeta v_1, \quad W = w. \tag{9.30}$$

Combining Eqs. (9.19), (9.20) and (9.30), we obtain

$$\mathbf{r} = \mathbf{r}^{(0)} + (u + \zeta u_1)\,\mathbf{a}^{(0)} + (v + \zeta v_1)\,\mathbf{b}^{(0)} + w\mathbf{n}^{(0)}. \tag{9.31}$$

It is seen that Eqs. (9.30) are natural extensions of Eqs. (8.99) of the thin plate to the thin shell.

Next, we shall obtain strain–displacement relations. From Eq. (9.31), we have the following relations:

$$\begin{aligned}
\frac{1}{A}\frac{\partial \mathbf{r}}{\partial \alpha} &= \left[\left(1 - \frac{\zeta}{R_\alpha}\right) + l_{11} + m_{11}\zeta\right]\mathbf{a}^{(0)} + (l_{21} + m_{21}\zeta)\,\mathbf{b}^{(0)} \\
&\quad + (l_{31} + m_{31}\zeta)\,\mathbf{n}^{(0)}, \\
\frac{1}{B}\frac{\partial \mathbf{r}}{\partial \beta} &= (l_{12} + m_{12}\zeta)\,\mathbf{a}^{(0)} + \left[\left(1 - \frac{\zeta}{R_\beta}\right) + l_{22} + m_{22}\zeta\right]\mathbf{b}^{(0)} \\
&\quad + (l_{32} + m_{32}\zeta)\,\mathbf{n}^{(0)}, \\
\frac{\partial \mathbf{r}}{\partial \zeta} &= u_1\mathbf{a}^{(0)} + v_1\mathbf{b}^{(0)} + \mathbf{n}^{(0)},
\end{aligned} \tag{9.32}$$

where it is defined that

$$\begin{aligned}
l_{11} &= \frac{1}{A}\frac{\partial u}{\partial \alpha} + \frac{v}{AB}\frac{\partial A}{\partial \beta} - \frac{w}{R_\alpha}, & l_{12} &= \frac{1}{B}\frac{\partial u}{\partial \beta} - \frac{v}{AB}\frac{\partial B}{\partial \alpha}, \\
l_{21} &= \frac{1}{A}\frac{\partial v}{\partial \alpha} - \frac{u}{AB}\frac{\partial A}{\partial \beta}, & l_{22} &= \frac{1}{B}\frac{\partial v}{\partial \beta} + \frac{u}{AB}\frac{\partial B}{\partial \alpha} - \frac{w}{R_\beta}, \\
l_{31} &= \frac{1}{A}\frac{\partial w}{\partial \alpha} + \frac{u}{R_\alpha}, & l_{32} &= \frac{1}{B}\frac{\partial w}{\partial \beta} + \frac{v}{R_\beta}.
\end{aligned} \tag{9.33}$$

and

$$m_{11} = \frac{1}{A}\frac{\partial u_1}{\partial \alpha} + \frac{v_1}{AB}\frac{\partial A}{\partial \beta}, \quad m_{12} = \frac{1}{B}\frac{\partial u_1}{\partial \beta} - \frac{v_1}{AB}\frac{\partial B}{\partial \alpha},$$
$$m_{21} = \frac{1}{A}\frac{\partial v_1}{\partial \alpha} - \frac{u_1}{AB}\frac{\partial A}{\partial \beta}, \quad m_{22} = \frac{1}{B}\frac{\partial v_1}{\partial \beta} + \frac{u_1}{AB}\frac{\partial B}{\partial \alpha}, \tag{9.34}$$
$$m_{31} = \frac{u_1}{R_\alpha}, \qquad m_{32} = \frac{v_1}{R_\beta}.$$

By the use of Eqs. (9.9) and (9.31), and with the aid of the above relations, the strains $f_{\lambda\mu}$ can be calculated. The strains thus obtained are then linearized with respect to the displacements and substituted into Eqs. (9.21) to obtain following strain–displacement relations for a small displacement theory including the effect of transverse shear deformation:

$$\varepsilon_\alpha = \frac{\varepsilon_{\alpha o} - \zeta k_\alpha}{1 - \zeta/R_\alpha}, \quad \varepsilon_\beta = \frac{\varepsilon_{\beta o} - \zeta k_\beta}{1 - \zeta/R_\beta},$$
$$\gamma_{\alpha\beta} = \frac{\gamma_{\alpha\beta o} - 2\zeta k_{\alpha\beta} + \zeta^2 k^*_{\alpha\beta}}{(1 - \zeta/R_\alpha)(1 - \zeta/R_\beta)}, \tag{9.35}$$
$$\gamma_{\alpha\zeta} = \frac{\gamma_{\alpha\zeta o}}{1 - \zeta/R_\alpha}, \quad \gamma_{\beta\zeta} = \frac{\gamma_{\beta\zeta o}}{1 - \zeta/R_\beta},$$

where

$$\varepsilon_{\alpha o} = l_{11}, \quad \varepsilon_{\beta o} = l_{22}, \quad \gamma_{\alpha\beta o} = l_{12} + l_{21}, \tag{9.36}$$

$$k_\alpha = -m_{11}, \quad k_\beta = -m_{22},$$
$$2k_{\alpha\beta} = -m_{21} - m_{12} + \frac{l_{12}}{R_\alpha} + \frac{l_{21}}{R_\beta}, \tag{9.37}$$
$$k^*_{\alpha\beta} = -\frac{m_{12}}{R_\alpha} - \frac{m_{21}}{R_\beta},$$

$$\gamma_{\alpha\zeta o} = u_1 + l_{31}, \quad \gamma_{\beta\zeta o} = v_1 + l_{32}. \tag{9.38}$$

9.3. Analysis of Strain under the Kirchhoff–Love Hypothesis

The analysis of strains including the effect of transverse shear deformation has been made in the last section. We shall now proceed to the analysis of strain under the Kirchhoff–Love hypothesis that the straight fibres of the shell which are perpendicular to the middle surface before deformation remain straight and perpendicular to the deformed middle surface and suffer no extensions.† This is an extension of the Kirchhoff hypothesis for thin plates to thin shells. We observe that a shell theory under the hypothesis is a special case of the theory based on Eqs. (9.25) and (9.28).

† The Kirchhoff–Love hypothesis is usually understood to include the assumption $\sigma_\zeta = 0$ as well. However, only the hypothesis described here will be called the Kirchhoff–Love in this book.

We consider an arbitrary point of the middle surface having the coordinates (α, β, 0) before deformation, and denote its position vectors before and after deformation by $\mathbf{r}_0^{(0)}$ and $\mathbf{r}_0$, respectively, which are related to the displacement vector $\mathbf{u}_0$ introduced in Eq. (9.25) by

$$\mathbf{r}_0 = \mathbf{r}_0^{(0)} + \mathbf{u}_0. \tag{9.39}$$

Then, it is seen that the hypothesis allows us to express $\mathbf{r}$ as

$$\mathbf{r} = \mathbf{r}_0 + \zeta\mathbf{n}, \tag{9.40}$$

where $\mathbf{n}$ is a unit normal to the deformed middle surface and is given by

$$\mathbf{n} = \frac{\partial \mathbf{r}_0}{\partial \alpha} \times \frac{\partial \mathbf{r}_0}{\partial \beta} \Bigg/ \left| \frac{\partial \mathbf{r}_0}{\partial \alpha} \times \frac{\partial \mathbf{r}_0}{\partial \beta} \right|. \tag{9.41}$$

Since

$$\mathbf{r}_0 = \mathbf{r}_0^{(0)} + u\mathbf{a}^{(0)} + v\mathbf{b}^{(0)} + w\mathbf{n}^{(0)}, \tag{9.42}$$

we may express $\mathbf{n}$ in terms of u, v and w as follows:

$$\mathbf{n} = \frac{L\mathbf{a}^{(0)} + M\mathbf{b}^{(0)} + N\mathbf{n}^{(0)}}{\sqrt{L^2 + M^2 + N^2}}, \tag{9.43}$$

where

$$\begin{aligned} L &= -l_{31} + l_{21}l_{32} - l_{22}l_{31}, \\ M &= -l_{32} + l_{12}l_{31} - l_{11}l_{32}, \\ N &= 1 + l_{11} + l_{22} + l_{11}l_{22} - l_{12}l_{21}. \end{aligned} \tag{9.44}$$

From Eqs. (9.9), (9.19), (9.39) and (9.40), we obtain

$$\mathbf{u} = \mathbf{u}_0 + \zeta(\mathbf{n} - \mathbf{n}_0). \tag{9.45}$$

We observe from Eqs. (9.25) and (9.45) that the hypothesis imposes the following condition on $\mathbf{u}_1$:

$$\mathbf{u}_1 = \mathbf{n} - \mathbf{n}^{(0)} \tag{9.46}$$

and reduces the freedom of shell deformation to u, v and w only.

When a shell problem is confined to a small displacement theory, Eq. (9.46) reduces to

$$u_1 = -l_{31}, \quad v_1 = -l_{32}. \tag{9.47}$$

Thus, the hypothesis allows us to express the displacement components as

$$U = u - l_{31}\zeta, \quad V = v - l_{32}\zeta, \quad W = w, \tag{9.48}$$

and the strain–displacement relations as

$$\begin{aligned} \varepsilon_\alpha &= \frac{\varepsilon_{\alpha 0} - \zeta\varkappa_\alpha}{1 - \zeta/R_\alpha}, \quad \varepsilon_\beta = \frac{\varepsilon_{\beta 0} - \zeta\varkappa_\beta}{1 - \zeta/R_\beta}, \\ \gamma_{\alpha\beta} &= \frac{\gamma_{\alpha\beta 0} - 2\zeta\varkappa_{\alpha\beta} + \zeta^2\varkappa^*_{\alpha\beta}}{(1 - \zeta/R_\alpha)(1 - \zeta/R_\beta)}, \end{aligned} \tag{9.49}$$

where

$$\varkappa_\alpha = -\hat{m}_{11}, \quad \varkappa_\beta = -\hat{m}_{22},$$
$$2\varkappa_{\alpha\beta} = -\hat{m}_{21} - \hat{m}_{12} + \frac{l_{12}}{R_\alpha} + \frac{l_{21}}{R_\beta}, \tag{9.50}$$
$$\varkappa^*_{\alpha\beta} = -\frac{\hat{m}_{12}}{R_\alpha} - \frac{\hat{m}_{21}}{R_\beta},$$

$$-\hat{m}_{11} = \frac{1}{A}\frac{\partial l_{31}}{\partial\alpha} + \frac{l_{32}}{AB}\frac{\partial A}{\partial\beta}, \quad -\hat{m}_{12} = \frac{1}{B}\frac{\partial l_{31}}{\partial\beta} - \frac{l_{32}}{AB}\frac{\partial B}{\partial\alpha},$$
$$-\hat{m}_{21} = \frac{1}{A}\frac{\partial l_{32}}{\partial\alpha} - \frac{l_{31}}{AB}\frac{\partial A}{\partial\beta}, \quad -\hat{m}_{22} = \frac{1}{B}\frac{\partial l_{32}}{\partial\beta} + \frac{l_{31}}{AB}\frac{\partial B}{\partial\alpha}. \tag{9.51}$$

We note here several formulae which will be useful in subsequent formulations:

$$\varkappa_{\alpha\beta} = -\hat{m}_{21} + \frac{l_{12}}{R_\alpha} = -\hat{m}_{12} + \frac{l_{21}}{R_\beta}. \tag{9.52}$$
$$\gamma_{\alpha\beta} = \frac{l_{21} + \zeta\hat{m}_{21}}{1 - \zeta/R_\alpha} + \frac{l_{12} + \zeta\hat{m}_{12}}{1 - \zeta/R_\beta}, \tag{9.53}$$
$$\varkappa^*_{\alpha\beta} = \left(\frac{1}{R_\alpha} + \frac{1}{R_\beta}\right)\varkappa_{\alpha\beta} - \frac{1}{R_\alpha R_\beta}\gamma_{\alpha\beta 0}, \tag{9.54}$$

where the conditions of Codazzi are used for the proof of Eqs. (9.52).

9.4. A Linearized Thin Shell Theory under the Kirchhoff–Love Hypothesis

We shall begin this section by prescribing the following thin shell problem. The body forces, together with the forces applied on the upper and lower surfaces of the shell, are defined per unit area of the middle surface S_m, and their components are defined by

$$\bar{\mathbf{Y}} = \bar{Y}_\alpha \mathbf{a}^{(0)} + \bar{Y}_\beta \mathbf{b}^{(0)} + \bar{Y}_n \mathbf{n}^{(0)}. \tag{9.55}$$

The external force $\bar{\mathbf{F}}$ is applied on the part S_1 of the side boundary S and its components are defined by

$$\bar{\mathbf{F}} = \bar{F}_\alpha \mathbf{a}^{(0)} + \bar{F}_\beta \mathbf{b}^{(0)} + \bar{F}_n \mathbf{n}^{(0)}. \tag{9.56}$$

The displacement components are prescribed on the remaining part S_2 of the side boundary S.

We shall derive a linearized thin shell theory for this problem under the Kirchhoff–Love hypothesis. The principle of virtual work for this problem may be written as:

$$\iiint_V (\sigma_\alpha\,\delta\varepsilon_\alpha + \sigma_\beta\,\delta\varepsilon_\beta + \tau_{\alpha\beta}\,\delta\gamma_{\alpha\beta})\,dV$$
$$- \iint_{S_m} (\bar{Y}_\alpha\,\delta u + \bar{Y}_\beta\,\delta v + \bar{Y}_n\,\delta w)\,AB\,d\alpha\,d\beta$$
$$- \iint_{S_1} (\bar{F}_\alpha\,\delta U + \bar{F}_\beta\,\delta V + \bar{F}_n\,\delta W)\,dS = 0, \tag{9.57}$$ †

† See Eq. (4.84).

where Eqs. (9.48) and (9.49) as well as the geometrical relations (9.14) and (9.16) have been substituted. Before proceeding to the reduction of Eqs. (9.57), we shall introduce the following notation for the stress resultants:

$$N_\alpha = \int \sigma_\alpha\left(1 - \frac{\zeta}{R_\beta}\right) d\zeta, \quad N_\beta = \int \sigma_\beta\left(1 - \frac{\zeta}{R_\alpha}\right) d\zeta,$$

$$N_{\alpha\beta} = \int \tau_{\alpha\beta}\left(1 - \frac{\zeta}{R_\beta}\right) d\zeta, \quad N_{\beta\alpha} = \int \tau_{\beta\alpha}\left(1 - \frac{\zeta}{R_\alpha}\right) d\zeta, \tag{9.58}$$

$$M_\alpha = \int \sigma_\alpha\left(1 - \frac{\zeta}{R_\beta}\right) \zeta d\zeta, \quad M_\beta = \int \sigma_\beta\left(1 - \frac{\zeta}{R_\alpha}\right) \zeta\, d\zeta,$$

$$M_{\alpha\beta} = \int \tau_{\alpha\beta}\left(1 - \frac{\zeta}{R_\beta}\right) \zeta\, d\zeta, \quad M_{\beta\alpha} = \int \tau_{\beta\alpha}\left(1 - \frac{\zeta}{R_\alpha}\right) \zeta\, d\zeta, \tag{9.59}$$

$$AB\tilde{Q}_\alpha = \frac{\partial}{\partial\alpha}(BM_\alpha) + \frac{\partial}{\partial\beta}(AM_{\beta\alpha}) + \frac{\partial A}{\partial\beta} M_{\alpha\beta} - \frac{\partial B}{\partial\alpha} M_\beta,$$

$$AB\tilde{Q}_\beta = \frac{\partial}{\partial\beta}(AM_\beta) + \frac{\partial}{\partial\alpha}(BM_{\alpha\beta}) + \frac{\partial B}{\partial\alpha} M_{\beta\alpha} - \frac{\partial A}{\partial\beta} M_\alpha, \tag{9.60}$$

$$N_{\alpha\nu} = N_\alpha l + N_{\beta\alpha} m, \quad N_{\beta\nu} = N_{\alpha\beta} l + N_\beta m, \tag{9.61}$$

$$M_{\alpha\nu} = M_\alpha l + M_{\beta\alpha} m, \quad M_{\beta\nu} = M_{\alpha\beta} l + M_\beta m,$$

$$M_\nu = M_{\alpha\nu} l + M_{\beta\nu} m, \quad M_{\nu s} = -M_{\alpha\nu} m + M_{\beta\nu} l, \tag{9.62}$$

$$V_n = \tilde{Q}_\alpha l + \tilde{Q}_\beta m \tag{9.63}$$

and

$$\bar{N}_{\alpha\nu} = \int \bar{F}_\alpha H\, d\zeta, \quad \bar{N}_{\beta\nu} = \int \bar{F}_\beta H\, d\zeta, \quad \bar{V}_n = \int \bar{F}_n H\, d\zeta,$$

$$\bar{M}_{\alpha\nu} = \int \bar{F}_\alpha H\zeta\, d\zeta, \quad \bar{M}_{\beta\nu} = \int \bar{F}_\beta H\zeta\, d\zeta, \tag{9.64}$$

$$\bar{M}_\nu = \bar{M}_{\alpha\nu} l + \bar{M}_{\beta\nu} m, \quad \bar{M}_{\nu s} = -\bar{M}_{\alpha\nu} m + \bar{M}_{\beta\nu} l.$$

Here $H(\zeta)$ is obtained from Eq. (9.16) as

$$H(\zeta) = \sqrt{[m(1 - \zeta/R_\alpha)]^2 + [l(1 - \zeta/R_\beta)]^2}. \tag{9.65}$$

In Eqs. (9.58), (9.59), and (9.64), as well as throughout the present section, the integrations with respect to ζ extend from $\zeta = -h/2$ to $\zeta = h/2$, where h denotes the thickness of the shell. The quantities defined by Eqs. (9.58) and (9.59) are stress resultants and moments per unit length of the coordinate curves α and β of the middle surface as shown in Fig. 9.5. The quantities N_α, N_β, $N_{\alpha\beta}$ and $N_{\beta\alpha}$ are in-plane stress resultants, while M_α, M_β, $M_{\alpha\beta}$ and $M_{\beta\alpha}$ are bending and twisting moments. The quantities $\tilde{Q}_\alpha$ and $\tilde{Q}_\beta$ defined by Eqs. (9.60) are proved equal to shearing forces Q_α and Q_β defined per unit length of the curves α and β of the middle surface by considering the equilibrium conditions with respect to moments of the shell element in the figure.† The quantities defined by Eqs. (9.64) are prescribed external forces and

† See the footnote of Section 7.2 and Section 8.2 for similar developments.

moments per unit length along the boundary. It is seen that $\bar{V}_n$ is the shearing force acting in the direction of the normal $\mathbf{n}^{(0)}$, while $\bar{M}_\nu$ and $\bar{M}_{\nu s}$ are bending and twisting moments on the boundary. With the aid of these relations, we have

$$\iiint_V (\sigma_\alpha\, \delta\varepsilon_\alpha + \sigma_\beta\, \delta\varepsilon_\beta + \tau_{\alpha\beta}\, \delta\gamma_{\alpha\beta})\, AB\left(1 - \frac{\zeta}{R_\alpha}\right)\left(1 - \frac{\zeta}{R_\beta}\right) d\alpha\, d\beta\, d\zeta$$

$$= \iint_{S_m} [N_\alpha\, \delta l_{11} + N_{\alpha\beta}\, \delta l_{21} + \tilde{Q}_\alpha\, \delta l_{31} + N_{\beta\alpha}\, \delta l_{12} + N_\beta\, \delta l_{22}$$
$$+ \tilde{Q}_\beta\, \delta l_{32}]\, AB\, d\alpha\, d\beta$$
$$- \int_{C_1+C_2} [M_{\alpha\nu}\, \delta l_{31} + M_{\beta\nu}\, \delta l_{32}]\, ds. \tag{9.66}$$

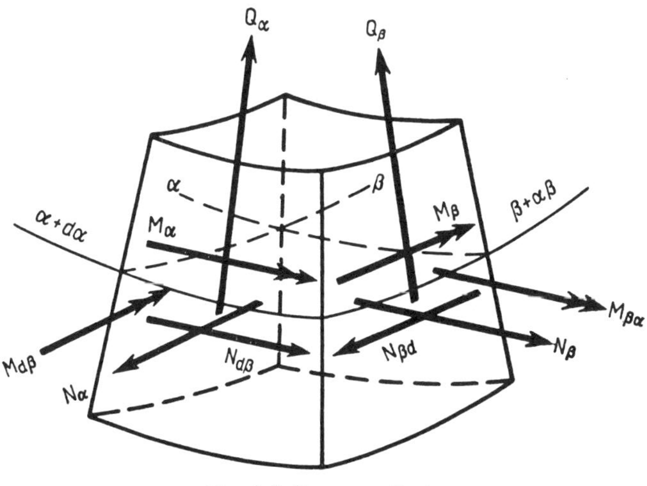

FIG. 9.5. Stress resultants.

Substituting Eq. (9.66) into Eq. (9.57), and using Eqs. (9.18), we obtain finally,

$$- \iint_{S_m} \left\{\left[(BN_\alpha)_{,\alpha} + (AN_{\beta\alpha})_{,\beta} + \frac{\partial A}{\partial \beta} N_{\alpha\beta} - \frac{\partial B}{\partial \alpha} N_\beta - \frac{AB}{R_\alpha} \tilde{Q}_\alpha + \bar{Y}_\alpha AB\right] \delta u\right.$$
$$\left. + [\cdots]\, \delta v + \left[(B\tilde{Q}_\alpha)_{,\alpha} + (A\tilde{Q}_\beta)_{,\beta} + AB\left(\frac{N_\alpha}{R_\alpha} + \frac{N_\beta}{R_\beta}\right) + \bar{Y}_n AB\right] \delta w\right\} d\alpha\, d\beta$$
$$+ \int_{C_1} \left\{\left[\left(N_{\alpha\nu} - \frac{M_{\alpha\nu}}{R_\alpha}\right) - \left(\bar{N}_{\alpha\nu} - \frac{\bar{M}_{\alpha\nu}}{R_\alpha}\right)\right] \delta u + [\cdots]\, \delta v\right.$$
$$\left. + [(V_n + M_{\nu s,s}) - (\bar{V}_n + \bar{M}_{\nu s,s})]\, \delta w - (M_\nu - \bar{M}_\nu)\, \delta w_{,\nu}\right\} ds$$
$$+ (\text{integrals on } C_2) = 0. \tag{9.67}$$

Consequently, the equations of equilibrium are:

$$\frac{\partial}{\partial\alpha}(BN_\alpha) + \frac{\partial}{\partial\beta}(AN_{\beta\alpha}) + \frac{\partial A}{\partial\beta}N_{\alpha\beta} - \frac{\partial B}{\partial\alpha}N_\beta - \frac{AB}{R_\alpha}\tilde{Q}_\alpha + \bar{Y}_\alpha AB = 0,$$
$$\frac{\partial}{\partial\beta}(AN_\beta) + \frac{\partial}{\partial\alpha}(BN_{\alpha\beta}) + \frac{\partial B}{\partial\alpha}N_{\beta\alpha} - \frac{\partial A}{\partial\beta}N_\alpha - \frac{AB}{R_\beta}\tilde{Q}_\beta + \bar{Y}_\beta AB = 0, \quad (9.68)$$
$$\frac{\partial}{\partial\alpha}(B\tilde{Q}_\alpha) + \frac{\partial}{\partial\beta}(A\tilde{Q}_\beta) + AB\left(\frac{N_\alpha}{R_\alpha} + \frac{N_\beta}{R_\beta}\right) + \bar{Y}_n AB = 0.$$

The boundary conditions on C_1 become:

$$N_{\alpha\nu} - \frac{M_{\alpha\nu}}{R_\alpha} = \bar{N}_{\alpha\nu} - \frac{\bar{M}_{\alpha\nu}}{R_\alpha}, \quad N_{\beta\nu} - \frac{M_{\beta\nu}}{R_\beta} = \bar{N}_{\beta\nu} - \frac{\bar{M}_{\beta\nu}}{R_\beta},$$
$$V_n + \frac{\partial M_{\nu s}}{\partial s} = \bar{V}_n + \frac{\partial \bar{M}_{\nu s}}{\partial s}, \quad M_\nu = \bar{M}_\nu. \quad (9.69\text{a, b, c, d})$$

Equation (9.69c) shows that under the Kirchhoff–Love hypothesis, the actions of the twisting moments $M_{\nu s}$ and $\bar{M}_{\nu s}$ distributed along the boundary are replaced by that of the shearing forces V_n and $\bar{V}_n$, respectively. This result is similar to the result we encountered in Section 8.2. Equation (9.67) also suggests that in this approximate theory, the boundary conditions on C_2 are specified as follows:

$$u = \bar{u}, \quad v = \bar{v}, \quad w = \bar{w}, \quad \frac{\partial w}{\partial \nu} = \overline{\frac{\partial w}{\partial \nu}}. \quad (9.70)$$

Next, we shall obtain relations between the stress resultants and strains. We may combine Eqs. (9.23), (9.49), (9.58) and (9.59) to yield the following relations:

$$N_\alpha = \frac{E}{(1-\nu^2)}\int\left[\frac{\varepsilon_{\alpha 0} - \zeta\varkappa_\alpha}{1-\zeta/R_\alpha} + \nu\frac{\varepsilon_{\beta 0} - \zeta\varkappa_\beta}{1-\zeta/R_\beta}\right]\left(1 - \frac{\zeta}{R_\beta}\right)d\zeta, \ldots,$$
$$M_{\beta\alpha} = G\int\frac{\left(1 - \frac{\zeta^2}{R_\alpha R_\beta}\right)\gamma_{\alpha\beta 0} - 2\left[1 - \frac{\zeta}{2}\left(\frac{1}{R_\alpha} + \frac{1}{R_\beta}\right)\right]\zeta\varkappa_{\alpha\beta}}{1 - \zeta/R_\beta}\zeta\, d\zeta. \quad (9.71)$$

Since the exact evaluation of these integrals lead to very cumbersome stress resultant–strain relations, Lur'e proposed to expand the integrands of the integrals in Eqs. (9.71) in power series of ζ and discard terms higher than ζ^2 before integration, thus obtaining:[(1)]

$$N_\alpha = \frac{Eh}{(1-\nu^2)}\left[\varepsilon_{\alpha 0} + \nu\varepsilon_{\beta 0} + \frac{h^2}{12}\left(\frac{1}{R_\alpha} - \frac{1}{R_\beta}\right)\left(\frac{\varepsilon_{\alpha 0}}{R_\alpha} - \varkappa_\alpha\right)\right],$$
$$N_\beta = \frac{Eh}{(1-\nu^2)}\left[\varepsilon_{\beta 0} + \nu\varepsilon_{\alpha 0} + \frac{h^2}{12}\left(\frac{1}{R_\beta} - \frac{1}{R_\alpha}\right)\left(\frac{\varepsilon_{\beta 0}}{R_\beta} - \varkappa_\beta\right)\right],$$
$$N_{\alpha\beta} = Gh\left[\gamma_{\alpha\beta 0} + \frac{h^2}{12}\left(\frac{1}{R_\alpha} - \frac{1}{R_\beta}\right)\left(\frac{\gamma_{\alpha\beta 0}}{R_\alpha} - \varkappa_{\alpha\beta}\right)\right],$$
$$N_{\beta\alpha} = Gh\left[\gamma_{\alpha\beta 0} + \frac{h^2}{12}\left(\frac{1}{R_\beta} - \frac{1}{R_\alpha}\right)\left(\frac{\gamma_{\alpha\beta 0}}{R_\beta} - \varkappa_{\alpha\beta}\right)\right], \quad (9.72)$$

$$M_\alpha = -D\left[\varkappa_\alpha + \nu\varkappa_\beta - \left(\frac{1}{R_\alpha} - \frac{1}{R_\beta}\right)\varepsilon_{\alpha 0}\right],$$

$$M_\beta = -D\left[\varkappa_\beta + \nu\varkappa_\alpha - \left(\frac{1}{R_\beta} - \frac{1}{R_\alpha}\right)\varepsilon_{\beta 0}\right],$$

$$M_{\alpha\beta} = -D(1-\nu)\left(\varkappa_{\alpha\beta} - \frac{\gamma_{\alpha\beta 0}}{2R_\alpha}\right),$$

$$M_{\beta\alpha} = -D(1-\nu)\left(\varkappa_{\alpha\beta} - \frac{\gamma_{\alpha\beta 0}}{2R_\beta}\right), \tag{9.73}$$

where $D = Eh^3/12(1-\nu^2)$ is the bending rigidity of the shell. The strain energy of the shell is calculable with the aid of Eqs. (9.49) and Eq. (3) of Appendix B, and is expressed in the accuracy of Lur'e's approximation as follows:

$$U = \tfrac{1}{2}\iint_{S_m} [N_\alpha\varepsilon_{\alpha 0} + N_\beta\varepsilon_{\beta 0} + S^*_{\alpha\beta}\gamma_{\alpha\beta 0} - M_\alpha\varkappa_\alpha - M_\beta\varkappa_\beta - 2M^*_{\alpha\beta}\varkappa_{\alpha\beta}]\, AB\, d\alpha\, d\beta, \tag{9.74}$$

where Eqs. (9.72), (9.73) and

$$\begin{bmatrix} S^*_{\alpha\beta} \\ M^*_{\alpha\beta} \end{bmatrix} = \begin{bmatrix} C_{11} & C_{12} \\ C_{21} & C_{22} \end{bmatrix} \begin{bmatrix} \gamma_{\alpha\beta 0} \\ -2\varkappa_{\alpha\beta} \end{bmatrix}, \tag{9.75}$$

$$C_{11} = Gh\left[1 + \frac{h^2}{12}\left(\frac{1}{R_\alpha^2} - \frac{1}{R_\alpha R_\beta} + \frac{1}{R_\beta^2}\right)\right],$$

$$C_{12} = C_{21} = \frac{1}{4}D(1-\nu)\left(\frac{1}{R_\alpha} + \frac{1}{R_\beta}\right), \tag{9.76}$$

$$C_{22} = \tfrac{1}{2}D(1-\nu),$$

have been substituted to express the stress resultants in terms of the displacements. We note that the accuracy of Eqs. (9.72) and (9.73) is found an apparent one if we consider the assumptions and hypothesis on which the present thin shell theory is based:[1, 2] the terms containing $h^2/12$ in Eqs. (9.72) and those containing $\varepsilon_{\alpha 0}$, $\varepsilon_{\beta 0}$ or $\gamma_{\alpha\beta 0}$ in Eqs. (9.73) are usually very small and may be neglected in comparison with the preceding terms. Consequently, these relations have been used very rarely in their original form for practical purposes.

9.5. Simplified Formulations

Since the thin shell theory derived in Section 9.4 leads to rather cumbersome formulations of the shell problem, we shall be interested in deriving simpler formulations in the remainder of this chapter. We shall adopt a simplifying assumption that the shell is so thin that terms of smaller magnitude may be neglected in geometrical as well as strain–displacement relations. The quantities h/R_α and h/R_β may be considered negligible in comparison with unity under this simplifying assumption. To begin with, we admit that Eqs. (9.14) and (9.16) may reduce to

$$dV = AB\,d\alpha\,d\beta\,d\zeta, \tag{9.77}$$

$$dS = ds\,d\zeta, \tag{9.78}$$

respectively. The factor $H(\zeta)$ in Eqs. (9.64) may be taken equal to unity due to the same reason.

Next, we shall proceed to the reduction of the strain–displacement relations to simpler form and summarize results of some considerations as follows:

(a) *Linearized thin shell theory including the effect of transverse shear deformation.* We allow that Eqs. (9.35) reduce to

$$\begin{aligned} \varepsilon_\alpha &= \varepsilon_{\alpha 0} - \zeta k_\alpha, \quad \varepsilon_\beta = \varepsilon_{\beta 0} - \zeta k_\beta, \\ \gamma_{\alpha\beta} &= \gamma_{\alpha\beta 0} - 2\zeta k_{\alpha\beta}, \\ \gamma_{\alpha\zeta} &= \gamma_{\alpha\zeta 0}, \quad \gamma_{\beta\zeta} = \gamma_{\beta\zeta 0}, \end{aligned} \tag{9.79}$$

while the displacement components are still given by Eqs. (9.30).

(b) *Linearized thin shell theory under the Kirchhoff–Love hypothesis.* We allow that Eqs. (9.49) reduce to

$$\begin{aligned} \varepsilon_\alpha &= \varepsilon_{\alpha 0} - \zeta \varkappa_\alpha, \quad \varepsilon_\beta = \varepsilon_{\beta 0} - \zeta \varkappa_\beta, \\ \gamma_{\alpha\beta} &= \gamma_{\alpha\beta 0} - 2\zeta \varkappa_{\alpha\beta}, \end{aligned} \tag{9.80}$$

while the displacement components are still given by Eqs. (9.48).

We shall derive in this connection nonlinear strain–displacement relations under the Kirchhoff–Love hypothesis plus the simplifying assumption. The exact relations may be obtained by employing Eq. (9.45) as an expression for the displacements and calculating the strain tensors in a manner similar to the development in Section 9.2. However, we shall be satisfied with obtaining approximate strain–displacement relations: we shall retain nonlinear terms in the expressions of the strains of the middle surface, but we shall retain only linear terms in the formulation of curvatures. Thus, we allow that the nonlinear strain–displacement relations reduce to

$$\begin{aligned} e_{\alpha\alpha} &= e_{\alpha\alpha 0} - \zeta \varkappa_\alpha, \quad e_{\beta\beta} = e_{\beta\beta 0} - \zeta \varkappa_\beta, \\ e_{\alpha\beta} &= e_{\alpha\beta 0} - \zeta \varkappa_{\alpha\beta}, \end{aligned} \tag{9.81}$$

where

$$\begin{aligned} 2e_{\alpha\alpha 0} &= (1 + l_{11})^2 + l_{21}^2 + l_{31}^2 - 1, \\ 2e_{\beta\beta 0} &= l_{12}^2 + (1 + l_{22})^2 + l_{32}^2 - 1, \\ 2e_{\alpha\beta 0} &= (1 + l_{11})\,l_{12} + l_{21}(1 + l_{22}) + l_{31}l_{32}, \end{aligned} \tag{9.82}$$

while the displacements are still given by Eqs. (9.48). It is obvious that the linearization of the curvature terms in Eqs. (9.81) and the use of Eqs. (9.48) restrict the field of application of a nonlinear theory founded on these relations. However, this choice is considered useful in application to shell problems such as buckling or vibrational problems of shells where small displacement motions are executed about equilibrium configurations with initial membrane stresses. See Ref. 3 for more detailed considerations on finite displacement theories of thin shells.

9.6. A Simplified Linear Theory under the Kirchhoff–Love Hypothesis

We shall consider again the thin shell problem presented in Section 9.4 and derive for it a linear theory by the use of Eqs. (9.48), (9.77), (9.78) and (9.80). The principle of virtual work is employed for the derivation of governing equations. The principle suggests the adoption of the following definition of the stress resultants:

$$N_\alpha = \int \sigma_\alpha \, d\zeta, \quad N_\beta = \int \sigma_\beta \, d\zeta,$$

$$S_{\alpha\beta} = \int \tau_{\alpha\beta} \, d\zeta, \quad S_{\beta\alpha} = \int \tau_{\beta\alpha} \, d\zeta, \tag{9.83†}$$

$$M_\alpha = \int \sigma_\alpha \zeta \, d\zeta, \quad M_\beta = \int \sigma_\beta \zeta \, d\zeta,$$

$$M_{\alpha\beta} = \int \tau_{\alpha\beta} \zeta \, d\zeta, \quad M_{\beta\alpha} = \int \tau_{\beta\alpha} \zeta \, d\zeta, \tag{9.84†}$$

$$N_{\alpha\beta} = S_{\alpha\beta} - \frac{M_{\alpha\beta}}{R_\beta}, \quad N_{\beta\alpha} = S_{\beta\alpha} - \frac{M_{\beta\alpha}}{R_\alpha}. \tag{9.85†}$$

After some calculation, we find that in this simplified linear theory of the thin shell, the equations of equilibrium as well as the mechanical and geometrical boundary conditions are derived in the same form as those in Section 9.4. However, the stress resultant–strain relations and the expression of the strain energy of the shell are now given in simpler form as follows:

$$N_\alpha = \frac{Eh}{(1-\nu^2)} [\varepsilon_{\alpha 0} + \nu \varepsilon_{\beta 0}], \quad N_\beta = \frac{Eh}{(1-\nu^2)} [\nu \varepsilon_{\alpha 0} + \varepsilon_{\beta 0}],$$

$$S_{\alpha\beta} = S_{\beta\alpha} = Gh\gamma_{\alpha\beta 0}. \tag{9.86‡}$$

$$M_\alpha = -D[\varkappa_\alpha + \nu \varkappa_\beta], \quad M_\beta = -D[\nu \varkappa_\alpha + \varkappa_\beta],$$

$$M_{\alpha\beta} = M_{\beta\alpha} = -D(1-\nu)\, \varkappa_{\alpha\beta}. \tag{9.87‡}$$

$$N_{\alpha\beta} = S_{\alpha\beta} - M_{\alpha\beta}/R_\beta = Gh[\gamma_{\alpha\beta 0} + (h^2/6R_\beta)\, \varkappa_{\alpha\beta}],$$

$$N_{\beta\alpha} = S_{\beta\alpha} - M_{\beta\alpha}/R_\alpha = Gh[\gamma_{\alpha\beta 0} + (h^2/6R_\alpha)\, \varkappa_{\alpha\beta}]. \tag{9.88‡}$$

$$U = \frac{1}{2} \iint_{S_m} \Big\{ \frac{Eh}{(1-\nu^2)} \Big[(\varepsilon_{\alpha 0} + \varepsilon_{\beta 0})^2 + 2(1-\nu) \Big(\frac{1}{4} \gamma_{\alpha\beta 0}^2 - \varepsilon_{\alpha 0} \varepsilon_{\beta 0} \Big) \Big]$$

$$+ D[(\varkappa_\alpha + \varkappa_\beta)^2 + 2(1-\nu)(\varkappa_{\alpha\beta}^2 - \varkappa_\alpha \varkappa_\beta)] \Big\} AB \, d\alpha \, d\beta. \tag{9.89}$$

† Compare these equations with Eqs. (9.58) and (9.59).
‡ Compare these equations with Eqs. (9.72) and (9.73).

We note here that the second terms in the right-hand sides of Eqs. (9.88) are frequently neglected in comparison with the first terms to obtain the following simpler relations:

$$N_{\alpha\beta} = N_{\beta\alpha} = Gh\gamma_{\alpha\beta 0}. \qquad (9.90)\dagger$$

The use of Eqs. (9.90) for practical purposes may be considered justified if we remember the assumptions on which the present thin shell theory is based. However, Eqs. (9.88) are employed for theoretical presentations because this choice is consistent with the results derived from the principle of virtual work or the principle of minimum potential energy.

9.7. A Nonlinear Thin Shell Theory under the Kirchhoff–Love Hypothesis

We shall consider the thin shell problem presented in Section 9.4 and derive for it a nonlinear theory under the Kirchhoff–Love hypothesis by the use of Eqs. (9.81). The principle of virtual work for the present problem may be written as

$$\begin{aligned}\iiint_V (\sigma_\alpha\, \delta e_{\alpha\alpha} + \sigma_\beta\, \delta e_{\beta\beta} + 2\tau_{\alpha\beta}\, \delta e_{\alpha\beta})\, dV \\ - \iint_{S_m} (\bar{Y}_\alpha\, \delta u + \bar{Y}_\beta\, \delta v + \bar{Y}_n\, \delta w)\, AB\, d\alpha\, d\beta \\ - \iint_{S_1} (\bar{F}_\alpha\, \delta U + \bar{F}_\beta\, \delta V + \bar{F}_n\, \delta W)\, dS = 0,\end{aligned} \qquad (9.91)\ddagger$$

where Eqs. (9.48), (9.77), (9.78) and (9.81) have been substituted. With the aid of the stress resultants defined by Eqs. (9.83), (9.84) and (9.85), we have

$$\begin{aligned}&\iiint_V (\sigma_\alpha\, \delta e_{\alpha\alpha} + \sigma_\beta\, \delta e_{\beta\beta} + 2\tau_{\alpha\beta}\, \delta e_{\alpha\beta})\, AB\, d\alpha\, d\beta\, d\zeta \\ &\quad = \iint_{S_m} [(N_\alpha + N_\alpha l_{11} + S_{\alpha\beta} l_{12})\, \delta l_{11} + (N_{\beta\alpha} + S_{\beta\alpha} l_{11} + N_\beta l_{12})\, \delta l_{12} \\ &\qquad + (N_{\alpha\beta} + N_\alpha l_{21} + S_{\alpha\beta} l_{22})\, \delta l_{21} + (N_\beta + S_{\beta\alpha} l_{21} + N_\beta l_{22})\, \delta l_{22} \\ &\qquad + (\tilde{Q}_\alpha + N_\alpha l_{31} + S_{\alpha\beta} l_{32})\, \delta l_{31} + (\tilde{Q}_\beta + S_{\beta\alpha} l_{31} + N_\beta l_{32})\, \delta l_{32}]\, AB\, d\alpha\, d\beta \\ &\qquad - \int_{C_1 + C_2} [M_{\alpha\nu}\, \delta l_{31} + M_{\beta\nu}\, \delta l_{32}]\, ds.\end{aligned} \qquad (9.92)$$

Comparing Eq. (9.92) with Eq. (9.66), we find that the following replacements yield the desired equations of equilibrium and mechanical boundary conditions for the nonlinear theory:

$$\begin{aligned}&N_\alpha \text{ by } N_\alpha + N_\alpha l_{11} + S_{\alpha\beta} l_{12}, \quad N_{\beta\alpha} \text{ by } N_{\beta\alpha} + S_{\beta\alpha} l_{11} + N_\beta l_{12}, \\ &N_{\alpha\beta} \text{ by } N_{\alpha\beta} + N_\alpha l_{21} + S_{\alpha\beta} l_{22}, \quad N_\beta \text{ by } N_\beta + S_{\beta\alpha} l_{21} + N_\beta l_{22}, \\ &\tilde{Q}_\alpha \text{ by } \tilde{Q}_\alpha + N_\alpha l_{31} + S_{\alpha\beta} l_{32}, \quad \tilde{Q}_\beta \text{ by } \tilde{Q}_\beta + S_{\beta\alpha} l_{31} + N_\beta l_{32}.\end{aligned} \qquad (9.93)$$

† Compare these equations with Eqs. (9.72) and (9.73).

‡ See Eq. (4.84).

Thus, we have for the equations of equilibrium in S_m,

$$\begin{aligned}
&\frac{\partial}{\partial\alpha}\{B[N_\alpha + N_\alpha l_{11} + S_{\alpha\beta} l_{12}]\} + \frac{\partial}{\partial\beta}\{A[N_{\beta\alpha} + S_{\beta\alpha} l_{11} + N_\beta l_{12}]\} \\
&\quad + \frac{\partial A}{\partial\beta}[N_{\alpha\beta} + N_\alpha l_{21} + S_{\alpha\beta} l_{22}] - \frac{\partial B}{\partial\alpha}[N_\beta + S_{\beta\alpha} l_{21} + N_\beta l_{22}] \\
&\quad - \frac{AB}{R_\alpha}[\tilde{Q}_\alpha + N_\alpha l_{31} + S_{\alpha\beta} l_{32}] + \bar{Y}_\alpha AB = 0, \\
&\frac{\partial}{\partial\beta}\{A[N_\beta + S_{\beta\alpha} l_{21} + N_\beta l_{32}]\} + \frac{\partial}{\partial\alpha}\{B[N_{\alpha\beta} + N_\alpha l_{21} + S_{\alpha\beta} l_{22}]\} \\
&\quad + \frac{\partial B}{\partial\alpha}[N_{\beta\alpha} + S_{\beta\alpha} l_{11} + N_\beta l_{12}] - \frac{\partial A}{\partial\beta}[N_\alpha + N_\alpha l_{11} + S_{\alpha\beta} l_{12}] \\
&\quad - \frac{AB}{R_\beta}[\tilde{Q}_\beta + S_{\beta\alpha} l_{31} + N_\beta l_{32}] + \bar{Y}_\beta AB = 0, \\
&\frac{\partial}{\partial\alpha}\{B[\tilde{Q}_\alpha + N_\alpha l_{31} + S_{\alpha\beta} l_{32}]\} + \frac{\partial}{\partial\beta}\{A[\tilde{Q}_\beta + S_{\beta\alpha} l_{31} + N_\beta l_{32}]\} \\
&\quad + \frac{AB}{R_\alpha}[N_\alpha + N_\alpha l_{11} + S_{\alpha\beta} l_{12}] + \frac{AB}{R_\beta}[N_\beta + S_{\beta\alpha} l_{21} + N_\beta l_{22}] \\
&\quad + \bar{Y}_n AB = 0
\end{aligned} \tag{9.94}$$

and for the mechanical boundary conditions on C_1,

$$\begin{aligned}
&[N_\alpha + N_\alpha l_{11} + S_{\alpha\beta} l_{12}]\, l + [N_{\beta\alpha} + S_{\beta\alpha} l_{11} + N_\beta l_{12}]\, m - \frac{M_{\alpha\nu}}{R_\alpha} = \bar{N}_{\alpha\nu} - \frac{\bar{M}_{\alpha\nu}}{R_\alpha}, \\
&[N_{\alpha\beta} + N_\alpha l_{21} + S_{\alpha\beta} l_{22}]\, l + [N_\beta + S_{\beta\alpha} l_{21} + N_\beta l_{22}]\, m - \frac{M_{\beta\nu}}{R_\beta} = \bar{N}_{\beta\nu} - \frac{\bar{M}_{\beta\nu}}{R_\beta}, \\
&[\tilde{Q}_\alpha + N_\alpha l_{31} + S_{\alpha\beta} l_{32}]\, l + (\tilde{Q}_\beta + S_{\beta\alpha} l_{31} + N_\beta l_{32}]\, m + \frac{\partial M_{\nu s}}{\partial s} = \bar{V}_n + \frac{\partial \bar{M}_{\nu s}}{\partial s}, \\
&\qquad M_\nu = \bar{M}_\nu.
\end{aligned} \tag{9.95}$$

The geometrical boundary conditions are still given by Eqs. (9.70). The stress resultant–strain relations and the expression for the strain energy of the nonlinear theory are obtained from Eqs. (9.86), (9.87), (9.88) and (9.89) by replacing $\varepsilon_{\alpha 0}$, $\varepsilon_{\beta 0}$ and $\gamma_{\alpha\beta 0}$ with $e_{\alpha\alpha 0}$, $e_{\beta\beta 0}$ and $2e_{\alpha\beta 0}$, respectively.†

9.8. A Linearized Thin Shell Theory Including the Effect of Transverse Shear Deformations

We shall consider again the problem prescribed in Section 9.4 and derive for it a linearized thin shell theory including the effect of transverse shear deformation in which all the prescribed forces and boundary conditions are

† As mentioned at the end of the last section, we may replace $S_{\alpha\beta}$ and $S_{\beta\alpha}$ in Eqs. (9.94) and (9.95) by $N_{\alpha\beta}$ and $N_{\beta\alpha}$, respectively, and employ the simpler relations $N_{\alpha\beta} = N_{\beta\alpha} = S_{\alpha\beta} = S_{\beta\alpha} = 2Ghe_{\alpha\beta 0}$ for practical purposes.

now assumed to be time-dependent. The principle of virtual work for the present dynamical problem may be written as

$$\int_{t_1}^{t_2}\Big\{\iiint_V [\sigma_\alpha\,\delta\varepsilon_\alpha + \sigma_\beta\,\delta\varepsilon_\beta + \tau_{\alpha\beta}\,\delta\gamma_{\alpha\beta} + \tau_{\alpha\zeta}\,\delta\gamma_{\alpha\zeta} + \tau_{\beta\zeta}\,\delta\gamma_{\beta\zeta}]\,dV$$
$$-\,\delta\iiint_V \tfrac{1}{2}\varrho(\dot{U}^2 + \dot{V}^2 + \dot{W}^2)\,dV$$
$$-\iint_{S_m}(\bar{Y}_\alpha\,\delta u + \bar{Y}_\beta\,\delta v + \bar{Y}_n\,\delta w)\,AB\,d\alpha\,d\beta$$
$$-\iint_{S_1}[\bar{F}_\alpha\,\delta U + \bar{F}_\beta\,\delta V + \bar{F}_n\,\delta W]\,dS\Big\}\,dt = 0. \qquad (9.96)†$$

We shall adopt the simplifying assumption introduced in Section 9.5 and employ Eqs. (9.30), (9.77), (9.78) and (9.79) for the formulation. Here, we introduce the following new definitions in addition to Eqs. (9.83), (9.84) and (9.85):

$$Q_\alpha = \int \tau_{\alpha\zeta}\,d\zeta, \quad Q_\beta = \int \tau_{\beta\zeta}\,d\zeta, \qquad (9.97)$$

$$m = \varrho hAB, \qquad I_m = \frac{1}{12}\varrho h^3 AB. \qquad (9.98)$$

The quantities defined by Eqs. (9.97) are shearing forces per unit length of the coordinate curves of the middle surface as shown in Fig. 9.5. The quantities defined by Eqs. (9.98) are related to the mass and mass moment of inertia of the shell element shown in the same figure. With the aid of the stress resultants thus defined, we have

$$\iiint_V [\sigma_\alpha\,\delta\varepsilon_\alpha + \sigma_\beta\,\delta\varepsilon_\beta + \tau_{\alpha\beta}\,\delta\gamma_{\alpha\beta} + \tau_{\alpha\zeta}\,\delta\gamma_{\alpha\zeta} + \tau_{\beta\zeta}\,\delta\gamma_{\beta\zeta}]\,dV$$
$$= \iint_{S_m}[N_\alpha\,\delta l_{11} + N_{\alpha\beta}\,\delta l_{21} + Q_\alpha\,\delta l_{31} + Q_\alpha\,\delta u_1$$
$$+ N_{\beta\alpha}\,\delta l_{12} + N_\beta\,\delta l_{22} + Q_\beta\,\delta l_{32} + Q_\beta\,\delta v_1$$
$$+ M_\alpha\,\delta m_{11} + M_{\alpha\beta}\,\delta m_{21} + M_{\beta\alpha}\,\delta m_{12} + M_\beta\,\delta m_{22}]\,AB\,d\alpha\,d\beta. \qquad (9.99)$$

Substituting Eq. (9.99) into Eqs. (9.96) and using Eqs. (9.30) for the displacement components, we have the equations of motion,

$$\frac{\partial}{\partial\alpha}(BN_\alpha) + \frac{\partial}{\partial\beta}(AN_{\beta\alpha}) + \frac{\partial A}{\partial\beta}N_{\alpha\beta} - \frac{\partial B}{\partial\alpha}N_\beta - \frac{AB}{R_\alpha}Q_\alpha + \bar{Y}_\alpha AB = m\ddot{u},$$
$$\frac{\partial}{\partial\beta}(AN_\beta) + \frac{\partial}{\partial\alpha}(BN_{\alpha\beta}) + \frac{\partial B}{\partial\alpha}N_{\beta\alpha} - \frac{\partial A}{\partial\beta}N_\alpha - \frac{AB}{R_\beta}Q_\beta + \bar{Y}_\beta AB = m\ddot{v},$$
$$\frac{\partial}{\partial\alpha}(BQ_\alpha) + \frac{\partial}{\partial\beta}(AQ_\beta) + AB\left(\frac{N_\alpha}{R_\alpha} + \frac{N_\beta}{R_\beta}\right) + \bar{Y}_n AB = m\ddot{w}, \qquad (9.100)$$
$$\frac{\partial}{\partial\alpha}(BM_\alpha) + \frac{\partial}{\partial\beta}(AM_{\beta\alpha}) + \frac{\partial A}{\partial\beta}M_{\alpha\beta} - \frac{\partial B}{\partial\alpha}M_\beta - ABQ_\alpha = I_m\ddot{u}_1,$$
$$\frac{\partial}{\partial\beta}(AM_\beta) + \frac{\partial}{\partial\alpha}(BM_{\alpha\beta}) + \frac{\partial B}{\partial\alpha}M_{\beta\alpha} - \frac{\partial A}{\partial\beta}M_\alpha - ABQ_\beta = I_m\ddot{v}_1,$$

† See Eq. (5.81).

and the boundary conditions on C_1,

$$N_{\alpha\nu} = \bar{N}_{\alpha\nu}, \quad N_{\beta\nu} = \bar{N}_{\beta\nu}, \quad Q_\alpha l + Q_\beta m = \bar{V}_n,$$
$$M_{\alpha\nu} = \bar{M}_{\alpha\nu}, \quad M_{\beta\nu} = \bar{M}_{\beta\nu}, \tag{9.101}$$

while it is suggested that the boundary conditions on C_2 are specified approximately by:

$$u = \bar{u}, \quad v = \bar{v}, \quad w = \bar{w}, \quad u_1 = \bar{u}_1, \quad v_1 = \bar{v}_1 \tag{9.102}$$

The stress resultant–strain relations and the expression of the strain energy are obtained from Eqs. (9.23), (9.79), (9.83), (9.84), (9.85) and (9.97), together with the aid of Eq. (3) in Appendix B, as follows:

$$N_\alpha = \frac{Eh}{(1-\nu^2)}(\varepsilon_{\alpha 0} + \nu\varepsilon_{\beta 0}), \quad N_\beta = \frac{Eh}{(1-\nu^2)}(\nu\varepsilon_{\alpha 0} + \varepsilon_{\beta 0}),$$
$$S_{\alpha\beta} = S_{\beta\alpha} = Gh\gamma_{\alpha\beta 0}, \tag{9.103}$$
$$M_\alpha = -D(k_\alpha + \nu k_\beta), \quad M_\beta = -D(\nu k_\alpha + k_\beta),$$
$$M_{\alpha\beta} = M_{\beta\alpha} = -D(1-\nu)\, k_{\alpha\beta}, \tag{9.104}$$
$$N_{\alpha\beta} = Gh\left[\gamma_{\alpha\beta 0} + \frac{h^2}{6R_\beta} k_{\alpha\beta}\right], \quad N_{\beta\alpha} = Gh\left[\gamma_{\alpha\beta 0} + \frac{h^2}{6R_\alpha} k_{\alpha\beta}\right], \tag{9.105}$$
$$Q_\alpha = Gkh\gamma_{\alpha\zeta 0}, \quad Q_\beta = Gkh\gamma_{\beta\zeta 0}, \tag{9.106}$$
$$U = \frac{1}{2}\iint_{S_m} \left\{ \frac{Eh}{(1-\nu^2)} \left[(\varepsilon_{\alpha 0} + \varepsilon_{\beta 0})^2 + 2(1-\nu)\left(\frac{1}{4}\gamma_{\alpha\beta 0}^2 - \varepsilon_{\alpha 0}\varepsilon_{\beta 0}\right)\right] \right.$$
$$+ D[(k_\alpha + k_\beta)^2 + 2(1-\nu)(k_{\alpha\beta}^2 - k_\alpha k_\beta)]$$
$$\left. + Gkh(\gamma_{\alpha\zeta 0}^2 + \gamma_{\beta\zeta 0}^2) \right\} AB\, d\alpha\, d\beta. \tag{9.107}$$

The factor k in Eqs. (9.106) and (9.107) has been appended to take account of the non-uniformity of the shearing strains $\gamma_{\alpha\zeta}$ and $\gamma_{\beta\zeta}$ over the cross section. For isotropic shell, the factor k may assume the same value as mentioned in Section 8.8.[4]

It is seen from the above formulation that we have five mechanical boundary conditions on C_1 and the same number of geometrical boundary conditions on C_2 and these are compatible with the assumed degree of freedom of the displacement components, i.e. u, v, w, u_1 and v_1. We have replaced the action of $M_{\nu s}$ and $\bar{M}_{\nu s}$ by that of V_n and $\bar{V}_n$ in the thin shell theories under the Kirchhoff–Love hypothesis. However, such replacements are no longer necessary in the thin shell theory including the effect of transverse shear deformation. This is similar to the result we encountered in Section 8.8.

9.9. Some Remarks

Since the appearance of Love's approximate theory, many books have been written on theories of thin shells (see Refs. 1 through 11, for example). Many papers concerning shell problems have been published; an extensive bibliography is given in Ref. 12, while Ref. 13 presents a survey of progress on this topic. Theories of thin shells have been proposed by many authors in the literature, and some discrepancies have been found among them. A comparison of various theories was made in Refs. 4, 9, 14, 15 and 16.

The thin shell theories developed in this chapter are based on the assumptions postulated in Section 9.2. As we remarked in Section 8.10, the simultaneous use of the first and second assumptions mentioned in Section 9.2 may lead to an inconsistency in the stress–strain relations. For the improvement of the accuracy of thin shell theories, as well as for the complete removal of the inconsistency, it is necessary to abandon the assumptions and to assume the displacement components as

$$U = \sum_{m=0}^{n} u_m(\alpha, \beta)\, \zeta^m, \quad V = \sum_{m=0}^{n} v_m(\alpha, \beta)\, \zeta^m, \quad W = \sum_{m=0}^{n} w_m(\alpha, \beta)\zeta^m, \tag{9.108}$$

where the number of terms must be chosen properly.

Hildebrand, Reissner and Thomas have presented a theory of shells[17] in which U, V and W are approximated by quadratic functions with respect to ζ, namely $m = 2$ in Eqs. (9.108). Naghdi has employed the approximation

$$U = u_0 + \zeta u_1, \quad V = v_0 + \zeta v_1, \quad W = w_0 + \zeta w_1 + \zeta^2 w_2 \tag{9.109}$$

and has developed a theory by the use of Reissner's variational principle.[18, 19] He applied his theory to problems of wave propagation in cylindrical shells[20] and concluded that the form of the displacements in Eqs. (9.30) needs no improvements, if a theory is sought in which only transverse shear deformation and rotary inertia effects are retained.[13] We note here two papers which are related to the variational formulation of thin shell theories.[21, 22]

Variational Principles in shell theories in combination with the Rayleigh–Ritz method have provided powerful tools for solving shell problems approximately (see Refs. 23 through 26, for example). A theory of thin cylindrical shells was proposed by Donnel,[27] and has been extensively used for analyzing problems of thin cylindrical shells. The problems of buckling and post-buckling behavior of shells have been two of the central problems in shell theory.[28, 29] A snap-through theory was proposed by Kármán and Tsien for the buckling of cylindrical and spherical shells.[30, 31, 32] As other problems of great engineering concern, we may mention thermal stresses and thermal buckling of shells,[33, 34] and shell vibrations.[16, 35, 36, 37]

Bibliography

1. V. V. Novozhilov, *The Theory of Thin Shells*, Translated by B. G. Lowe, P. Noordhoff Ltd., Groningen, Netherlands, 1959.
2. A. L. Gol'denveizer, *Theory of Elastic Thin Shells*, Translated by G. Hermann, Pergamon Press, 1961.
3. V. V. Novozhilov, *Foundations of the Nonlinear Theory of Elasticity*, Graylock, 1953.

4. P. M. Naghdi, Foundations of Elastic Shell Theory, *Progress in Solid Mechanics*, edited by I. N. Sneddon and R. Hill, Vol. IV, Chapter 1, North-Holland, 1963.
5. A. E. H. Love, *Mathematical Theory of Elasticity*, Cambridge University Press, 4th edition, 1927.
6. S. Timoshenko and S. Woinowsky-Krieger, *Theory of Plates and Shells*, McGraw-Hill, 1959.
7. W. Flügge, *Statik und Dynamik der Schalen*, Springer Verlag, 1934.
8. A. E. Green and W. Zerna, *Theoretical Elasticity*, Oxford University Press, 1954.
9. W. S. Wlassow, *Allgemeine Schalen-Theorie und ihre Anwendungen in der Technik*, Akademie-Verlag, 1958.
10. W. Flügge, *Stresses in Shells*, Springer Verlag, 1960.
11. Kh. M. Mustari and K. Z. Galimov, *Non-linear Theory of Thin Elastic Shells*, Translated by J. Morgenstern, J. J. Schorr-Kon and PST Staff, Israel Program for Scientific Translations Ltd., 1962.
12. W. A. Nash, *Bibliography on Shells and Shell-like Structures*, David Taylor Model Basin Report 863, 1954. *Bibliography on Shells and Shell-like Structures (1954–1956)*. Engineering and Industrial Experimental Station, University of Florida, 1957.
13. P. M. Naghdi, A Survey of Recent Progress in the Theory of Elastic Shells, *Applied Mechanics Reviews*, Vol. 9, No. 9, pp. 365–8, September 1956.
14. W. T. Koiter, A Consistent First Approximation in the General Theory of Thin Elastic Shells, *Proceedings of the Symposium on the Theory of Thin Elastic Shells*, I.U.T.A.M., Delft, pp. 12–33, North-Holland, Amsterdam, 1960.
15. D. S. Houghton and D. J. Johns, A Comparison of the Characteristic Equations in the Theory of Circular Cylindrical Shells, *The Aeronautical Quarterly*, Vol. 12, Part 3, pp. 228–36, August 1961.
16. R. L. Bisplinghoff and H. Ashley, *Principles of Aeroelasticity*, John Wiley, 1962.
17. F. B. Hildebrand, E. Reissner and G. B. Thomas, *Notes on the Foundations of the Theory of Small Displacements of Orthotropic Shells*, NACA TN 1833, 1949.
18. P. M. Naghdi, On the Theory of Thin Elastic Shells, *Quarterly of Applied Mathematics*, Vol. 14, No. 4, pp. 369–80, January 1957.
19. P. M. Naghdi, The Effect of Transverse Shear Deformation on the Bending of Elastic Shells of Revolution, *Quarterly of Applied Mathematics*, Vol. 15, No. 1, pp. 41–52, April 1957.
20. P. M. Naghdi and P. M. Cooper, Propagation of Elastic Waves in Cylindrical Shells, including the Effects of Transverse Shear and Rotary Inertia, *Journal of Acoustical Society of America*, Vol. 28, No. 1, pp. 56–63, January 1956.
21. E. Trefftz, Ableitung der Schalenbiegungsgleichungen mit dem Castiglianoschen Prinzip, *Zeitschrift für Angewandte Mathematik und Mechanik*, Vol. 15, No. 1/2, pp. 101–8, February 1935.
22. E. Reissner, Variational Considerations for Elastic Beams and Shells, *Journal of the Engineering Mechanics Division, Proceedings of the American Society of Civil Engineers*, Vol. 88, No. EM 1, pp. 23–57, February 1962.
23. R. Schmidt and G. A. Wempner, The Nonlinear Conical Spring, *Transactions of the American Society for Mechanical Engineers*, Series E, Vol. 26, No. 4, pp. 681–2, December 1959.
24. N. C. Dahl, Toroidal-Shell Expansion Joints, *Journal of Applied Mechanics*, Vol. 20, No. 4, pp. 497–503, December 1953.
25. C. E. Turner and H. Ford, Stress and Deflection Studies of Pipeline Expansion Bellows, *Proceedings of the Institute of Mechanical Engineers*, Vol. 171, No. 15, pp. 526–52, 1957.
26. P. G. Kafka and M. B. Dunn, Stiffness of Curved Circular Tubes with Internal Pressure, *Journal of Applied Mechanics*, Vol. 23, No. 2, pp. 247–54, June 1956.

27. L. H. DONNEL, A New Theory for the Buckling of Thin Cylinders under Axial Compression and Bending, *Transactions of American Society for Mechanical Engineers*, Vol. 56, No. 11, pp. 795–806, November 1934.
28. S. TIMOSHENKO, *Theory of Elastic Stability*, McGraw-Hill, 1936.
29. H. L. LANGHAAR, General Theory of Buckling, *Applied Mechanics Review*, Vol. 11, No. 11, pp. 585–8, November 1958.
30. T. VON KÁRMÁN and H. S. TSIEN, The Buckling of Spherical Shells by External Pressure, *Journal of the Aeronautical Sciences*, Vol. 7, No. 2, pp. 43–50, December 1939.
31. T. VON KÁRMÁN and H. S. TSIEN, The Buckling of Thin Cylindrical Shells Under Axial Compression, *Journal of the Aeronautical Sciences*, Vol. 8, No. 8, pp. 303–12, June 1941.
32. H. S. TSIEN, A Theory for the Buckling of Thin Shells, *Journal of the Aeronautical Sciences*, Vol. 9, No. 10, pp. 373–83, August 1942.
33. N. J. HOFF, Buckling of Thin Cylindrical Shell under Hoop Stresses Varying in Axial Direction, *Journal of Applied Mechanics*, Vol. 24, No. 3, pp. 405–12, September 1957.
34. D. J. JOHNS, D. S. HOUGHTON and J. P. H. WEBBER, *Buckling due to Thermal Stress of Cylindrical Shells subjected to Axial Temperature Distributions*, College of Aeronautics, Cranfield, CoA Report No. 147, 1961.
35. R. N. ARNOLD and G. B. WARBURTON, The Flexural Vibrations of Thin Cylinders, *Proceedings of the Institute of Mechanical Engineers*, Vol. 167, No. 1, pp. 62–74, 1953.
36. J. B. BERRY and E. REISSNER, The Effect of an Internal Compressible Fluid Column on the Breathing Vibrations of a Thin Pressurized Cylindrical Shell, *Journal of the Aeronautical Sciences*, Vol. 25, No. 5, pp. 288–94, May 1958.
37. J. S. MIXSON and R. W. HERR, *An Investigation of the Vibration Characteristics of Pressurized Thin-walled Circular Cylinders Partly Filled with Liquid*, NASA TR R-145, 1962.

CHAPTER 10

STRUCTURES

10.1. Finite Redundancy

Thus far, variational formulations have been developed for simply connected, continuous bodies, the torsion of a bar with a hole treated in Section 6.3 being the only exception. It will be shown in the present chapter that these formulations are applicable, with slight modifications, to structures: multiply connected continuous bodies built up from basic members or components. For the sake of simplicity, we shall restrict the investigation to the small displacement theory of structures.

We shall assume that a structure under consideration can be fictitiously split into a number of simply connected members, the deformation characteristics of which have been derived with the aid of methods of analysis for simply connected bodies. Problems involving the entire structure are then reduced to the determination of internal forces existing at the joints of these members and at the points at which the structure is supported.

A structure is called redundant or statically indeterminate if the equations of equilibrium are not sufficient for the determination of all the internal forces: the degree of redundancy is then the difference between the number of unknown internal forces and the number of independent equations of equilibrium for the structure. According to this terminology, structures should be treated, in general, as multiply connected, continuous bodies with infinite redundancy. The analysis of such structures would lead to formidable calculations. However, experimental evidence and design experience have shown that we are justified in simplifying our analysis of structures by approximating the deformations of the members by finite degree of freedom systems. In other words, structures may be treated as bodies with finite redundancy under special circumstances.

Both trusses and frames are examples of structures in which such simplifications are permitted. All the members of a truss are assumed to be pin-jointed and capable of transmitting tensile or compressional axial loads only, while frame members are assumed to be capable of transmitting axial, bending and torsional forces and moments. In order to make such simplifications valid, each member must be slender and judiciously connected at the joints, and external forces must be carefully applied. Structures such as trusses or frames are sometimes called lumped parameter circuits by analogy with their electric counterparts.

The principle of virtual work and related variational principles have been found extremely effective in analyses of such simplified structures. An approach using the principle of minimum potential energy is usually called a displacement method,† while another using the principle of minimum complementary energy is called a force method.† These two methods have been guiding principles for analyses of structures. Due to the limited space available, we shall concentrate mainly on the analysis of trusses and frames, and put emphasis on the variational formulations. For practical details of numerical examples and applications to other structures, we shall be satisfied with listing related books in the bibliography for the reader's reference (Refs. 1 through 14).

10.2. Deformation Characteristics of a Truss Member and Presentation of a Truss Problem

We shall consider a truss member under end forces P, as shown in Fig. 10.1, and assume that the end force–elongation relation has been obtained:

$$P = P(\delta), \tag{10.1}$$

or conversely

$$\delta = \delta(P). \tag{10.2}$$

The elongation δ may be considered to be the displacement of one end of the member in the direction of the end force P while the other end is fixed. The strain and complementary energies stored in the member are given by

$$U = \int_0^\delta P(\delta)\, d\delta \tag{10.3}$$

and

$$V = \int_0^P \delta(P)\, dP, \tag{10.4}$$

respectively. For an elastic member with uniform cross sectional area A_0 and original length l, we have

$$P = \frac{EA_0}{l}\delta, \tag{10.5}$$

$$\delta = \frac{l}{EA_0}P, \tag{10.6}$$

$$U \doteq \frac{EA_0}{2l}\delta^2, \tag{10.7}$$

$$V = \frac{l}{2EA_0}P^2. \tag{10.8}$$

† They are also called the stiffness method and the flexibility method, respectively.

It is obvious that we have the following relations from Eqs. (10.3) and (10.4):

$$\frac{\partial U}{\partial \delta} = P, \qquad \frac{\partial V}{\partial P} = \delta. \qquad (10.9), (10.10)$$

When the bar is oriented arbitrarily with respect to a set of Cartesian reference axes, we require a relation between its elongation and its displacements. We shall denote the position vectors of the two ends of the bar before and after deformation by $\mathbf{r}_1^{(0)}$, $\mathbf{r}_2^{(0)}$ and $\mathbf{r}_1$, $\mathbf{r}_2$ respectively, which are related to the displacement vectors of the two ends of the bar, $\mathbf{u}_1$ and $\mathbf{u}_2$, by

$$\mathbf{r}_1 = \mathbf{r}_1^{(0)} + \mathbf{u}_1, \quad \mathbf{r}_2 = \mathbf{r}_2^{(0)} + \mathbf{u}_2. \qquad (10.11)$$

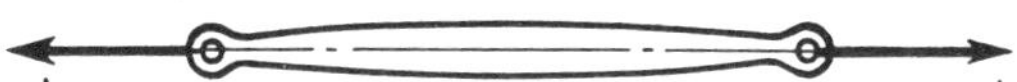

FIG. 10.1. Truss member loaded by end forces.

Then, the elongation of the bar, δ_{12}, is given in terms of the displacements as follows:

$$\begin{aligned}\delta_{12} &= |\mathbf{r}_2 - \mathbf{r}_1| - |\mathbf{r}_2^{(0)} - \mathbf{r}_1^{(0)}| \\ &= (\mathbf{u}_2 - \mathbf{u}_1) \cdot (\mathbf{r}_2^{(0)} - \mathbf{r}_1^{(0)})/|\mathbf{r}_2^{(0)} - \mathbf{r}_1^{(0)}|, \end{aligned} \qquad (10.12)$$

where higher order terms are neglected due to the assumption of small displacement.

Let us now consider a truss structure consisting of m members and n joints in three-dimensional space, where rectangular Cartesian coordinates will be used as a reference system. Let the truss joints be denoted by i; $i = 1, 2, \ldots, n$ and let a member which connects two joints i and j be represented by a double suffix ij; $ij = 1, 2, \ldots, m$. The direction cosines of the ij-th member before deformation are denoted by $\lambda_{ij}, \mu_{ij}, \nu_{ij}$, where the direction from the j-th joint to the i-th point is taken positive. Obviously:

$$\lambda_{ij} = -\lambda_{ji}, \quad \mu_{ij} = -\mu_{ji}, \quad \nu_{ij} = -\nu_{ji}. \qquad (10.13)$$

Denoting the elongation of the ij-th member by δ_{ij} and the displacement component of the i-th and j-th joints by u_i, v_i, w_i and u_j, v_j, w_j respectively, and using Eq. (10.12), we have the following elongation–displacement relations:

$$\delta_{ij} = (u_i - u_j)\,\lambda_{ij} + (v_i - v_j)\,\mu_{ij} + (w_i - w_j)\,\nu_{ij}. \qquad (10.14)†$$

We shall specify the boundary conditions for the truss structure as follows. All the external forces acting on the truss structure are applied at k joints

† The symbol δ_{ij} used throughout the present chapter should not be confused with the Kronecker symbol defined in the preceding chapters.

of the total n:

$$X_i = \bar{X}_i, \quad Y_i = \bar{Y}_i, \quad Z_i = \bar{Z}_i, \quad (i = 1, 2, \ldots, k), \tag{10.15}$$ †

as shown in Fig. 10.2, where

$$X_i = \sum_j P_{ij}\lambda_{ij}, \quad Y_i = \sum_j P_{ij}\mu_{ij}, \quad Z_i = \sum_j P_{ij}\nu_{ij}. \tag{10.16}$$

In Eqs. (10.16), P_{ij} is the internal end force of the ij-th member, and the summation with respect to j is taken over all the members which are directly connected to the i-th joint. For the sake of simplicity, we assume that the truss structure is rigidly fixed at the remaining $(n - k)$ joints:

$$u_i = 0, \quad v_i = 0, \quad w_i = 0, \quad (i = k + 1, \ldots, n). \tag{10.17}$$ ‡

In order to simplify the problem further, we assume that the geometrical boundary conditions are sufficient to let the external forces be independent of each other.

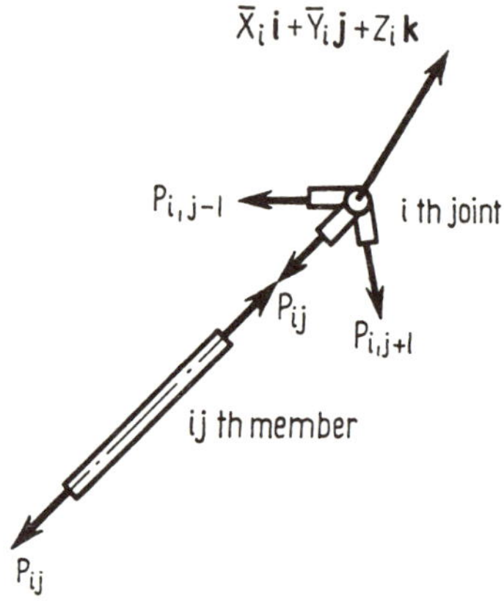

FIG. 10.2. i-th joint and ij-th member.

We assume that the internal force–elongation relations have been obtained for each member as

$$P_{ij} = P_{ij}(\delta_{ij}), \tag{10.18}$$

or conversely

$$\delta_{ij} = \delta_{ij}(P_{ij}). \tag{10.19}$$

† It is not necessary that all the three components of the external force or of the displacement be prescribed at a joint. Only the complementary relations between $\bar{X}_i$ and $\bar{u}_i$, $\bar{Y}_i$ and $\bar{v}_i$, $\bar{Z}_i$ and w_i are necessary. However, we have prescribed our problem as above to simplify the subsequent formulations.

‡ An extension to a truss problem in which the geometrical boundary conditions are given by $u_i = \bar{u}_i$, $v_i = \bar{v}_i$, $w_i = \bar{w}_i$ $(i = k + 1, \ldots, n)$ is straightforward.

Combining Eqs. (10.14) and (10.18) with the boundary conditions (10.15) and (10.17), we have the necessary and sufficient number of equations for determining $(2m + 3k)$ unknown quantities P_{ij}, δ_{ij}, u_i, v_i and w_i where $ij = 1, 2, \ldots, m$ and $i = 1, 2, \ldots, k$.

10.3. Variational Formulations of the Truss Problem

We shall consider the principle of virtual work for the truss problem. Denoting the virtual displacements of the i-th joint by δu_i, δv_i and δw_i, and using Eqs. (10.15), we have

$$\sum_{i=1}^{k} [(X_i - \bar{X}_i)\,\delta u_i + (Y_i - \bar{Y}_i)\,\delta v_i + (Z_i - \bar{Z}_i)\,\delta w_i] = 0. \tag{10.20}$$

Then recalling Eqs. (10.17), we can transform Eq. (10.20) into:

$$\sum_{ij=1}^{m} P_{ij}\,\delta\delta_{ij} - \sum_{i=1}^{k} (\bar{X}_i\,\delta u_i + \bar{Y}_i\,\delta v_i + \bar{Z}_i\,\delta w_i) = 0, \tag{10.21}$$

where Eq. (10.14) has been substituted. This is the principle of virtual work for the truss problem.

The principle (10.21) suggests that the function for the principle of minimum potential energy of the truss problem is given by

$$\Pi = \sum_{ij=1}^{m} U_{ij}(\delta_{ij}) - \sum_{i=1}^{k} (\bar{X}_i u_i + \bar{Y}_i v_i + \bar{Z}_i w_i), \tag{10.22}$$

where

$$U_{ij}(\delta_{ij}) = \int_0^{\delta ij} P_{ij}(\delta_{ij})\, d\delta_{ij}, \tag{10.23}$$

and where Eq. (10.14) has been substituted. In the above, the quantities subject to variation are u_i, v_i and w_i under the subsidiary conditions (10.17). The function (10.22) may be transformed through familiar procedure to the generalized form:

$$\begin{aligned} \Pi_{\mathrm{I}} = & \sum_{ij=1}^{m} U_{ij}(\delta_{ij}) - \sum_{i=1}^{k} (\bar{X}_i u_i + \bar{Y}_i u_i + \bar{Z}_i w_i) \\ & - \sum_{ij=1}^{m} P_{ij}\{\delta_{ij} - [(u_i - u_j)\,\lambda_{ij} + (v_i - v_j)\,\mu_{ij} + (w_i - w_j)\nu_{ij}]\} \\ & - \sum_{i=k+1}^{n} (X_i u_i + Y_i v_i + Z_i w_i), \end{aligned} \tag{10.24}$$

where the quantities subject to variation are δ_{ij}, P_{ij}, u_i, v_i and w_i; $ij = 1, 2, \ldots, m$ and $i = 1, 2, \ldots, n$, under no subsidiary conditions.

The function for the principle of minimum complementary energy may be derived from Eq. (10.24) in the usual manner as follows:

$$\Pi_c = \sum_{ij=1}^{m} V_{ij}(P_{ij}), \tag{10.25}$$

where

$$V_{ij}(P_{ij}) = \int_0^{P_{ij}} \delta_{ij}(P_{ij})\, dP_{ij}. \tag{10.26}$$

The independent variables subject to variation in Eq. (10.25) are P_{ij}; $ij = 1, 2, \ldots, m$ under the subsidiary conditions (10.15).

Thus, we have derived the principle of minimum complementary energy from the principle of minimum potential energy. However, it is obvious that the principle of minimum complementary energy is derivable alternatively by use of the principle of complementary virtual work

$$\sum_{ij=1}^{m} \delta_{ij}\, \delta P_{ij} = 0, \tag{10.27}$$

where δ_{ij} and δP_{ij} are so chosen as to satisfy Eqs. (10.14) and (10.17), and

$$\delta X_i = 0, \quad \delta Y_i = 0, \quad \delta Z_i = 0, \quad (i = 1, 2, \ldots, k) \tag{10.28}$$

respectively. We note that if Eq. (10.27) holds for any combination of δP_{ij} which satisfy Eqs. (10.28), the elongation δ_{ij} must be derived from u_i, v_i and w_i as given by Eqs. (10.14) and (10.17).

We shall now derive equations for the displacement components of the joints where the external forces are applied. We assume that the truss problem has been solved and the applied external forces at the i-th joint are increased by the amounts $d\bar{X}_i$, $d\bar{Y}_i$ and $d\bar{Z}_i$; $i = 1, 2, \ldots, k$, while the geometrical boundary conditions are kept unchanged. Then, in a manner similar to the development in Section 2.6, we have

$$\sum_{ij=1}^{m} \delta_{ij}\, dP_{ij} = \sum_{i=1}^{k} (u_i\, d\bar{X}_i + v_i\, d\bar{Y}_i + w_i\, d\bar{Z}_i). \tag{10.29}$$

Equation (10.29) is equivalent to what is known as Castigliano's theorem applied to the truss structure.

10.4. The Force Method Applied to the Truss Problem

We observe that Eqs. (10.15) consist of $3k$ equations and are generally insufficient for the determination of the m unknowns P_{ij}. In other words, the truss structure is redundant, and the degree of redundancy is $R = m - 3k$ by definition. We shall obtain the remaining R equations from the principle of minimum complementary energy (10.25). To begin with, we obtain the general solution of Eqs. (10.15):

$$P_{ij} = \sum_{p=1}^{R} a_{ijp}\chi_p + \sum_{l=1}^{k} (\alpha_{ijl}\bar{X}_l + \beta_{ijl}\bar{Y}_l + \gamma_{ijl}\bar{Z}_l), \quad (ij = 1, 2, \ldots, m). \tag{10.30}$$

The first terms on the right-hand side of Eqs. (10.30) constitute the general solution of the homogeneous equations,

$$\sum_j P_{ij}\lambda_{ij} = 0, \quad \sum_j P_{ij}\mu_{ij} = 0, \quad \sum_j P_{ij}\nu_{ij} = 0, \qquad (i = 1, 2, \ldots, k), \tag{10.31}$$

thus defining a self-equilibrating system of the internal end forces. Eqs. (10.30) can be written in matrix form as follows:

$$\{P\} = [a]\{\chi\} + [\alpha]\{\bar{X}\}, \tag{10.32}$$

where the notations { } and [ ] denote column and rectangular matrices, respectively, and

$$\{P\} = \begin{bmatrix} P_{12} \\ \vdots \\ P_{n-1,n} \end{bmatrix}, \quad [a] = \begin{bmatrix} a_{12,1} & \cdots\cdots & a_{12,R} \\ \vdots & & \vdots \\ a_{n-1,n,1} & \cdots\cdots & a_{n-1,n,R} \end{bmatrix}, \quad \{\chi\} = \begin{bmatrix} \chi_1 \\ \vdots \\ \chi_R \end{bmatrix},$$

$$[\alpha] = \begin{bmatrix} \alpha_{12,1} & \cdots & \alpha_{12,k} & \beta_{12,1} & \cdots & \beta_{12,k} & \gamma_{12,1} & \cdots & \gamma_{12,k} \\ \vdots & & \vdots & \vdots & & \vdots & \vdots & & \vdots \\ \alpha_{n-1,n,1} & \cdots & \alpha_{n-1,n,k} & \beta_{n-1,n,1} & \cdots & \beta_{n-1,n,k} & \gamma_{n-1,n,1} & \cdots & \gamma_{n-1,n,k} \end{bmatrix},$$

$$\{\bar{X}\} = \{\bar{X}_1, \ldots, \bar{X}_k, \ \bar{Y}_1, \ldots, \bar{Y}_k, \ \bar{Z}_1, \ldots, \bar{Z}_k\}. \tag{10.33†}$$

Introducing Eq. (10.30) into the principle of minimum complementary energy (10.25), and taking variations with respect to χ_p, we obtain:

$$\sum_{ij=1}^{m} a_{ijp}\,\delta_{ij} = 0 \quad (p = 1, 2, \ldots, R) \tag{10.34}$$

or in matrix form

$$[a]'\{\delta\} = 0, \tag{10.35}$$

where [ ]′ denotes the transpose of the matrix [ ], and $\{\delta\} = \{\delta_{12}, \ldots, \delta_{n-1,n}\}$. It is obvious that relations equivalent to Eqs. (10.34) are obtainable from the principle of complementary virtual work (10.27). Equations (10.34) provide conditions which must exist between the elongations of the members in order to ensure that none of the connections between the members of the truss are broken after deformation. They are geometrical relations and

† A column matrix is denoted by either one of the following symbols:

$$\{x_1, x_2, \ldots, x_n\}, \quad \begin{bmatrix} x_1 \\ x_2 \\ \vdots \\ x_n \end{bmatrix},$$

of which the former is frequently used for the sake of saving space.

must hold irrespective of the internal end force–elongation relationships employed. Namely, they are *the compatibility conditions in the large* for the truss structure, and have the same geometrical meaning as Eq. (6.45), which was derived for the torsion of a bar with a hole. By the use of Eqs. (10.30), Eqs. (10.34) are reduced to simultaneous algebraic equations with respect to χ_p; $p = 1, 2, \ldots, R$. By solving these equations, we obtain the values of the χ_p, which, in turn, determine P_{ij} and δ_{ij} through Eqs. (10.30) and (10.19).

When all the members behave elastically, the relations between the internal end forces and elongations are given in matrix form by:

$$\{\delta\} = [C]\{P\}. \tag{10.36}$$

In the truss structure, $[C]$ is a diagonal matrix as suggested by Eq. (10.6). By the use of Eq. (10.32) and (10.36), we find that Eq. (10.35) provides

$$\{\chi\} = -[G]^{-1}[H]\{\bar{X}\} \tag{10.37}$$

for the determination of χ_p, where

$$\begin{aligned} [G] &= [a]'\,[C]\,[a], \\ [H] &= [a]'\,[C]\,[\alpha]. \end{aligned} \tag{10.38}$$

Substituting Eq. (10.37) into Eq. (10.32), we obtain

$$\{P\} = [[\alpha] - [a]\,[G]^{-1}\,[H]]\,\{\bar{X}\}. \tag{10.39}$$

By the use of Eqs. (10.36) and (10.39), we can determine the elongations of all the members of the truss structure.

Next, we shall derive equations for the displacement components of the joints where the external forces are applied. Introducing Eqs. (10.32) into Eq. (10.29), and remembering that the external forces are assumed independent of each other, we obtain:

$$u_l = \sum_{ij=1}^{m} \alpha_{ijl}\,\delta_{ij}, \quad v_l = \sum_{ij=1}^{m} \beta_{ijl}\,\delta_{ij}, \quad w_l = \sum_{ij=1}^{m} \gamma_{ijl}\,\delta_{ij},$$

$$(l = 1, 2, \ldots, k) \tag{10.40}$$

or in matrix form as

$$\{u\} = [\alpha]'\,\{\delta\}, \tag{10.41}$$

where

$$\{u\} = \{u_1, \ldots, u_k,\ v_1, \ldots, v_k,\ w_1, \ldots, w_k\}.$$

By the use of Eqs. (10.36), (10.39) and (10.41), we obtain the displacement components of the joints where the external forces are applied, and matrices of structural influence coefficients can be derived. The above method constitutes the main part of the force method.

A note is made here on the displacement method applied to the truss problem. A well-known procedure is followed by substituting Eqs. (10.14), (10.17) and (10.18) into Eqs. (10.15) and solving these $3k$ equations to determine the unknown displacement components u_i, v_i and w_i; $i = 1, 2, \ldots, k$. Once the displacement components have been obtained, the deformation and internal forces of the truss structure can be determined by the use of Eqs. (10.14) and (10.18).

10.5. A Simple Example of a Truss Structure

As an application of the preceding formulations, we consider the plane truss structure shown in Fig. 10.3. The truss consists of six members and four joints, namely $m = 6$ and $n = 4$. The external forces are applied at the joints ① and ② in the directions of the x- and y-axes, while at the joint ③,

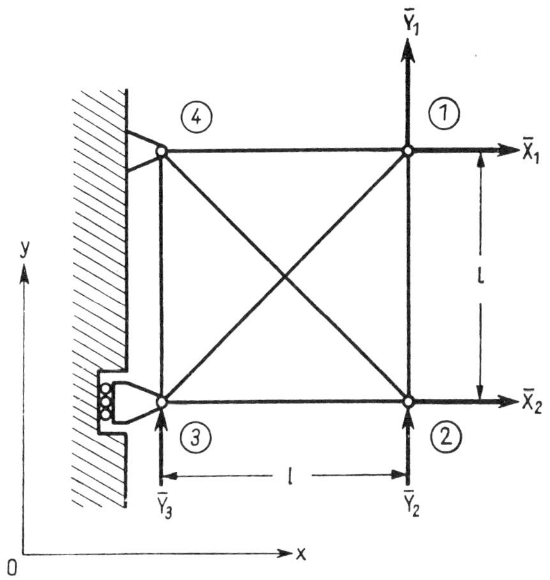

FIG. 10.3. A truss structure.

the force is applied in the y-direction only. The equations of equilibrium at these joints are:

$$P_{14} + (1/\sqrt{2})\,P_{13} = \bar{X}_1, \quad P_{12} + (1/\sqrt{2})\,P_{13} = \bar{Y}_1,$$
$$P_{23} + (1/\sqrt{2})\,P_{24} = \bar{X}_2, \quad P_{12} + (1/\sqrt{2})\,P_{24} = -\bar{Y}_2,$$
$$P_{34} + (1/\sqrt{2})\,P_{13} = -\bar{Y}_3. \tag{10.42}$$

The geometrical boundary conditions are prescribed at the joints ③ and ④ as

$$u_3 = 0, \quad u_4 = 0, \quad v_4 = 0. \tag{10.43}$$

Since we have five equations of equilibrium (10.42) for six unknown internal end forces, the redundancy of the truss is 6—5 = 1, and Eqs. (10.42) can be written in matrix form as:

$$\begin{bmatrix} P_{12} \\ P_{13} \\ P_{14} \\ P_{23} \\ P_{24} \\ P_{34} \end{bmatrix} = \begin{bmatrix} -1/\sqrt{2} & 0 & 0 & 1 & 0 & 0 \\ 1 & 0 & 0 & 0 & 0 & 0 \\ -1/\sqrt{2} & 1 & 0 & 0 & 0 & 0 \\ -1/\sqrt{2} & 0 & 1 & 1 & 1 & 0 \\ 1 & 0 & 0 & -\sqrt{2} & -\sqrt{2} & 0 \\ -1/\sqrt{2} & 0 & 0 & 0 & 0 & -1 \end{bmatrix} \begin{bmatrix} P_{13} \\ \bar{X}_1 \\ \bar{X}_2 \\ \bar{Y}_1 \\ \bar{Y}_2 \\ \bar{Y}_3 \end{bmatrix}. \tag{10.44}$$

We observe that the unknown end force P_{13} in Eq. (10.44) plays the role of χ_1 defined in Eqs. (10.30). In the notation defined in Eq. (10.33), the right-hand side of Eq. (10.44) provides:

$$[a] = \begin{bmatrix} -1/\sqrt{2} \\ 1 \\ -1/\sqrt{2} \\ -1/\sqrt{2} \\ 1 \\ -1/\sqrt{2} \end{bmatrix}, \quad [\alpha] = \begin{bmatrix} 0 & 0 & 1 & 0 & 0 \\ 0 & 0 & 0 & 0 & 0 \\ 1 & 0 & 0 & 0 & 0 \\ 0 & 1 & 1 & 1 & 0 \\ 0 & 0 & -\sqrt{2} & -\sqrt{2} & 0 \\ 0 & 0 & 0 & 0 & -1 \end{bmatrix}. \tag{10.45}$$

Consequently, noticing $\{\delta\} = \{\delta_{12}, \delta_{13}, \delta_{14}, \delta_{23}, \delta_{24}, \delta_{34}\}$, we obtain from Eq. (10.35) the compatibility condition in the large:

$$-\sqrt{2}(\delta_{13} + \delta_{24}) + \delta_{12} + \delta_{14} + \delta_{23} + \delta_{34} = 0. \tag{10.46}$$

The displacement components at the joints ①, ② and ③ are obtained from Eqs. (10.41):

$$\begin{aligned} u_1 &= \delta_{14}, \quad v_1 = \delta_{12} + \delta_{23} - \sqrt{2}\,\delta_{24}, \\ u_2 &= \delta_{23}, \quad v_2 = \delta_{23} - \sqrt{2}\,\delta_{24}, \\ & \qquad v_3 = -\delta_{34}. \end{aligned} \tag{10.47}$$

10.6. Deformation Characteristics of a Frame Member

Next, we shall deal with a frame structure. To begin with, we consider the deformation characteristics of a frame member. For the sake of simplicity, we shall take a beam which, as shown in Fig. 10.4, is rigidly fixed at one end ① and is subjected to end forces and moment, N_{12}, Q_{12}, M_{12}, at the other end ② and to a concentrated load $\bar{P}$ acting in the middle of the span. We denote components of deformation under the application of these forces and moment by:

δ_{12}^N = the displacement of the end ② in the direction of N_{12},
δ_{12}^Q = the displacement of the end ② in the direction of Q_{12},
δ_{12}^M = the rotational angle of the end ② in the direction of M_{12},
δ_{12}^P = the displacement of the point of application in the direction of the external force $\bar{P}$.

These quantities may be calculated from Castigliano's theorem:

$$\delta_{12}^{N} = \frac{\partial V_{12}}{\partial N_{12}}, \quad \delta_{12}^{Q} = \frac{\partial V_{12}}{\partial Q_{12}}, \quad \delta_{12}^{M} = \frac{\partial V_{12}}{\partial M_{12}}, \quad \delta_{12}^{\bar{P}} = \frac{\partial V_{12}}{\partial \bar{P}}. \tag{10.48}$$

In the above, the quantity V_{12} is the complementary energy stored in the beam ⑫. If we employ the elementary beam theory for the analysis of the frame member, it is given by

$$V_{12} = \frac{1}{2}\int_0^l \left(\frac{N^2}{EA_0} + \frac{M^2}{EI} \right) dx, \tag{10.49}$$

where

$$N = N_{12}, \tag{10.50a}$$

and

$$M = \begin{cases} M_{12} - (l - x)\,Q_{12} + \left(\frac{l}{2} - x\right)\bar{P}, & 0 \leqq x \leqq \frac{l}{2}, \\ M_{12} - (l - x)\,Q_{12}, & \frac{l}{2} \leqq x \leqq l. \end{cases} \tag{10.50b}$$

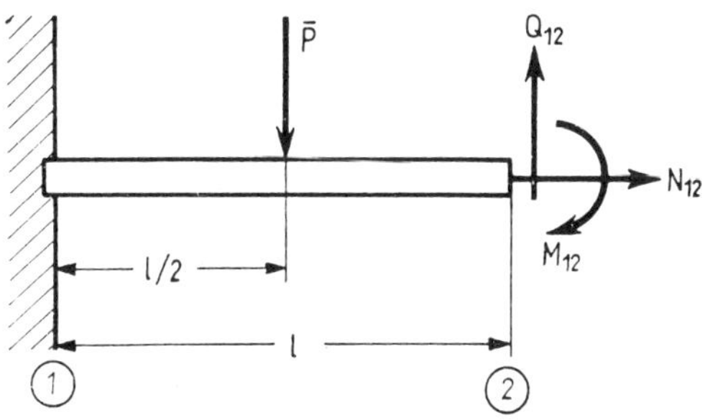

FIG. 10.4. A cantilever beam.

When the beam is of uniform cross section along the span, we obtain the flexibility matrix of the beam ⑫ as follows:

$$\begin{bmatrix} \delta_{12}^{N} \\ \delta_{12}^{Q} \\ \delta_{12}^{M} \\ \delta_{12}^{\bar{P}} \end{bmatrix} = \begin{bmatrix} \frac{l}{EA_0} & 0 & 0 & 0 \\ 0 & \frac{1}{3}\frac{l^3}{EI} & -\frac{1}{2}\frac{l^2}{EI} & -\frac{5}{48}\frac{l^3}{EI} \\ 0 & -\frac{1}{2}\frac{l^2}{EI} & \frac{l}{EI} & \frac{1}{8}\frac{l^2}{EI} \\ 0 & -\frac{5}{48}\frac{l^3}{EI} & \frac{1}{8}\frac{l^2}{EI} & \frac{1}{24}\frac{l^3}{EI} \end{bmatrix} \begin{bmatrix} N_{12} \\ Q_{12} \\ M_{12} \\ \bar{P} \end{bmatrix}. \tag{10.51}$$

For later convenience, we note that Eqs. (10.51) can be inverted into the stiffness matrix form as follows:

$$\begin{bmatrix} N_{12}^{①} \\ Q_{12}^{①} \\ M_{12}^{①} \\ N_{12}^{②} \\ Q_{12}^{②} \\ M_{12}^{②} \end{bmatrix} = \begin{bmatrix} \frac{EA_0}{l} & 0 & 0 & -\frac{EA_0}{l} & 0 & 0 & 0 \\ 0 & \frac{12EI}{l^3} & -\frac{6EI}{l^2} & 0 & -\frac{12EI}{l^3} & -\frac{6EI}{l^2} & \frac{1}{2}\bar{P} \\ 0 & -\frac{6EI}{l^2} & \frac{4EI}{l} & 0 & \frac{6EI}{l^2} & \frac{2EI}{l} & -\frac{1}{8}\bar{P}l \\ -\frac{EA_0}{l} & 0 & 0 & \frac{EA_0}{l} & 0 & 0 & 0 \\ 0 & -\frac{12EI}{l^3} & \frac{6EI}{l^3} & 0 & \frac{12EI}{l^3} & \frac{6EI}{l^2} & \frac{1}{2}\bar{P} \\ 0 & -\frac{6EI}{l^2} & \frac{2EI}{l} & 0 & \frac{6EI}{l^2} & \frac{4EI}{l} & \frac{1}{8}\bar{P}l \end{bmatrix} \begin{bmatrix} u_1 \\ v_1 \\ \theta_1 \\ u_2 \\ v_2 \\ \theta_2 \\ 1 \end{bmatrix} \tag{10.52}$$

where u_1, v_1, θ_1 and u_2, v_2, θ_2 are the displacement components in the x- and y-directions and the rotational angle in the clockwise direction of the joints ① and ② respectively, as shown in Fig. 10.5. Since the beam is in static equilibrium, we have the following relations among the external forces and moments:

$$N_{12}^{①} + N_{12}^{②} = 0, \quad Q_{12}^{①} + Q_{12}^{②} - \bar{P} = 0,$$

$$M_{12}^{①} + M_{12}^{②} - Q_{12}^{②} l + \tfrac{1}{2}\bar{P}l = 0. \tag{10.53}$$

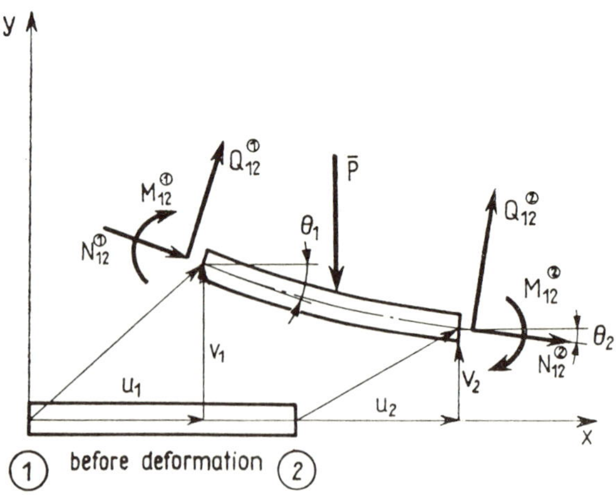

FIG. 10.5. A beam element.

When a curved beam is oriented arbitrarily with respect to a set of Cartesian reference axes and is subject to combined actions of axial, shearing, bending and torsional end forces and moments together with external loads distributed along the span, the relations between the external forces and the resulting deformation become complicated. The relations for a straight beam have been obtained in the flexibility matrix and stiffness matrix forms and are widely used in structural analysis.†

10.7. The Force Method Applied to a Frame Problem

With the above preliminaries, we shall now proceed to the analysis of a frame structure. Rather than attempting to develop general formulations for a three-dimensional frame structure, we shall consider only a simple

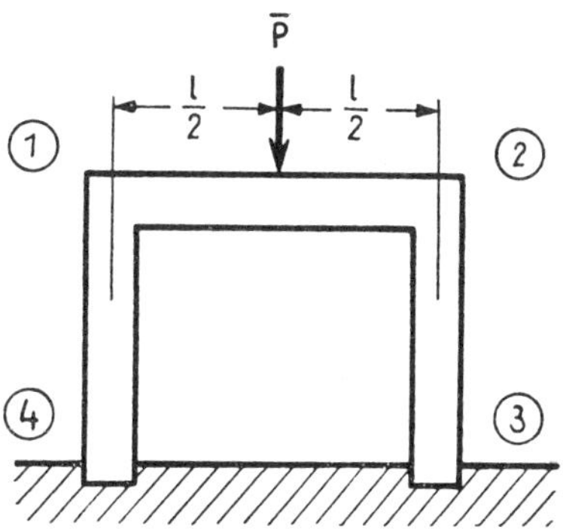

FIG. 10.6. A frame structure.

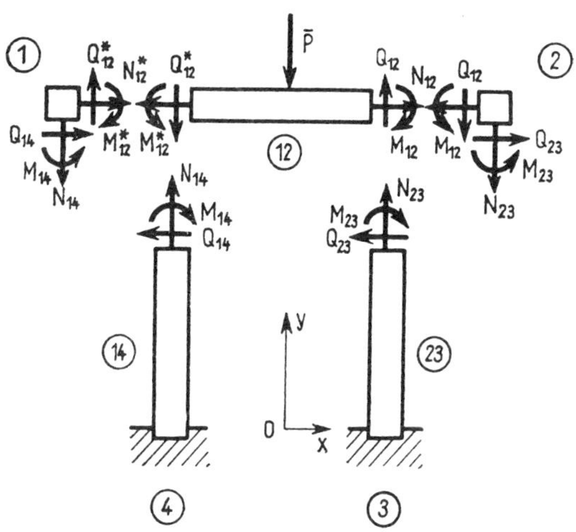

FIG. 10.7. Free body diagram of the frame structure.

† For the stiffness matrix, see Ref. 15.

example of the plane frame illustrated in Fig. 10.6 and discover that variational formulations proceed in a manner similar to that for the truss structure.

We shall be interested in the force method applied to the frame problem. To begin with, let the frame structure be cut fictitiously into several members and the internal end forces and moments be defined properly for the separate members. For example, the structure may be cut into three members, ⑫, ㉓ and ⑭, as shown in Fig. 10.7, and the internal end forces and moments: $N_{12}, Q_{12}, M_{12}; N_{23}, Q_{23}, M_{23}; N_{14}, Q_{14}, M_{14}$ may be defined on one end of each of these three members, while the asterisked quantities N_{12}^*, Q_{12}^*, M_{12}^* are defined on the opposite end of member ⑫. From equilibrium conditions for member ⑫, we have the following relations among these quantities:

$$N_{12}^* = N_{12}, \quad Q_{12}^* = Q_{12} - \bar{P}, \quad M_{2}^* = M_{12} - Q_{12}l + \tfrac{1}{2}\bar{P}l. \quad (10.54)$$

We define geometrical quantities of deformation for these members as follows. With respect to ⑫, the end on which the asterisked quantities are defined is assumed to be held fixed, and components of deformation denoted by δ_{12}^N, δ_{12}^Q, δ_{12}^M and $\delta_{12}^{\bar{P}}$ are defined as given in the preceding section. In a similar manner, the quantities δ_{23}^N, δ_{23}^Q and δ_{23}^M are defined with respect to ㉓ in the directions of N_{23}, Q_{23} and M_{23}, respectively, and the quantities δ_{14}^N, δ_{14}^Q and δ_{14}^M with respect to ⑭ in the directions of N_{14}, Q_{14} and M_{14}, respectively. Then, denoting the complementary energies of the members ⑫, ㉓ and ⑭ by $V_{12}(N_{12}, Q_{12}, M_{12}, \bar{P})$, $V_{23}(N_{23}, Q_{23}, M_{23})$ and $V_{14}(N_{14}, Q_{14}, M_{14})$, respectively, we have

$$\delta_{12}^N = \frac{\partial V_{12}}{\partial N_{12}}, \ldots, \delta_{14}^M = \frac{\partial V_{14}}{\partial M_{14}}. \quad (10.55)$$

Let us now consider reassembling these three members into a frame structure. To begin with, the equilibrium conditions at the joints ① and ② should be satisfied (see Fig. 10.7):

$$-N_{14} + Q_{12}^* = 0, \quad Q_{14} + N_{12}^* = 0, \quad -M_{14} + M_{12}^* = 0, \quad (10.56)$$

$$-N_{12} + Q_{23} = 0, \quad -Q_{12} - N_{23} = 0, \quad -M_{12} - M_{23} = 0. \quad (10.57)$$

Eliminating the asterisked quantities from Eqs. (10.56) by the use of Eqs. (10.54), we obtain:

$$N_{12} + Q_{14} = 0, \quad Q_{12} - N_{14} - \bar{P} = 0,$$
$$M_{12} - M_{14} - Q_{12}l + \tfrac{1}{2}\bar{P}l = 0. \quad (10.58)$$

Equations (10.57) and (10.58) comprise six equations of equilibrium for the nine internal end forces and moments, thus showing that the redundancy is 9 − 6 = 3.† These equilibrium equations can be written in matrix form as

† Due to the symmetry property of the problem, we may reduce the redundancy further. However, this property will not be taken into account here, since our purpose is to show the procedure of the force method.

follows:

$$\begin{bmatrix} N_{12} \\ Q_{12} \\ M_{12} \\ N_{23} \\ Q_{23} \\ M_{23} \\ N_{14} \\ Q_{14} \\ M_{14} \end{bmatrix} = \begin{bmatrix} 1 & 0 & 0 & 0 \\ 0 & 1 & 0 & 0 \\ 0 & 0 & 1 & 0 \\ 0 & -1 & 0 & 0 \\ 1 & 0 & 0 & 0 \\ 0 & 0 & -1 & 0 \\ 0 & 1 & 0 & -1 \\ -1 & 0 & 0 & 0 \\ 0 & -l & 1 & \frac{1}{2}l \end{bmatrix} \begin{bmatrix} N_{12} \\ Q_{12} \\ M_{12} \\ \bar{P} \end{bmatrix}, \tag{10.59}$$

where N_{12}, Q_{12} and M_{12} are chosen as the independent end forces and moments. Using the notation defined in Section 10.4, we may write

$$[a] = \begin{bmatrix} 1 & 0 & 0 \\ 0 & 1 & 0 \\ 0 & 0 & 1 \\ 0 & -1 & 0 \\ 1 & 0 & 0 \\ 0 & 0 & -1 \\ 0 & 1 & 0 \\ -1 & 0 & 0 \\ 0 & -l & 1 \end{bmatrix}, \quad [\alpha] = \begin{bmatrix} 0 \\ 0 \\ 0 \\ 0 \\ 0 \\ 0 \\ -1 \\ 0 \\ \frac{1}{2}l \end{bmatrix}. \tag{10.60}$$

We must introduce three conditions of compatibility in the large in order to solve the present problem. They are given by the stationary conditions of Π_c defined by

$$\Pi_c = V_{12} + V_{23} + V_{14}, \tag{10.61}$$

where the independent quantities subject to variation are N_{12}, ..., and M_{14} under the subsidiary conditions (10.57) and (10.58). A careful consideration shows that the conditions of compatibility in the large are given by

$$[a]' \{\delta\} = 0, \tag{10.62}$$

where

$$\{\delta\} = \{\delta_{12}^N, \delta_{12}^Q, \delta_{12}^M, \delta_{23}^N, \delta_{23}^Q, \delta_{23}^M, \delta_{14}^N, \delta_{14}^Q, \delta_{14}^M\}.$$

When written in explicit form, Eq. (10.62) becomes:

$$\delta_{12}^N + \delta_{23}^Q - \delta_{14}^Q = 0, \quad \delta_{12}^Q - \delta_{23}^N + \delta_{14}^N - \delta_{14}^M l = 0,$$
$$\delta_{12}^M - \delta_{23}^M + \delta_{14}^M = 0. \tag{10.63}$$

These are the conditions of compatibility in the large for the frame structure. By combining Eqs. (10.55), (10.59) and (10.63), we have necessary and sufficient equations for the determination of the nine unknown internal end forces and moments.

Next, we shall determine the displacement $\delta_{\bar{P}}$ of the point of application in the direction of the external force $\bar{P}$. We assume that the frame problem has been solved and the external force $\bar{P}$ is increased by an amount $d\bar{P}$. Then we have

$$\delta_{12}^{N}\, dN_{12} + \delta_{12}^{Q}\, dQ_{12} + \cdots + \delta_{12}^{\bar{P}}\, d\bar{P} = \delta_{\bar{P}}\, d\bar{P} \tag{10.64}$$

in a manner similar to the development in Section 2.6. From Eq. (10.64) we obtain

$$\begin{aligned} \delta_{\bar{P}} &= [\alpha]' \{\delta\} + \delta_{12}^{\bar{P}} \\ &= -\delta_{14}^{N} + \tfrac{1}{2}\delta_{14}^{M} l + \delta_{12}^{\bar{P}}. \end{aligned} \tag{10.65}$$

We note here that, if the subsidiary conditions (10.57) and (10.58) are introduced into the framework of Π_c via Lagrange multipliers, the expression (10.61) is transformed into Π_c^* defined as

$$\begin{aligned} \Pi_c^* = {} & V_{12} + V_{23} + V_{14} + (N_{12} + Q_{14})\, u_1 \\ & + (Q_{12} - N_{14} - \bar{P})\, v_1 + (M_{12} - M_{14} - Q_{12} l + \tfrac{1}{2}\bar{P} l)\, \theta_1 \\ & - (N_{12} - Q_{23})\, u_2 - (Q_{12} + N_{23})\, v_2 - (M_{12} + N_{23})\, \theta_2, \end{aligned} \tag{10.66}$$

where the quantities subject to variation are N_{12}, Q_{12}, M_{12}, N_{14}, Q_{14}, M_{14}, N_{23}, Q_{23}, M_{23}, u_1, v_1, θ_1, u_2, v_2 and θ_2 with no subsidiary conditions. The physical meanings of u_1, v_1, θ_1 and u_2, v_2, θ_2 are the displacement components in the x- and y-directions and the rotational angle in the clockwise direction of the joints ① and ② respectively. The expression (10.66) may be considered as an extension of the Hellinger–Reissner principle to the frame structure. When the mechanical quantities N_{12}, Q_{12}, ... and M_{23} are eliminated from the expression (10.66) by the use of the stationary conditions such that

$$\frac{\partial V_{12}}{\partial N_{12}} + u_1 - u_2 = 0, \ldots, \tag{10.67}$$

we have the function for the principle of minimum potential energy:

$$\begin{aligned} \Pi = {} & U_{12} + U_{14} + U_{23} - \frac{1}{384}\frac{\bar{P}^2 l^3}{EI} \\ & + \bar{P}\left[\frac{1}{2}(v_1 + v_2) + \frac{l}{8}(\theta_2 - \theta_1)\right], \end{aligned} \tag{10.68}$$

where $U_{12}(u_1, v_1, \theta_1, u_2, v_2, \theta_2)$, $U_{14}(u_1, v_1, \theta_1)$ and $U_{23}(u_2, v_2, \theta_2)$ are the strain energies stored in these members, and the quantities subject to variation are u_1, v_1, θ_1, u_2, v_2 and θ_2 with no subsidiary conditions.†

We observe that the inversion of Eqs. (10.67) provides a stiffness matrix in the same form as given by Eqs. (10.52). We observe also that by the use of the notations defined before, Eqs. (10.67) can be written as

$$\begin{aligned} &\delta_{12}^{N} = u_2 - u_1, \quad \delta_{12}^{Q} = v_2 + \theta_1 l - v_1, \quad \delta_{12}^{M} = \theta_2 - \theta_1, \\ &\delta_{23}^{N} = v_2, \quad \delta_{23}^{Q} = -u_2, \quad \delta_{23}^{M} = \theta_2, \\ &\delta_{14}^{N} = v_1, \quad \delta_{14}^{Q} = -u_1, \quad \delta_{14}^{M} = \theta_1, \end{aligned} \tag{10.69}$$

and the compatibility conditions in the large (10.63) hold among them.

† See Problems 3 and 4 of Chapter 10 in Appendix H for the explicit expressions of U_{12}, U_{14} and U_{23}.

A mention is made here of the stiffness matrix method applied to the analysis of the frame structure.[14] A formal procedure begins with deriving the deformation characteristics of all the frame members in stiffness matrix forms. Then, a transformation of coordinates relating to member coordinates and absolute coordinates is applied to express all the stiffness matrices in absolute coordinates. Next, equations of equilibrium are derived for all the joints with respect to forces and moments, and, by the use of the transformed stiffness matrices, these equations are expressed in terms of the deformation quantities such as displacements and rotational angles belonging to the joints. It is seen that the equations of equilibrium thus derived are equivalent to those obtained by applying the principle of minimum potential energy. By solving these equations, we can determine all the deformation quantities of the joints. Then, by the use of the stiffness matrices, all the internal forces and moments at the joints, and consequently, the deformation and stress of the frame structure can be determined.

As is easily seen, one must struggle with a large number of linear simultaneous algebraic equations with the deformation quantities at the joints as unknowns in the application of the the stiffness matrix method. The amazing advance in the development of the high-speed digital computer has made such computations a routine calculation.[15]

10.8. Notes on the Force Method Applied to Semi-monocoque Structures

Semi-monocoque construction has wide applications in light structures such as airplanes, ships and so forth. A semi-monocoque structure usually consists of panels and stringers, where the panels are used as transmitters of in-plane forces, especially as shear members, and the stringers as transmitters of axial forces. Variational principles have been formulated extensively for analyses of these structures, reducing them to finite degree-of-deformation or finite redundancy systems.

We shall have some considerations on the force method applied to the analysis of a semi-monocoque structure,† and review briefly its variational background. In the force method, a semi-monocoque structure is usually split fictitiously into an assemblage of a number of elements consisting of stringers and rectangular panels. For the sake of simplicity, we assume that each stringer has uniform cross section and each panel uniform thickness. One of the simplest assumptions on the stress distribution of these members is as follows: A stringer is assumed to be subjected to end forces P and P^* together with uniformly distributed load q as shown in Fig. 10.8. From static equilibrium, we have

$$P^* = P + ql. \tag{10.70}$$

† For the force method applied to semi-monocoque structures, see the books of Refs. 1 through 13 and the papers of Refs. 16 through 24.

The complementary energy of the stringer, V_s, is

$$V_s = \int_0^l \frac{1}{2EA_0} [P + (l - x) q]^2 \, dx. \tag{10.71}$$

For a stringer with constant EA_0, we obtain

$$V_s = \frac{1}{2EA_0} \left[P^2 l + Pql^2 + \frac{1}{3} q^2 l^3 \right]. \tag{10.72}$$

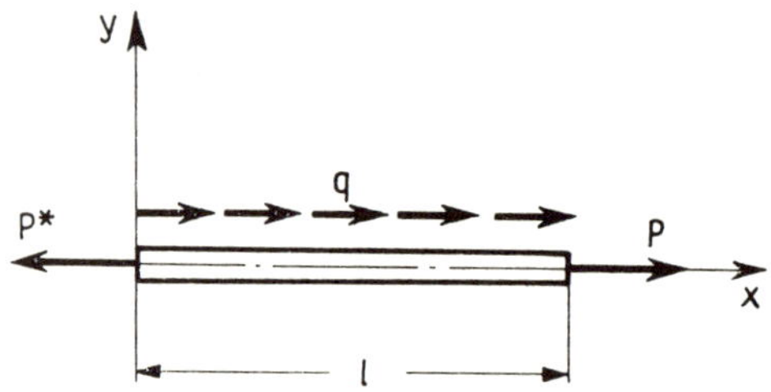

FIG. 10.8. A stringer under axial forces and distributed shear.

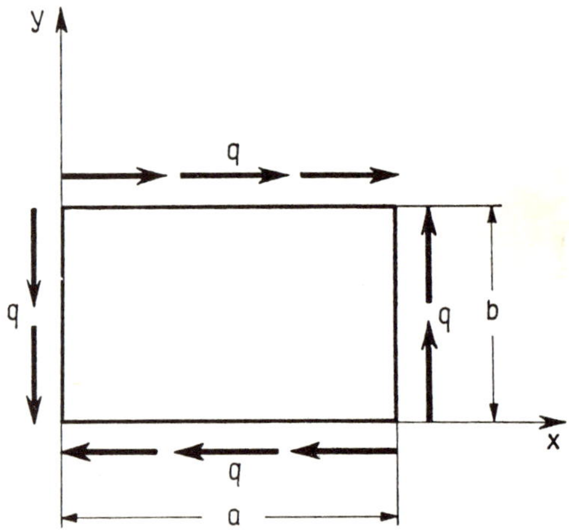

FIG. 10.9. A shear panel.

A rectangular panel is assumed to be in a state of uniform shearing stress under a shear flow q, uniformly distributed along the four edges as shown in Fig. 10.9, where $q = \tau t$, τ is the shearing stress and t is the thickness of the panel. The complementary energy of the panel, V_p, is

$$V_p = \int_0^a \int_0^b \frac{1}{2G} \tau^2 t \, dx \, dy = \frac{1}{2Gt} q^2 ab. \tag{10.73}$$

By summing the complementary energies of all the members, we obtain the total complementary energy of the semi-monocoque structure. Then, the force method requires that the total complementary energy assumes an absolute minimum with respect to the variation of the internal edge forces under the subsidiary conditions that the internal edge forces are continous between adjacent members and the equilibrium conditions at the joints or nodes must be satisfied. The above procedure constitutes the main part of the force method.

We have seen that in trusses and frames, the deformation quantities such as elongations and rotational angles are associated with internal end forces and moments through the complementary energy of the members as shown by Eqs. (10.10) and (10.48), and that the principle of minimum complementary energy provides the compatibility conditions in the large existing among them. However, for structural members under more complicated internal loadings, the geometrical meanings of deformation quantities associated with generalized internal forces through complementary energy of the members become less clear, although the general process of formulation remains the same. For example, let us consider a stringer shown in Fig. 10.8, of which the complementary energy is given by Eq. (10.71). Denoting the displacement in the direction of the x-axis by $u(x)$, and employing the relation

$$P + (l - x)\, q = EA_0(du/dx), \tag{10.74}$$

we obtain:

$$\frac{\partial V_s}{\partial P} = \int_0^l \frac{du}{dx}\, dx = u(l) - u(0), \tag{10.75}$$

$$\frac{\partial V_s}{\partial q} = \int_0^l \frac{du}{dx}(l - x)\, dx = \int_0^l [u(x) - u(0)]\, dx, \tag{10.76}$$

thus realizing the physical meaning of these derivatives. It can be said that although geometrical meanings of deformation quantities derived from the complementary energy may become obscure, a sequence of approximate solutions obtained via the force method must converge to the actual solution if the degree of redundancy is increased without limit.

It should be noted here that, although a solution can be obtained by applying one of the variational principles, it is in general an approximate solution. For example, by applying a force method, we can obtain compatibility conditions in the large which are consistent with the degrees of simplification in establishing the total complementary energy. However, they are generally approximations to the exact conditions of compatibility. Local continuity of displacement between members is generally violated in the approximate solution. We consider, for example, the force method applied to a plane semi-monocoque structure consisting of panels and stringers as shown in Fig. 10.10. We assume the distribution of internal edge forces of these members as shown in Figs. 10.8 and 10.9, and determine the values of these internal edge forces by the force method. Since a rectangular panel subjected to a uniform shear flow q is deformed as shown in Fig. 10.11, in which γ is the shearing strain given by

$$\gamma = q/Gt, \tag{10.77}$$

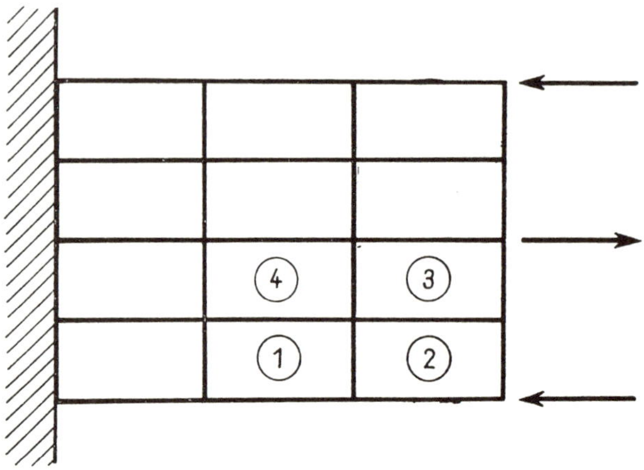

FIG. 10.10. A semi-monocoque plane structure consisting of rectangular panels and stringers.

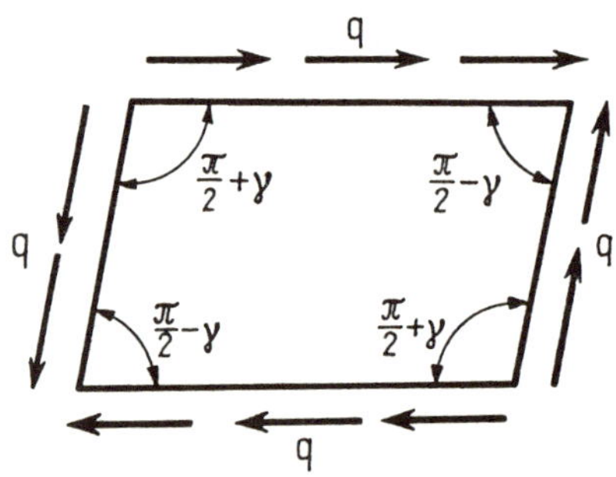

FIG. 10.11. Shear strain γ.

it would be required that a geometrical relation

$$\gamma_{①} - \gamma_{②} + \gamma_{③} - \gamma_{④} = 0 \tag{10.78}$$

must hold for a joint formed by the four panels ①, ②, ③ and ④ in Fig. 10.10 in order to ensure that the connections between these panels are not broken after deformation. However, conditions of this kind are not satisfied in general. Thus, if we were to calculate the displacement components of the members independently, using the values of the internal forces obtained from the force method, we would find discontinuities of displacements on the boundaries between the members. For an improvement of the accuracy of the approximate solution, we must assume more complicated internal edge forces instead of the uniformly distributed shear flow. It is obvious that the principle of minimum complementary energy provides an effective tool for achieving the improvement.

10.9. Notes on the Stiffness Matrix Method Applied to Semi-monocoque Structures

Next, we shall have some considerations on the stiffness matrix method applied to semi-monocoque structures following the pioneering work by Turner, Clough, Martin and Topp,[25] and review briefly its variational background.† In the stiffness matrix method, a semi-monocoque structure

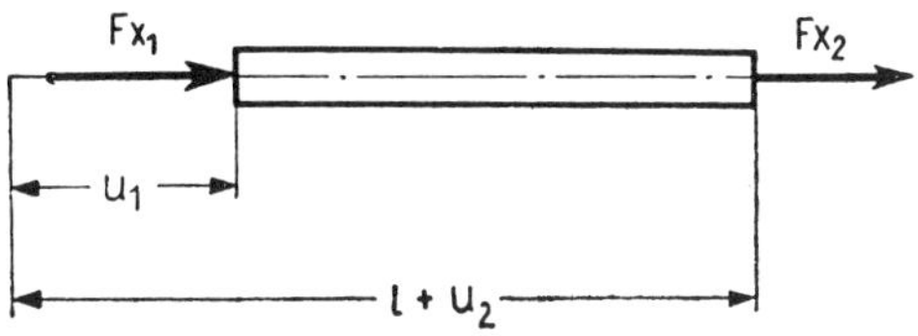

FIG. 10.12. A stringer element.

is usually split fictitiously into an assemblage of a number of elements consisting of stringers and triangular panels. For the sake of simplicity, we assume that each stringer has uniform cross section and each panel uniform thickness. One of the simplest assumptions on the deformation of these members is as follows: A stringer is in a state of uniform strain and its strain energy, U_s, is

$$U_s = \frac{1}{2}\frac{EA_0}{l}(u_2 - u_1)^2, \tag{10.79}$$

where u_1 and u_2 are the displacements of both ends in the axial direction, as shown in Fig. 10.12. Defining the end forces by

$$F_{x1} = \frac{\partial U_s}{\partial u_1}, \quad F_{x2} = \frac{\partial U_s}{\partial u_2}, \tag{10.80}$$

we obtain the stiffness matrix of the stringer:

$$\begin{bmatrix} F_{x1} \\ F_{x2} \end{bmatrix} = \frac{EA_0}{l}\begin{bmatrix} -1 & 1 \\ 1 & -1 \end{bmatrix}\begin{bmatrix} u_1 \\ u_2 \end{bmatrix}. \tag{10.81}$$

† For the stiffness matrix method applied to semi-monocoque structures, see also the books of Refs. 10 through 13.

A triangular panel is assumed in a state of uniform strain:

$$\varepsilon_x = \frac{\partial u}{\partial x} = a, \quad \varepsilon_y = \frac{\partial v}{\partial y} = b,$$

$$\gamma_{xy} = \frac{\partial u}{\partial y} + \frac{\partial v}{\partial x} = c, \tag{10.82}$$

from which we obtain

$$u = ax + Ay + B,$$

$$v = by + (c - A)\, x + C, \tag{10.83}$$

A, *B* and *C* being constants of integration which define rigid body translation and rotation of the triangle. Denoting the displacement components

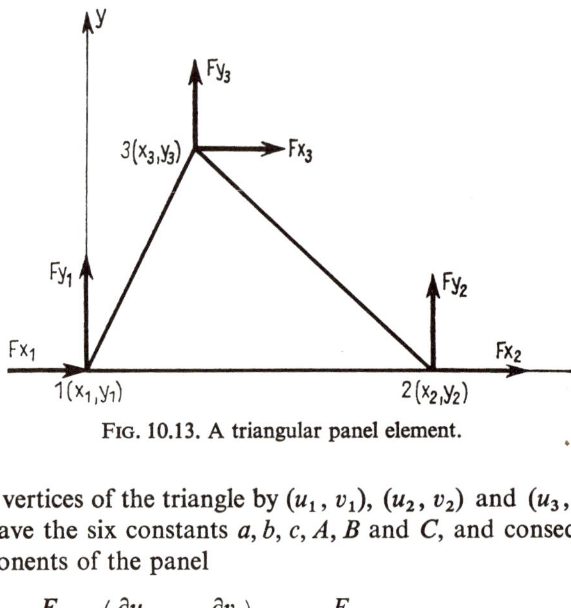

FIG. 10.13. A triangular panel element.

of the three vertices of the triangle by (u_1, v_1), (u_2, v_2) and (u_3, v_3) respectively, we have the six constants a, b, c, A, B and C, and consequently, the stress components of the panel

$$\sigma_x = \frac{E}{(1-\nu^2)}\left(\frac{\partial u}{\partial x} + \nu\frac{\partial v}{\partial y}\right) = \frac{E}{(1-\nu^2)}(a + \nu b),$$

$$\sigma_y = \frac{E}{(1-\nu^2)}\left(\nu\frac{\partial u}{\partial x} + \frac{\partial v}{\partial y}\right) = \frac{E}{(1-\nu^2)}(\nu a + b), \tag{10.84}$$

$$\tau_{xy} = G\gamma_{xy} = Gc,$$

and the strain energy of the panel

$$U_p = \frac{1}{2}\iint\left[\frac{E}{(1-\nu^2)}(\varepsilon_x + \varepsilon_y)^2 + G(\gamma_{xy}^2 - 4\varepsilon_x\varepsilon_y)\right] t\, dx\, dy$$

$$= \frac{1}{4}\left[\frac{E}{(1-\nu^2)}(a + b)^2 + G(c^2 - 4ab)\right] t\, x_2 y_3, \tag{10.85}$$

in terms of u_1, v_1, u_2, v_2, u_3 and v_3, where t is the thickness of the plate. It is seen that the stress components σ_x, σ_y and τ_{xy} given by Eqs. (10.84) satisfy the equations of equilibrium:

$$\frac{\partial \sigma_x}{\partial x} + \frac{\partial \tau_{xy}}{\partial y} = 0, \qquad \frac{\partial \tau_{xy}}{\partial x} + \frac{\partial \sigma_y}{\partial y} = 0. \tag{10.86}$$

Defining the node forces at the vertices by

$$F_{x_1} = \frac{\partial U_p}{\partial u_1}, \quad F_{y_1} = \frac{\partial U_p}{\partial v_1}, \quad \ldots, \quad F_{y_3} = \frac{\partial U_p}{\partial v_3}, \tag{10.87}$$

we obtain

$$\begin{bmatrix} F_{x_1} \\ F_{y_1} \\ F_{x_2} \\ F_{y_2} \\ F_{x_3} \\ F_{y_3} \end{bmatrix} = [K] \begin{bmatrix} u_1 \\ v_1 \\ u_2 \\ v_2 \\ u_3 \\ v_3 \end{bmatrix}, \tag{10.88}$$

where $[K]$ is a symmetric matrix given by

$$[K] = \frac{Et}{2(1 - \nu^2)} \times$$

$$\begin{bmatrix} \frac{y_3}{x_2} + \frac{\lambda_1 x_{23}^2}{x_2 y_3} & & & & & \\ -\frac{\lambda_2 x_{32}}{x_2} & \frac{x_{23}^2}{x_2 y_3} + \frac{\lambda_1 y_3}{x_2} & & & & \\ -\frac{y_3}{x_2} + \frac{\lambda_1 x_3 x_{23}}{x_2 y_3} & \frac{\nu x_{32}}{x_2} + \frac{\lambda_1 x_3}{x_2} & \frac{y_3}{x_2} + \frac{\lambda_1 x_3^2}{x_2 y_3} & & & \\ \frac{\nu x_3}{x_2} + \frac{\lambda_1 x_{32}}{x_2} & \frac{x_3 x_{23}}{x_2 y_3} - \frac{\lambda_1 y_3}{x_2} & -\frac{\lambda_2 x_3}{x_2} & \frac{x_3^2}{x_2 y_3} + \frac{\lambda_1 y_3}{x_2} & & \\ -\frac{\lambda_1 x_{23}}{y_3} & -\lambda_1 & -\frac{\lambda_1 x_3}{y_3} & \lambda_1 & \frac{\lambda_1 x_2}{y_3} & \\ -\nu & -\frac{x_{23}}{y_3} & \nu & -\frac{x_3}{y_3} & 0 & \frac{x_2}{y_3} \end{bmatrix}, \tag{10.89}$$

with $x_{ij} = x_i - x_j$, $\lambda_1 = (1 - \nu)/2$ and $\lambda_2 = (1 + \nu)/2$.

With these preliminaries, the analysis of a semi-monocoque structure by the use of the stiffness matrix method proceeds as follows: First, a transformation from the member axes to the absolute axes is applied to express all the stiffness matrices in the absolute coordinates. Next, by the use of these stiffness matrices, we obtain equilibrium conditions of all the nodes in terms of the displacement components of the nodes. Since continuity of the displacements along the fictitiously cut edges between the elements has been satisfied by restricting the displacements between the nodes to the linear variation, we find that the equilibrium equations thus derived are equivalent to those obtained by applying the principle of minimum potential energy. By solving these equations, we can determine the displacements of all the nodes. Then, the stress of the stringers and triangular panels can be computed. The above procedure constitutes the main part of the stiffness method. It is observed that the stress components (σ_x, σ_y, τ_{xy}) are uniform in each triangle and change discontinuously from one triangle to another. Similar discontinuities exist between neighboring triangles and stringers. For a method of smoothing the stress discontinuity, the reader is advised to read a paper by Turner and Martin.[12]

We remember that although a solution can be obtained by an application of a displacement method, it is in general an approximate solution. By applying the stiffness matrix method, we can obtain equations of equilibrium which are consistent with the degree of simplification employed in establishing the total potential energy. However, they are generally approximations of the exact equations of equilibrium. Local continuity of the internal forces along the fictitiously cut edges or surfaces between elements is generally violated as mentioned above. We shall consider, for example, a part of a panel which is split fictitiously into several sub-elements as shown in Fig. 10.14 and obtain an interpretation of the equations of equilibrium provided by the stiffness matrix method. For the sake of simplicity, we assume that no external forces are applied at the i-th node.

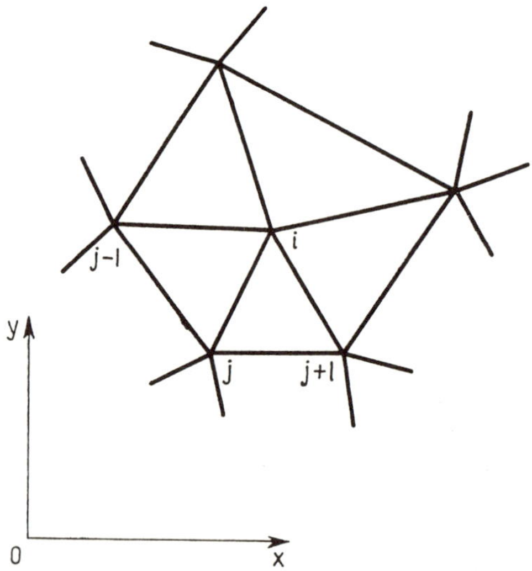

FIG. 10.14. Assembly of triangular panel elements.

The equation of equilibrium of the i-th node in the direction of the x-axis is given by

$$\frac{\partial \Pi}{\partial u_i} = 0, \qquad (10.90)$$

where Π is the total potential energy and u_i is the displacement component of the i-th node in the x-direction. It is obvious that a result equivalent to Eq. (10.90) is obtainable from the principle of virtual work for the problem

$$\sum_{\Delta} \iint (\sigma_x \,\delta\varepsilon_x + \sigma_y \,\delta\varepsilon_y + \tau_{xy} \,\delta\gamma_{xy}) \, t \, dx \, dy + \cdots = 0, \tag{10.91}$$

by requiring that the coefficient of δu_i must vanish, where the notation $\sum_{\Delta}$ means that the summation is taken with respect to all the triangles, and ε_x, ε_y and γ_{xy} are derived from continuous functions u and v such that

$$\varepsilon_x = \frac{\partial u}{\partial x}, \quad \varepsilon_y = \frac{\partial v}{\partial y}, \quad \gamma_{xy} = \frac{\partial u}{\partial y} + \frac{\partial v}{\partial x}. \tag{10.92}$$

Since σ_x, σ_y and τ_{xy} are so chosen as to satisfy Eqs. (10.86), the contribution due to δu_i in the principle (10.91) reduces, via integrations by parts, to

$$\sum_{j} \int_{i}^{j} [X_\nu^{(i,j,j-1)} + X_\nu^{(i,j,j+1)}] \, \delta u_{ij} t \, ds = 0, \tag{10.93}$$

where $X_\nu^{(i,j,j-1)}$ and $X_\nu^{(i,j,j+1)}$ are the components of the internal stresses distributed on the ij-th edge of the triangles $(i, j, j-1)$ and $(i, j, j+1)$ in the direction of the x-axis respectively, and u_{ij} is the displacement components of the ij-th edge in the x-direction. In Eq. (10.93), the summation with respect to j is taken over all the edges which are directly connected to the i-th node. Since u_{ij} is chosen to vary linearly along the ij-th edge, we have

$$\delta u_{ij} = \left(1 - \frac{s_{ij}}{l_{ij}}\right) \delta u_i, \tag{10.94}$$

where l_{ij} is the length of the ij-th edge, and s_{ij} is a coordinate measured from the i-th to the j-th node. Then, Eq. (10.93) reduces to

$$\sum_{j} \int_{i}^{j} [X_\nu^{(i,j,j-1)} + X_\nu^{(i,j,j+1)}] \left(1 - \frac{s_{ij}}{l_{ij}}\right) t \, ds = 0, \tag{10.95}$$

which requires that the weighted mean of the unbalanced internal stresses along the edges directly connected to the i-th node should vanish. Equation (10.95) provides an interpretation of Eq. (10.90).

Thus, the deformation and stress obtained by the stiffness matrix method is approximate. For an improvement of the accuracy of the approximate solution, we may have two approaches.[25] First, the triangular stiffness matrix may be used and the desired accuracy may be obtained by using a sufficient number of sub-elements, or, second, a more general panel stiffness matrix may be used with fewer sub-elements. It is obvious that the principle of minimum potential energy provides an effective tool for the second approach.

Bibliography

1. A. S. Niles and J. S. Newell, *Airplane Structures*, John Wiley, 1943.
2. D. J. Peery, *Aircraft Structures*, McGraw-Hill, 1949.
3. N. J. Hoff, *The Analysis of Structures*, John Wiley, 1956.
4. P. Kuhn, *Stresses in Aircraft and Shell Structures*, McGraw-Hill, 1956.
5. R. L. Bisplinghoff, H. Ashley, and R. L. Halfman, *Aeroelasticity*, Addison-Wesley, 1955.
6. W. S. Hemp, *Methods for the Theoretical Analysis of Aircraft Structures*, AGARD Lecture Course April, 1957.
7. J. H. Argyris, On the Analysis of Complex Elastic Structures, *Applied Mechanics Reviews*, Vol. 11, No. 7, pp. 331–8, July 1958.
8. J. H. Argyris and S. Kelsey, *Energy Theorems and Structural Analysis*, Butterworth, 1960.
9. E. C. Pestel and F. A. Leckie, *Matrix Methods in Elastomechanics*, McGraw-Hill, 1963.
10. J. H. Argyris, *Recent Advances in Matrix Methods of Structural Analysis*, Pergamon Press 1964.
11. R. H. Gallagher, *A Correlation Study of Methods of Matrix Structural Analysis*, Pergamon Press 1964.
12. F. de Veubeke, Editor, *Matrix Methods of Structural Analysis*, Pergamon Press 1964.
13. O. C. Zienkiewicz and G. S. Holister, Editor, *Stress Analysis*, John Wiley, 1965.
14. H. C. Martin, *Introduction to Matrix Methods of Structural Analysis*, McGraw-Hill, 1966.
15. *IBM 7090/7094 FRAN Framed Structure Analysis Program*, International Business Machine Corporation, August 21, 1964.
16. H. Ebner and H. Köller, Über die Einleitung von Längskraften in Versteiften Zylinderschalen, *Jahrbuch der Deutschen Luftfahrtforschung*, pp. 464–73, 1937.
17. E. Ebner and H. Köller, Zur Berechnung des Kraftverlaufs in Versteiften Zylinderschalen, *Luftfahrtforschung*, Vol. 14, No. 12, pp. 607–26, December 1937.
18. E. Ebner and H. Köller, Über den Kraftverlauf in Längs-und-querversteiften Scheiben, *Luftfahrtforschung*, Vol. 15, No. 11, pp. 527–42, October 1938.
19. S. Levy, Computations of Influence Coefficients for Aircraft Structures with Discontinuities and Sweepback, *Journal of the Aeronautical Sciences*, Vol. 14, No. 10, pp. 547–60, October 1947.
20. A. L. Lang and R. L. Bisplinghoff, Some Results of Sweptback Wing Structural Studies, *Journal of the Aeronautical Sciences*, Vol. 18, No. 11, pp. 705–17, November 1951.
21. B. Langefors, *Structural Analysis of Swept-back Wings by Matrix-Transformations*, SAAB Aircraft Company, Linköping, Sweden, SAAB TN 3, 1951.
22. B. Langefors, Analysis of Elastic Structures by Matrix Transformation with Special Regard to Semi-monocoque Structures, *Journal of the Aeronautical Sciences*, Vol. 19, No. 8, pp. 451–8, July 1952.
23. T. Rand, An Approximate Method for the Calculation of Stresses in Sweptback Wings, *Journal of the Aeronautical Sciences*, Vol. 18, No. 1, pp. 61–3, January 1951.
24. L. B. Wehle and W. A. Lansing, A Method for Reducing the Analysis of Complex Redundant Structures to a Routine Procedure, *Journal of the Aeronautical Sciences*, Vol. 19, No. 10, pp. 677–84, October 1952.
25. M. J. Turner, R. W. Clough, H. C. Martin and L. J. Topp, Stiffness and Deflection Analysis of Complex Structures, *Journal of the Aeronautical Sciences*, Vol. 23, No. 9, pp. 805–23, September 1956.

CHAPTER 11

THE DEFORMATION THEORY OF PLASTICITY

11.1. The Deformation Theory of Plasticity

This chapter will discuss variational principles based on the deformation theory of plasticity.† The deformation theory is characterized among theories of plasticity as the one in which relations between instantaneous states of stress and strain are postulated in such a way, that when the strain is given, the stress is uniquely determined or vice versa. However, this determination may or may not be unique in both directions. For example, if the stress is given in terms of the strain as

$$\sigma^{\lambda\mu} = \sigma^{\lambda\mu}(e_{\alpha\beta}), \tag{11.1}$$

the inverse relations may or may not be unique in determining the strain in terms of the stress.

In deriving variational principles in this chapter, we shall assume that the stress–strain relations do not change during the loading process. This assumption restricts the deformation theory problems which we may formulate to those in which the loading increases monotonically. Consequently, the above assumption in effect renders the deformation theory of plasticity undistinguishable from the nonlinear theory of elasticity discussed in Chapter 3, except for materials which obey a yield condition. Furthermore, we shall employ the assumption of small displacements and define a problem of the deformation theory of plasticity as follows:‡

(1) *Equations of equilibrium*

$$\sigma_{ij,j} = 0, \tag{11.2}$$

where body forces are assumed absent for the sake of simplicity;

† It is well established that the deformation theory of plasticity is unsuitable for describing completely the plastic behavior of a metal and should be replaced by the flow theory of plasticity, which follows in Chapter 12. However, this brief chapter is devoted to the deformation theory of plasticity because of historical interest and its frequent use due to mathematical simplicity.

‡ The summation convention is employed throughout Chapters 11 and 12. Thus, a repeated Roman subscript means summation over the values (1, 2, 3).

(2) *Strain–displacement relations*

$$2\varepsilon_{ij} = u_{i,j} + u_{j,i}; \tag{11.3}$$

(3) *Stress–strain relations*

$$\sigma_{ij} = \sigma_{ij}(\varepsilon_{mn}), \tag{11.4}$$

or conversely

$$\varepsilon_{mn} = \varepsilon_{mn}(\sigma_{ij}); \tag{11.5}$$

(4) *Boundary conditions*

$$\sigma_{ij}n_j = F_i \quad \text{on} \quad S_1, \tag{11.6}$$

$$u_i = \bar{u}_i \quad \text{on} \quad S_2. \tag{11.7}$$

Then, by taking the same steps we took in Chapter 1, we obtain the following expressions for the principles of virtual work and complementary virtual work:

$$\iiint_V \sigma_{ij}\, \delta\varepsilon_{ij}\, dV - \iint_{S_1} F_i\, \delta u_i\, dS = 0, \tag{11.8}$$

and

$$\iiint_V \varepsilon_{ij} \delta\sigma_{ij} dV - \iint_{S_2} \delta\sigma_{ij} n_j\, \bar{u}_i dS = 0. \tag{11.9}$$

We repeat that these two principles hold independently of the stress–strain relations.

If Eqs. (11.4) are analytic functions which assure the existence of the state function A defined by

$$\delta A = \sigma_{ij}\, \delta\varepsilon_{ij}, \tag{11.10†}$$

the principle (11.8) leads to the principle of stationary potential energy:‡

$$\delta\Pi = 0, \tag{11.11}$$

where

$$\Pi = \iiint_V A(u_i)\, dV - \iint_{S_1} F_i u_i\, dS. \tag{11.12}$$

On the other hand, if Eqs. (11.5) are analytic functions which assure the existence of the state function B defined by

$$\delta B = \varepsilon_{ij}\, \delta\sigma_{ij}, \tag{11.13†}$$

the principle (11.9) leads to the principle of stationary complementary energy:‡

$$\delta\Pi_c = 0, \tag{11.14}$$

† When the stress system under consideration is uniaxial, such as in a bar under tension, the existence of the state functions A and B is assured for the deformation theory of plasticity. This suggests that variational procedures will be extremely powerful in analyzing structures, if the stresses in the structures can be assumed uniaxial and the deformation theory of plasticity can be applied.[1, 2, 3]

‡ The appellations "potential energy" and "complementary energy" seem to be misleading in the theory of plasticity. However, we shall employ them because their mathematical definitions are the same as in the theory of elasticity.

where

$$\Pi_c = \iiint_V B(\sigma_{ij})\, dV - \iint_{S_2} \sigma_{ij} n_j \bar{u}_i \, dS. \tag{11.15}$$

If it is assumed further that Eqs. (11.5) are unique inverse relations of Eqs. (11.4), and vice versa, we can transform the principle of stationary potential energy into the principle of stationary complementary energy, and vice versa, in a manner similar to the development in Chapter 2.

Thus, the stationary property of the two functionals (11.12) and (11.15) is assured under the assumptions mentioned above. However, the maximum or minimum properties of these functionals cannot be guaranteed unless the stress–strain relations are specified more in detail. Following Ref. 4, we shall review some of the variational principles related to the deformation theory of plasticity.

11.2. Strain-hardening Material

A type of deformation theory called the secant modulus theory in which the stress–strain relations are given by

$$\sigma'_{ij} = \mu \varepsilon'_{ij} \tag{11.16}$$

will be discussed in this section. Here, σ'_{ij} and ε'_{ij} are the stress and strain deviators defined as $\sigma'_{ij} = \sigma_{ij} - \sigma\delta_{ij}$ and $\varepsilon'_{ij} = \varepsilon_{ij} - e\delta_{ij}$, where $\sigma = (1/3)\,\sigma_{ii}$, $e = (1/3)\,\varepsilon_{ii}$ and δ_{ij} is the Kronecker symbol. The quantity μ appearing in Eq. (11.16) is assumed to be a positive quantity which depends in general on the state of strain. It follows immediately from Eq. (11.16) that

$$S = \mu\Gamma, \tag{11.17}$$

where

$$S = \sqrt{\sigma'_{ij}\sigma'_{ij}}, \quad \Gamma = \sqrt{\varepsilon'_{ij}\varepsilon'_{ij}}, \tag{11.18}$$

and consequently

$$S\, dS = \sigma'_{ij}\, d\sigma'_{ij}, \quad \Gamma\, d\Gamma = \varepsilon'_{ij}\, d\varepsilon'_{ij}. \tag{11.19}$$

It is assumed that S is a single-valued continuous function of Γ, as shown in Fig. 11.1, i.e.

$$S = S(\Gamma), \tag{11.20}$$

and that the relations

$$S/\Gamma = \mu > 0, \quad dS/d\Gamma > 0 \tag{11.21}$$

hold throughout the regions of Γ and S under consideration. By combining Eqs. (11.16) and (11.17), we have

$$\sigma'_{ij} = (S(\Gamma)/\Gamma)\, \varepsilon'_{ij}, \tag{11.22}$$

and

$$\varepsilon'_{ij} = (\Gamma(S)/S)\, \sigma'_{ij}. \tag{11.23}$$

Only five of the relations in Eqs. (11.22) or (11.23) are independent. Therefore, a sixth relation, compressibility, will be added, i.e.

$$\sigma = 3Ke, \tag{11.24}$$

where K is the bulk modulus of the material. Under the assumption that the plastic deformation causes no volume change, K can be expressed in terms of Young's modulus E and Poisson's ratio ν as follows:

$$3K = \frac{E}{1-2\nu}. \tag{11.25}$$

With the above preliminaries, we obtain the following expressions for A and B under the secant modulus theory:

$$A = \frac{3E}{2(1-2\nu)} e^2 + \int_0^{\Gamma} S(\Gamma)\, d\Gamma, \tag{11.26}$$

$$B = \frac{3(1-2\nu)}{2E} \sigma^2 + \int_0^{S} \Gamma(S)\, dS. \tag{11.27}$$

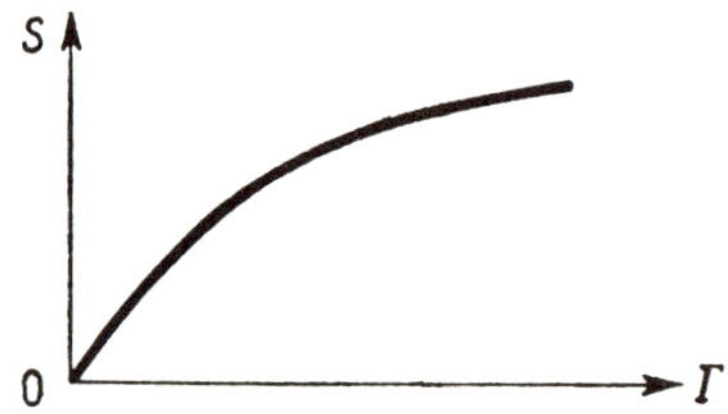

FIG. 11.1. $S - \Gamma$ relation for a strain-hardening material.

By substituting Eqs. (11.26) and (11.27) into Eqs. (11.12) and (11.15) respectively, we have the expressions of the functionals Π and Π_c of the problem for a material obeying the secant modulus theory. We then have two variational principles which are called Kachanov's principles [4, 5, 6] and stated as follows:

Kachanov principle 1. The exact solution of the problem renders the functional Π a minimum with respect to admissible displacement variations.†

Kachanov principle 2. The exact solution of the problem renders the functional Π_c a minimum with respect to admissible stress variation.‡

† Since $2\delta^2 A = [3E/(1-2\nu)](\delta e)^2 + (S/\Gamma^3)[\Gamma^2 \delta\varepsilon'_{ij}\delta\varepsilon'_{ij} - (\varepsilon'_{ij}\delta\varepsilon'_{ij})^2] + (1/\Gamma^2)(dS/d\Gamma) \times (\varepsilon'_{ij}\delta\varepsilon'_{ij})^2$, $\Gamma^2\delta\varepsilon'_{ij}\delta\varepsilon'_{ij} \geqq (\varepsilon'_{ij}\delta\varepsilon'_{ij})^2$ by Schwarz's inequality and $dS/d\Gamma > 0$ from Eq. (11.21), we have $\delta^2\Pi \geqq 0$.

‡ Since $2\delta^2 B = [3(1-2\nu)/E](\delta\sigma)^2 + (\Gamma/S^3)[S^2\delta\sigma'_{ij}\delta\sigma'_{ij} - (\sigma'_{ij}\delta\sigma'_{ij})^2] + (1/S^2)(d\Gamma/dS) \times (\sigma'_{ij}\delta\sigma'_{ij})^2$, $S^2\delta\sigma'_{ij}\delta\sigma'_{ij} \geqq (\sigma'_{ij}\delta\sigma'_{ij})^2$ by Schwarz's inequality and $d\Gamma/dS > 0$ from Eq. (11.21), we have $\delta^2\Pi_c \geqq 0$.

11.3. Perfectly Plastic Material

The secant modulus theory will now be specialized to the case of a perfectly plastic material obeying the Mises yield condition. The $S - \Gamma$ relation will be assumed as shown in Fig. 11.2: the material behaves elastically for $S < \sqrt{2}\,k$ and flows for $S = \sqrt{2}\,k$, where k is the yield limit in simple shear. The expressions for A and B, and the stress–strain relations for the perfectly plastic material may be formally obtained as follows. We replace the $S - \Gamma$ curve in Fig. 11.1 with a broken line such that

$$\begin{aligned} S &= 2G\Gamma && \text{for} \quad \Gamma < \Gamma_0, \\ S &= S_0 + 2\beta(\Gamma - \Gamma_0) && \text{for} \quad \Gamma \geqq \Gamma_0, \end{aligned} \tag{11.28}$$

where $S_0 = \sqrt{2}\,k = 2G\Gamma_0$ and β is a positive constant. We calculate the expressions for A and B for the relation given by Eqs. (11.28), and drive the stress–strain relations from

$$\sigma_{ij} = \partial A/\partial \varepsilon_{ij}, \quad \varepsilon_{ij} = \partial B/\partial \sigma_{ij}. \tag{11.29}$$

We then let β approach zero. Thus, we obtain the following expressions for A and B for the perfectly plastic material:

$$\begin{aligned} A &= \frac{3E}{2(1-2\nu)} e^2 + G\Gamma^2 && \text{for} \quad \Gamma < \Gamma_0, \\ A &= \frac{3E}{2(1-2\nu)} e^2 + G\Gamma_0^2 + \sqrt{2}\,k(\Gamma - \Gamma_0) && \text{for} \quad \Gamma > \Gamma_0, \end{aligned} \tag{11.30}$$

$$B = \frac{3(1-2\nu)}{2E} \sigma^2 + \frac{1}{4G} S^2. \tag{11.31}$$

The resulting stress–strain relations obtained by the above limiting procedures are as follows:

$$\begin{aligned} \sigma_{ij} &= \frac{E}{(1-2\nu)} e\delta_{ij} + 2G\varepsilon'_{ij} && \text{for} \quad \Gamma < \Gamma_0, \\ \sigma_{ij} &= \frac{E}{(1-2\nu)} e\delta_{ij} + \frac{\sqrt{2}\,k}{\Gamma} \varepsilon'_{ij} && \text{for} \quad \Gamma \geqq \Gamma_0, \end{aligned} \tag{11.32}$$

$$\begin{aligned} \varepsilon_{ij} &= \frac{(1-2\nu)}{E} \sigma\delta_{ij} + \frac{1}{2G} \sigma'_{ij} && \text{for} \quad S < \sqrt{2}\,k, \\ \varepsilon_{ij} &= \frac{(1-2\nu)}{E} \sigma\delta_{ij} + \frac{1}{2G} \sigma'_{ij} + \lambda\sigma'_{ij} && \text{for} \quad S = \sqrt{2}\,k, \end{aligned} \tag{11.33}$$

where λ is a positive, indeterminate and finite quantity defined by

$$\lim_{\beta \to 0,\, S \to S_0} (S - S_0)/2\beta S = \lambda. \tag{11.34}$$

A material which exhibits the stress–strain relation of Eqs. (11.32), or equivalently (11.33), is called a Hencky material for which the Haar–Kármán principle holds.[7] This principle may be stated as follows:

Among arbitrary sets of admissible stress components which satisfy the equations of equilibrium, the mechanical boundary condition on S_1 and the condition $\sigma'_{ij}\sigma'_{ij} \leqq 2k^2$, the exact solution renders

$$\Pi_c = \iiint_V \left[\frac{3(1-2\nu)}{2E}\sigma^2 + \frac{1}{4G}\sigma'_{ij}\sigma'_{ij}\right] dV - \iint_{S_2} \sigma_{ij} n_j \bar{u}_i \, dS \tag{11.35}$$

an absolute minimum.

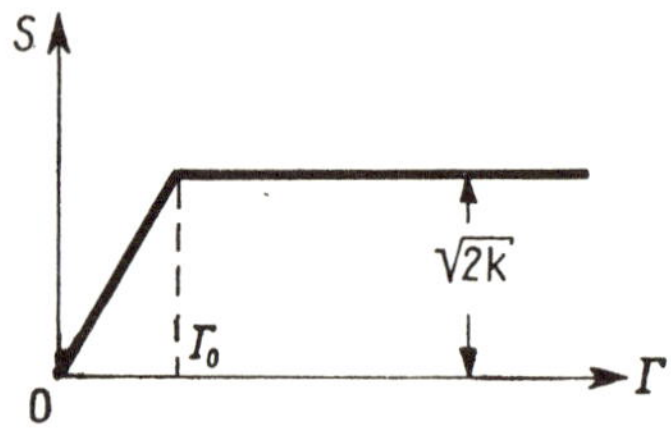

FIG. 11.2. $S - \Gamma$ relation for a perfectly plastic material.

The proof of the Haar–Kármán principle given by Greenberg will be followed here.[4] Let σ_{ij}, ε_{ij} and u_i represent the stresses, strains and displacements of the exact solution and let the stress components of an admissible solution be denoted by σ^*_{ij}. In addition, the elastic and plastic portions of the strain components will be separated by writing

$$\varepsilon_{ij} = \varepsilon^e_{ij} + \varepsilon^p_{ij}. \tag{11.36}$$

Then, observing that the body consists of a plastic region V_p and an elastic region V_e, we may write the first variation of Π_c as

$$\begin{aligned}
\delta\Pi_c &= \iiint_V \varepsilon^e_{ij}\,\delta\sigma_{ij}\, dV - \iint_{S_2} \delta\sigma_{ij} n_j \bar{u}_i \, dS \\
&= \iiint_V (\varepsilon_{ij} - \varepsilon^p_{ij})\,\delta\sigma_{ij}\, dV - \iint_{S_2} \delta\sigma_{ij} n_j \bar{u}_i \, dS \\
&= -\iiint_V (\delta\sigma_{ij})_{,j} u_i \, dV + \iint_{S_1} \delta\sigma_{ij} n_j u_i \, dS \\
&\quad + \iint_{S_2} \delta\sigma_{ij} n_j (u_i - \bar{u}_i)\, dS - \iiint_{V_p} \varepsilon^p_{ij}\,\delta\sigma_{ij}\, dV \\
&= -\iiint_{V_p} \varepsilon^p_{ij}\,\delta\sigma_{ij}\, dV.
\end{aligned} \tag{11.37}$$

Since we have

$$\varepsilon^p_{ij} = \lambda\sigma'_{ij}, \quad \lambda > 0 \tag{11.38}$$

from Eq. (11.33), we can write

$$\delta \Pi_c = -\iiint_{V_p} \lambda \sigma'_{ij}\, \delta\sigma_{ij}\, dV. \tag{11.39}$$

The exact solution satisfies $\sigma'_{ij}\sigma'_{ij} = 2k^2$ in the plastic region, while the admissible solution has been chosen so that $\sigma^{*\prime}_{ij}\sigma^{*\prime}_{ij} \leqq 2k^2$. Since Schwarz's inequality proves the following relations:

$$\sigma'_{ij}\sigma^{*\prime}_{ij} \leqq \sqrt{\sigma'_{ij}\sigma'_{ij}}\,\sqrt{\sigma^{*\prime}_{ij}\sigma^{*\prime}_{ij}} \leqq 2k^2, \tag{11.40}$$

we have

$$\begin{aligned} \sigma'_{ij}\,\delta\sigma_{ij} &= \sigma'_{ij}(\sigma^*_{ij} - \sigma_{ij}) \\ &= \sigma'_{ij}\sigma^{*\prime}_{ij} - \sigma'_{ij}\sigma'_{ij} \leqq 0, \end{aligned} \tag{11.41}$$

where the equality holds only when $\sigma^*_{ij} = \sigma_{ij}$. Consequently, we conclude that

$$\delta \Pi_c \geqq 0. \tag{11.42}$$

The functional Π_c is a quadratic form with respect to the stress components and the second variation can be proved to be positive. Consequently, the functional Π_c is rendered an absolute minimum for the exact solution.

The above proof shows that, when a subsidiary condition is given in the form of an inequality, it is no longer necessary that the first variation must vanish for an absolute maximum or minimum of a functional. An example noted in Ref. 4 is that a maximum of the parabola $y = x^2$ under a subsidiary condition $0 \leqq x \leqq 1$ is attained at $x = 1$ at which point, however, $y'(1) \neq 0$. A brief mention is made in Appendix F concerning variational formulations of a problem with a subsidiary condition in the form of an inequality.

11.4. A Special Case of Hencky Material

Variational principles will now be treated for a special case of a Hencky material, i.e. one which is assumed to be incompressible and everywhere plastic. The $S - \Gamma$ relation is as shown in Fig. 11.3. We observe that for the

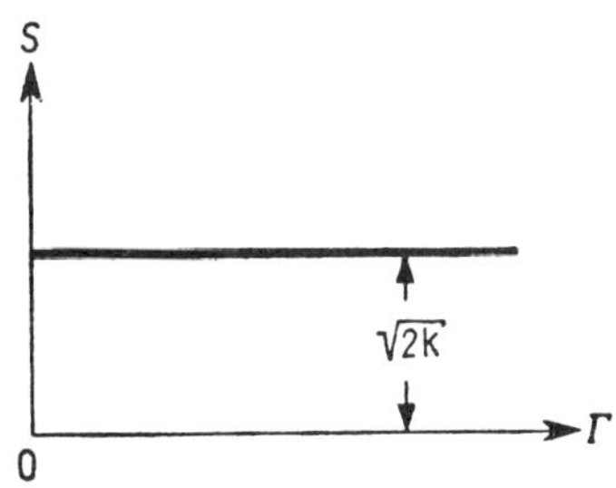

FIG. 11.3. $S - \Gamma$ relation for a special case of Hencky material.

special case, the expressions (11.30) and (11.31) are reduced to

$$A = \sqrt{2}\, k\Gamma, \tag{11.43}$$

$$B = 0, \tag{11.44}$$

and the corresponding stress–strain relations are

$$\sigma_{ij} = \frac{\sqrt{2}\, k}{\sqrt{\varepsilon_{mn}\varepsilon_{mn}}}\varepsilon_{ij}. \tag{11.45}$$

The principle of minimum potential energy then holds as follows:

Among admissible solutions which satisfy the conditions of compatibility, the geometrical boundary conditions on S_2 and the incompressibility condition, the actual solution† renders

$$\Pi = \sqrt{2}\, k \iiint_V \sqrt{\varepsilon_{ij}\varepsilon_{ij}}\, dV - \iint_{S_1} F_i \bar{u}_i\, dS \tag{11.46}$$

an absolute minimum.

This principle is analogous to Markov's principle for the Saint-Venant–Levy–Mises material in the flow theory of plasticity. On the other hand, the principle of minimum complementary energy is expressed as follows:

Among admissible solutions which satisfy the equations of equilibrium, the yield condition $\sigma'_{ij}\sigma'_{ij} = 2k^2$ and the mechanical boundary conditions on S_1, the actual solution† renders

$$\Pi_c = -\iint_{S_2} \sigma_{ij} n_j \bar{u}_i\, dS \tag{11.47}$$

an absolute minimum.

This principle is equivalent to Sadowsky's principle of maximum plastic work which states that among admissible solutions, the actual solution renders

$$\iint_S \sigma_{ij} n_j \bar{u}_i\, dS$$

an absolute maximum.[8] Sadowsky's principle is analogous to Hill's principle for the Saint-Venant–Levy–Mises material in the flow theory of plasticity. Proofs of the above two principles are found in Ref. 4 (see also Section 12.5 of this book).

Bibliography

1. N. J. Hoff, *The Analysis of Structures*, John Wiley, 1956.
2. J. H. Argyris and S. Kelsey, *Energy Theorems and Structural Analysis*, Butterworth, 1960.
3. H. Langhaar, *Energy Methods in Applied Mechanics*, John Wiley, 1962.

† Except for a possible indeterminate uniform hydrostatic stress.

4. H. J. GREENBERG, *On the Variational Principles of Plasticity*, Brown University, ONR, NR-041-032, March 1949.
5. L. M. KACHANOV, Variational Principles for Elastic-Plastic Solids, *Prikladnaia Mathematika i Mekhanika*, Vol. 6, pp. 187–96, 1942. (Translation prepared at Brown University for the Taylor Model Basin in 1946.)
6. A. A. ILYUSHIN, Some Problems in the Theory of Plastic Deformations, *Prikladnaia Matematika i Mekhanika*, Vol. 7, pp. 245–72, 1943. (Translation prepared at Brown University for the Taylor Model Basin in 1946).
7. A. HAAR and TH. V. KÁRMÁN, Zur Theorie der Spannungszustände in Plastischen und Sandartigen Medien, *Nach. der Wiss. zu Göttingen*, pp. 204–18, 1909.
8. M. A. SADOWSKY, A Principle of Maximum Plastic Resistance, *Journal of Applied Mechanics*, Vol. 10, No. 2, pp. 65–8, June 1943.

CHAPTER 12

THE FLOW THEORY OF PLASTICITY

12.1. The Flow Theory of Plasticity

It is well established that unique relations do not exist in general between stress and strain components in the plastic region; the strain depends not only on the final state of stress, but also on the loading history. Therefore, the stress–strain relations which have been familiar to us in the theory of elasticity must be replaced by relations between increments of stress and strain in developing theories of plasticity. This avenue of the theory of plasticity is called the incremental strain theory or flow theory of plasticity.† The deformation theory of plasticity treated in the last chapter is only a special case of the flow theory and has been found unsuitable for a complete description of the plastic behavior of a metal.

We begin by observing that the flow theory of plasticity employs the Eulerian descriptive technique. Namely, a set of values of the rectangular Cartesian coordinates which an arbitrary point of a body under consideration occupies at the generic time is employed for specifying the point during subsequent incremental deformations. The stress components σ_{ij} at the generic time are defined with respect to these coordinates in a manner similar to the definition of initial stresses in Section 5.1.

Following Prager, let us define a problem in the flow theory of plasticity. At a given instant of time t, a body is assumed to be in a state of static equilibrium, and the state of stress σ_{ij} and its loading history are assumed to be known throughout the body. Now, external force increments $\overline{dF}_i$, $i = 1, 2, 3$ are prescribed on S_1 and displacement increments $d\bar{u}_i$, $i = 1, 2, 3$ are prescribed on S_2. Our problem is then to determine increments of the stress $d\sigma_{ij}$ and displacement du_i induced in the body under the assumptions that the increments are infinitesimal and all the governing equations may be linearized. Thus, we have:[1]

(1) *Equations of equilibrium*

$$d\sigma_{ij,j} = 0; \tag{12.1}‡$$

† Refs. 1 through 6.

‡ Here, $d\sigma_{ij}$ is the change of σ_{ij} in a given element of the body (Lagrangian), and differs from the change of σ_{ij} at a fixed point (Eulerian), denoted by $d^*\sigma_{ij}$, by the amount $d\sigma_{ij} - d^*\sigma_{ij} = \sigma_{ij,k}\, du_k$. Both the original and incremental stress distributions satisfy the equations of equilibrium: $\sigma_{ij,j} = 0$ and $d^*\sigma_{ij,j} = 0$. Consequently, we have $d\sigma_{ij,j} - \sigma_{ij,k}\, du_{k,j} = 0$, which leads to Eqs. (12.1) under the assumption that increments of plastic strain are constrained to be of order $(1/E) \times$ (the stress-increments), where E is the Young's modulus of the material.[1]

(2) *Strain–displacement relations*

$$2d\varepsilon_{ij} = du_{i,j} + du_{j,i}; \tag{12.2}$$

(3) *Linear relations between stress increments* $d\sigma_{ij}$ *and strain increments* $d\varepsilon_{kl}$; (12.3)

(4) *Boundary conditions*

$$d\sigma_{ij}n_j = d\bar{F}_i \quad \text{on} \quad S_1, \tag{12.4}$$

$$du_i = d\bar{u}_i \quad \text{on} \quad S_2. \tag{12.5}$$

It is seen that the above problem is defined in a manner similar to an elasticity problem of the small displacement theory, except for the stress–strain relations. Once the problem in flow theory has been thus formulated, problems of finite plastic deformation can be analyzed by integrating the resulting relations along the prescribed loading path.

It is apparent from the above relations that the principles of virtual work and complementary virtual work may be written for the present problem as

$$\iiint_V d\sigma_{ij}\,\delta d\varepsilon_{ij}\,dV - \iint_{S_1} d\bar{F}_i\,\delta du_i\,dS = 0 \tag{12.6}$$

and

$$\iiint_V d\varepsilon_{ij}\delta d\sigma_{ij}\,dV - \iint_{S_2} \overline{du}_i\delta d\sigma_{ij}n_j\,dS = 0, \tag{12.7}$$

respectively.

The above problem may be considered as quasi-static and stated in terms of rate as follows. At a given instant of time, a body is assumed to be in a state of quasi-static equilibrium. Now, the rates of application of the external forces $\dot{\bar{F}}_i$, $i = 1, 2, 3$ are prescribed on S_1, while the surface velocity $\bar{v}_i$, $i = 1, 2, 3$ are prescribed on S_2. Our problem is then to find the stress rates $\dot{\sigma}_{ij}$ and velocities v_i induced in the body. Here, a dot sign denotes a derivative with respect to time, while v_i denotes a component of the velocity with respect to the rectangular Cartesian coordinates.

The governing equations for the problem expressed in terms of rate are obtained from Eqs. (12.1) through (12.5) by replacing $d\sigma_{ij}$, $d\varepsilon_{ij}$, du_i, $d\bar{F}_i$ and $\overline{du}_i$ with $\dot{\sigma}_{ij}$, $\dot{\varepsilon}_{ij}$, v_i, $\dot{\bar{F}}_i$ and $\bar{v}_i$, respectively. Two principles corresponding to Eqs. (12.6) and (12.7) may then be written as

$$\iiint_V \dot{\sigma}_{ij}\,\delta\dot{\varepsilon}_{ij}\,dV - \iint_{S_1} \dot{\bar{F}}_i\,\delta v_i\,dS = 0 \tag{12.8}$$

and

$$\iiint_V \dot{\varepsilon}_{ij}\,\delta\dot{\sigma}_{ij}\,dV - \iint_{S_2} \bar{v}_i\,\delta\dot{\sigma}_{ij}n_j\,dS = 0. \tag{12.9}$$

Since the state of stress σ_{ij} at the given instant forms a self-equilibrating system, we may add the following two principles:

$$\iiint_V \sigma_{ij}\,\delta\dot{\varepsilon}_{ij}\,dV - \iint_{S_1} F_i\,\delta v_i\,dS = 0, \tag{12.10}$$

and

$$\iiint_V \dot{\varepsilon}_{ij}\,\delta\sigma_{ij}\,dV - \iint_{S_2} \bar{v}_i\,\delta\sigma_{ij}n_j\,dS = 0, \tag{12.11}$$

where $F_i = \sigma_{ij}n_j$ on S_1.

With these preliminaries, we shall review some of variational principles related to the flow theory of plasticity.†

12.2. Strain-hardening Material

Following Ref. 1, we shall adopt:

$$d\varepsilon_{ij} = \frac{(1-2\nu)}{E}\,d\sigma\,\delta_{ij} + \frac{d\sigma'_{ij}}{2G} + \alpha^{**}h\frac{\partial f}{\partial\sigma_{ij}}\,df \tag{12.12}$$

as the incremental stress–strain relations for a strain-hardening material, and call

$$d\varepsilon^e_{ij} = \frac{(1-2\nu)}{E}\,d\sigma\,\delta_{ij} + \frac{d\sigma'_{ij}}{2G}, \tag{12.13}$$

$$d\varepsilon^p_{ij} = \alpha^{**}h\frac{\partial f}{\partial\sigma_{ij}}\,df, \tag{12.14}$$

the elastic and plastic strain components, respectively. Here, $d\sigma$ and $d\sigma'_{ij}$ are the increment of the average hydrostatic stress and the incremental stress deviator: $d\sigma = (1/3)\,d\sigma_{ii}$ and $d\sigma'_{ij} = d\sigma_{ij} - d\sigma\,\delta_{ij}$. The function h is a positive definite function of σ_{ij}. The function $f(\sigma_{ij})$ is called the yield condition and the surface

$$f = c \tag{12.15}$$

the yield surface. The parameter c in the above specifies the final state of strain-hardening, and its value may vary from point to point throughout the body. Since $df = (\partial f/\partial\sigma_{ij})\,d\sigma_{ij}$, we may specify the following loading terminology concerning a set of incremental stresses $d\sigma_{ij}$:

$$\begin{aligned} &\text{loading} && \text{if } df > 0,\\ &\text{neutral} && \text{if } df = 0,\\ &\text{unloading} && \text{if } df < 0. \end{aligned} \tag{12.16}$$

† Refs. 1 through 6.

The quantity α^{**} in Eqs. (12.12) is defined via the above relations in the following manner:

$$\begin{aligned}
&\alpha^{**} = 1 \quad \text{where} \quad f(\sigma_{ij}) = c \quad \text{and} \quad df \geqq 0, \\
&\alpha^{**} = 0 \quad \text{where} \quad f(\sigma_{ij}) < c, \quad \text{or where} \\
&\qquad\qquad f(\sigma_{ij}) = c \quad \text{and} \quad df < 0.
\end{aligned} \tag{12.17}$$

The parameter c may be given as a function of the total plastic work:

$$c = F\left(\int \sigma_{ij}\, d\varepsilon_{ij}^{p}\right) \tag{12.18}$$

where F is a monotonically increasing positive function and the integral is taken along the loading path. From Eqs. (12.14), (12.15) and (12.18), we find that the three functions f, h and F are related by $hF'(\partial f/\partial \sigma_{ij})\, \sigma_{ij} = 1$.

By multiplying both sides of Eq. (12.12) by $\partial f/\partial \sigma_{ij}$, and taking summations with respect to i and j, we obtain

$$\frac{\partial f}{\partial \sigma_{ij}}\, d\varepsilon_{ij} = \left(\frac{1}{2Gh} + \frac{\partial f}{\partial \sigma_{ij}} \frac{\partial f}{\partial \sigma_{ij}}\right) h\, df \tag{12.19}$$

for $\alpha^{**} = 1$. This suggests that $df > 0$ corresponds to $(\partial f/\partial \sigma_{ij})\, d\varepsilon_{ij} > 0$. With this preliminary, we may obtain from Eq. (12.12) the following inverse relations expressing $d\sigma_{ij}$ in terms of $d\varepsilon_{ij}$:

$$d\sigma_{ij} = \frac{E}{(1 - 2\nu)}\, de\, \delta_{ij} + 2G\, d\varepsilon'_{ij} - \alpha^{*} \frac{2G\left(\dfrac{\partial f}{\partial \sigma_{kl}}\, d\varepsilon_{kl}\right)}{\left(\dfrac{1}{2Gh} + \dfrac{\partial f}{\partial \sigma_{pq}} \dfrac{\partial f}{\partial \sigma_{pq}}\right)} \frac{\partial f}{\partial \sigma_{ij}} \tag{12.20}$$

where $de = (1/3)\, d\varepsilon_{ii}$ and $d\varepsilon'_{ij} = d\varepsilon_{ij} - de\, \delta_{ij}$. The notation α^{*} is defined in the following manner:

$$\begin{aligned}
&\alpha^{*} = 1 \quad \text{where} \quad f(\sigma_{ij}) = c \quad \text{and} \quad (\partial f/\partial \sigma_{kl})\, d\varepsilon_{kl} \geqq 0, \\
&\alpha^{*} = 0 \quad \text{where} \quad f(\sigma_{ij}) < c \quad \text{or where} \\
&\qquad\qquad f(\sigma_{ij}) = c \quad \text{and} \quad (\partial f/\partial \sigma_{kl})\, d\varepsilon_{kl} < 0.
\end{aligned} \tag{12.21}$$

Equations (12.12) or (12.20) are linear and homogeneous in terms of the strain increments $d\varepsilon_{ij}$ and the stress increments $d\sigma_{ij}$. In that sense, they are similar to the stress–strain relations in the linear theory of elasticity, except that they are given in pairs, one set for loading and one set for unloading, and that the coefficients which correspond to the elastic constants are dependent on the state of stress and loading history of the element just before the incremental deformation occurs.

Before proceeding with variational formulations of the problem, we must ascertain whether or not the incremental stress–strain relations, (12.12) and (12.20), assure the existence of state functions $\mathscr{A}$ and $\mathscr{B}$ defined as:

$$\delta \mathscr{A} = d\sigma_{ij}\, \delta d\varepsilon_{ij} \tag{12.22}$$

and

$$\delta\mathscr{B} = d\varepsilon_{ij}\,\delta d\sigma_{ij}. \tag{12.23}$$

Indeed, we find that they are given by

$$\mathscr{A} = \frac{3E}{2(1-2\nu)}(de)^2 + G\,d\varepsilon'_{ij}\,d\varepsilon'_{ij} - \alpha^* \frac{G\left(\dfrac{\partial f}{\partial \sigma_{kl}}\,d\varepsilon_{kl}\right)^2}{\left(\dfrac{1}{2Gh} + \dfrac{\partial f}{\partial \sigma_{pq}}\dfrac{\partial f}{\partial \sigma_{pq}}\right)}, \tag{12.24}$$

$$\mathscr{B} = \frac{3(1-2\nu)}{2E}(d\sigma)^2 + \frac{d\sigma'_{ij}\,d\sigma'_{ij}}{4G} + \frac{1}{2}\alpha^{**}h(df)^2. \tag{12.25}$$

Consequently, two variational principles may be created for the work-hardening material. The first principle states:

Among admissible solutions which satisfy the conditions of compatibility and the geometrical boundary conditions, the exact solution renders the functional

$$\Pi = \iiint_V \mathscr{A}\,dV - \iint_{S_1} d\bar{F}_i\,du_i\,dS \tag{12.26}$$

an absolute minimum.

On the other hand, the second principle states:

Among admissible solutions which satisfy the equations of equilibrium and the mechanical boundary conditions, the exact solution renders the functional

$$\Pi_c = \iiint_V \mathscr{B}\,dV - \iint_S d\sigma_{ij}n_j\,d\bar{u}_i\,dS \tag{12.27}$$

an absolute minimum.

The proof of these principles is given in Ref. 1, together with reference to the pioneers who contributed to the establishment of the principles.

12.3. Perfectly Plastic Material

By substituting $h = 1/\beta$ into Eq. (12.12), where β is assumed to be a positive constant, and letting β approach zero in such a way that

$$\lim_{\beta\to 0,\, df\to 0} \frac{df}{\beta} = d\lambda > 0 \tag{12.28}$$

where $d\lambda$ is a positive, indeterminate and finite quantity, we find the following incremental stress–strain relations for a perfectly plastic material:

$$d\varepsilon_{ij} = \frac{1-2\nu}{E}\,d\sigma\,\delta_{ij} + \frac{1}{2G}\,d\sigma'_{ij} + \alpha^{**}\frac{\partial f}{\partial \sigma_{ij}}\,d\lambda. \tag{12.29}$$

Here,

$$\begin{aligned} &\alpha^{**} = 1 \quad \text{where} \quad f(\sigma_{ij}) = c \quad \text{and} \quad df = 0, \\ &\alpha^{**} = 0 \quad \text{where} \quad f(\sigma_{ij}) < c \quad \text{or where} \\ &\qquad\qquad f(\sigma_{ij}) = c \quad \text{and} \quad df < 0, \end{aligned} \tag{12.30}$$

and c is a material constant defining the yield conditions of the material under consideration.

Simultaneous solution of Eqs. (12.29) yields the following inverse relations:

$$d\sigma_{ij} = \frac{E}{(1-2\nu)}\, de\, \delta_{ij} + 2G\, d\varepsilon'_{ij} - \alpha^* \frac{2G\left(\dfrac{\partial f}{\partial \sigma_{kl}} d\varepsilon_{kl}\right)}{\left(\dfrac{\partial f}{\partial \sigma_{pq}} \dfrac{\partial f}{\partial \sigma_{pq}}\right)} \frac{\partial f}{\partial \sigma_{ij}}. \tag{12.31}$$

Here,

$$\alpha^* = 1 \quad \text{where} \quad f(\sigma_{ij}) = c \quad \text{and} \quad (\partial f/\partial \sigma_{kl})\, d\varepsilon_{kl} \geqq 0.$$

$$\alpha^* = 0 \quad \text{where} \quad f(\sigma_{ij}) < c, \quad \text{or where}$$

$$f(\sigma_{ij}) = c \quad \text{and} \quad (\partial f/\partial \sigma_{kl})\, d\varepsilon_{kl} < 0. \tag{12.32}$$

The expressions for $\mathscr{A}$ and $\mathscr{B}$ now become:

$$\mathscr{A} = \frac{3E}{2(1-2\nu)} (de)^2 + G\, d\varepsilon'_{ij}\, d\varepsilon'_{ij} - \alpha^* \frac{G\left(\dfrac{\partial f}{\partial \sigma_{kl}} d\varepsilon_{kl}\right)^2}{\left(\dfrac{\partial f}{\partial \sigma_{pq}} \dfrac{\partial f}{\partial \sigma_{pq}}\right)}, \tag{12.33}$$

$$\mathscr{B} = \frac{3(1-2\nu)}{2E} (d\sigma)^2 + \frac{1}{4G}\, d\sigma'_{ij}\, d\sigma'_{ij}, \tag{12.34}$$

for the perfectly plastic material.

By the use of the expressions for $\mathscr{A}$ and $\mathscr{B}$ thus derived, we can obtain two variational principles for the perfectly plastic material in a manner similar to the development in the last section, except that admissible solutions in the second principle are now subject to an additional subsidiary condition in the form of an inequality,† namely, $df \leqq 0$. The proof of these principles is given in Ref. 1, together with reference to the pioneers who contributed to the establishment of the principles.

12.4. The Prandtl–Reuss Equation

The Prandtl–Reuss equation is a special case of the stress–strain relations of Eqs. (12.12) and (12.29) and is based on the assumption that

$$f = \bar{\sigma} \tag{12.35}$$

where

$$\bar{\sigma} = \sqrt{\frac{3}{2}}\{\sigma'_{ij}\sigma'_{ij}\}^{1/2}. \tag{12.36‡}$$

† This is the same situation which we encountered in the Haar–Kármán principle of the deformation theory of plasticity.

‡ An overbar does not indicate that the barred quantity is prescribed in the notation $\bar{\sigma}$.

Introducing the notation $d\bar{\varepsilon}^p = \sqrt{(2/3)}\,\{d\varepsilon^p_{ij}\,d\varepsilon^p_{ij}\}^{1/2}$, Eq. (12.18) is found to be replaced by

$$\bar{\sigma} = H\left(\int d\bar{\varepsilon}^p\right) \tag{12.37†}$$

where H is a monotonically increasing positive function. Since the relation $hH' = 1$ is obtained from Eqs. (12.12), (12,35) and (12.37), we may employ the following equation in place of Eq. (12.12):

$$d\varepsilon_{ij} = \frac{(1-2\nu)}{E}\,d\sigma\,\delta_{ij} + \frac{d\sigma'_{ij}}{2G} + \alpha^{**}\,\frac{3\sigma'_{ij}\,d\bar{\sigma}}{2\bar{\sigma}H'}\,. \tag{12.38}$$

Here

$$\begin{aligned} \alpha^{**} &= 1 \quad \text{where} \quad \bar{\sigma} = c \quad \text{and} \quad d\bar{\sigma} \geqq 0, \\ \alpha^{**} &= 0 \quad \text{where} \quad \bar{\sigma} < c \quad \text{or where} \\ &\qquad \bar{\sigma} = c \quad \text{and} \quad d\bar{\sigma} < 0. \end{aligned} \tag{12.39}$$

The inverse form of Eq. (12.38) can be shown to be

$$d\sigma_{ij} = \frac{E}{(1-2\nu)}\,de\,\delta_{ij} + 2G\,d\varepsilon'_{ij} - \alpha^*\,\frac{3G(\sigma'_{kl}\,d\varepsilon_{kl})}{\bar{\sigma}^2\left(\dfrac{H'}{3G}+1\right)}\,\sigma'_{ij}\,. \tag{12.40}$$

Here

$$\begin{aligned} \alpha^{*} &= 1 \quad \text{where} \quad \bar{\sigma} = c \quad \text{and} \quad \sigma'_{ij}\,d\varepsilon_{ij} \geqq 0, \\ \alpha^{*} &= 0 \quad \text{where} \quad \bar{\sigma} < c \quad \text{or where} \\ &\qquad \bar{\sigma} = c \quad \text{and} \quad \sigma'_{ij}\,d\varepsilon_{ij} < 0. \end{aligned} \tag{12.41}$$

Equations (12.38) and (12.40) are called the Prandtl–Reuss equations for a strain-hardening material. If both sides of Eqs. (12.38) and (12.40) are divided by dt, we have the relations in the form of rate:

$$\dot{\varepsilon}_{ij} = \frac{(1-2\nu)}{E}\,\dot{\sigma}\delta_{ij} + \frac{\dot{\sigma}'_{ij}}{2G} + \alpha^{**}\,\frac{3\sigma'_{ij}\dot{\bar{\sigma}}}{2\bar{\sigma}H'}\,, \tag{12.42}$$

$$\dot{\sigma}_{ij} = \frac{E}{(1-2\nu)}\,\dot{e}\,\delta_{ij} + 2G\dot{\varepsilon}_{ij} - \alpha^*\,\frac{3G(\sigma'_{kl}\dot{\varepsilon}_{kl})}{\bar{\sigma}^2\left(\dfrac{H'}{3G}+1\right)}\,\sigma'_{ij} \tag{12.43}$$

where the definitions of α^{**} and α^* are obtained from Eqs. (12.39) and (12.41) by replacing $d\bar{\sigma}$ and $d\varepsilon_{ij}$ with $\dot{\bar{\sigma}}$ and $\dot{\varepsilon}_{ij}$, respectively.

The above relations may be extended to a perfectly plastic material for which the yield condition is given by:

$$\sigma'_{ij}\sigma'_{ij} = 2k^2, \tag{12.44}$$

where k is a constant dependent on the material. By putting

$$\lim_{d\bar{\sigma}\to 0,\,H'\to 0} \frac{3d\bar{\sigma}}{2\bar{\sigma}H'} = d\lambda > 0 \tag{12.45}$$

† An overbar does not indicate that the barred quantity is prescribed in the notation $d\bar{\varepsilon}^P$.

in Eq. (12.38), where $d\lambda$ is a positive, indeterminate and finite quantity, we obtain

$$d\varepsilon_{ij} = \frac{(1-2\nu)}{E} d\sigma\, \delta_{ij} + \frac{d\sigma'_{ij}}{2G} + \alpha^{**}\sigma'_{ij}\, d\lambda. \tag{12.46}$$

Here,

$$\begin{aligned} &\alpha^{**} = 1 \quad \text{where} \quad \sigma'_{ij}\sigma'_{ij} = 2k^2 \quad \text{and} \quad \sigma'_{ij}\, d\sigma'_{ij} = 0, \\ &\alpha^{**} = 0 \quad \text{where} \quad \sigma'_{ij}\sigma'_{ij} < 2k^2, \quad \text{or where} \\ &\qquad \sigma'_{ij}\sigma'_{ij} = 2k^2 \quad \text{and} \quad \sigma'_{ij}\, d\sigma'_{ij} < 0. \end{aligned} \tag{12.47}$$

The inverse forms of Eqs. (12.46) can be shown to be:

$$d\sigma_{ij} = \frac{E}{(1-2\nu)} de\, \delta_{ij} + 2G\, d\varepsilon'_{ij} - \alpha^* \frac{G}{k^2} (\sigma'_{kl}\, d\varepsilon_{kl})\, \sigma'_{ij}. \tag{12.48}$$

Here,

$$\begin{aligned} &\alpha^* = 1 \quad \text{where} \quad \sigma'_{ij}\sigma'_{ij} = 2k^2 \quad \text{and} \quad \sigma'_{ij}\, d\varepsilon_{ij} \geqq 0, \\ &\alpha^* = 0 \quad \text{where} \quad \sigma'_{ij}\sigma'_{ij} < 2k^2, \quad \text{or where} \\ &\qquad \sigma'_{ij}\sigma'_{ij} = 2k^2 \quad \text{and} \quad \sigma'_{ij}\, d\varepsilon_{ij} < 0. \end{aligned} \tag{12.49}$$

Equations (12.46) and (12.48) are called the Prandtl–Reuss equations for a perfectly plastic material. The rate forms of these equations are given by

$$\dot{\varepsilon}_{ij} = \frac{(1-2\nu)}{E} \dot{\sigma}\delta_{ij} + \frac{\dot{\sigma}'_{ij}}{2G} + \alpha^{**}\mu\sigma'_{ij}, \tag{12.50}$$

and

$$\dot{\sigma}_{ij} = \frac{E}{(1-2\nu)} \dot{e}\, \delta_{ij} + 2G\dot{\varepsilon}'_{ij} - \alpha^* \frac{G}{k^2} (\sigma'_{kl}\dot{\varepsilon}_{kl})\, \sigma'_{ij}, \tag{12.51}$$

respectively, where $\mu = d\lambda/dt > 0$. The definitions of α^{**} and α^* in these equations are obtained from Eqs. (12.47) and (12.49) by replacing $d\sigma'_{ij}$ and $d\varepsilon_{ij}$ with $\dot{\sigma}'_{ij}$ and $\dot{\varepsilon}_{ij}$, respectively.

Variational principles similar to those given in Sections 12.2 and 12.3 have been derived for materials which obey the Prandtl–Reuss equations.[1]

12.5. The Saint-Venant–Levy–Mises Equations

If the elastic strain rates in Eqs. (12.50) are assumed to be negligible compared to the plastic strain rates, we have

$$\dot{\varepsilon}_{ij} = \mu\sigma'_{ij} \quad \text{where} \quad \sigma'_{ij}\sigma'_{ij} = 2k^2 \quad \text{and} \quad \sigma'_{ij}\dot{\sigma}'_{ij} = 0, \tag{12.52a}$$

$$\begin{aligned} &\dot{\varepsilon}_{ij} = 0 \quad \text{where} \quad \sigma'_{ij}\sigma'_{ij} < 2k^2, \quad \text{or where} \\ &\qquad \sigma'_{ij}\sigma'_{ij} = 2k^2 \quad \text{and} \quad \sigma'_{ij}\dot{\sigma}'_{ij} < 0. \end{aligned} \tag{12.52b}$$

The inverse relations are

$$\sigma'_{ij} = \frac{\sqrt{2}\, k}{\sqrt{\dot{\varepsilon}_{kl}\dot{\varepsilon}_{kl}}} \dot{\varepsilon}_{ij} \tag{12.53}$$

for Eq. (12.52a) only. Materials governed by the above relations are called rigid-plastic materials. Equations (12.52a, b) are called the Saint-Venant–Levy–Mises equations for rigid-plastic materials.

We shall consider variational principles for a body composed of rigid-plastic material under the assumption that the entire body is plastic. The problem in this section will be defined in a slightly different manner from the previous problems:

(1) *Equations of equilibrium*

$$\sigma_{ij,j} = 0; \tag{12.54}$$

(2) *Yield condition*

$$\sigma'_{ij}\sigma'_{ij} = 2k^2; \tag{12.55}$$

(3) *Stress–strain rate relations*: Eq. (12.53);

(4) *Rate of strain–velocity relation*

$$2\dot{\varepsilon}_{ij} = v_{i,j} + v_{j,i}; \tag{12.56}$$

(5) *Condition of incompressibility*

$$\dot{\varepsilon}_{ii} = 0; \tag{12.57}$$

(6) *Boundary conditions*

$$\sigma_{ij}n_j = \bar{F}_i \quad \text{on} \quad S_1, \tag{12.58}$$

$$v_i = \bar{v}_i \quad \text{on} \quad S_2. \tag{12.59}$$

There result two variational principles, the first of which may be stated as follows:

> Among admissible solutions which satisfy the conditions of compatibility and incompressibility, as well as the geometrical boundary conditions on S_2, the actual solution† renders
>
> $$\Pi = \sqrt{2}\, k \iiint_V \sqrt{\dot{\varepsilon}_{ij}\dot{\varepsilon}_{ij}}\, dV - \iint_{S_1} \bar{F}_i v_i \, dS \tag{12.60}$$
>
> an absolute minimum.

This is called Markov's principle.[7] The proof is as follows. Let the stress, strain rate and velocity of the exact solution be denoted by σ_{ij}, $\dot{\varepsilon}_{ij}$ and v_i, and the strain rate and velocity of an admissible solution by $\dot{\varepsilon}^*_{ij}$ and v^*_i. Then, since

$$\sigma'_{ij}\dot{\varepsilon}^*_{ij} \leqq \sqrt{\sigma'_{ij}\sigma'_{ij}}\sqrt{\dot{\varepsilon}^*_{ij}\dot{\varepsilon}^*_{ij}} \tag{12.61}$$

by Schwarz's inequality, and

$$\sigma'_{ij}\dot{\varepsilon}^*_{ij} = \sigma_{ij}\dot{\varepsilon}^*_{ij} \tag{12.62}$$

by the incompressibility condition, we obtain from Eqs. (12.55), (12.61) and (12.62) the following relation:

$$\sigma_{ij}\dot{\varepsilon}^*_{ij} \leqq \sqrt{2}\, k \sqrt{\dot{\varepsilon}^*_{ij}\dot{\varepsilon}^*_{ij}}\,. \tag{12.63}$$

† Except for a possible indeterminate uniform hydrostatic pressure.

On the other hand, Eqs. (12.53) and (12.57) provide:

$$\sigma_{ij}\dot{\varepsilon}_{ij} = \sqrt{2}\, k \sqrt{\dot{\varepsilon}_{ij}\dot{\varepsilon}_{ij}}\,. \tag{12.64}$$

Combining Eqs. (12.63) and (12.64), we obtain

$$\sqrt{2}\, k \left(\sqrt{\dot{\varepsilon}_{ij}^{*}\dot{\varepsilon}_{ij}^{*}} - \sqrt{\dot{\varepsilon}_{ij}\dot{\varepsilon}_{ij}}\right) \geqq \sigma_{ij}(\dot{\varepsilon}_{ij}^{*} - \dot{\varepsilon}_{ij}). \tag{12.65}$$

Integration of Eq. (12.65) through the entire body and integrations by parts yield:

$$\sqrt{2}\, k \iiint_V \sqrt{\dot{\varepsilon}_{ij}^{*}\dot{\varepsilon}_{ij}^{*}}\, dV - \iint_{S_1} \bar{F}_i v_i^{*}\, dS \geqq$$

$$\sqrt{2}\, k \iiint \sqrt{\dot{\varepsilon}_{ij}\dot{\varepsilon}_{ij}}\, dV - \iint_{S_2} \bar{F}_i v_i\, dS. \tag{12.66}$$

Since v_i^* is an arbitrary admissible velocity, Eq. (12.65) proves Markov's principle. The second principle may be stated as follows:

Among admissible solutions which satisfy the equations of equilibrium, yield condition and the mechanical boundary conditions on S_2, the actual solution† renders

$$\Pi_c = -\iint_{S_2} \sigma_{ij} n_j \bar{v}_i\, dS \tag{12.67}$$

an absolute minimum.

This is equivalent to Hill's principle of maximum plastic work which states that among admissible solutions, the actual solution renders

$$\iiint_{S_2} \sigma_{ij} n_j \bar{v}_i\, dS$$

an absolute maximum.[1, 8] The proof is as follows. Let the stress, strain rate and velocity of the actual solution be denoted by σ_{ij}, $\dot{\varepsilon}_{ij}$ and v_i, and the stress of an admissible solution by σ_{ij}^*. Then, from

$$\sigma_{ij}'\sigma_{ij}' = 2k^2, \quad \sigma_{ij}^{*\prime}\sigma_{ij}^{*\prime} = 2k^2 \tag{12.68}$$

and

$$\sigma_{ij}'\sigma_{ij}^{*\prime} \leqq \sqrt{\sigma_{ij}'\sigma_{ij}'}\sqrt{\sigma_{ij}^{*\prime}\sigma_{ij}^{*\prime}} = 2k^2 \tag{12.69}$$

we have

$$(\sigma_{ij}^{*\prime} - \sigma_{ij}')\,\sigma_{ij}' \leqq 0. \tag{12.70}$$

Substitution of Eqs. (12.53) and (12.57) into Eq. (12.70) yields:

$$(\sigma_{ij}^{*} - \sigma_{ij})\,\dot{\varepsilon}_{ij} \leqq 0. \tag{12.71}$$

Integration of Eq. (12.71) throughout the body and integrations by parts provide:

$$\iint_{S_2} \sigma_{ij} n_j \bar{v}_i\, dS \geqq \iint_{S_2} \sigma_{ij}^{*} n_j \bar{v}_i\, dS. \tag{12.72}$$

Since σ_{ij}^* is an arbitrary admissible stress, Eq. (12.72) proves Hill's principle.

† Except for a possible indeterminate uniform hydrostatic pressure.

A weak statement can be made on the above two principles that the functionals (12.60) and (12.67) are rendered stationary with respect to admissible velocity and stress variations, respectively. For example, we can show

$$\delta \Pi = 0 \tag{12.73}$$

for the exact solution with respect to admissible displacement variations. The principle can then be generalized into:

$$\begin{aligned} \Pi_I = \sqrt{2}\, k \iiint_V \sqrt{\dot{\varepsilon}_{ij}\dot{\varepsilon}_{ij}}\, dV - \iint_{S_1} \bar{F}_i v_i \, dS \\ - \iiint_V \{[\dot{\varepsilon}_{ij} - (\tfrac{1}{2})(v_{i,j} + v_{j,i})]\, \sigma_{ij} - \dot{\varepsilon}_{ij}\, \delta_{ij} \sigma\}\, dV \\ - \iint_{S_2} \sigma_{ij} n_j (v_i - \bar{v}_i)\, dS \end{aligned} \tag{12.74}$$

where σ_{ij} and σ are Lagrange multipliers which introduce the conditions (12.56), (12.57) and (12.59) into the variational expression. The stationary condition of the functional (12.74) with respect to $\dot{\varepsilon}_{ij}$ can be shown to be

$$\sigma_{ij} - \sigma\, \delta_{ij} = \frac{\sqrt{2}\, k}{\sqrt{\dot{\varepsilon}_{kl}\dot{\varepsilon}_{kl}}}\, \dot{\varepsilon}_{ij} \tag{12.75}$$

and the expression of the functional (12.67) can be derived through the elimination of u_i and ε_{ij} in the usual manner.

12.6. Limit Analysis

One of the most successful applications of variational formulations in the flow theory of plasticity is undoubtedly the theory of limit analysis.[2] Consider a continuum or structure, hereafter called a body, which consists of a material obeying the perfectly plastic Prandtl–Reuss equations (12.50). Surface tractions $\bar{F}_i$; $i = 1, 2, 3$ are prescribed on S_1, and displacements are prescribed on S_2 such that $u_i = 0$; $i = 1, 2, 3$. We assume that the surface tractions are applied in proportional loading, that is, the external traction is assumed to be given by $\varkappa F_i$: $i = 1, 2, 3$ where $\varkappa$ is a monotonically increasing parameter. When the value of $\varkappa$ is sufficiently small, the body behaves elastically. As $\varkappa$ increases, a point in the body reaches the plastic state; beyond this value of $\varkappa$ the theory of elasticity is no longer applicable. As $\varkappa$ increases further, the plastic region of the body spreads gradually, although larger parts of the body may still be in the elastic state. If the value of $\varkappa$ continues to increase, a state of impending plastic flow will be reached in such a way that an increase of plastic strain under constant surface tractions becomes possible for the first time during the loading process. The set of surface tractions which correspond to the impending plastic flow is called the collapse load of the body, and the ratio of the collapse load to the design load is called the safety factor and is denoted by S. Thus, the safety factor is the value of $\varkappa$ at the collapse load. The problem in limit analysis is to determine the safety factor of the body under the prescribed surface tractions.

We observe that at the collapse load the elastic strain rates and stress rates are identically zero and the body behaves as rigid-plastic.[2] Consequently, the equations governing the state of impending plastic flow may be given as follows:

(1) *Equations of equilibrium*

$$\sigma_{ij,j} = 0; \tag{12.76}$$

(2) *Yield condition*

$$\sigma'_{ij}\sigma'_{ij} \leqq 2k^2; \tag{12.77}$$

(3) *Stress–strain rate relations*

$$\dot{\varepsilon}_{ij} = \mu\sigma'_{ij} \quad \text{where} \quad \sigma'_{ij}\sigma'_{ij} = 2k^2, \tag{12.78a}$$

$$\dot{\varepsilon}_{ij} = 0 \quad \text{where} \quad \sigma'_{ij}\sigma'_{ij} < 2k^2; \tag{12.78b}$$

(4) *Rate of strain–velocity relation*

$$2\dot{\varepsilon}_{ij} = v_{i,j} + v_{j,i}; \tag{12.79}$$

(5) *Condition of incompressibility*

$$\dot{\varepsilon}_{ii} = 0; \tag{12.80}$$

(6) *Boundary conditions*

$$\sigma_{ij}n_j = S\bar{F}_i \quad \text{on} \quad S_1, \tag{12.81}$$

$$v_i = 0 \quad \text{on} \quad S_2. \tag{12.82}$$

These equations constitute an eigenvalue problem in which the value of S is determined as an eigenvalue.

The theory of limit analysis puts emphasis on the derivation of upper and lower bound formulae for the safety factor. We shall restrict the problem under consideration to continuous stress and velocity fields for the sake of brevity† and introduce the following nomenclature. A set of stress components σ^*_{ij} will be called statically admissible if it satisfies Eqs. (12.76), (12.77) and

$$\sigma^*_{ij}n_j = m_s\bar{F}_i \quad \text{on} \quad S_1, \tag{12.83}$$

where m_s is a number called a statically admissible multiplier. A set of velocity components u^*_i will be called kinematically admissible if it satisfies Eqs. (12.80) and (12.82), and the condition

$$\int_{S_1} \bar{F}_i v^*_i \, dS > 0. \tag{12.84}$$

† An extension to discontinuous velocity fields is found in Ref. 9.

A number defined by

$$m_k = \frac{\sqrt{2}k \iiint\limits_V \sqrt{\dot{\varepsilon}_{ij}^* \dot{\varepsilon}_{ij}^*}\, dV}{\int\limits_{S_1} \bar{F}_i v_i^*\, dS} \tag{12.85}$$

will be called a kinematically admissible multiplier, where $2\dot{\varepsilon}_{ij}^* = v_{i,j}^* + v_{j,i}^*$. We then obtain the following upper and lower bound formulae for the safety factor:

$$m_s \leqq S \leqq m_k. \tag{12.86}$$

The proof is given as follows: First, we observe that Eq. (12.63) is still valid for the present problem. Integration of Eq. (12.63) through the entire body and integrations by parts yield:

$$S \iint\limits_{S_1} \bar{F}_i v_i^*\, dS \leqq \sqrt{2}k \iiint\limits_V \sqrt{\dot{\varepsilon}_{ij}^* \dot{\varepsilon}_{ij}^*}\, dV \tag{12.87}$$

which, under the assumption (12.84), proves $S \leqq m_k$. Second, we observe that Eq. (12.71) is still valid for the present problem. Integration of Eq. (12.71) through the entire body and integrations by parts yield:

$$(m_s - S) \iint\limits_{S_1} \bar{F}_i v_i\, dS \leqq 0. \tag{12.88}$$

Since

$$\iint\limits_{S_1} \bar{F}_i v_i\, dS \geqq 0$$

for the exact solution, Eq. (12.88) proves $m_s \leqq S$.

Thus, the upper and lower bound formulae for the safety factor have been obtained by the simultaneous use of the two variational principles. In that sense, Eqs. (12.86) are analogous to the upper and lower bound formulae for torsional rigidity, which were derived in Section 6.5 from the principles of minimum potential and complementary energy, although the determination of torsional rigidity is a boundary value problem and not an eigenvalue problem. A variational consideration is given in Ref. 10 on the bound formulae, Eq. (12.86).

The theory of limit analysis has been formulated for plane strain problems where detailed investigations have been made on the discontinuity of velocity and stress fields (see Ref. 2). An excellent example of the plane strain problem is shown in Ref. 11: a prismatic cylinder having a square section and a circular hole at the center is under an uniform internal pressure, and upper and lower bounds for the safety factor are obtained by assuming discontinuous velocity as well as stress fields. Limit theory has also proved very powerful in the analysis of plates, shells and multi-component structures (see Refs. 12 through 16).

12.7. Some Remarks

It has been assumed throughout Chapters 11 and 12 that the displacement components are given in terms of three continuous functions. However, deformation in the plastic region is known to consist of infinitesimal slips.[1] This means that the representation of displacements by three continuous functions is only an approximation, and suggests that the theory of plasticity may be improved by taking the discontinuous character of the displacement into account. One of the advances in this direction is known as the theory of dislocation, an excellent description of which is given in Ref. 17. A brief mention of variational principles in the theory of creep is made in Appendix G.

Bibliography

1. R. Hill, *Mathematical Theory of Plasticity*, Oxford, 1950.
2. W. Prager and P. G. Hodge, Jr., *Theory of Perfectly Plastic Solids*, John Wiley, 1951.
3. W. Prager, *An Introduction to Plasticity*, Addison-Wesley, 1959.
4. P. G. Hodge, Jr., The Mathematical Theory of Plasticity, in *Elasticity and Plasticity* by J. N. Goodier and P. G. Hodge, Jr., pp. 51–152, John Wiley, 1958.
5. D. C. Drucker, Variational Principles in the Mathematical Theory of Plasticity, *Proceedings of Symposia in Applied Mathematics*, Vol. 8, pp. 7–22, McGraw-Hill, 1958.
6. W. T. Koiter, General Theorems for Elastic-Plastic Solids, *Progress in Solid Mechanics*, Vol. 1, Chapter IV, pp. 167–221, North-Holland, Amsterdam, Interscience, New York, 1960.
7. A. A. Markov, On Variational Principles in the Theory of Plasticity, *Prikladnaia Matematika i Mekhanika*, Vol. 11, pp. 339–50, 1947. (Translation prepared at Brown University.)
8. R. Hill, A Variational Principle of Maximum Plastic Work in Classical Plasticity, *Quarterly Journal of Mechanics and Applied Mathematics*, Vol. 1, pp. 18–28, 1948.
9. D. C. Drucker, W. Prager and H. J. Greenberg, Extended Limit Design Theorems for Continuous Media, *Quarterly of Applied Mathematics*, Vol. 9, No. 4, pp. 381–9, 1952.
10. T. Mura and S. L. Lee, Application of Variational Principles to Limit Analysis, *Quarterly of Applied Mathematics*, Vol. XXI, No. 3, pp. 244–8, October 1963.
11. D. C. Drucker, H. J. Greenberg and W. Prager, The Safety Factor of an Elastic–Plastic Body in Plane Strain, *Journal of Applied Mechanics*, Vol. 18, No. 4, pp. 371–8, December 1951.
12. Van den Broeck, *Theory of Limit Design*, John Wiley, 1948.
13. B. G. Neal, *The Plastic Methods of Structural Analysis*, John Wiley, 1956.
14. L. S. Beedle, *Plastic Design of Steel Frames*, John Wiley, 1958.
15. P. G. Hodge, Jr., *Plastic Analysis of Structures*, McGraw-Hill, 1959.
16. P. G. Hodge, Jr., *Limit Analysis of Rotationally Symmetric Plates and Shells*, Prentice-Hall, 1963.
17. B. A. Bilby, Continuous Distributions of Dislocations, *Progress in Solid Mechanics*, Vol. 1, Chapter VII, pp. 329–98, North-Holland, Amsterdam, Interscience, New York, 1960.

APPENDIX A

EXTREMUM OF A FUNCTION WITH A SUBSIDIARY CONDITION

WE SHALL consider a problem of finding the extremum of a function under a subsidiary condition. For the sake of simplicity, we shall take a simple example to illustrate the procedure.

PROBLEM: Find the extremum of the function:

$$z = f(x, y) = x^2 + y^2 - 2x - 4y + 6, \tag{1}$$

under the subsidiary condition:

$$g(x, y) = 2x + y - 1 = 0. \tag{2}$$

Geometrically speaking, the problem is one of finding the extremal value of z on the curve of intersection of $z = f(x, y)$ and $g(x, y) = 0$. One of the ways of solving the problem is to eliminate one of the variables, say y, from Eq. (1) by the use of Eq. (2), thus obtaining

$$z = f(x, y(x)) = f^*(x) = 5x^2 + 2x + 3, \tag{3}$$

and then finding the extremum of z by the condition

$$\frac{df^*}{dx} = 10x + 2 = 0, \tag{4}$$

By solving Eq. (4), we obtain

$$x = -\tfrac{1}{5}, \tag{5}$$

and find that

$$z_{\text{ext}} = \tfrac{14}{5}, \quad y = \tfrac{7}{5}. \tag{6}$$

It is observed that this extremal value proves to be the absolute minimum of $f^*(x)$.

The method of Lagrange multiplier asserts that the above problem is equivalent to finding the stationary value of the function z_I defined by

$$z_I = x^2 + y^2 - 2x - 4y + 6 + \lambda(2x + y - 1), \tag{7}$$

where the independent variables are x, y and λ and the stationary conditions are given by

$$\partial z_I/\partial x = 2x - 2 + 2\lambda = 0, \tag{8}$$

$$\partial z_I/\partial y = 2y - 4 + \lambda = 0, \tag{9}$$

$$\partial z_I/\partial \lambda = 2x + y - 1 = 0. \tag{10}$$

Trevithick Libraries

Items on loan 25/11/2010 20:44

XXXXX48776

Item Details	Due Date
Many-particle physics / Gera	06/12/2010
RF power amplifiers for wire	29/11/2010
Introduction to molecular-ort	04/01/2011
Parallel MATLAB for multico	03/02/2011
Art of molecular dynamics si	07/01/2011
Cellular solids : structure an	03/02/2011
* Variational methods in elas	24/02/2011
Professional C / Simon Rob	16/12/2010
Introduction to lattice dynam	24/01/2011
Quantum theory of solids / C	16/12/2010
Molecular modelling : princip	29/11/2010
Atomistic modeling of materi	19/01/2011

* Indicates items borrowed today

Overdue items will incur charges.

Renew items at: http://library.cf.ac.uk

By solving these equations, we obtain

$$x = -\tfrac{1}{5}, \quad y = \tfrac{7}{5}, \quad \lambda = \tfrac{6}{5}, \tag{11}$$

and find the stationary value of z_I as follows:

$$z_{I_{st}} = \tfrac{14}{5} = z_{\text{ext}}. \tag{12}$$

If one of the independent variables, say x, is eliminated from z_I through the use of Eq. (8), the function is transformed into

$$z_{II} = y^2 + \lambda y - \lambda^2 - 4y + 5, \tag{13}$$

where there remain only two independent variables, namely λ and y. The stationary conditions are then given by:

$$\partial z_{II}/\partial y = 2y + \lambda - 4 = 0, \tag{14}$$

$$\partial z_{II}/\partial \lambda = y - 2\lambda + 1 = 0, \tag{15}$$

yielding immediately

$$y = \tfrac{7}{5}, \quad \lambda = \tfrac{6}{5}, \tag{16}$$

and consequently

$$z_{II_{st}} = \tfrac{14}{5} = z_{\text{ext}}, \quad x = -\tfrac{1}{5}. \tag{17}$$

Going further, we shall eliminate x and y from z_I through the use of Eqs. (8) and (9). Then, the function is transformed into

$$z_{III} = -\tfrac{5}{4}\lambda^2 + 3\lambda + 1, \tag{18}$$

where λ is the only remaining independent variable. The stationary condition is then given by

$$dz_{III}/d\lambda = -\tfrac{5}{2}\lambda + 3 = 0, \tag{19}$$

which gives

$$\lambda = \tfrac{6}{5}, \tag{20}$$

and consequently

$$z_{III_{st}} = \tfrac{14}{5} = z_{\text{ext}}, \quad x = -\tfrac{1}{5}, \quad y = \tfrac{7}{5}. \tag{21}$$

It is observed that $z_{III_{st}}$ is the absolute maximum of the function z_{III} with respect to the single variable λ. It is obvious that if Eq. (10) is employed as a subsidiary condition, the function z_I reduces to the original function.

Thus, it has been shown that the stationary values are the same for all the transformed formulations. The extremal value obtained as the minimum in the function (3) is given as the maximum in the function (18). However, $z_{I_{st}}$ and $z_{II_{st}}$ are no longer either maximum or minimum of the functions z_I and z_{II}, respectively.

Bibliography

1. R. Courant and D. Hilbert, *Methods of Mathematical Physics*, Vol. 1, Interscience, New York, 1953.
2. C. Lanczos, *The Variational Principles of Mechanics*, University of Toronto Press, 1949.

APPENDIX B

STRESS-STRAIN RELATIONS FOR A THIN PLATE

It is a common practice to assume that the transverse normal stress σ_z may be neglected in the stress–strain relations when formulating a thin plate problem approximately. This assumption reduces Eqs. (1.10) and (1.11) to

$$
\begin{aligned}
\sigma_x &= \frac{E}{(1-\nu^2)}(\varepsilon_x + \nu\varepsilon_y), & \tau_{yz} &= G\gamma_{yz}, \\
\sigma_y &= \frac{E}{(1-\nu^2)}(\nu\varepsilon_x + \varepsilon_y), & \tau_{zx} &= G\gamma_{zx}, \\
\sigma_z &= 0, & \tau_{xy} &= G\gamma_{xy},
\end{aligned} \tag{1}
$$

and

$$
\begin{aligned}
E\varepsilon_x &= \sigma_x - \nu\sigma_y, & G\gamma_{yz} &= \tau_{yz}, \\
E\varepsilon_y &= -\nu\sigma_x + \sigma_y, & G\gamma_{zx} &= \tau_{zx}, \\
E\varepsilon_z &= -\nu(\sigma_x + \sigma_y), & G\gamma_{xy} &= \tau_{xy},
\end{aligned} \tag{2}
$$

respectively. The expressions for the strain and complementary energy functions for the above stress–strain relations can be shown to be

$$A = \frac{E}{2(1-\nu^2)}(\varepsilon_x + \varepsilon_y)^2 + \frac{G}{2}(\gamma_{yz}^2 + \gamma_{zx}^2 + \gamma_{xy}^2 - 4\varepsilon_x\varepsilon_y), \tag{3}$$

$$B = \frac{1}{2E}[(\sigma_x + \sigma_y)^2 + 2(1+\nu)(\tau_{yz}^2 + \tau_{zx}^2 + \tau_{xy}^2 - \sigma_x\sigma_y)]. \tag{4}$$

When Eqs. (3.38) are employed as the stress–strain relations to take account of nonlinear strain–displacement relations, we obtain relations corresponding to Eqs. (1), (2) and (3) above by replacing ε_x, ε_y, ε_z, γ_{yz}, γ_{zx} and γ_{xy} with e_{xx}, e_{yy}, e_{zz}, $2e_{yz}$, $2e_{zx}$ and $2e_{xy}$, respectively.

The assumption that the transverse normal stress may be neglected in the stress–strain relations is frequently employed in thermal stress problems of a thin plate and reduces Eqs. (5.50) and (5.51) to

$$
\begin{aligned}
\sigma_x &= \frac{E}{(1-\nu^2)}(e_{xx} + \nu e_{yy}) - \frac{Ee^\theta}{(1-\nu)}, & \tau_{yz} &= 2Ge_{yz}, \\
\sigma_y &= \frac{E}{(1-\nu^2)}(\nu e_{xx} + e_{yy}) - \frac{Ee^\theta}{(1-\nu)}, & \tau_{zx} &= 2Ge_{zx}, \\
\sigma_z &= 0, & \tau_{xy} &= 2Ge_{xy},
\end{aligned} \tag{5}
$$

and

$$
\begin{aligned}
e_{xx} &= \frac{1}{E}(\sigma_x - \nu\sigma_y) + e^\theta, & e_{yz} &= \frac{1}{2G}\tau_{yz}, \\
e_{yy} &= \frac{1}{E}(-\nu\sigma_x + \sigma_y) + e^\theta, & e_{zx} &= \frac{1}{2G}\tau_{zx}, \\
e_{zz} &= -\frac{\nu}{E}(\sigma_x + \sigma_y) + e^\theta, & e_{xy} &= \frac{1}{2G}\tau_{xy},
\end{aligned} \tag{6}
$$

respectively. The expressions for the strain and complementary energy functions for the above stress–strain relations can be shown to be

$$
\begin{aligned}
A = {} & \frac{E}{2(1-\nu^2)}(e_{xx} + e_{yy})^2 + 2G(e_{yz}^2 + e_{zx}^2 + e_{xy}^2 - e_{xx}e_{yy}) \\
& - \frac{Ee^\theta}{(1-\nu)}(e_{xx} + e_{yy}),
\end{aligned} \tag{7}
$$

$$
\begin{aligned}
B = {} & \frac{1}{2E}[(\sigma_x + \sigma_y)^2 + 2(1+\nu)(\tau_{yz}^2 + \tau_{zx}^2 + \tau_{xy}^2 - \sigma_x\sigma_y)] \\
& + e^\theta(\sigma_x + \sigma_y).
\end{aligned} \tag{8}
$$

When linear strain–displacement relations are employed, we obtain relations corresponding to Eqs. (5), (6) and (7) above by replacing e_{xx}, e_{yy}, e_{zz}, $2e_{yz}$, $2e_{zx}$ and $2e_{xy}$ with ε_x, ε_y, ε_z, γ_{yz}, γ_{zx} and γ_{xy}, respectively.

APPENDIX C

A BEAM THEORY INCLUDING THE EFFECT OF TRANSVERSE SHEAR DEFORMATION

WE SHALL derive an approximate beam theory, including the effect of transverse shear deformation, by employing the generalized principle of minimum complementary energy (2.41). Consider a beam of uniform cross section which is clamped at $x = 0$ and in static equilibrium under terminal loads at the other end $x = l$. It is assumed that body forces and surface forces on the side boundary are absent; thus a torsion free bending is realized in the (x, z) plane. The principle (2.41) is written for the present problem as follows:

$$-\Pi_R^* = \iiint\limits_V \left[B(\sigma_x, \sigma_y, \ldots, \tau_{xy}) + \left(\frac{\partial \sigma_x}{\partial x} + \frac{\partial \tau_{xy}}{\partial y} + \frac{\partial \tau_{xz}}{\partial z}\right) u \right.$$
$$\left. + \left(\frac{\partial \tau_{xy}}{\partial x} + \frac{\partial \sigma_y}{\partial y} + \frac{\partial \tau_{yz}}{\partial z}\right) v + \left(\frac{\partial \tau_{xz}}{\partial x} + \frac{\partial \tau_{yz}}{\partial y} + \frac{\partial \sigma_z}{\partial z}\right) w \right] dx\, dy\, dz$$
$$+ \text{(terms on the boundary surfaces)}, \tag{1}$$

where u, v and w are displacement components and functions of (x, y, z). We assume:

$$\sigma_x = \frac{N(x)}{A_0} + \frac{M(x)}{I} z, \tag{2}$$

$$\tau_{xy} = Q(x)\, \Theta_{xy}(y, z), \quad \tau_{xz} = Q(x)\, \Theta_{xz}(y, z), \tag{3}$$

$$\sigma_y = \sigma_z = \tau_{yz} = 0. \tag{4}$$

It is seen that Eq. (2) is the same as Eq. (7.29). The two functions Θ_{xy} and Θ_{xz} in Eq. (3) are chosen to satisfy

$$\frac{\partial \Theta_{xy}}{\partial y} + \frac{\partial \Theta_{xz}}{\partial z} = -\frac{z}{I} \tag{5}$$

on the cross section, and

$$\Theta_{xy} m + \Theta_{xz} n = 0 \tag{6}$$

on the side boundary, where m and n are the direction cosines of the normal ν drawn outwards on the side boundary, namely $m = \cos(y, \nu)$ and $n = \cos(z, \nu)$. It is required that the functions Θ_{xy} and Θ_{xz} so chosen are good approximations of the stress components τ_{xy} and τ_{xz} induced in the beam. Substituting Eqs. (2), (3) and (4) into Eq. (1) and employing Eq. (2.21) for the expression of B, we have

$$-\Pi_R^* = \int_0^l \left[\frac{N^2}{2EA_0} + \frac{M^2}{2EI} + \frac{Q^2}{2GkA_0} + N'U_0 + (M' - Q)\, U_1 + Q'(V_0 + W_0) \right] dx + (\text{terms on the both ends}), \tag{7}$$

where it is defined that

$$\frac{1}{kA_0} = \iint (\Theta_{xy}^2 + \Theta_{xz}^2)\, dy\, dz, \tag{8}$$

and

$$U_0 A_0 = \iint u\, dy\, dz, \qquad U_1 I = \iint uz\, dy\, dz,$$
$$V_0 = \iint v\Theta_{xy}\, dy\, dz, \quad W_0 = \iint w\Theta_{xz}\, dy\, dz, \tag{9}$$

integrations being extended over the cross section of the beam. In the functional (7), the quantities subject to variation are N, M, Q, U_0, U_1, V_0 and W_0. We have the following stationary conditions:

$$N = EA_0 U_0', \quad M = EIU_1', \quad Q = GkA_0[(V_0 + W_0)' + U_1], \tag{10}$$
$$N' = 0 \quad M' - Q = 0, \quad Q' = 0. \tag{11}$$

Comparing Eqs. (10) with Eqs. (7.112), (7.113) and (7.114), we observe that, if the quantities u, u_1 and w are interpreted as

$$u = U_0, \quad u_1 = U_1, \quad w = V_0 + W_0, \tag{12}$$

the two approaches provide equivalent formulations except for the value of k as far as the present static problem is concerned. The values of the transverse shear rigidity are shown below for three cross sections.

(1) *Rectangular cross section.* Let the breadth and height of the section be b and h, respectively, as shown in Fig. C 1. We then have[1]

$$\Theta_{xy} = 0, \quad \Theta_{xz} = (1/2I)\,[(h/2)^2 - z^2],$$
$$A_0 = bh, \quad I = \tfrac{1}{12}bh^3, \quad k = \tfrac{5}{6}. \tag{13}$$

(2) *Circular cross section.* Let the radius of the section be a as shown in Fig. C 2. We then have[1]

$$\Theta_{xy} = -\frac{(1 + 2\nu)}{4(1 + \nu)} \frac{1}{I} yz, \quad \Theta_{xz} = \frac{(3 + 2\nu)}{8(1 + \nu)} \frac{1}{I} \left(a^2 - z^2 - \frac{1 - 2\nu}{3 + 2\nu} y^2 \right),$$

$$A_0 = \pi a^2, \quad I = \frac{\pi}{4} a^4, \quad k = 0.851 \quad \text{for} \quad \nu = 0.3. \tag{14}$$

(3) *A single-celled, closed, thin-walled circular tube.* In a thin-walled tube, the shearing stress τ on the cross section is assumed to be in the direction of the

periphery of the wall.[2, 3] For a tube with the radius a and constant thickness t shown in Fig. C 3, we have

$$\tau = (Q/\pi at)\cos\theta,$$
$$\Theta_{xy} = -(1/\pi at)\cos\theta\sin\theta, \quad \Theta_{xz} = (1/\pi at)\cos^2\theta,$$
$$A_0 = 2\pi at, \quad k = \tfrac{1}{2}. \tag{15}$$

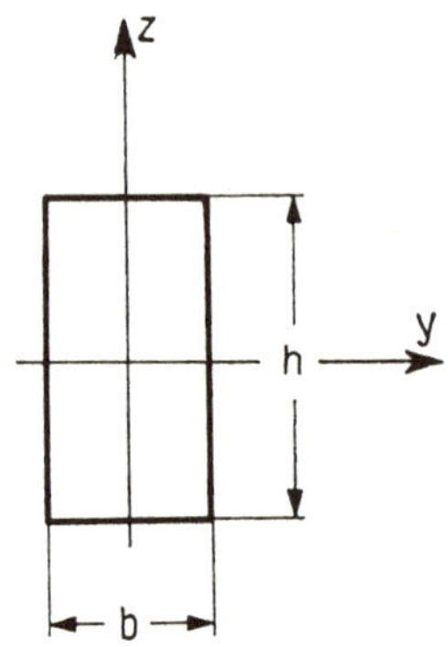

FIG. C1. A rectangular section.

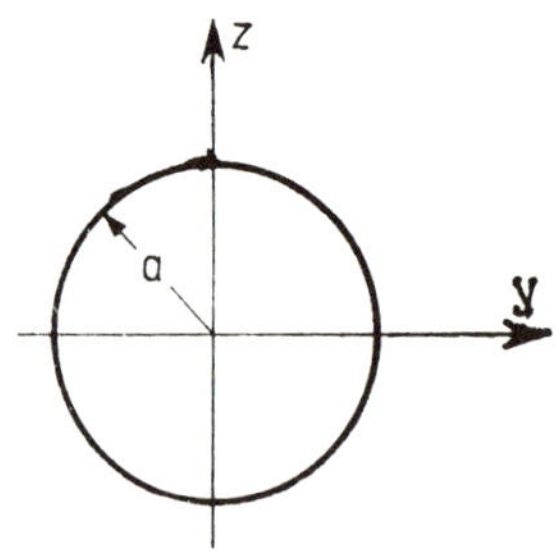

FIG. C2. A circular section.

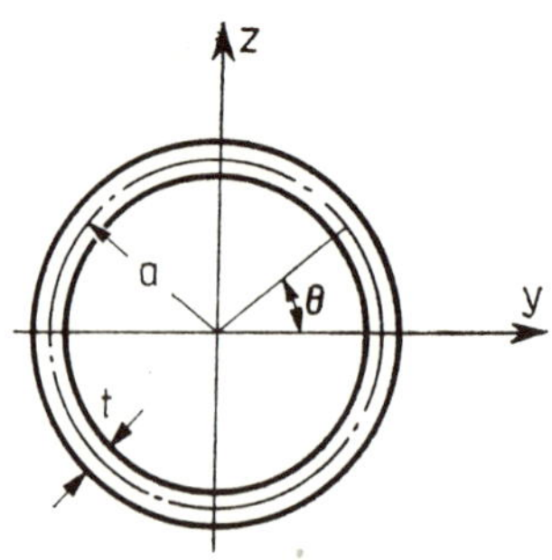

FIG. C3. A thin-walled, circular tube.

Bibliography

1. S. Timoshenko and J. N. Goodier, *Theory of Elasticity*, McGraw-Hill, 1951.
2. Y. C. Fung, *An Introduction to the Theory of Aeroelasticity*, John Wiley, 1955.
3. D. J. Peery, *Aircraft Structures*, McGraw-Hill, 1949.

APPENDIX D

A THEORY OF PLATE BENDING INCLUDING THE EFFECT OF TRANSVERSE SHEAR DEFORMATION

WE SHALL derive an approximate theory of plate bending including the effect of transverse shear deformation, employing the generalized principle of minimum complementary energy (2.41). We prescribe that the thin plate is in static equilibrium under mechanical boundary conditions

$$\bar{F}_x, \quad \bar{F}_y, \quad \bar{F}_z \tag{1}$$

on S_1 and geometrical boundary conditions

$$\bar{U}, \quad \bar{V}, \quad \bar{W} \tag{2}$$

on S_2, while the external forces applied to the upper and lower surfaces are given by:

$$\begin{aligned} \tau_{xz} &= 0, \quad \tau_{yz} = 0, \quad \sigma_z = \bar{p}, \quad \text{on} \quad z = h/2, \\ \tau_{xz} &= 0, \quad \tau_{yz} = 0, \quad \sigma_z = 0, \quad \text{on} \quad z = -h/2. \end{aligned} \tag{3}$$

The body forces are assumed absent. The principle (2.41) can be written for the present problem as follows:

$$\begin{aligned} -\Pi_R^* = \iiint\limits_V \Big[& B(\sigma_x, \ldots, \tau_{xy}) + \left(\frac{\partial \sigma_x}{\partial x} + \frac{\partial \tau_{xy}}{\partial y} + \frac{\partial \tau_{xz}}{\partial z} \right) U \\ & + \left(\frac{\partial \tau_{xy}}{\partial x} + \frac{\partial \sigma_y}{\partial y} + \frac{\partial \tau_{yz}}{\partial z} \right) V + \left(\frac{\partial \tau_{xz}}{\partial x} + \frac{\partial \tau_{yz}}{\partial y} + \frac{\partial \sigma_z}{\partial z} \right) W \Big] \, dx \, dy \, dz \\ & - \iint\limits_{S_2} [(\sigma_x l + \tau_{xy} m) \bar{U} + (\tau_{xy} l + \sigma_y m) \bar{V} + (\tau_{xz} l + \tau_{yz} m) \bar{W}] \, dz \, ds \\ & + (\text{integrals on } S_1 \text{ and } z = \pm h/2), \end{aligned} \tag{4}$$

where $l = \cos(x, \nu)$ and $m = \cos(y, \nu)$. Following Refs. 1, 2, 3 and 4, we may choose,

$$\begin{aligned} \sigma_x &= \frac{M_x}{(h^2/6)} \frac{z}{(h/2)}, \quad \sigma_y = \frac{M_y}{(h^2/6)} \frac{z}{(h/2)}, \quad \tau_{xy} = \frac{M_{xy}}{(h^2/6)} \frac{z}{(h/2)}, \\ \tau_{xz} &= \frac{Q_x}{(2h/3)} \left[1 - \left(\frac{z}{h/2} \right)^2 \right], \quad \tau_{yz} = \frac{Q_y}{(2h/3)} \left[1 - \left(\frac{z}{h/2} \right)^2 \right], \\ \sigma_z &= \frac{3}{4} \bar{p} \left[\frac{z}{(h/2)} - \frac{1}{3} \left(\frac{z}{h/2} \right)^3 + \frac{2}{3} \right]. \end{aligned} \tag{5}$$

Substituting these equations into Eq. (4) and employing Eq. (2.21) for the expression of B, we obtain

$$-\Pi_R^* = \iint_{S_m} \Big\{ \frac{1}{2}\frac{12}{Eh^3}\Big[(M_x + M_y)^2 + 2(1+\nu)(M_{xy}^2 - M_xM_y) - \frac{\nu}{5}h^2\bar{p}(M_x+M_y)\Big] + \frac{1}{2Gkh}(Q_x^2+Q_y^2) + \Big(\frac{\partial M_x}{\partial x} + \frac{\partial M_{xy}}{\partial y} - Q_x\Big)u_1 + \Big(\frac{\partial M_{xy}}{\partial x} + \frac{\partial M_y}{\partial y} - Q_y\Big)v_1 + \Big(\frac{\partial Q_x}{\partial x} + \frac{\partial Q_y}{\partial y} + \bar{p}\Big)w_0 \Big\} dx\,dy - \int_{C_2} [(M_xl + M_{xy}m)\bar{u}_1 + (M_{xy}l + M_ym)\bar{v}_1 + (Q_xl + Q_ym)\bar{w}_0]\,ds + (\text{integral on } C_1), \tag{6}$$

where $k = 5/6$ and it is defined that

$$u_1 = \Big(\frac{12}{h^3}\Big)\int uz\,dz, \quad v_1 = \Big(\frac{12}{h^3}\Big)\int vz\,dz, \quad w_0 = \Big(\frac{3}{2h}\Big)\int w\Big[1 - \Big(\frac{z}{h/2}\Big)^2\Big]dz,$$

$$\bar{u}_1 = \Big(\frac{12}{h^3}\Big)\int \bar{u}z\,dz, \quad \bar{v}_1 = \Big(\frac{12}{h^3}\Big)\int \bar{v}z\,dz, \quad \bar{w}_0 = \Big(\frac{3}{2h}\Big)\int \bar{w}\Big[1 - \Big(\frac{z}{h/2}\Big)^2\Big]dz.$$

(7a, b, c, d, e, f)

The quantities subject to variation in the functional (6) are M_x, M_y, M_{xy}, Q_x, Q_y, u_1, v_1 and w_0. The stationary conditions can be shown to be the equations of equilibrium,

$$\frac{\partial M_x}{\partial x} + \frac{\partial M_{xy}}{\partial y} - Q_x = 0, \quad \frac{\partial M_{xy}}{\partial x} + \frac{\partial M_y}{\partial y} - Q_y = 0,$$

$$\frac{\partial Q_x}{\partial x} + \frac{\partial Q_y}{\partial y} + \bar{p} = 0, \tag{8a, b, c}$$

and the stress resultant–displacement relations,

$$M_x = D\Big(\frac{\partial u_1}{\partial x} + \nu\frac{\partial v_1}{\partial y}\Big) + \frac{\nu}{10(1-\nu)}h^2\bar{p}, \tag{9a}$$

$$M_y = D\Big(\nu\frac{\partial u_1}{\partial x} + \frac{\partial v_1}{\partial y}\Big) + \frac{\nu}{10(1-\nu)}h^2\bar{p}, \tag{9b}$$

$$M_{xy} = \frac{Gh^3}{12}\Big(\frac{\partial u_1}{\partial y} + \frac{\partial v_1}{\partial x}\Big), \tag{9c}$$

$$Q_x = Gkh\Big(\frac{\partial w_0}{\partial x} + u_1\Big), \quad Q_y = Gkh\Big(\frac{\partial w_0}{\partial y} + v_1\Big), \tag{9d, e}$$

together with the geometrical boundary conditions,

$$u_1 = \bar{u}_1, \quad v_1 = \bar{v}_1, \quad w_0 = \bar{w}_0, \quad \text{on} \quad C_2. \tag{10}$$

The above results suggest that the mechanical boundary conditions can be specified approximately as follows:

$$M_x l + M_{xy} m = \int \bar{F}_x z \, dz, \quad M_{xy} l + M_y m = \int \bar{F}_y z \, dz,$$
$$Q_x l + Q_y m = \int \bar{F}_z \, dz \quad \text{on} \quad C_1. \tag{11}$$

Unless the surface load $\bar{p}$ is highly concentrated, the last terms in Eqs. (9a) and (9b) may be neglected in comparison to the preceding terms. Then, comparing the above equations with those derived in Section 8.8, we observe that, if the quantities u_1, v_1 and w are interpreted as

$$u_1 = U_1, \quad v_1 = V_1, \quad w = W_0, \tag{12}$$

the two approaches provide equivalent formulations except for the value of k as far as the present static problem is concerned.

Bibliography

1. E. Reissner, On the Theory of Bending of Elastic Plates, *Journal of Mathematics and Physics*, Vol. 23, No. 4, pp. 184–91, November 1944.
2. E. Reissner, The Effect of Transverse-Shear Deformation on the Bending of Elastic Plates, *Journal of Applied Mechanics*, Vol. 12, No. 2, pp. 69–77, June 1945.
3. E. Reissner, On Bending of Elastic Plates, *Quarterly of Applied Mathematics*, Vol. 5, No. 1, pp. 55–68, April 1947.
4. E. Reissner, On a Variational Theorem in Elasticity, *Journal of Mathematics and Physics*, Vol. XXIX, No. 2, pp. 90–5, July 1950.

APPENDIX E

SPECIALIZATIONS TO SEVERAL KINDS OF SHELLS

EXPLICIT expressions of the geometrical quantities defined in Chapter 9 are shown for several kinds of shells in the following:

1. *Flat plate* (see Fig. E1)

$$(ds_0^{(0)})^2 = (dx)^2 + (dy)^2 .$$

$$\alpha = x, \quad \beta = y, \quad A = 1, \quad B = 1, \quad R_\alpha = \infty, \quad R_\beta = \infty .$$

$$l_{11} = \frac{\partial u}{\partial x}, \quad l_{12} = \frac{\partial u}{\partial y},$$

$$l_{21} = \frac{\partial v}{\partial x}, \quad l_{22} = \frac{\partial v}{\partial y},$$

$$l_{31} = \frac{\partial w}{\partial x}, \quad l_{32} = \frac{\partial w}{\partial y} .$$

$$\varepsilon_x = \frac{\partial u}{\partial x}, \quad \varepsilon_y = \frac{\partial v}{\partial y}, \quad \gamma_{xy} = \frac{\partial u}{\partial y} + \frac{\partial v}{\partial x},$$

$$\varkappa_x = \frac{\partial^2 w}{\partial x^2}, \quad \varkappa_y = \frac{\partial^2 w}{\partial y^2}, \quad \varkappa_{xy} = \frac{\partial^2 w}{\partial x \, \partial y} .$$

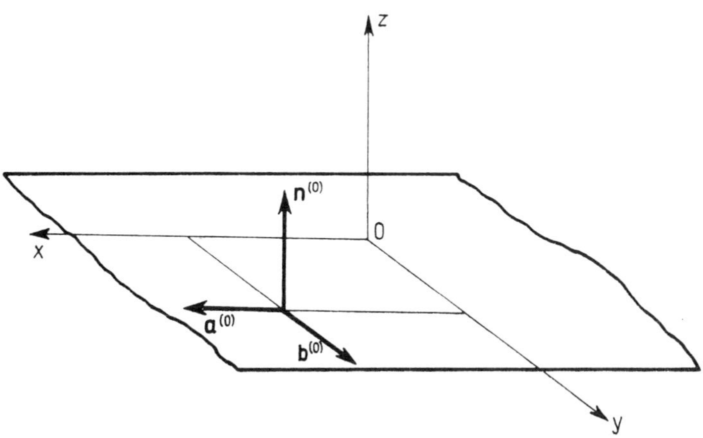

FIG. E1. Flat plate.

2. *Cylindrical shell* (see Fig. E2)

$$x = x, \quad y = a\cos\varphi, \quad z = a\sin\varphi.$$

$$(ds_0^{(0)})^2 = (dx)^2 + (a\,d\varphi)^2.$$

$$\alpha = x, \quad \beta = \varphi, \quad A = 1, \quad B = a, \quad R_\alpha = \infty, \quad R_\beta = a.$$

$$l_{11} = \frac{\partial u}{\partial x}, \quad l_{12} = \frac{1}{a}\frac{\partial u}{\partial \varphi},$$

$$l_{21} = \frac{\partial v}{\partial x}, \quad l_{22} = \frac{1}{a}\left(\frac{\partial v}{\partial \varphi} - w\right),$$

$$l_{31} = \frac{\partial w}{\partial x}, \quad l_{32} = \frac{1}{a}\left(\frac{\partial w}{\partial \varphi} + v\right).$$

$$\varepsilon_x = \frac{\partial u}{\partial x}, \quad \varepsilon_\varphi = \frac{1}{a}\left(\frac{\partial v}{\partial \varphi} - w\right), \quad \gamma_{x\varphi} = \frac{1}{a}\frac{\partial u}{\partial \varphi} + \frac{\partial v}{\partial x}.$$

$$\varkappa_x = \frac{\partial^2 w}{\partial x^2}, \quad \varkappa_\varphi = \frac{1}{a^2}\left(\frac{\partial^2 w}{\partial \varphi^2} + \frac{\partial v}{\partial \varphi}\right), \quad \varkappa_{x\varphi} = \frac{1}{a}\left(\frac{\partial^2 w}{\partial x\,\partial \varphi} + \frac{\partial v}{\partial x}\right).$$

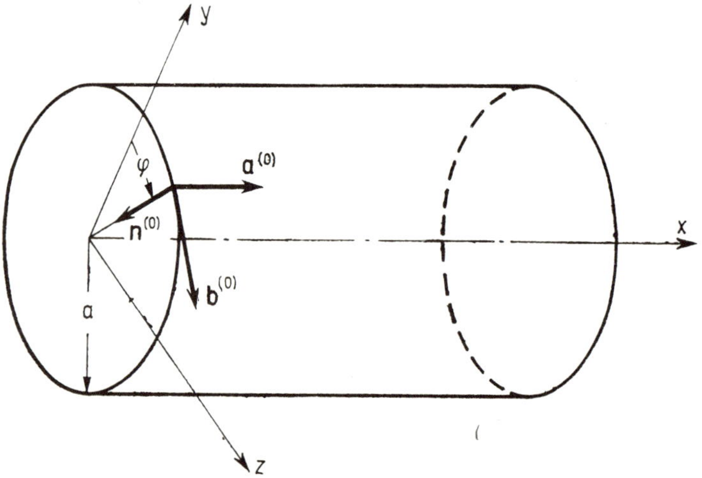

FIG. E2. Cylindrical shell.

3. *Spherical shell* (see Fig. E3)

$$x = a\sin\varphi\cos\theta, \quad y = a\sin\varphi\sin\theta, \quad z = a\cos\varphi.$$

$$(ds_0^{(0)})^2 = (a\sin\varphi\,d\theta)^2 + (a\,d\varphi)^2.$$

$$\alpha = \theta, \quad \beta = \varphi, \quad A = a\sin\varphi, \quad B = a, \quad R_\alpha = a, \quad R_\beta = a.$$

$$l_{11} = \frac{1}{a}\left(\frac{1}{\sin\varphi}\frac{\partial u}{\partial\theta} + v\cot\varphi - w\right), \quad l_{12} = \frac{1}{a}\frac{\partial u}{\partial\varphi},$$

$$l_{21} = \frac{1}{a}\left(\frac{1}{\sin\varphi}\frac{\partial v}{\partial\theta} - u\cot\varphi\right), \quad l_{32} = \frac{1}{a}\left(\frac{\partial v}{\partial\varphi} - w\right),$$

$$l_{31} = \frac{1}{a}\left(\frac{1}{\sin\varphi}\frac{\partial w}{\partial\theta} + u\right), \quad l_{32} = \frac{1}{a}\left(\frac{\partial w}{\partial\varphi} + v\right).$$

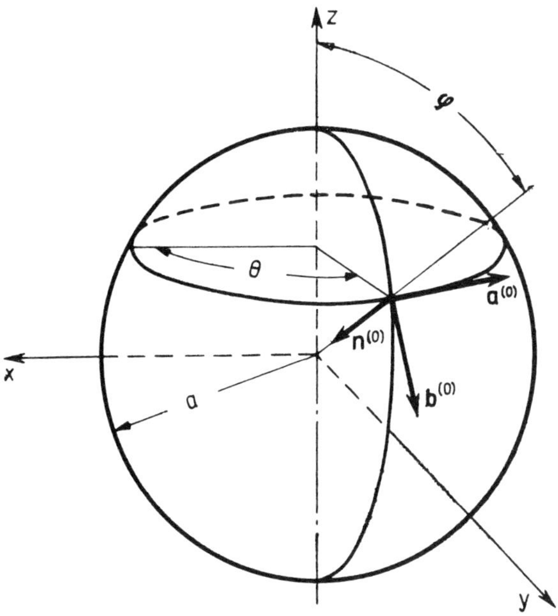

FIG. E3. Spherical shell.

4. *Rotationally symmetric shell* (see Fig. E4)

$$(ds_0^{(0)})^2 = (R_\theta \sin\varphi \, d\theta)^2 + (R_\varphi \, d\varphi)^2 .$$

$$\alpha = \theta, \quad \beta = \varphi, \quad A = R_\theta \sin\varphi, \quad B = R_\varphi .$$

$$l_{11} = \frac{1}{R_\theta}\left(\frac{1}{\sin\varphi}\frac{\partial u}{\partial\theta} + v\cot\varphi - w\right), \quad l_{12} = \frac{1}{R_\varphi}\frac{\partial u}{\partial\varphi}.$$

$$l_{21} = \frac{1}{R_\theta}\left(\frac{1}{\sin\varphi}\frac{\partial v}{\partial\theta} - u\cot\varphi\right), \quad l_{22} = \frac{1}{R_\varphi}\left(\frac{\partial v}{\partial\varphi} - w\right),$$

$$l_{31} = \frac{1}{R_\theta}\left(\frac{1}{\sin\varphi}\frac{\partial w}{\partial\theta} + u\right), \quad l_{32} = \frac{1}{R_\varphi}\left(\frac{\partial w}{\partial\varphi} + v\right).$$

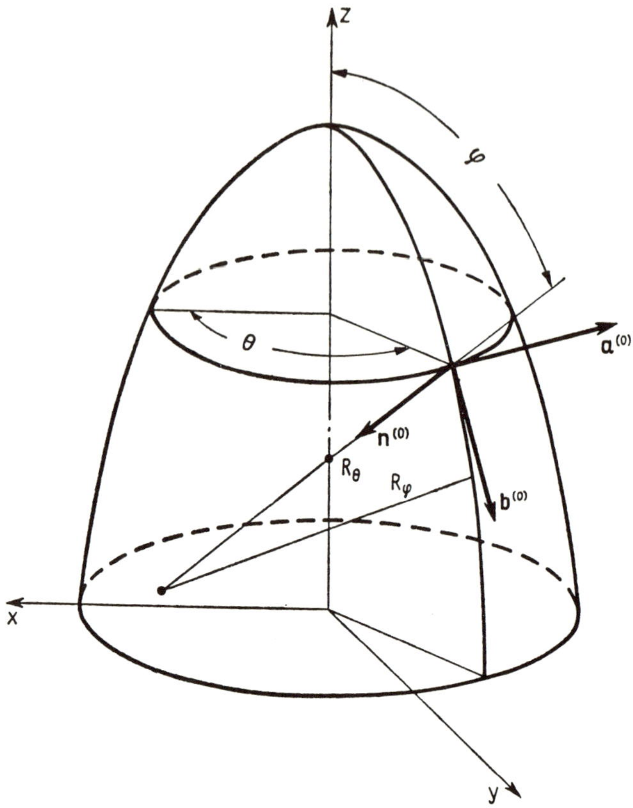

FIG. E4. Rotationally symmetric shell.

APPENDIX F

A NOTE ON THE HAAR-KÁRMÁN PRINCIPLE

EQUATION (11.42) shows that the Haar–Kármán principle does not possess a stationary property in the conventional sense. However, if the subsidiary condition in the form of inequality:

$$\sigma'_{ij}\sigma'_{ij} - 2k^2 \leqq 0 \tag{1}$$

is written as

$$2k^2 - \sigma'_{ij}\sigma'_{ij} - z^2 = 0, \tag{2}$$

where z is a real variable, the functional (11.35) can be generalized as follows:[1, 2]

$$\begin{aligned} \Pi_R^* = &- \iiint_V \left[\frac{3(1-2\nu)}{2E}\sigma^2 + \frac{1}{4G}\sigma'_{ij}\sigma'_{ij}\right] dV + \iint_{S_2} \sigma_{ij} n_j \bar{u}_i \, dS \\ &+ \iiint_V [-\sigma_{ij,j} u_i + (\lambda/2)(2k^2 - \sigma'_{ij}\sigma'_{ij} - z^2)] \, dz \\ &+ \iint_{S_1} (\sigma_{ij} n_j - \bar{F}_i) \, u_i \, dS, \end{aligned} \tag{3}$$

where u_i and λ are Lagrange multipliers which introduce the equations of equilibrium, mechanical boundary conditions and the yield condition into the variational expression. It is interesting to observe that the stationary conditions of the functional (3) with respect to σ_{ij} and z provide:

$$\frac{1}{2}(u_{i,j} + u_{j,i}) = \frac{(1-2\nu)}{E}\sigma\,\delta_{ij} + \frac{1}{2G}\sigma'_{ij} + \lambda\sigma'_{ij}, \tag{4}$$

$$z\lambda = 0. \tag{5}$$

The solutions of Eq. (5) are

$$z = 0, \quad \lambda = 0. \tag{6a, 6b}$$

The first solution corresponds to the plastic state, while the second one corresponds to the elastic state.

Bibliography

1. F. A. VALENTINE, The Problem of Lagrange with Differential Inequalities as Added Side Conditions, in, *Contributions to the Calculus of Variations, 1933–1937*, University of Chicago Press, 1937.
2. A. MIELE, The Calculus of Variations in Applied Aerodynamics and Flight Mechanics, in, *Optimization Techniques with Applications to Aerospace Systems*, edited by G. Leitmann, Academic Press, 1962.

APPENDIX G

VARIATIONAL PRINCIPLES IN THE THEORY OF CREEP

DEFORMATIONS of materials consist not only of elastic and plastic strains, but also of a time-dependent portion, especially at elevated temperature. This portion of the deformation proceeds with the lapse of time, even under constant external loads, and is known as the phenomenon of creep.† Creep deformations in structures cause changes in shape, changes in stress distribution and such instabilities as creep buckling. Consequently, creep is considered to be one of the decisive factors in the analysis of structures exposed to high temperatures.

Several proposals have been made on the establishment of variational principles in the theory of creep. Wang and Prager have formulated variational principles for a boundary value problem defined (using the notation of Chapter 12) as follows.(6) A body of work-hardening plastic material is assumed to have been deformed, including creep behavior, and at the time t occupies a region V bounded by a surface S. It is also assumed that the temperature θ, the stress σ_{ij} and the state of strain-hardening f are known throughout V. We now prescribe an infinitesimal temperature change $d\theta$ throughout V, infinitesimal changes $d\bar{F}_i$ of the surface tractions on S_1 and infinitesimal changes $d\bar{u}_i$ of the surface displacements on S_2. Given relations between incremental components of elastic, plastic, thermal and creep strain, denoted by $d\varepsilon_{ij}^e$, $d\varepsilon_{ij}^p$, $d\varepsilon_{ij}^\theta$ and $d\varepsilon_{ij}^c$, respectively, and incremental components of stress, temperature and time, the problem is then to find the stress increments $d\sigma_{ij}$ and displacement increments du_i induced in the body. It is understood that the sum of the thermal and creep strains, $d\varepsilon_{ij}^\theta + d\varepsilon_{ij}^c$, can be taken as initial strain increments, and the problem is thus reduced to a boundary value problem of a body with initial strain increments in the flow theory of plasticity.(7)

Sanders, McComb and Schlechte have formulated another variational principle for a boundary value problem which may be defined (using the notation in Section 5.5) as follows.(8) Consider that the stresses $\sigma^{\lambda\mu}$ and the displacements u^λ are known at the time t. Given the surface force rates $\dot{\bar{F}}^\lambda$, the surface displacement rates $\dot{\bar{u}}^\lambda$, the body forces rates $\dot{\bar{P}}^\lambda$, together with

† Refs. 1 through 5.

relations between stress rates and strain rates, find the stress rates $\dot{\sigma}^{\lambda\mu}$ and the displacement rates u^{λ} induced in the body. It is understood that the creep strain rates $\dot{e}^{c}_{\lambda\mu}$ can be taken as initial strain rates,[(9)] and the principles derived in Section 5.5 may be employed for the establishment of variational principles.

Bibliography

1. F. K. G. ODGVIST, Recent Advances in Theories of Creep of Engineering Materials, *Applied Mechanics Reviews*, pp. 517–19, December 1954.
2. N. J. HOFF, Approximate Analysis of Structures in the Presence of Moderately Large Creep Deformations, *Quarterly of Applied Mathematics* Vol. 12, No. 1, pp. 49–55, April 1954.
3. T. H. H. PIAN, *Stress Distribution and Deformation Due to Creep*, Aerodynamic Heating of Aircraft Structures in High Speed Flight, Notes for a Special Summer Program, Department of Aeronautical Engineering, Massachusetts Institute of Technology, pp. 15–1 to 15–34, June 25–July 6, 1956.
4. W. PRAGER, Total Creep and Varying Loads, *Journal of the Aeronautical Sciences*, Vol. 24, No. 2, pp. 153–5, February 1957.
5. N. J. HOFF, editor, *High Temperature Effects in Aircraft Structures*, AGARDograph 28, Pergamon Press, 1958.
6. A. J. WANG and W. PRAGER, Thermal and Creep Effects in Work-Hardening Elastic-Plastic Solids, *Journal of the Aeronautical Sciences*, Vol. 21, No. 5, pp. 343–4, May 1954.
7. K. WASHIZU, *Variational Principles in Elasticity and Plasticity*, Aeroelastic and Structures Research Laboratory, Massachusetts Institute of Technology, Technical Report 25-18, March 1955.
8. J. L. SANDERS, JR., H. G. MCCOMB, JR., and F. R. SCHLECHTE, *A Variational Theorem for Creep with Application to Plates and Columns*, NACA TN 4003, 1957.
9. T. H. H. PIAN, On the Variational Theorem for Creep, *Journal of the Aeronautical Sciences*, Vol. 24, No. 11, pp. 846–7, November 1957.

APPENDIX H

PROBLEMS

CHAPTER 1

Problems Related to Sections 1.1 and 1.2

1. Show that by use of Eqs. (1.5) and (1.10), we may express Eqs. (1.4) and (1.12) in terms of displacements as follows:

$$\begin{aligned} \Delta u + \frac{1}{1-2\nu}\frac{\partial e}{\partial x} + \frac{\bar{X}}{G} &= 0, \\ \Delta v + \frac{1}{1-2\nu}\frac{\partial e}{\partial y} + \frac{\bar{Y}}{G} &= 0, \\ \Delta w + \frac{1}{1-2\nu}\frac{\partial e}{\partial z} + \frac{\bar{Z}}{G} &= 0, \end{aligned} \tag{i}$$

and

$$\begin{aligned} G\left[\left(2\frac{\partial u}{\partial x} + \frac{2\nu}{1-2\nu}e\right)l + \left(\frac{\partial u}{\partial y} + \frac{\partial v}{\partial x}\right)m + \left(\frac{\partial u}{\partial z} + \frac{\partial w}{\partial x}\right)n\right] &= \bar{X}_\nu, \\ G\left[\left(\frac{\partial v}{\partial x} + \frac{\partial u}{\partial y}\right)l + \left(2\frac{\partial v}{\partial y} + \frac{2\nu}{1-2\nu}e\right)m + \left(\frac{\partial v}{\partial z} + \frac{\partial w}{\partial y}\right)n\right] &= \bar{Y}_\nu, \\ G\left[\left(\frac{\partial w}{\partial x} + \frac{\partial u}{\partial z}\right)l + \left(\frac{\partial w}{\partial y} + \frac{\partial v}{\partial z}\right)m + \left(2\frac{\partial w}{\partial z} + \frac{2\nu}{1-2\nu}e\right)n\right] &= \bar{Z}_\nu, \end{aligned} \tag{ii}$$

respectively, where $\Delta(\ ) = (\ )_{,xx} + (\ )_{,yy} + (\ )_{,zz}$, $e = u_{,x} + v_{,y} + w_{,z}$, $(\ )_{,x} = \partial(\ )/\partial x$, $(\ )_{,y} = \partial(\ )/\partial y$, and $(\ )_{,z} = \partial(\ )/\partial z$. Show also that the elasticity problem is reduced to solving equations (i) under the boundary conditions; equation (ii) and Eqs. (1.14).

2. Show that if the body forces are absent, the conditions of compatibility, Eqs. (1.15), can be transformed, by the use of Eqs. (1.11) and (1.20), into

$$\begin{aligned} \Delta\sigma_x + \frac{1}{1+\nu}\frac{\partial^2\Theta}{\partial x^2} &= 0, & \Delta\tau_{yz} + \frac{1}{1+\nu}\frac{\partial^2\Theta}{\partial y\,\partial z} &= 0, \\ \Delta\sigma_y + \frac{1}{1+\nu}\frac{\partial^2\Theta}{\partial y^2} &= 0, & \Delta\tau_{zx} + \frac{1}{1+\nu}\frac{\partial^2\Theta}{\partial z\,\partial x} &= 0, \\ \Delta\sigma_z + \frac{1}{1+\nu}\frac{\partial^2\Theta}{\partial z^2} &= 0, & \Delta\tau_{xy} + \frac{1}{1+\nu}\frac{\partial^2\Theta}{\partial x\,\partial y} &= 0, \end{aligned} \tag{i}$$

where $\Delta(\ ) = (\ )_{,xx} + (\ )_{,yy} + (\ )_{,zz}$, and $\Theta = \sigma_x + \sigma_y + \sigma_z$. Show also that if it is further assumed that the boundary conditions are given entirely in terms of forces, the elasticity problem is reduced to solving equations (i) and Eqs. (1.20) under the boundary conditions $X_\nu = \bar{X}_\nu$, $Y_\nu = \bar{Y}_\nu$, and $Z_\nu = \bar{Z}_\nu$ on S.

3. We consider two sets of rectangular Cartesian coordinate systems (x, y, z) and $(\bar{x}, \bar{y}, \bar{z})$, and denote strain and stress components defined with respect to these systems by $\varepsilon_x, \varepsilon_y, \ldots, \gamma_{xy}$; $\sigma_x, \sigma_y, \ldots, \tau_{xy}$ and $\bar{\varepsilon}_x, \bar{\varepsilon}_y, \ldots, \bar{\gamma}_{xy}$; $\bar{\sigma}_x, \bar{\sigma}_y, \ldots, \bar{\tau}_{xy}$, respectively, where an overbar is used to distinguish between the two coordinate systems. For the sake of brevity, we shall also employ the following notations frequently:

$$\begin{aligned}
&x = x_1, \quad y = x_2, \quad z = x_3, \quad \bar{x} = \bar{x}_1, \ldots, \quad \bar{z} = \bar{x}_3;\\
&\varepsilon_x = \varepsilon_{11}, \quad \varepsilon_y = \varepsilon_{22}, \quad \varepsilon_z = \varepsilon_{33},\\
&\tfrac{1}{2}\gamma_{yz} = \varepsilon_{23} = \varepsilon_{32}, \quad \tfrac{1}{2}\gamma_{zx} = \varepsilon_{31} = \varepsilon_{13}, \quad \tfrac{1}{2}\gamma_{xy} = \varepsilon_{12} = \varepsilon_{21},\\
&\bar{\varepsilon}_x = \bar{\varepsilon}_{11}, \ldots, \quad \tfrac{1}{2}\bar{\gamma}_{xy} = \bar{\varepsilon}_{12} = \bar{\varepsilon}_{21};\\
&\sigma_x = \sigma_{11}, \quad \sigma_y = \sigma_{22}, \quad \sigma_z = \sigma_{33}, \quad \tau_{yz} = \sigma_{23}, \quad \tau_{zy} = \sigma_{32},\\
&\tau_{zx} = \sigma_{31}, \quad \tau_{xz} = \sigma_{13}, \quad \tau_{xy} = \sigma_{12}, \quad \tau_{yx} = \sigma_{21},\\
&\bar{\sigma}_x = \bar{\sigma}_{11}, \ldots, \quad \bar{\tau}_{yx} = \bar{\sigma}_{21}.
\end{aligned}$$

(1) Show that the following relations hold:

$$\begin{aligned}
\bar{\varepsilon}_{ij} &= \sum_{m=1}^{3}\sum_{n=1}^{3} \cos(\bar{x}_i, x_m)\cos(\bar{x}_j, x_n)\,\varepsilon_{mn},\\
\bar{\sigma}_{ij} &= \sum_{m=1}^{3}\sum_{n=1}^{3} \cos(\bar{x}_i, x_m)\cos(\bar{x}_j, x_n)\,\sigma_{mn}.
\end{aligned} \tag{i}$$

(2) Show that the following quantities are invariant with respect to transformations from one set of rectangular Cartesian coordinates to another:

$$\begin{aligned}
&\varepsilon_x + \varepsilon_y + \varepsilon_z,\\
&\varepsilon_y\varepsilon_z + \varepsilon_z\varepsilon_x + \varepsilon_x\varepsilon_y - \tfrac{1}{4}(\gamma_{yz}^2 + \gamma_{zx}^2 + \gamma_{xy}^2),\\
&\varepsilon_x\varepsilon_y\varepsilon_z + \tfrac{1}{4}(\gamma_{yz}\gamma_{zx}\gamma_{xy} - \varepsilon_x\gamma_{yz}^2 - \varepsilon_y\gamma_{zx}^2 - \varepsilon_z\gamma_{xy}^2),\\
&\sigma_x + \sigma_y + \sigma_z,\\
&\sigma_y\sigma_z + \sigma_z\sigma_x + \sigma_x\sigma_y - (\tau_{yz}^2 + \tau_{zx}^2 + \tau_{xy}^2),\\
&\sigma_x\sigma_y\sigma_z + 2\tau_{yz}\tau_{zx}\tau_{xy} - (\sigma_x\tau_{yz}^2 + \sigma_y\tau_{zx}^2 + \sigma_z\tau_{xy}^2).
\end{aligned} \tag{ii}$$

Show also that these quantities may be written as follows:

$$\begin{aligned}
&\sum_{i=1}^{3}\varepsilon_{ii}, \quad \frac{1}{2}\left[\left(\sum_{i=1}^{3}\varepsilon_{ii}\right)^2 - \sum_{i,j=1}^{3}\sum \varepsilon_{ij}\,\varepsilon_{ij}\right],\\
&\frac{1}{3!}\sum_{i,j,k,r,s,t=1}^{3}\sum\sum\sum\sum\sum e^{ijk}\,e^{rst}\,\varepsilon_{ir}\,\varepsilon_{js}\,\varepsilon_{kt},\\
&\sum_{i=1}^{3}\sigma_{ii}, \quad \frac{1}{2}\left[\left(\sum_{i=1}^{3}\sigma_{ii}\right)^2 - \sum_{i,j=1}^{3}\sum \sigma_{ij}\sigma_{ij}\right],\\
&\frac{1}{3!}\sum_{i,j,k,r,s,t=1}^{3}\sum\sum\sum\sum\sum e^{ijk}\,e^{rst}\sigma_{ir}\,\sigma_{js}\,\sigma_{kt},
\end{aligned} \tag{iii}$$

respectively, where,

$$e^{ijk} = 0 \quad \text{when any two of } i, j, k \text{ are equal},$$
$$= +1 \quad \text{when } i, j, k \text{ are an even permutaion of 1, 2, 3}, \qquad \text{(iv)}$$
$$= -1 \quad \text{when } i, j, k \text{ are an odd permutation of 1 2, 3}.$$

(3) Show that there exist only two independent elastic constants for an isotropic elastic body.

Problems Related to the Conditions of Compatibility

4. Show that

$$u(Q) = u(P) + \int_P^Q \left\{ \varepsilon_x\, dx + \left(\frac{1}{2}\gamma_{xy} - \varpi_z\right) dy + \left(\frac{1}{2}\gamma_{zx} + \varpi_y\right) dz \right\},$$

$$v(Q) = v(P) + \int_P^Q \left\{ \left(\frac{1}{2}\gamma_{xy} + \varpi_z\right) dx + \varepsilon_y\, dy + \left(\frac{1}{2}\gamma_{yz} - \varpi_x\right) dz \right\}, \qquad \text{(i)}$$

$$w(Q) = w(P) + \int_P^Q \left\{ \left(\frac{1}{2}\gamma_{zx} - \varpi_y\right) dx + \left(\frac{1}{2}\gamma_{yz} + \varpi_x\right) dy + \varepsilon_z\, dz \right\},$$

and

$$\varpi_x(Q) = \varpi_x(P) + \int_P^Q \left\{ \frac{1}{2}\left(\frac{\partial \gamma_{zx}}{\partial y} - \frac{\partial \gamma_{xy}}{\partial z}\right) dx + \left(\frac{1}{2}\frac{\partial \gamma_{yz}}{\partial y} - \frac{\partial \varepsilon_y}{\partial z}\right) dy + \left(\frac{\partial \varepsilon_z}{\partial y} - \frac{1}{2}\frac{\partial \gamma_{yz}}{\partial z}\right) dz \right\},$$

$$\bar{\varpi}_y(Q) = \varpi_y(P) + \int_P^Q \left\{ \left(\frac{\partial \varepsilon_x}{\partial z} - \frac{1}{2}\frac{\partial \gamma_{zx}}{\partial x}\right) dx + \frac{1}{2}\left(\frac{\partial \gamma_{xy}}{\partial z} - \frac{\partial \gamma_{yz}}{\partial x}\right) dy + \left(\frac{1}{2}\frac{\partial \gamma_{zx}}{\partial z} - \frac{\partial \varepsilon_z}{\partial x}\right) dz \right\}, \qquad \text{(ii)}$$

$$\varpi_z(Q) = \varpi_z(P) + \int_P^Q \left\{ \left(\frac{1}{2}\frac{\partial \gamma_{xy}}{\partial x} - \frac{\partial \varepsilon_x}{\partial y}\right) dx + \left(\frac{\partial \varepsilon_y}{\partial x} - \frac{1}{2}\frac{\partial \gamma_{xy}}{\partial y}\right) dy + \frac{1}{2}\left(\frac{\partial \gamma_{yz}}{\partial x} - \frac{\partial \gamma_{zx}}{\partial y}\right) dz \right\},$$

where ϖ_x, ϖ_y and ϖ_z are components of rotation defined by

$$2\varpi_x = \frac{\partial w}{\partial y} - \frac{\partial v}{\partial z}, \quad 2\varpi_y = \frac{\partial u}{\partial z} - \frac{\partial w}{\partial x}, \quad 2\varpi_z = \frac{\partial v}{\partial x} - \frac{\partial u}{\partial y}, \qquad \text{(iii)}$$

while P and Q are two arbitrary points in the body and integrations are taken along an arbitrary path between two points P and Q.

Next, by using these relations, show that the conditions of compatibility are given by Eqs. (1.15) for a simply-connected body.

5. Consider a doubly-connected body as shown in Fig. H 1, and reduce it to a simply-connected body by means of a barrier surface Ω. Take an arbitrary closed circuit C which has initial and final points on Ω and cannot

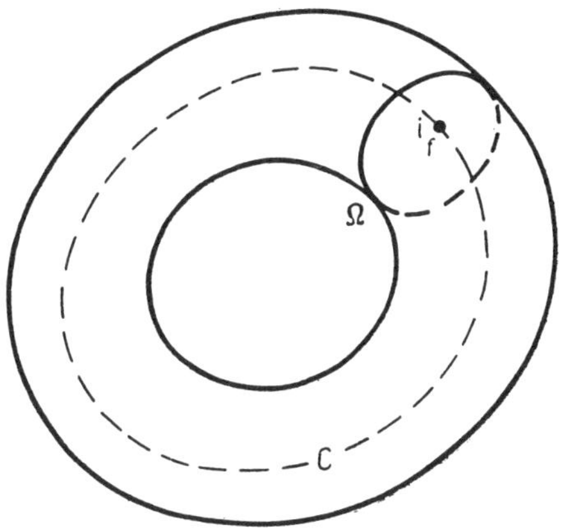

FIG. H 1.

be contracted to a point without passing out of the body. Applying equations (i) and (ii) of Problem 4 to the circuit C, show that even if strains of the body are continuous and satisfy the conditions of compatibility, Eqs. (1.15), we have

$$u_f - u_i = l_1 + p_2 z - p_3 y,$$
$$v_f - v_i = l_2 + p_3 x - p_1 z,$$
$$w_f - w_i = l_3 + p_1 y - p_2 x,$$

where l_1, l_2, l_3 and p_1, p_2, p_3 are constant, and the suffixes f and i are referred to the final and initial values, respectively. Note: see Ref. 1.1, pp. 221–8, and Ref. 1.20, pp. 99–110.†

Problems Related to Sections 1.6 and 1.7

6. Show that the two-dimensional elasticity problem treated in Section 1.7 is reduced to solving the following equations:

† Ref. 1.1 denotes Ref. 1 in the bibliography of Chapter 1.

(1) Displacement method: solve the differential equations

$$\Delta u + \frac{1+\nu}{1-\nu}\frac{\partial e}{\partial x} = 0, \quad \Delta v + \frac{1+\nu}{1-\nu}\frac{\partial e}{\partial y} = 0 \quad \text{in} \quad S \tag{i}$$

under the boundary conditions

$$\begin{aligned} \frac{E}{(1-\nu^2)}\left[\left(\frac{\partial u}{\partial x} + \nu\frac{\partial v}{\partial y}\right)l + \frac{1-\nu}{2}\left(\frac{\partial v}{\partial x} + \frac{\partial u}{\partial y}\right)m\right] &= \bar{X}_\nu, \\ \frac{E}{(1-\nu^2)}\left[\frac{1-\nu}{2}\left(\frac{\partial v}{\partial x} + \frac{\partial u}{\partial y}\right)l + \left(\nu\frac{\partial u}{\partial x} + \frac{\partial v}{\partial y}\right)m\right] &= \bar{Y}_\nu \quad \text{on} \quad C, \end{aligned} \tag{ii}$$

where $\Delta(\ ) = \partial^2(\ )/\partial x^2 + \partial^2(\ )/\partial y^2$ and $e = u_{,x} + v_{,y}$.

(2) Force method: solve the differential equation

$$\Delta\Delta F = 0 \quad \text{in} \quad S \tag{iii}$$

under the boundary conditions

$$\frac{d}{ds}\left(\frac{\partial F}{\partial y}\right) = \bar{X}_\nu, \quad -\frac{d}{ds}\left(\frac{\partial F}{dx}\right) = \bar{Y}_\nu \quad \text{on} \quad C, \tag{iv}$$

where $\Delta\Delta(\ ) = \partial^4(\ )/\partial x^4 + 2\partial^4(\ )/\partial x^2\,\partial y^2 + \partial^4(\ )/\partial y^4$.

7. Show that the principle of complementary virtual work for the two-dimensional problem treated in Section 1.7 may be given by

$$\iint_S (\varepsilon_x\,\delta\sigma_x + \varepsilon_y\,\delta\sigma_y + \gamma_{xy}\,\delta\tau_{xy})\,dx\,dy - \int_C (u\,\delta X_\nu + v\delta Y_\nu)\,ds = 0,$$

where σ_x, σ_y, τ_{xy}, X_ν and Y_ν have been expressed in terms of F by use of Eqs. (1.25) and (1.57), and u, v are Lagrange multipliers. Show also that we may derive from the above principle the following equations:

$$\varepsilon_{x,yy} + \varepsilon_{y,xx} - \gamma_{xy,xy} = 0$$

in S and

$$u(s) = \int_0^s [\varepsilon_x\,dx + (\tfrac{1}{2}\gamma_{xy} - \varpi_z)\,dy] + ay + b,$$

$$v(s) = \int_0^s [(\tfrac{1}{2}\gamma_{xy} + \varpi_z)\,dx + \varepsilon_y\,dy] - ax + c$$

on C, where

$$\varpi_z = \int_0^s [(\tfrac{1}{2}\gamma_{xy,x} - \varepsilon_{x,y})\,dx + (\varepsilon_{y,x} - \tfrac{1}{2}\gamma_{xy,y})\,dy],$$

while a, b, c are arbitrary constants and s is measured along the boundary C.

Problem Related to Section 1.9

8. We consider, as an example of Eq. (1.77), a two-dimensional problem in which the stress field is continuous, while the displacement field has a line of discontinuity as shown in Fig. H 2. To begin with, it is assumed

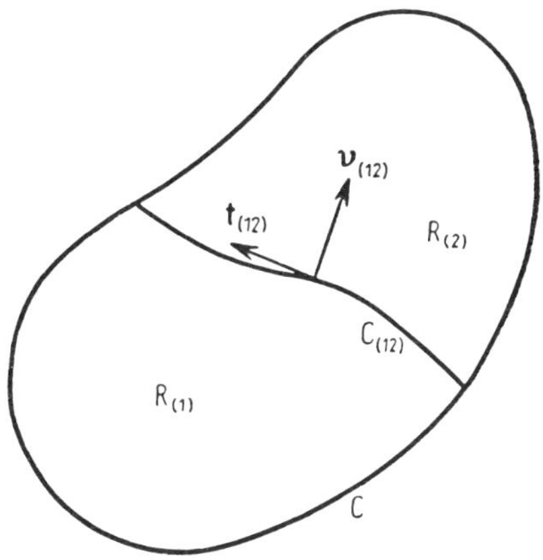

FIG. H 2.

that the line of discontinuity, denoted by $C_{(12)}$, divides the two-dimensional body R into two subregions $R_{(1)}$ and $R_{(2)}$. Two unit vectors, $\boldsymbol{\nu}_{(12)}$ and $\mathbf{t}_{(12)}$, are defined on $C_{(12)}$ such that $\boldsymbol{\nu}_{(12)}$ is the unit normal drawn from $R_{(1)}$ to $R_{(2)}$, and $\mathbf{t}_{(12)}$ is obtained by rotating $\boldsymbol{\nu}_{(12)}$ in a counter-clockwise direction through 90°. The stress components (σ_x, σ_y, τ_{xy}) are assumed continuous throughout the body R and to satisfy Eqs. (1.24) and (1.53). The tangential stress transmitted across the $C_{(12)}$ line from $R_{(2)}$ to $R_{(1)}$ is denoted by $T_{(12)}$ and taken positive when it is acting in the direction of $\mathbf{t}_{(12)}$. The displacement components (u, v) are assumed continuous in each of the subregions. The displacement components on $C_{(12)}$ of the subregions $R_{(1)}$ and $R_{(2)}$, taken in the directions of $\boldsymbol{\nu}_{(12)}$ and $\mathbf{t}_{(12)}$, are denoted by $v_{\nu(1)}$, $v_{\nu(2)}$ and $v_{t(1)}$, $v_{t(2)}$, respectively, and continuity of the normal components, i.e. $v_{\nu(1)} = v_{\nu(2)}$, is assumed. Then show that we have the divergence theorem as follows:

$$\iint_R (\sigma_x \varepsilon_x + \sigma_y \varepsilon_y + \tau_{xy}\gamma_{xy})\, dx\, dy$$
$$= \int_C (\bar{X}_\nu u + \bar{Y}_\nu v)\, ds + \int_{C_{(12)}} T_{(12)}\, [v_{t(1)} - v_{t(2)}]\, ds_{(12)},$$

where Eqs. (1.52) are assumed to hold in each of the subregions. Show also that the above relation holds even if the line $C_{(12)}$ does not extend between two points on the boundary, but is a line segment contained in the region R. Note: see Ref. 1.21, pp. 209–13.

CHAPTER 2

Problems Related to Sections 2.1 and 2.2

1. Prove Kirchhoff's theorem that the solution of the elasticity problem presented in Section 1.1 is unique.

2. Show that the stationary conditions of Π defined by Eq. (2.9) are coincident for an isotropic body with equations (i) and (ii) in Problem 1 of Chapter 1.

3. Show that, for the two-dimensional problem treated in Section 1.7, the functional for the principle of minimum complementary energy is given by

$$\Pi_C = \iint \left\{ \frac{1}{2E} \left\{ \left(\frac{\partial^2 F}{\partial x^2} + \frac{\partial^2 F}{\partial y^2} \right)^2 + 2(1+\nu) \left[\left(\frac{\partial^2 F}{\partial x \, \partial y} \right)^2 - \frac{\partial^2 F}{\partial x^2} \frac{\partial^2 F}{\partial y^2} \right] \right\} \right\} dx\, dy$$

and that the stationary condition is coincident with equation (iii) in Problem 6 of Chapter 1.

Problems Related to Quadratic Functions

4. We consider a quadratic function with n variables $x_1, x_2, \ldots, x_n$:

$$f(x_1, x_2, \ldots, x_n) = \{x\}' [A] \{x\}$$

where $[A]$ is a symmetric matrix, and $\{x\}$ is a column matrix:

$$[A] = \begin{bmatrix} a_{11} & \cdots & a_{1n} \\ \vdots & & \vdots \\ a_{n1} & \cdots & a_{nn} \end{bmatrix}, \qquad \{x\} = \begin{bmatrix} x_1 \\ \vdots \\ x_n \end{bmatrix},$$

and $\{\ \}'$ denotes the transposed matrix of $\{\ \}$. Show that the function f is positive definite if and only if

$$D_1 > 0, \quad D_2 > 0, \quad \ldots, \quad D_n > 0 \tag{i}$$

where $D_1, D_2, \ldots, D_n$ are the principal minors of the matrix A defined by

$$D_1 = a_{11}, \quad D_2 = \begin{vmatrix} a_{11} & a_{12} \\ a_{21} & a_{22} \end{vmatrix}, \quad \ldots, \quad D_n = \begin{vmatrix} a_{11} & \cdots\cdots & a_{1n} \\ \vdots & & \vdots \\ a_{n1} & \cdots\cdots & a_{nn} \end{vmatrix} \tag{ii}$$

Note: see Ref. 2.42, pp. 304–8. The relations (i) are useful, for example, in deriving some relations of inequality among the elastic constants from the assumption that the strain energy function is a positive definite function of the strain components.

5. We consider a function with n variables $x_1, x_2, \ldots, x_n$:

$$f(x_1, x_2, \ldots, x_n) = \tfrac{1}{2}\{x\}'[A]\{x\} - \{b\}'\{x\} \tag{i}$$

where $[A]$ is a positive definite symmetric matrix, $\{x\}$ is a column matrix, and $\{b\}'$ is a row matrix:

$$\{b\}' = [b_1, b_2, \ldots, b_n].$$

Show that the stationary conditions of f are given by

$$[A]\{x\} = \{b\} \tag{ii}$$

and that the minimum value of f is given by

$$f_{\min} = -\tfrac{1}{2}\{x_{st}\}'[A]\{x_{st}\} = -\tfrac{1}{2}\{b\}'\{x_{st}\} \tag{iii}$$

where $\{x_{st}\}$ denotes the solution of equations (ii).

Problems Related to the Concept of Function Space

Here we consider the elasticity problem defined in Section 1.1, assuming, however, that body forces are absent for the sake of simplicity.

6. Show that the principles of minimum complementary energy and minimum potential energy are given by

$$\tfrac{1}{2}(\mathbf{S}', \mathbf{S}') - [\bar{\mathbf{S}}, \mathbf{S}']_2 \geqq \tfrac{1}{2}(\mathbf{S}, \mathbf{S}) - [\bar{\mathbf{S}}, \mathbf{S}]_2 \tag{i}$$

and

$$\tfrac{1}{2}(\mathbf{S}'', \mathbf{S}'') - [\mathbf{S}'', \bar{\mathbf{S}}]_1 \geqq \tfrac{1}{2}(\mathbf{S}, \mathbf{S}) - [\mathbf{S}, \bar{\mathbf{S}}]_1 \tag{ii}$$

in vectorial notations, respectively, where

- $\mathbf{S}$ is the exact solution,
- $\mathbf{S}'$ satisfies Eqs. (1.20) and (1.12),
- $\mathbf{S}''$ satisfies Eqs. (1.5) and (1.14).

The brackets $[\ ]_1$ and $[\ ]_2$ denote surface integrals on S_1 ans S_2, respectively such that

$$[\mathbf{S}'', \bar{\mathbf{S}}]_1 = \iint_{S_1} (u''\bar{X}_\nu + v''\bar{Y}_\nu + w''\bar{Z}_\nu)\, dS,$$

$$[\bar{\mathbf{S}}, \mathbf{S}']_2 = \iint_{S_2} (\bar{u}X' + \bar{v}Y'_\nu + \bar{w}Z'_\nu)\, dS.$$

The bracket is so defined that it contains the displacement components of the first vector and the stress components of the second one. Note: see Ref. 2.20.

7. We consider a special case of the elasticity problem where the boundary conditions are given by

$$X_\nu = \bar{X}_\nu, \quad Y_\nu = \bar{Y}_\nu, \quad Z_\nu = \bar{Z}_\nu \quad \text{on} \quad S_1, \tag{i}$$

and

$$u = v = w = 0 \quad \text{on} \quad S_2. \tag{ii}$$

Let us take

$$\mathbf{S}' = \mathbf{S}'_0 + \sum_{p=1}^{m} a_p \mathbf{I}'_p, \qquad \mathbf{S}'' = \sum_{q=1}^{n} b_q \mathbf{I}''_q \tag{iii}$$

and determine a_p $(p = 1, 2, \ldots, m)$ and b_q $(q = 1, 2, \ldots, n)$ so that they make

$$\tfrac{1}{2}(\mathbf{S}', \mathbf{S}') \tag{iv}$$

and

$$\tfrac{1}{2}(\mathbf{S}'', \mathbf{S}'') - [\mathbf{S}'', \bar{\mathbf{S}}]_1, \tag{v}$$

minimum, respectively, where

$\mathbf{S}'_0$ satisfies Eqs. (1.20) and equations (i),

$\mathbf{I}'_p$ satisfies Eqs. (1.20) and homogeneous boundary conditions on S_1, namely,

$$X_\nu = Y_\nu = Z_\nu = 0,$$

$\mathbf{I}''_q$ satisfies Eqs. (1.5) and equations (ii).

Then, show that we have the following inequalities:

$$(\mathbf{S}'', \mathbf{S}'') \leqq (\mathbf{S}, \mathbf{S}) \leqq (\mathbf{S}', \mathbf{S}'). \tag{vi}$$

8. We consider another special case of the elasticity problem where the boundary conditions are given by

$$X_\nu = Y_\nu = Z_\nu = 0 \quad \text{on} \quad S_1, \tag{i}$$

and

$$u = \bar{u}, \quad v = \bar{v}, \quad w = \bar{w} \quad \text{on} \quad S_2.. \tag{ii}$$

Let us take

$$\mathbf{S}' = \sum_{p=1}^{m} a_p \mathbf{I}'_p, \qquad \mathbf{S}'' = \mathbf{S}''_0 + \sum_{q=1}^{n} b_q \mathbf{I}''_q \tag{iii}$$

and determine a_p $(p = 1, 2, \ldots, m)$ and b_q $(q = 1, 2, \ldots, n)$ so that they make

$$\tfrac{1}{2}(\mathbf{S}', \mathbf{S}') - [\bar{\mathbf{S}}, \mathbf{S}']_2 \tag{iv}$$

and

$$\tfrac{1}{2}(\mathbf{S}'', \mathbf{S}'') \tag{v}$$

minimum, respectively, where

$\mathbf{I}'_p$ satisfies Eqs. (1.20) and equations (i),

$\mathbf{S}''_0$ satisfies Eqs. (1.5) and equations (ii),

$\mathbf{I}''_q$ satisfies Eqs. (1.5) and homogeneous boundary conditions on S_2, namely,

$$u = v = w = 0.$$

Then, show that we have the following inequalities:

$$(\mathbf{S}', \mathbf{S}') \leqq (\mathbf{S}, \mathbf{S}) \leqq (\mathbf{S}'', \mathbf{S}''). \tag{vi}$$

Note: From equations (vi) of Problem 7 and equations (vi) of Problem 8, bounds formulae for some scalar quantities are obtainable as exemplified in Section 6.5. See also Ref. 2.43.

9. Obtain the following vectorial equations:

$$(\mathbf{S}'', \mathbf{S}) = [\mathbf{S}'', \overline{\mathbf{S}}]_1 + [\mathbf{S}'', \mathbf{S}]_2, \tag{i}$$

and

$$(\mathbf{S}, \mathbf{S}') = [\mathbf{S}, \mathbf{S}']_1 + [\overline{\mathbf{S}}, \mathbf{S}']_2, \tag{ii}$$

where **S**, **S**′, **S**″ are defined in the same manner as in Problem 6. Discuss relations between equation (i) and the unit displacement method, and also relations between equation (ii) and the unit load method. Note: See Ref. 2.14 for the unit displacement method and the unit load method.

10. We choose a vector S^*, having displacement components

$$\begin{aligned} u^* &= a_{11}x + a_{12}y + a_{13}z, \\ v^* &= a_{21}x + a_{22}y + a_{23}z, \\ w^* &= a_{31}x + a_{32}y + a_{33}z, \end{aligned} \tag{i}$$

where a_{ik} $(i, k = 1, 2, 3)$ are constants. Show that we have

$$\begin{aligned} (\mathbf{S}, \mathbf{S}^*) &= \iiint_V [\sigma_x a_{11} + \sigma_y a_{22} + \cdots + \tau_{xy}(a_{21} + a_{12})]\, dV \\ &= \iint_{S_1+S_2} [u^* X_\nu + v^* Y_\nu + w^* Z_\nu]\, dS \end{aligned} \tag{ii}$$

or

$$\begin{aligned} (\mathbf{S}, \mathbf{S}^*) &= \iiint_V [\sigma_x^* \varepsilon_x + \sigma_y^* \varepsilon_y + \cdots + \tau_{xy}^* \gamma_{xy}]\, dV \\ &= \iint_{S_1+S_2} [u X_\nu^* + v Y_\nu^* + w Z_\nu^*]\, dS \end{aligned} \tag{iii}$$

where **S** is the exact solution. Note: The above relations show that if the boundary conditions are given either entirely in terms of forces or entirely in terms of displacements, we can calculate the average value of stresses or strains of the exact solution.

Problems Related to Section 2.6

11. We consider an elastic body which is held fixed on S_2. We apply two systems of body forces plus surface forces on S_1:

$$\bar{X}, \bar{Y}, \bar{Z}, \quad \bar{X}_\nu, \bar{Y}_\nu, \bar{Z}_\nu; \quad \bar{X}^*, \bar{Y}^*, \bar{Z}^*, \quad \bar{X}_\nu^*, \bar{Y}_\nu^*, \bar{Z}_\nu^*$$

to the elastic body independently, and denote displacement components due to these forces by

$$u, v, w; \quad u^*, v^*, w^*,$$

respectively. Then show that Maxwell–Betti's theorem

$$\iiint_V (\bar{X}u^* + \bar{Y}v^* + \bar{Z}w^*)\, dV + \iint_{S_1} (\bar{X}_\nu u^* + \bar{Y}_\nu v^* + \bar{Z}_\nu w^*)\, dS$$
$$= \iiint_V (\bar{X}^* u + \bar{Y}^* v + \bar{Z}^* w)\, dV + \iint_{S_1} (\bar{X}_\nu^* u + \bar{Y}_\nu^* v + \bar{Z}_\nu^* w)\, dS$$

holds between them.

12. Show that for a concentrated moment $\bar{M}$ on S_1, Castigliano's theorem provides:

$$\frac{\partial V}{\partial \bar{M}} = \theta \qquad \text{(i)}$$

where θ is the rotational angle of the local surface (where $\bar{M}$ is applied) in the direction of $\bar{M}$.

13. Examine relations between the unit displacement method and Eq. (2.49). Examine also relations between the unit load method and Castigliano's theorem.

Problem Related to Variational Principles of Elasticity

14. We divide the elastic body treated in Section 1.1 into two parts $V_{(1)}$ and $V_{(2)}$ fictitiously, and denote their interface by $S_{(12)}$.

(1) Show that the functional for the principle of minimum potential energy, Eq. (2.12), can be written with the use of Lagrange multipliers p_x, p_y and p_z as follows:

$$\begin{aligned}
\Pi = &\iiint_{V_{(1)}} [A(u_{(1)}, v_{(1)}, w_{(1)}) - (\bar{X}u_{(1)} + \bar{Y}v_{(1)} + \bar{Z}w_{(1)})]\, dV \\
&+ \iiint_{V_{(2)}} [A(u_{(2)}, v_{(2)}, w_{(2)}) - (\bar{X}u_{(2)} + \bar{Y}v_{(2)} + \bar{Z}w_{(2)})]\, dV \\
&- \iint_{S_1} (\bar{X}_\nu u_{(1)} + \bar{Y}_\nu v_{(1)} + \bar{Z}_\nu w_{(1)})\, dS \\
&+ \iint_{S_{(12)}} [p_x(u_{(1)} - u_{(2)}) + p_y(v_{(1)} - v_{(2)}) + p_z(w_{(1)} - w_{(2)})]\, dS \qquad \text{(i)}
\end{aligned}$$

where it is assumed without loss of generality that S_1 belongs to $V_{(1)}$. The independent quantities subject to variation in the functional (i) are $u_{(1)}$, $v_{(1)}$, $w_{(1)}$, $u_{(2)}$, $v_{(2)}$, $w_{(2)}$, p_x, p_y, p_z under subsidiary conditions Eqs. (1.14). Derive also the stationary conditions of the functional (i).

(2) Show that, by the use of Lagrange multipliers q_x, q_y and q_z, the functional for the principle of minimum complementary energy, Eq. (2.23), can be written as follows:

$$\begin{aligned}\Pi_C = &\iiint_{V_{(1)}} B(\sigma_{x(1)}, \sigma_{y(1)}, \ldots, \tau_{xy(1)})\, dV \\ &+ \iiint_{V_{(2)}} B(\sigma_{x(2)}, \sigma_{y(2)}, \ldots, \tau_{xy(2)})\, dV \\ &- \iint_{S_2} (\bar{u}X_{\nu(2)} + \bar{v}Y_{\nu(2)} + \bar{w}Z_{\nu(2)})\, dS \\ &+ \iint_{S_{(12)}} [q_x(X_{\nu(1)} + X_{\nu(2)}) + q_y(Y_{\nu(1)} + Y_{\nu(2)}) \\ &\qquad + q_z(Z_{\nu(1)} + Z_{\nu(2)})]\, dS, \end{aligned} \tag{ii}$$

where it is assumed without loss of generality that S_2 belongs to $V_{(2)}$. In defining $X_{\nu(1)}$, ... and $Z_{\nu(2)}$ on the interface, the outward normals are employed: the unit normal drawn from $V_{(1)}$ to $V_{(2)}$ is used in defining $X_{\nu(1)}$, $Y_{\nu(1)}$ and $Z_{\nu(1)}$, while the unit normal drawn from $V_{(2)}$ to $V_{(1)}$ is used in defining $X_{\nu(2)}$, $Y_{\nu(2)}$ and $Z_{\nu(2)}$. The independent quantities subject to variation in the functional (ii) are $\sigma_{x(1)}, \ldots, \tau_{xy(1)}, \sigma_{x(2)}, \ldots, \tau_{xy(2)}, q_x, q_y$ and q_z, under subsidiary conditions Eqs. (1.4) and (1.12). Derive also the stationary conditions of the functional (ii).

Problems Related to Variational Formulation

15. We consider an eigenvalue problem of a function $u(x)$ defined in $a \leqq x \leqq b$:

$$-\frac{d}{dx}\left[p(x)\frac{du}{dx}\right] + r(x)\,u - \lambda u = 0 \tag{i}$$

with boundary conditions

$$u'(a) - \alpha u(a) = 0, \quad u'(b) + \beta u(b) = 0 \tag{ii}$$

where λ is a parameter related to the eigenvalues, and α and β are specified constants. Show that we have a variational expression for the eigenvalue problem as follows:

$$\begin{aligned}\Pi = &\frac{1}{2}\int_a^b \left[p\left(\frac{du}{dx}\right)^2 + ru^2 - \lambda u^2\right] dx \\ &+ \frac{1}{2}\alpha p(a)\,[u(a)]^2 + \frac{1}{2}\beta p(b)\,[u(b)]^2, \end{aligned} \tag{iii}$$

where the function subject to variation is $u(x)$.

16. We consider a heat conduction problem, the field equation of which is given by

$$Q_{x,x} + Q_{y,y} + Q_{z,z} = \bar{q}(x, y, z), \tag{i}$$

where (Q_x, Q_y, Q_z) is heat flux and $\bar{q}$ is the heat source. The relation between heat flux and temperature gradient is assumed to be

$$\begin{bmatrix} Q_x \\ Q_y \\ Q_z \end{bmatrix} = - \begin{bmatrix} c_{11} & c_{12} & c_{13} \\ c_{21} & c_{22} & c_{23} \\ c_{31} & c_{32} & c_{33} \end{bmatrix} \begin{bmatrix} \theta,_x \\ \theta,_y \\ \theta,_z \end{bmatrix} \tag{ii}$$

where θ is the temperature, and the c_{ij} are constant and symmetric:

$$c_{ij} = c_{ji}; \quad i, j = 1, 2, 3. \tag{iii}$$

The boundary conditions are assumed to be

$$Q_x l + Q_y m + Q_z n = -K(\bar{\theta}_a - \theta) \quad \text{on} \quad S_1 \tag{iv}$$

and

$$\theta = \bar{\theta} \quad \text{on} \quad S_2, \tag{v}$$

where (l, m, n) are direction cosines of the normal drawn outwards from the boundary, K is a constant and $\bar{\theta}_a$ and $\bar{\theta}$ are prescribed. Then show that we have the following variational expression for this problem:

$$\begin{aligned} \Pi = \tfrac{1}{2} \iiint_V [& c_{11}\theta,_x^2 + c_{22}\theta,_y^2 + c_{33}\,\theta,_z^2 \\ & + 2c_{23}\theta,_y\theta,_z + 2c_{31}\theta,_z\theta,_x + 2c_{12}\,\theta,_x\theta,_y]\, dV \\ & - \iiint_V \bar{q}\theta\, dV - \iint_{S_1} K(\bar{\theta}_a\theta - \tfrac{1}{2}\theta^2)\, dS, \end{aligned} \tag{vi}$$

where the function subject to variation is $\theta(x, y, z)$ under the subsidiary condition (v).

CHAPTER 4

Problems Related to Section 4.1

1. Vectors and tensors are systems of numbers or functions whose components obey a certain transformation law when the coordinate variables in the space undergo a transformation such that:

$$\bar{\alpha}^\lambda = \bar{\alpha}^\lambda(\alpha^1, \alpha^2, \alpha^3); \quad \lambda = 1, 2, 3. \tag{i}$$

An overbar is used to distinguish between two coordinate systems α^λ and $\bar{\alpha}^\lambda$. A system v^λ is called a contravariant vector if its components $\bar{v}^\lambda$ in the new variables satisfy the relations:

$$\bar{v}^{\lambda} = \frac{\partial \bar{\alpha}^{\lambda}}{\partial \alpha^{\mu}} v^{\mu}. \tag{ii}$$

Similarly, we define a covariant vector v_{λ} by

$$\bar{v}_{\lambda} = \frac{\partial \alpha^{\mu}}{\partial \bar{\alpha}^{\lambda}} v_{\mu}, \tag{iii}$$

a contravariant tensor of order two $a^{\alpha\beta}$ by

$$\bar{a}^{\alpha\beta} = \frac{\partial \bar{\alpha}^{\alpha}}{\partial \alpha^{\lambda}} \frac{\partial \bar{\alpha}^{\beta}}{\partial \alpha^{\mu}} a^{\lambda\mu}, \tag{iv}$$

a mixed tensor of order two $a^{\alpha\cdot}_{\cdot\beta}$ by

$$\bar{a}^{\alpha\cdot}_{\cdot\beta} = \frac{\partial \bar{\alpha}^{\alpha}}{\partial \alpha^{\lambda}} \frac{\partial \alpha^{\mu}}{\partial \bar{\alpha}^{\beta}} a^{\lambda\cdot}_{\cdot\mu} \tag{v}$$

and a covariant tensor of order two $a_{\alpha\beta}$ by

$$\bar{a}_{\alpha\beta} = \frac{\partial \alpha^{\lambda}}{\partial \bar{\alpha}^{\alpha}} \frac{\partial \alpha^{\mu}}{\partial \bar{\alpha}^{\beta}} a_{\lambda\mu}. \tag{vi}$$

In general, a system $a^{\lambda\cdot\cdot\cdots}_{\cdot\mu\nu\cdots}$ is called a tensor when its components $\bar{a}^{\alpha\cdot\cdot\cdots}_{\cdot\beta\gamma\cdots}$ in the new variables satisfy the relations:

$$\bar{a}^{\alpha\cdot\cdot\cdots}_{\cdot\beta\gamma\cdots} = \frac{\partial \bar{\alpha}^{\alpha}}{\partial \alpha^{\lambda}} \frac{\partial \alpha^{\mu}}{\partial \bar{\alpha}^{\beta}} \frac{\partial \alpha^{\nu}}{\partial \bar{\alpha}^{\gamma}} \cdots a^{\lambda\cdot\cdot\cdots}_{\cdot\mu\nu\cdots} \tag{vii}$$

(1) Show that the quantities v^{λ} defined by Eq. (4.15) and v_{λ} defined by Eq. (4.18) are contravariant and covariant vectors, respectively.

(2) Show that the quantities $g_{\lambda\mu}$ defined by Eq. (4.6) and $g^{\lambda\mu}$ defined by Eq. (4.7) are covariant and contravariant tensors of order two, respectively.

(3) Show that the quantities defined by

$$\varepsilon^{\lambda\mu\nu} = \frac{e^{\lambda\mu\nu}}{\sqrt{g}} \tag{viii}$$

are a contravariant tensor of order three, where $e^{\lambda\mu\nu}$ is defined by equations (iv) of Problem 3 of Chapter 1, and g is given by Eq. (4.28). Note: see Ref. 4.1, pp. 10–12.

(4) A tensor of order two can be given by any one of the following three forms: $a^{\lambda\mu}$, $a^{\lambda\cdot}_{\cdot\mu}$ and $a_{\lambda\mu}$, Show that any one of them can be changed into another form by use of the principle of raising and lowering an index of a component of the tensor such that

$$a^{\lambda\mu} = g^{\lambda\varkappa} a_{\varkappa\cdot}^{\cdot\mu}, \quad a_{\lambda\mu} = g_{\lambda\varkappa} a^{\varkappa\cdot}_{\cdot\mu}.$$

2. We employ the transformation represented by equation (i) of Problem 1 again. By use of the relations:

$$\mathbf{g}_\lambda = \frac{\partial \mathbf{r}^{(0)}}{\partial \alpha^\lambda}, \quad \mathbf{g}_{\bar{\lambda}} = \frac{\partial \mathbf{r}^{(0)}}{\partial \bar{\alpha}^\lambda} = \frac{\partial \alpha^\varkappa}{\partial \bar{\alpha}^\lambda} \mathbf{g}_\varkappa,$$

$$\mathbf{g}_{\bar{\lambda}, \bar{\mu}} = \frac{\partial^2 \mathbf{r}^{(0)}}{\partial \bar{\alpha}^\lambda \, \partial \bar{\alpha}^\mu} = \frac{\partial}{\partial \bar{\alpha}^\mu} \left(\frac{\partial \alpha^\varkappa}{\partial \bar{\alpha}^\lambda} \mathbf{g}_\varkappa \right),$$

show that we have

$$\frac{\partial \alpha^\varkappa}{\partial \bar{\alpha}^\lambda} \overline{\begin{Bmatrix} \lambda \\ \mu\nu \end{Bmatrix}} = \frac{\partial^2 \alpha^\varkappa}{\partial \bar{\alpha}^\lambda \, \partial \bar{\alpha}^\nu} + \frac{\partial \alpha^\varrho}{\partial \bar{\alpha}^\mu} \frac{\partial \alpha^\omega}{\partial \bar{\alpha}^\nu} \begin{Bmatrix} \varkappa \\ \varrho\omega \end{Bmatrix}, \tag{i}$$

where $\overline{\begin{Bmatrix} \lambda \\ \mu\nu \end{Bmatrix}}$ is the Christoffel 3-index symbol of the second kind in the $\bar{\alpha}^\lambda$ coordinate system. By use of Eq. (i), show that $v^\lambda{}_{;\nu}$ defined by Eq. (4.17) as well as $v_{\lambda;\nu}$ defined by Eq. (4.21) are tensors of order two. Show also that $T^{\lambda_1 \cdots\cdots \lambda_r}_{\mu_1 \cdots\cdots \mu_s; \nu}$ defined by Eq. (4.22) is a tensor.

3. Discuss geometrical relations between $\mathbf{g}_\lambda$ and $\mathbf{g}^\mu$ defined by Eqs. (4.5) and (4.8), respectively, and show that

$$\mathbf{g}_1 \times \mathbf{g}_2 = \sqrt{g}\, \mathbf{g}^3, \quad \mathbf{g}_2 \times \mathbf{g}_3 = \sqrt{g}\, \mathbf{g}^1, \quad \mathbf{g}_3 \times \mathbf{g}_1 = \sqrt{g}\, \mathbf{g}^2.$$

4. We consider a special case of curvilinear coordinate systems:

$$(ds^{(0)})^2 = g_{11} (d\alpha^1)^2 + 2g_{12}\, d\alpha^1\, d\alpha^2 + g_{22} (d\alpha^2)^2 + (d\alpha^3)^2,$$

where g_{11}, g_{12} and g_{22} are functions of α^1 and α^2 only. Obtain the following relations for the Christoffel symbols:

$$\begin{Bmatrix} 1 \\ 11 \end{Bmatrix} = \frac{1}{2} (g^{11} g_{11,1} + 2g^{12} g_{21,1} - g^{12} g_{11,2}),$$

$$\begin{Bmatrix} 2 \\ 11 \end{Bmatrix} = \frac{1}{2} (g^{21} g_{11,1} + 2g^{22} g_{21,1} - g^{22} g_{11,2}),$$

$$\begin{Bmatrix} 1 \\ 22 \end{Bmatrix} = \frac{1}{2} (g^{12}\, g_{22,2} + 2g^{11} g_{12,2} - g^{11} g_{22,1}),$$

$$\begin{Bmatrix} 2 \\ 22 \end{Bmatrix} = \frac{1}{2} (g^{22} g_{22,2} + 2g^{21} g_{12,2} - g^{12} g_{22,1}),$$

$$\begin{Bmatrix} 1 \\ 12 \end{Bmatrix} = \begin{Bmatrix} 1 \\ 21 \end{Bmatrix} = \frac{1}{2} (g^{11} g_{11,2} + g^{12} g_{22,1}),$$

$$\begin{Bmatrix} 2 \\ 12 \end{Bmatrix} = \begin{Bmatrix} 2 \\ 21 \end{Bmatrix} = \frac{1}{2} (g^{22} g_{22,1} + g^{21} g_{11,2})$$

and all the other $\begin{Bmatrix} \lambda \\ \mu\nu \end{Bmatrix}$ are zero.

Show also that if the variables (α^1, α^2) constitute an orthogonal curvilinear system, namely, $g_{12} = 0$, we have

$$\begin{Bmatrix} 1 \\ 11 \end{Bmatrix} = \frac{1}{A}\frac{\partial A}{\partial \alpha}, \qquad \begin{Bmatrix} 2 \\ 22 \end{Bmatrix} = \frac{1}{B}\frac{\partial B}{\partial \beta}.$$

$$\begin{Bmatrix} 2 \\ 11 \end{Bmatrix} = -\frac{A}{B^2}\frac{\partial A}{\partial \beta}, \qquad \begin{Bmatrix} 1 \\ 22 \end{Bmatrix} = -\frac{B}{A^2}\frac{\partial B}{\partial \alpha},$$

$$\begin{Bmatrix} 1 \\ 12 \end{Bmatrix} = \begin{Bmatrix} 1 \\ 21 \end{Bmatrix} = \frac{1}{A}\frac{\partial A}{\partial \beta}, \qquad \begin{Bmatrix} 2 \\ 12 \end{Bmatrix} = \begin{Bmatrix} 2 \\ 21 \end{Bmatrix} = \frac{1}{B}\frac{\partial B}{\partial \alpha}.$$

where $\alpha^1 = \alpha$, $\alpha^2 = \beta$, $g_{11} = A^2$ and $g_{22} = B^2$.

5. We consider an orthogonal curvilinear coordinate system defined by

$$(ds^{(0)})^2 = A^2\left(1 - \frac{\zeta}{R_\alpha}\right)^2 (d\alpha)^2 + B^2\left(1 - \frac{\zeta}{R_\beta}\right)^2 (d\beta)^2 + (d\zeta)^2.$$

where A, B, R_α and R_β are functions of (α, β) only. Choosing $\alpha^1 = \alpha$, $\alpha^2 = \beta$, $\alpha^3 = \zeta$, show that the Christoffel symbols at $\zeta = 0$ are obtained as follows:

$$\begin{Bmatrix} 1 \\ 11 \end{Bmatrix}_0 = \frac{1}{A}\frac{\partial A}{\partial \alpha}, \qquad \begin{Bmatrix} 2 \\ 11 \end{Bmatrix}_0 = -\frac{A}{B^2}\frac{\partial A}{\partial \beta},$$

$$\begin{Bmatrix} 1 \\ 12 \end{Bmatrix}_0 = \begin{Bmatrix} 1 \\ 21 \end{Bmatrix}_0 = \frac{1}{A}\frac{\partial A}{\partial \beta}, \qquad \begin{Bmatrix} 2 \\ 12 \end{Bmatrix}_0 = \begin{Bmatrix} 2 \\ 21 \end{Bmatrix}_0 = \frac{1}{B}\frac{\partial B}{\partial \alpha},$$

$$\begin{Bmatrix} 1 \\ 13 \end{Bmatrix}_0 = \begin{Bmatrix} 1 \\ 31 \end{Bmatrix}_0 = -\frac{1}{R_\alpha}, \qquad \begin{Bmatrix} 2 \\ 13 \end{Bmatrix}_0 = \begin{Bmatrix} 2 \\ 31 \end{Bmatrix}_0 = 0,$$

$$\begin{Bmatrix} 1 \\ 22 \end{Bmatrix}_0 = -\frac{B}{A^2}\frac{\partial B}{\partial \alpha}, \qquad \begin{Bmatrix} 2 \\ 22 \end{Bmatrix}_0 = \frac{1}{B}\frac{\partial B}{\partial \beta},$$

$$\begin{Bmatrix} 1 \\ 23 \end{Bmatrix}_0 = \begin{Bmatrix} 1 \\ 32 \end{Bmatrix}_0 = 0, \qquad \begin{Bmatrix} 2 \\ 23 \end{Bmatrix}_0 = \begin{Bmatrix} 2 \\ 32 \end{Bmatrix}_0 = -\frac{1}{R_\beta},$$

$$\begin{Bmatrix} 1 \\ 33 \end{Bmatrix}_0 = 0, \qquad \begin{Bmatrix} 2 \\ 33 \end{Bmatrix}_0 = 0,$$

$$\begin{Bmatrix} 3 \\ 11 \end{Bmatrix}_0 = \frac{A^2}{R_\alpha}, \qquad \begin{Bmatrix} 3 \\ 22 \end{Bmatrix}_0 = \frac{B^2}{R_\beta},$$

$$\begin{Bmatrix} 3 \\ 12 \end{Bmatrix}_0 = \begin{Bmatrix} 3 \\ 21 \end{Bmatrix}_0 = \begin{Bmatrix} 3 \\ 13 \end{Bmatrix}_0 = \begin{Bmatrix} 3 \\ 31 \end{Bmatrix}_0 = \begin{Bmatrix} 3 \\ 23 \end{Bmatrix}_0 = \begin{Bmatrix} 3 \\ 32 \end{Bmatrix}_0 = \begin{Bmatrix} 3 \\ 33 \end{Bmatrix}_0 = 0$$

where $\begin{Bmatrix} \lambda \\ \mu\nu \end{Bmatrix}_0$ denotes the value of $\begin{Bmatrix} \lambda \\ \mu\nu \end{Bmatrix}$ at $\zeta = 0$.

Problems Related to the Conditions of Compatibility and Stress Functions

6. We consider a simply-connected body, assuming that strain components $f_{\lambda\mu}$ are given as functions of $(\alpha^1, \alpha^2, \alpha^3)$, and $\left\{\begin{Bmatrix} \lambda \\ \mu\nu \end{Bmatrix}\right\}$ are expressed in terms of $f_{\alpha\beta}$ by use of Eqs. (4.34) and (4.36). Show that Eqs. (4.42) are necessary and sufficient conditions for the following equations to be integrable:

$$d\mathbf{r} = \mathbf{G}_\lambda \, d\alpha^\lambda, \quad d\mathbf{G}_\lambda = \left\{\left\{ {\varkappa \atop \lambda\mu} \right\}\right\} \mathbf{G}_\varkappa \, d\alpha^\mu, \tag{i}$$

That is, they are necessary and sufficient conditions for the existence of single-valued vector functions $\mathbf{r}$ and $\mathbf{G}_\lambda$. Obtain relations between equations (i) above and equations (i) and (ii) of Problem 4 of Chapter 1.

7. Show that the conditions of compatibility before deformation are given by

$$R^{(0)\lambda}{}_{\bullet\mu\nu\omega} = 0, \tag{i}$$

where $R^{(0)\lambda}{}_{\bullet\mu\nu\omega}$ is defined from Eqs. (4.43) by replacing $\left\{\left\{ {\lambda \atop \mu\nu} \right\}\right\}, \ldots$ with $\left\{ {\lambda \atop \mu\nu} \right\}, \ldots$, and discuss the physical meaning of these conditions.

8. We confine our problem to small displacement theory. Show that Eqs. (4.40) and (4.53) reduce to

$$\begin{aligned} f_{\lambda\mu} &= \tfrac{1}{2} (g_{\lambda\varkappa} v^\varkappa{}_{;\mu} + g_{\mu\varkappa} v^\varkappa{}_{;\lambda}) \\ &= \tfrac{1}{2} (v_{\lambda;\mu} + v_{\mu;\lambda}), \end{aligned} \tag{i}$$

and

$$\tau^{\lambda\varkappa}{}_{;\varkappa} + \bar{P}^\lambda_g = 0, \tag{ii}$$

respectively. Show also that equations (i) can be derived from equations (ii) by use of the principle of complementary virtual work. Note:

$$\left\{ {\lambda \atop \lambda\nu} \right\} = \frac{1}{\sqrt{g}} \frac{\partial \sqrt{g}}{\partial \alpha^\nu}.$$

9. We confine our problem to the small displacement theory. Show that the Riemann–Christoffel curvature tensor is reduced to

$$\begin{aligned} R_{\lambda\mu\nu\omega} &= f_{\lambda\omega,\mu\nu} + f_{\mu\nu,\lambda\omega} - f_{\lambda\nu,\mu\omega} - f_{\mu\omega,\lambda\nu} \\ &\quad - 2\left[\left\{ {\alpha \atop \mu\nu} \right\}\left\{ {\beta \atop \lambda\omega} \right\} - \left\{ {\alpha \atop \mu\omega} \right\}\left\{ {\beta \atop \lambda\nu} \right\}\right] f_{\alpha\beta} \\ &\quad + 2\left[\left\{ {\alpha \atop \mu\nu} \right\} \gamma_{\lambda\omega\alpha} + \left\{ {\alpha \atop \lambda\omega} \right\} \gamma_{\mu\nu\alpha} - \left\{ {\alpha \atop \mu\omega} \right\} \gamma_{\lambda\nu\alpha} - \left\{ {\alpha \atop \lambda\nu} \right\} \gamma_{\mu\omega\alpha}\right] \end{aligned} \tag{i}$$

or

$$R_{\lambda\mu\nu\omega} = f_{\lambda\omega;\mu\nu} + f_{\mu\nu;\lambda\omega} - f_{\lambda\nu;\mu\omega} - f_{\mu\omega;\lambda\nu} \tag{ii}$$

where, by definition,

$$\gamma_{\lambda\mu\nu} = \tfrac{1}{2} (f_{\nu\lambda,\mu} + f_{\nu\mu,\lambda} - f_{\lambda\mu,\nu}). \tag{iii}$$

Show also that if

$$S^{\lambda\mu} = \varepsilon^{\lambda\alpha\varkappa} \varepsilon^{\mu\beta\varrho} f_{\alpha\beta;\varkappa\varrho} \tag{iv}$$

the conditions of compatibility are given by

$$S^{\lambda\mu} = 0 \quad (\lambda, \mu = 1, 2, 3), \tag{v}$$

where $\varepsilon^{\lambda\mu\nu}$ is defined by equation (viii) of Problem 1. Notes: (1) $\gamma_{\lambda\mu\nu}$ as defined by equation (iii) is not a tensor (see Ref. 4.1, pp. 61–2). (2) $f_{\lambda\mu;\nu\omega} = [f_{\lambda\mu;\nu}]_{;\omega}$.

10. We confine our problem to the small displacement theory. Show that the principle of virtual work may be written as follows:

$$\iiint \tau^{\lambda\mu}\, \delta f_{\lambda\mu} \sqrt{g}\, d\alpha^1\, d\alpha^2\, d\alpha^3 - \iiint \psi_{\lambda\mu}\, \varepsilon^{\lambda\alpha\varkappa}\, \varepsilon^{\mu\beta\varrho}\, \delta f_{\alpha\beta;\varkappa\varrho} \sqrt{g}\, d\alpha^1\, d\alpha^2\, d\alpha^3 + \cdots = 0 \tag{i}$$

where $\psi_{\lambda\mu}$ are Lagrange multipliers. Show also that we obtain

$$\tau^{\lambda\mu} = \varepsilon^{\lambda\alpha\varkappa}\, \varepsilon^{\mu\beta\varrho}\, \psi_{\alpha\beta;\varkappa\varrho} \tag{ii}$$

from the principle, thus demonstrating that a symmetric covariant tensor $\psi_{\alpha\beta}$ plays the role of stress functions in the small displacement theory expressed in curvilinear coordinates. Notes: (1) $\varepsilon^{\lambda\mu\nu}{}_{;\alpha} = 0$. (2) see Section 1.8 for similar developments.

Problems Related to a Two-dimensional Skew Coordinate System†

11. We consider a two-dimensional skew coordinate system (ξ, η):

$$x = \xi + \eta \cos\alpha, \quad y = \eta \sin\alpha$$

as shown in Fig. H 3, where α is a constant. We confine subsequent formulations to small displacement theory and choose $\alpha^1 = \xi$, $\alpha^2 = \eta$.

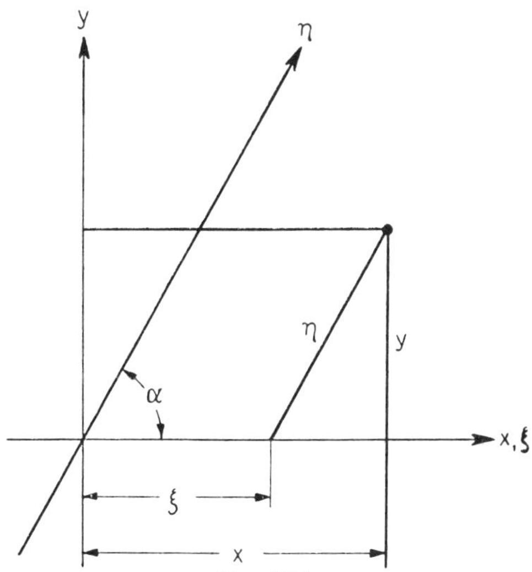

FIG. H 3.

† Ref. 4.14.

(1) Derive the following relations:

$$x_{,\xi} = 1, \quad x_{,\eta} = \cos\alpha, \qquad y_{,\xi} = 0, \quad y_{,\eta} = \sin\alpha,$$
$$\xi_{,x} = 1, \quad \xi_{,y} = -\cot\alpha, \quad \eta_{,x} = 0, \quad \eta_{,y} = \operatorname{cosec}\alpha,$$
$$g_{11} = 1, \quad g_{22} = 1, \quad g_{12} = g_{21} = \cos\alpha, \quad \sqrt{g} = \sin\alpha,$$
$$g^{11} = \operatorname{cosec}^2\alpha, \quad g^{22} = \operatorname{cosec}^2\alpha, \quad g^{12} = g^{21} = -\cos\alpha\operatorname{cosec}^2\alpha,$$
$$f_{11} = \varepsilon_x, \quad f_{22} = \varepsilon_x\cos^2\alpha + \varepsilon_y\sin^2\alpha + \gamma_{xy}\sin\alpha\cos\alpha,$$
$$2f_{12} = 2f_{21} = 2\varepsilon_x\cos\alpha + \gamma_{xy}\sin\alpha,$$
$$\tau^{11} = \sigma_x + \sigma_y\cot^2\alpha - 2\tau_{xy}\cot\alpha, \quad \tau^{22} = \sigma_y\operatorname{cosec}^2\alpha,$$
$$\tau^{12} = \tau^{21} = -\sigma_x\cot\alpha + \tau_{xy}\operatorname{cosec}\alpha.$$

(2) Denoting the displacement vector by

$$\mathbf{u} = u\mathbf{g}_1 + v\mathbf{g}_2,$$

where $\mathbf{g}_1$ and $\mathbf{g}_2$ are the unit vectors along the ξ- and η-axes, respectively, show that

$$f_{11} = (u + v\cos\alpha)_{,\xi}, \quad f_{22} = (u\cos\alpha + v)_{,\eta}$$
$$2f_{12} = (u + v\cos\alpha)_{,\eta} + (u\cos\alpha + v)_{,\xi}.$$

(3) Show that the condition of compatibility is given by

$$f_{11,\eta\eta} + f_{22,\xi\xi} - 2f_{12,\xi\eta} = 0.$$

(4) By use of the principle of virtual work, show that the equations of equilibrium are given by

$$\tau^{11}_{,\xi} + \tau^{21}_{,\eta} = 0, \quad \tau^{12}_{,\xi} + \tau^{22}_{,\eta} = 0.$$

(5) By use of the principle of virtual work combined with the condition of compatibility:

$$\iint [\tau^{\lambda\mu}\,\delta f_{\lambda\mu} - F(\delta f_{11,\eta\eta} + \delta f_{22,\xi\xi} - 2\delta f_{12,\xi\eta})]\sin\alpha\, d\xi\, d\eta + \cdots = 0,$$

show that the stress components are expressed in terms of the stress function F:

$$\tau^{11} = F_{,\eta\eta}, \quad \tau^{22} = F_{,\xi\xi}, \quad \tau^{12} = \tau^{21} = -F_{,\xi\eta}.$$

(6) Show that if the stress–strain relations are given in the (x, y) coordinate system as

$$\begin{bmatrix} \sigma_x \\ \sigma_y \\ \tau_{xy} \end{bmatrix} = \frac{E}{(1-\nu^2)} \begin{bmatrix} 1 & \nu & 0 \\ \nu & 1 & 0 \\ 0 & 0 & \dfrac{1-\nu}{2} \end{bmatrix} \begin{bmatrix} \varepsilon_x \\ \varepsilon_y \\ \gamma_{xy} \end{bmatrix}$$

or inversely

$$\begin{bmatrix} \varepsilon_x \\ \varepsilon_y \\ \gamma_{xy} \end{bmatrix} = \frac{1}{E} \begin{bmatrix} 1 & -\nu & 0 \\ -\nu & 1 & 0 \\ 0 & 0 & 2(1+\nu) \end{bmatrix} \begin{bmatrix} \sigma_x \\ \sigma_y \\ \tau_{xy} \end{bmatrix}$$

then from Eqs. (4.76) we have stress–strain relations in the (ξ, η) coordinate system as follows:

$$\begin{bmatrix} \tau^{11} \\ \tau^{22} \\ \tau^{12} \end{bmatrix} = \frac{E \operatorname{cosec}^4 \alpha}{(1-\nu^2)} \begin{bmatrix} 1 & \cos^2\alpha + \nu \sin^2\alpha & -\cos\alpha \\ \cos^2\alpha + \nu\sin^2\alpha & 1 & -\cos\alpha \\ -\cos\alpha & -\cos\alpha & \dfrac{1+\cos^2\alpha - \nu\sin^2\alpha}{2} \end{bmatrix} \begin{bmatrix} f_{11} \\ f_{22} \\ 2f_{12} \end{bmatrix}$$

or inversely

$$\begin{bmatrix} f_{11} \\ f_{22} \\ 2f_{12} \end{bmatrix} = \frac{1}{E} \begin{bmatrix} 1 & \cos^2\alpha - \nu\sin^2\alpha & 2\cos\alpha \\ \cos^2\alpha - \nu\sin^2\alpha & 1 & 2\cos\alpha \\ 2\cos\alpha & 2\cos\alpha & 2(1+\cos^2\alpha + \nu\sin^2\alpha) \end{bmatrix} \begin{bmatrix} \tau^{11} \\ \tau^{22} \\ \tau^{12} \end{bmatrix}$$

Problems Related to Orthogonal Curvilinear Coordinates

12. Derive the following relations:

$$\frac{\partial \mathbf{j}_1}{\partial \alpha^1} = -\frac{1}{\sqrt{g_{22}}}\frac{\partial\sqrt{g_{11}}}{\partial\alpha^2}\mathbf{j}_2 - \frac{1}{\sqrt{g_{33}}}\frac{\partial\sqrt{g_{11}}}{\partial\alpha^3}\mathbf{j}_3,$$

$$\frac{\partial \mathbf{j}_1}{\partial \alpha^2} = \frac{1}{\sqrt{g_{11}}}\frac{\partial\sqrt{g_{22}}}{\partial\alpha^1}\mathbf{j}_2, \quad \frac{\partial \mathbf{j}_1}{\partial \alpha^3} = \frac{1}{\sqrt{g_{11}}}\frac{\partial\sqrt{g_{33}}}{\partial\alpha^1}\mathbf{j}_3,$$

$$\frac{\partial \mathbf{j}_2}{\partial \alpha^1} = \frac{1}{\sqrt{g_{22}}}\frac{\partial\sqrt{g_{11}}}{\partial\alpha^2}\mathbf{j}_1, \quad \frac{\partial \mathbf{j}_2}{\partial \alpha^2} = -\frac{1}{\sqrt{g_{33}}}\frac{\partial\sqrt{g_{22}}}{\partial\alpha^3}\mathbf{j}_3 - \frac{1}{\sqrt{g_{11}}}\frac{\partial\sqrt{g_{22}}}{\partial\alpha^1}\mathbf{j}_1,$$

$$\frac{\partial \mathbf{j}_2}{\partial \alpha^3} = \frac{1}{\sqrt{g_{22}}}\frac{\partial\sqrt{g_{33}}}{\partial\alpha^2}\mathbf{j}_3,$$

$$\frac{\partial \mathbf{j}_3}{\partial \alpha^1} = \frac{1}{\sqrt{g_{33}}}\frac{\partial\sqrt{g_{11}}}{\partial\alpha^3}\mathbf{j}_1, \quad \frac{\partial \mathbf{j}_3}{\partial \alpha^2} = \frac{1}{\sqrt{g_{33}}}\frac{\partial\sqrt{g_{22}}}{\partial\alpha^3}\mathbf{j}_2,$$

$$\frac{\partial \mathbf{j}_3}{\partial \alpha^3} = -\frac{1}{\sqrt{g_{11}}}\frac{\partial\sqrt{g_{33}}}{\partial\alpha^1}\mathbf{j}_1 - \frac{1}{\sqrt{g_{22}}}\frac{\partial\sqrt{g_{33}}}{\partial\alpha^2}\mathbf{j}_2,$$

where $\mathbf{j}_i$; $i = 1, 2, 3$ are defined by Eq. (4.95). Write these relations for cylindrical and polar coordinate systems.

13. Show that for the small displacement theory expressed in the cylindrical coordinates r, θ and z ($x = r\cos\theta$, $y = r\sin\theta$, $z = z$), we have

$$\alpha^1 = r, \quad \alpha^2 = \theta, \quad \alpha^3 = z,$$

$$g_{11} = 1, \quad g_{22} = r^2, \quad g_{33} = 1,$$

$$\varepsilon_r = \frac{\partial u_r}{\partial r}, \quad \varepsilon_\theta = \frac{1}{r}\left(\frac{\partial u_\theta}{\partial\theta} + u_r\right), \quad \varepsilon_z = \frac{\partial u_z}{\partial z},$$

$$\gamma_{r\theta} = \frac{\partial u_\theta}{\partial r} + \frac{1}{r}\frac{\partial u_r}{\partial\theta} - \frac{u_\theta}{r},$$

$$\gamma_{rz} = \frac{\partial u_z}{\partial r} + \frac{\partial u_r}{\partial z}, \quad \gamma_{z\theta} = \frac{1}{r}\frac{\partial u_z}{\partial\theta} + \frac{\partial u_\theta}{\partial z},$$

$$\frac{\partial \sigma_r}{\partial r} + \frac{1}{r}\frac{\partial \tau_{r\theta}}{\partial \theta} + \frac{\partial \tau_{rz}}{\partial z} + \frac{1}{r}(\sigma_r - \sigma_\theta) + \bar{Y}_r = 0,$$

$$\frac{\partial \tau_{r\theta}}{\partial r} + \frac{1}{r}\frac{\partial \sigma_\theta}{\partial \theta} + \frac{\partial \tau_{\theta z}}{\partial z} + \frac{2}{r}\tau_{r\theta} + \bar{Y}_\theta = 0,$$

$$\frac{\partial \tau_{rz}}{\partial r} + \frac{1}{r}\frac{\partial \tau_{\theta z}}{\partial \theta} + \frac{\partial \sigma_z}{\partial z} + \frac{\tau_{rz}}{r} + \bar{Y}_z = 0.$$

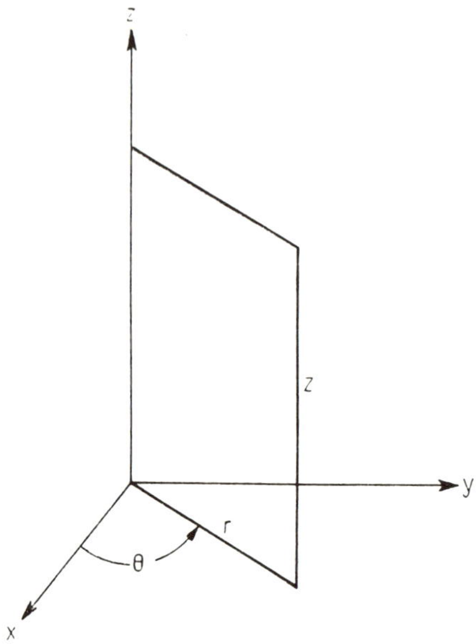

FIG. H 4.

14. Show that for the small displacement theory expressed in the polar coordinates r, θ and φ ($x = r \sin\varphi \cos\theta$, $y = r \sin\varphi \sin\theta$, $z = r\cos\varphi$), we have

$$\alpha^1 = r, \quad \alpha^2 = \theta, \quad \alpha^3 = \varphi,$$

$$g_{11} = 1, \quad g_{22} = (r\sin\varphi)^2, \quad g_{33} = r^2,$$

$$\varepsilon_r = \frac{\partial u_r}{\partial r},$$

$$\varepsilon_\theta = \frac{1}{r\sin\varphi}\frac{\partial u_\theta}{\partial \theta} + \frac{u_\varphi}{r}\cos\varphi + \frac{u_r}{r},$$

$$\varepsilon_\varphi = \frac{1}{r}\frac{\partial u_\varphi}{\partial \varphi} + \frac{u_r}{r},$$

$$\gamma_{\varphi\theta} = \frac{1}{r\sin\varphi}\frac{\partial u_\varphi}{\partial \theta} + \frac{1}{r}\frac{\partial u_\theta}{\partial \varphi} - \frac{u_\theta}{r}\cot\varphi.$$

$$\gamma_{r\varphi} = \frac{1}{r}\frac{\partial u_r}{\partial \varphi} - \frac{u_\varphi}{r} + \frac{\partial u_\varphi}{\partial r},$$

$$\gamma_{r\theta} = \frac{1}{r\sin\varphi}\frac{\partial u_r}{\partial \theta} - \frac{u_\theta}{r} + \frac{\partial u_\theta}{\partial r}.$$

$$\frac{\partial \sigma_r}{\partial r} + \frac{1}{r\sin\varphi}\frac{\partial \tau_{r\theta}}{\partial \theta} + \frac{1}{r}\frac{\partial \tau_{r\varphi}}{\partial \varphi} + \frac{2\sigma_r - \sigma_\theta - \sigma_\varphi + \tau_{r\varphi}\cot\varphi}{r} + \bar{Y}_r = 0,$$

$$\frac{\partial \tau_{r\theta}}{\partial r} + \frac{1}{r\sin\varphi}\frac{\partial \sigma_\theta}{\partial \theta} + \frac{1}{r}\frac{\partial \tau_{\varphi\theta}}{\partial \varphi} + \frac{3\tau_{r\theta} + 2\tau_{\varphi\theta}\cot\varphi}{r} + \bar{Y}_\theta = 0,$$

$$\frac{\partial \tau_{r\varphi}}{\partial r} + \frac{1}{r\sin\varphi}\frac{\partial \tau_{\theta\varphi}}{\partial \theta} + \frac{1}{r}\frac{\partial \sigma_\varphi}{\partial \varphi} + \frac{(\sigma_\varphi - \sigma_\theta)\cot\varphi + 3\tau_{r\varphi}}{r} + \bar{Y}_\varphi = 0.$$

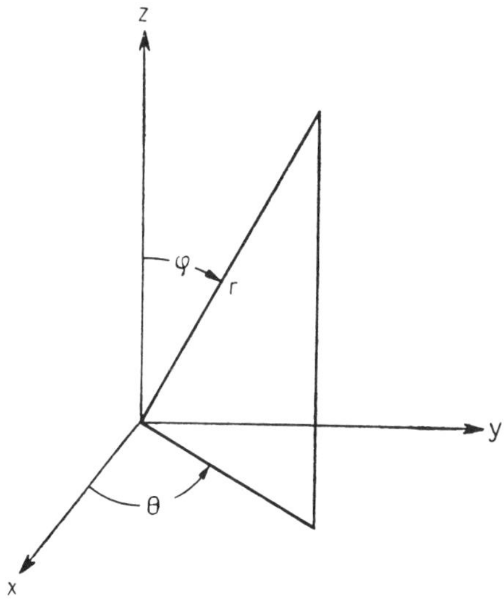

FIG. H 5.

CHAPTER 5

Problems Related to Sections 5.1 and 5.2

1. Show that the principle (5.5) can be expressed in a curvilinear coordinate system as follows:

$$\iiint_V (\tau^{(0)\lambda\mu} + \tau^{\lambda\mu})\, \delta f_{\lambda\mu} \sqrt{g}\, d\alpha^1\, d\alpha^2\, d\alpha^3 + \cdots = 0,$$

where $\tau^{(0)\lambda\mu}$ and $\tau^{\lambda\mu}$ are initial and incremental stresses referred to the curvilinear coordinate system, respectively, and Eq. (4.40) or (4.41) has been substituted.

2. Show that another expression for the functional of the principle of stationary potential energy is derived from Eq. (5.5) as

$$\Pi = \iiint_V [\sigma^{(0)\lambda\mu} e_{\lambda\mu} + A(e_{\lambda\mu}; \sigma^{(0)\lambda\mu}) - \bar{P}^{(0)\lambda} u^\lambda + \Phi(u^\lambda)]\, dV$$
$$+ \iint_{S_1} [-\bar{F}^{(0)\lambda} u^\lambda + \Psi(u^\lambda)]\, dS$$

for the initial stress problem, where $A(e_{\lambda\mu}; \sigma^{(0)\lambda\mu})$ is given by Eq. (5.10), and Eq. (5.6) has been substituted.

3. We have formulated stability problems in Sections 3.10, 3.11 and 5.2. Discuss relations between these formulations.

Problems Related to Sections 5.3 and 5.4

4. Show that if Eqs. (5.32) and (5.33) are given by

$$\sigma^{\lambda\mu} = \sigma^{\lambda\mu}(\epsilon_{\alpha\beta}),$$

and

$$\epsilon_{\lambda\mu} = \epsilon_{\lambda\mu}(\sigma^{\alpha\beta}),$$

respectively, we may have

$$dA = \sigma^{\lambda\mu}\, d\epsilon_{\lambda\mu}$$
$$dB = (e^{(0)}_{\lambda\mu} + \epsilon_{\lambda\mu})\, d\sigma^{\lambda\mu}$$

and consequently,

$$A = \int \sigma^{\lambda\mu}\, d\epsilon_{\lambda\mu}$$
$$B = e^{(0)}_{\lambda\mu}\sigma^{\lambda\mu} + \int \epsilon_{\lambda\mu}\, d\sigma^{\lambda\mu}$$

for the initial strain problem treated in Section 5.3. Compare these relations with Eqs. (5.43) through (5.53).

5. Show that if we confine the initial strain problem to the small displacement theory, we can prove that the actual solution is given by the minimum property of the total potential energy as well as total complementary energy.

6. We consider a thermal stress problem of an isotropic elastic body in the small displacement theory. Show that the functional of the principle of minimum potential energy is given by

$$\Pi = \iiint_V \left[A(u, v, w) - \Theta\left(\frac{\partial u}{\partial x} + \frac{\partial v}{\partial y} + \frac{\partial w}{\partial z}\right)\right] dx\, dy\, dz$$

for a body with free boundary surface, where $\Theta = E\alpha\theta/(1 - 2\nu)$. Show also that by the use of Green's theorem, the above equation is transformed into

$$\Pi = \iiint_V \left[A(u, v, w) + \left(\frac{\partial \Theta}{\partial x} u + \frac{\partial \Theta}{\partial y} v + \frac{\partial \Theta}{\partial z} \right) \right] dx\, dy\, dz - \iint_S (\Theta l u + \Theta m v + \Theta n w)\, dS$$

which indicates that the problem is equivalent to that of an elastic body under the body forces $(-\partial\Theta/\partial x,\ -\partial\Theta/\partial y,\ -\partial\Theta/\partial z)$ and hydraulic pressure $-\Theta$ distributed over the whole surface of the body.

Problems Related to Section 5.7

7. We denote the direction cosines between two rectangular Cartesian coordinate system (x^1, x^2, x^3) and (X^1, X^2, X^3) by

	X^1	X^2	X^3
x^1	l_1	m_1	n_1
x^2	l_2	m_2	n_2
x^3	l_3	m_3	n_3

and define the direction cosine matrix $[L]$ by

$$[L] = \begin{bmatrix} l_1 & m_1 & n_1 \\ l_2 & m_2 & n_2 \\ l_3 & m_3 & n_3 \end{bmatrix}.$$

Show that if the (x^1, x^2, x^3) system is rotated around the x^1-, x^2-, and x^3-axes by the angle of ϕ, θ and ψ, respectively, the direction cosine matrix of the new (x^1, x^2, x^3) system is given by

$$[\varrho_1(\phi)]\,[L], \quad [\varrho_2(\theta)]\,[L], \quad [\varrho_3(\psi)]\,[L],$$

respectively, where

$$[\varrho_1(\phi)] = \begin{bmatrix} 1 & 0 & 0 \\ 0 & \cos\phi & \sin\phi \\ 0 & -\sin\phi & \cos\phi \end{bmatrix}, \quad [\varrho_2(\theta)] = \begin{bmatrix} \cos\theta & 0 & -\sin\theta \\ 0 & 1 & 0 \\ \sin\theta & 0 & \cos\theta \end{bmatrix},$$

$$[\varrho_3(\psi)] = \begin{bmatrix} \cos\psi & \sin\psi & 0 \\ -\sin\psi & \cos\psi & 0 \\ 0 & 0 & 1 \end{bmatrix}.$$

Show also that Eq. (5.102) is obtained from the following matrix multiplication:

$$[\varrho_1(\phi)]\,[\varrho_2(\theta)]\,[\varrho_3(\psi)].$$

8. We have chosen the vector $\delta\mathbf{r}_G$ and three sclars $\delta\phi$, $\delta\theta$ and $\delta\psi$ as independent quantities in order to derive Eqs. (5.113) and (5.114) from Eq. (5.112). Show that the same equations may be obtained by resolving the vector $\mathbf{r}$ either as

$$\mathbf{r}_G = x_G\mathbf{i}_1 + y_G\,\mathbf{i}_2 + z_G\,\mathbf{i}_3$$

or as

$$\mathbf{r}_G = \xi\,\mathbf{k}_1 + \eta\,\mathbf{k}_2 + \zeta\,\mathbf{k}_3,$$

treating $(x_G, y_G, z_G, \phi, \theta, \psi)$ or $(\xi, \eta, \xi, \phi, \theta, \psi)$ respectively as a system of generalized coordinates.

CHAPTER 6

Problems Related to Section 6.1

1. Show that if the equations

$$u = -\theta yz, \quad v = \theta xz, \quad w = \theta\varphi(x, y) \tag{i}$$

are the solution of the Saint-Venant torsion problem, then a family of displacements

$$\begin{aligned} u &= -\theta z(y - y_0), \quad v = \theta z(x - x_0), \\ w &= \theta\varphi(x, y) - \theta(y_0 x - x_0 y) - \theta\varphi_0 \end{aligned} \tag{ii}$$

(where x_0, y_0 and φ_0 are arbitrary constants) are also the solution of the torsion problem, and that as far as the Saint-Venant torsion problem is concerned, the center of twist remains undetermined.

2. Show that

$$J = I_p - D$$

and consequently

$$J \leqq I_p,$$

where J is defined by Eq. (6.20), I_p is the polar moment of inertia, i.e. $I_p = \iint_S (x^2 + y^2)\,dx\,dy$, and $D = \iint_S [(\varphi_{,x})^2 + (\varphi_{,y})^2]\,dx\,dy$.

3. We consider a doubly-symmetric cross-section. The x- and y-axes are taken to coincide with the principal axes through the centroid of the cross-section, and the z-axis is taken as the axis of rotation. An additive constant of the Saint-Venant warping function is so determined that $\iint_S \varphi\,dx\,dy = 0$ (see equations (ii) of Problem 1). Then, show that the warping function thus determined has the following property:

$$\varphi(x, y) = -\varphi(-x, y) = -\varphi(x, -y) = \varphi(-x, -y).$$

4. We consider an approximate determination of the Saint-Venant warping function of a thin-walled open section as shown in Fig. H 6. The middle line of the wall is denoted by C. A coordinate s is taken along C and is measured from one end of the middle line. Two unit vectors **t** and **n**, are taken to be tangential to and normal to the middle line, respectively,

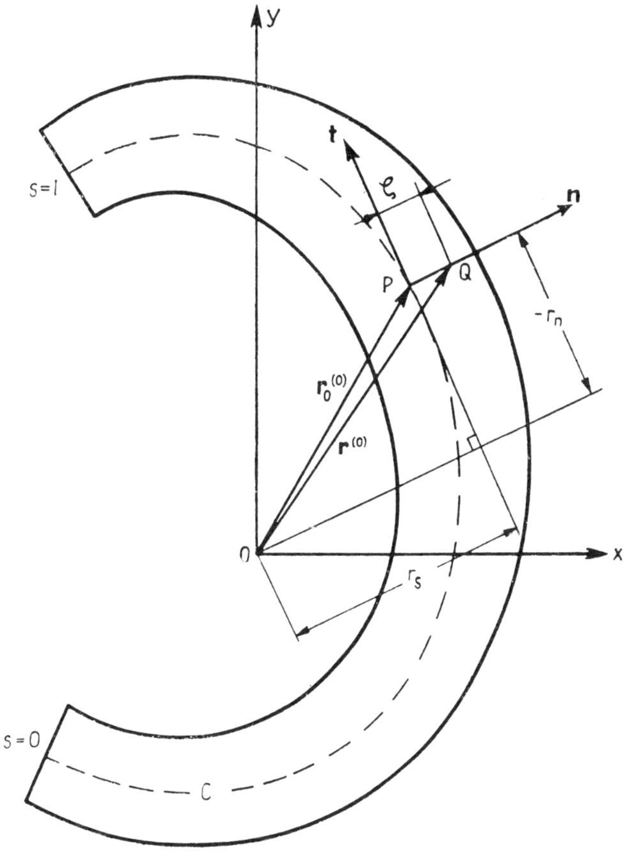

FIG. H 6.

so that the three unit vectors **n**, **t** and $\mathbf{i}_3$ constitute a right-handed system. Denoting the position vector of an arbitrary point P on C by $\mathbf{r}_0^{(0)}$, and that of an arbitrary point Q on the normal drawn at P by $\mathbf{r}^{(0)}$, we may write

$$\mathbf{r}^{(0)} = \mathbf{r}_0^{(0)} + \zeta \mathbf{n}, \tag{i}$$

where ζ is measured from the middle line. The equation (i) suggests that a set of parameters (s, ζ) may be taken as a curvilinear coordinate system defining the section. Denoting the shearing stress in the direction of the tangent on the middle line by τ_t, and the shearing stress in the direction of the normal by τ_n, and using the relations

$$\tau_{xz} = G\theta\left(\frac{\partial\varphi}{\partial x} - y\right), \quad \tau_{yz} = G\theta\left(\frac{\partial\varphi}{\partial y} + x\right), \tag{ii}$$

we have

$$\int_0^s \tau_t\, ds = G\theta \int_0^s \left(\frac{\partial\varphi}{\partial x}\, dx + \frac{\partial\varphi}{\partial y}\, dy\right) + G\theta \int_0^s r_s\, ds \tag{iii}$$

along the middle line from $s = 0$ to P, and

$$\int_0^\zeta \tau_n\, d\zeta = G\theta \int_0^\zeta \left(\frac{\partial\varphi}{\partial x}\, dx + \frac{\partial\varphi}{\partial y}\, dy\right) + G\theta \int_0^\zeta r_n\, d\zeta \tag{iv}$$

along the normal from P to Q, where

$$r_s = \mathbf{r}_0^{(0)} \cdot \mathbf{n}, \quad r_n = -\mathbf{r}_0^{(0)} \cdot \mathbf{t}. \qquad \text{(v), (vi)}$$

The geometrical interpretation of r_s and r_n which belong to the point P is shown in Fig. H 6. With these preliminaries, show that since τ_t and τ_n may be taken approximately equal to zero in the thin-walled open section, we have from equations (iii) and (iv) the value of φ at Q as follows:

$$\varphi = -\int_0^s r_s\, ds - \int_0^\zeta r_n\, d\zeta + \varphi_0, \tag{vii}$$

where φ_0 is an arbitrary constant. Equation (vii) determines the Saint-Venant warping function of the section. Consider also the shearing stress distribution of the thin-walled open section due to the Saint-Venant torsion. Note: see Ref. 6.2, pp. 272–5, and Refs. 6.7, 6.8, 6.19.

Problem Related to Section 6.2

5. Show that Eq. (6.32) can be derived from Eqs. (6.7), by eliminating w, and then expressing γ_{xz} and γ_{yz} in terms of ϕ by use of Eqs. (6.8) and (6.27).

Problems Related to Section 6.3

6. By use of the relation:

$$M = \iint_S (\tau_{yz}x - \tau_{xz}y)\, dx\, dy = -\iint_S \left(\frac{\partial\phi}{\partial x}\, x + \frac{\partial\phi}{\partial y}\, y\right) dx\, dy$$

show that we have

$$M = 2 \iint_S \phi \, dx \, dy$$

for a simply-connected cross-section, and

$$M = 2 \iint_S \phi \, dx \, dy + 2 \sum_{k=1}^{n} c_k A_k$$

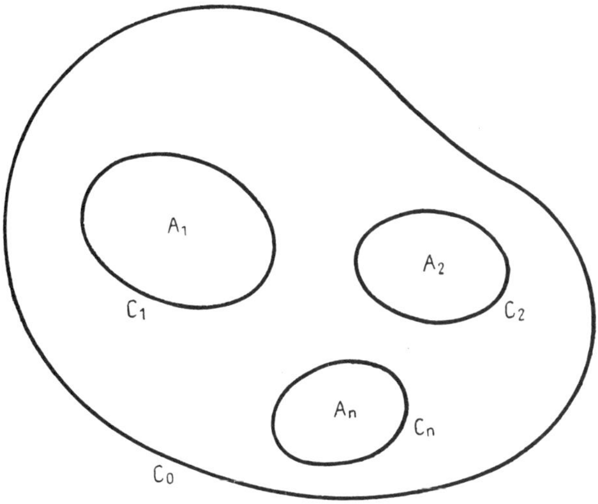

FIG. H 7.

for a multiply-connected cross-section consisting of an exterior boundary C_0 and interior boundaries $C_1, C_2, \ldots, C_n$, where c_k is the value of ϕ on the boundary C_k, and A_k is the area enclosed by the curve C_k. The value of ϕ on the exterior boundary C_0 is taken equal to zero.

7. Show that for a thin-walled closed section as shown in Fig. H 8, the shearing stress τ and the torsional rigidity GJ are given by

$$\tau = \frac{\bar{M}}{2A_0 t}, \tag{i}$$

and

$$GJ = \frac{4A_0^2 G}{\oint_C \frac{ds}{t}}, \tag{ii}$$

respectively, where A_0 is the area enclosed by the curve C (which is the mean of the outer and inner boundaries), s is measured along C, $t(s)$ is the thickness of the wall, and $\oint_C$ is the integral along the closed path C. Note: see Ref. 6.2, pp. 298–9.

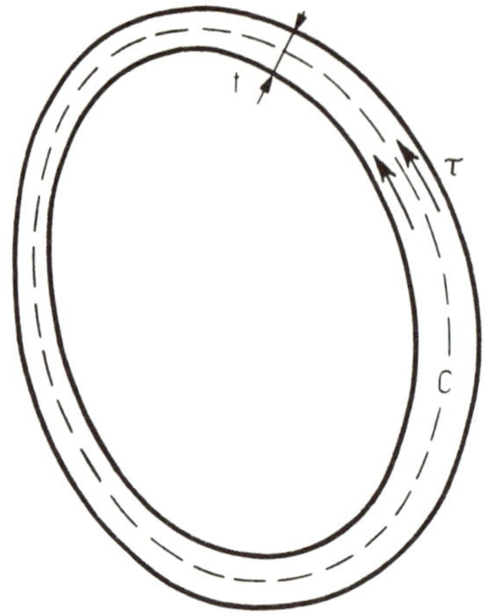

FIG. H 8.

8. We consider an approximate determination of the Saint-Venant warping function of the thin-walled closed section as shown in Fig. H 8. By use of equations (iii) and (iv) of Problem 4 plus equations (i) and (ii) of Problem 7, show that we have

$$\varphi = \frac{2A_0}{\oint_C \frac{ds}{t}} \int_0^s \frac{ds}{t} - \int_0^s r_s\, ds - \int_0^\zeta r_n\, d\zeta + \varphi_0 \tag{i}$$

which determines the Saint-Venant warping function of the section, where φ_0 is an arbitrary constant.

9. Consider two thin-walled circular cross-sections, of which one is closed and the other is open, as shown in Fig. H 9. Show that the torsional rigidities are given by

$$GJ = 2\pi a^3 tG \quad \text{for closed section}$$

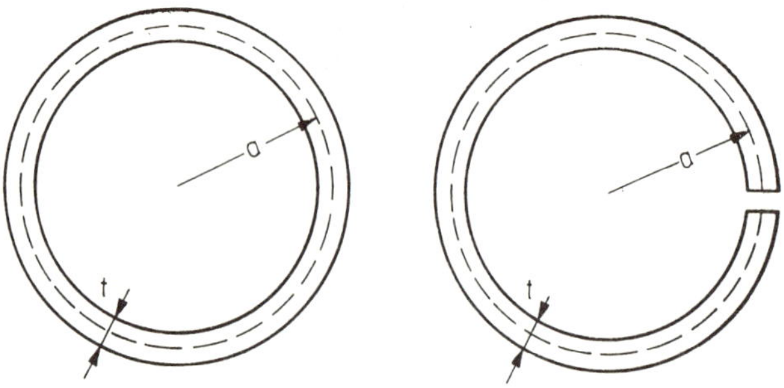

FIG. H 9.

and

$$GJ = \tfrac{2}{3}\pi a t^3 G \quad \text{for open section.}$$

Calculate the ratio $(GJ)_{closed}/(GJ)_{open}$ for $a/t = 10$, and discuss why the torsional rigidity of the open section is so drastically lower than that of the closed section. Note: see Ref. 6.2, pp. 272–5 and pp. 298–9.

10. Show that the Saint-Venant torsion problem of a thin-walled section with an inner wall as shown in Fig. H 10 can be solved by determining

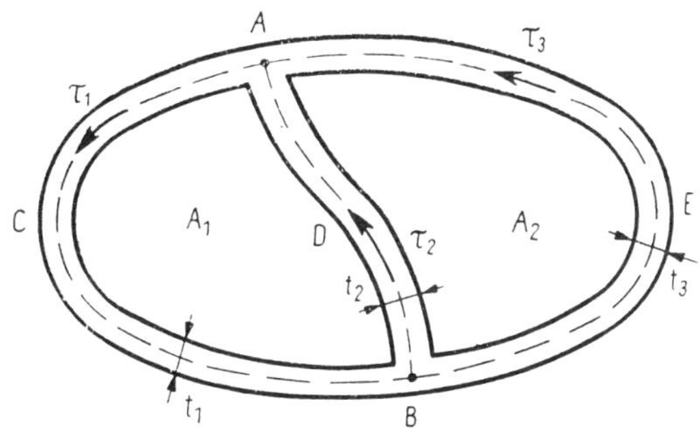

FIG. H 10.

the shearing stress τ_1, τ_2, τ_3 and the twist angle θ from the following equations:

$$\begin{aligned} t_1\tau_1 - t_2\tau_2 - t_3\tau_3 &= 0, \\ 2A_1t_1\tau_1 + 2A_2t_3\tau_3 &= \bar{M}, \\ \tau_1 s_1 + \tau_2 s_2 &= 2G\theta A_1, \\ \tau_3 s_3 - \tau_2 s_2 &= 2G\theta A_2, \end{aligned}$$

where the thickness t_1, t_2, t_3 are assumed constant along *ACB*, *ADB*, *AEB*, respectively. A_1 and A_2 are the areas enclosed by the closed curves *ACBD* and *ADBE*, respectively, and s_1, s_2, s_3 are the length of the curves *ACB*, *ADB*, *AEB* respectively. Note: see Ref. 6.2, pp. 301–2.

Problems Related to Section 6.5

11. Show that for a multiply-connected cross-section consisting of an exterior boundary C_0 and interior boundaries $C_1, C_2, \ldots, C_n$, the bounds formulae for the torsional rigidity can be formulated in a manner similar to those developed in Section 6.5, by replacing Eqs. (6.72) and (6.73) with

$$\bar{M} = -\iint \left(\frac{\partial \phi^*}{\partial x} x + \frac{\partial \phi^*}{\partial y} y \right) dx\, dy$$

and

$$\phi^* = 0 \quad \text{on} \quad C_0$$
$$\phi^* = c_k \quad \text{on} \quad C_k; \quad k = 1, 2, \ldots, n,$$

respectively, where c_k is some constant.

12. Consider a hollow square section as shown in Fig. H 11. Remembering the symmetric property of ϕ and w, we consider only the region $ABCD$

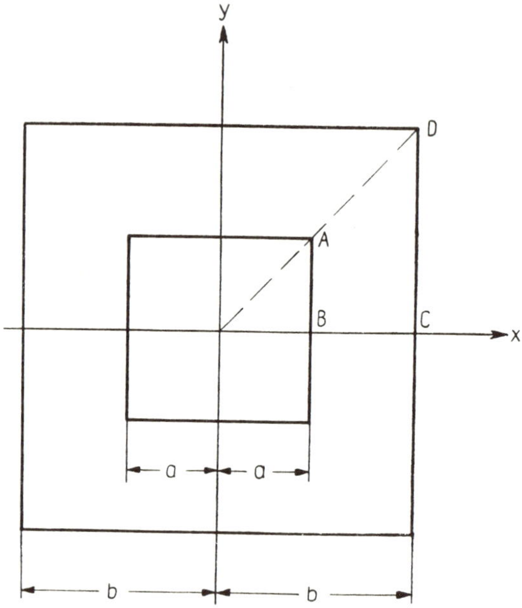

FIG. H 11.

and choose

$$\phi^* = a_1\phi_1(x, y) + a_2\phi_2(x, y)$$
$$\phi_1(x, y) = b(x - b)$$
$$\phi_2(x, y) = (x - b)^2$$

and

$$w^{**} = b_1 w_1(x, y)$$
$$w_1(x, y) = x^3y - xy^3 .$$

Then show that we have the following bounds for the torsional rigidity:

$$2(b^4 - a^4) \leqq J \leqq \frac{8}{3}(b^4 - a^4) - \frac{56}{135}\frac{(b^6 - a^6)^2}{b^8 - a^8}.$$

Note: see Ref. 6.14.

Problems Related to Non-uniform Torsion

13. We consider a torsion problem of a bar which is clamped at one end $z = 0$ and is subjected to a twisting moment $\bar{M}$ at the other end ($z = l$) as shown in Fig. H 12. The bar is assumed to have doubly symmetric

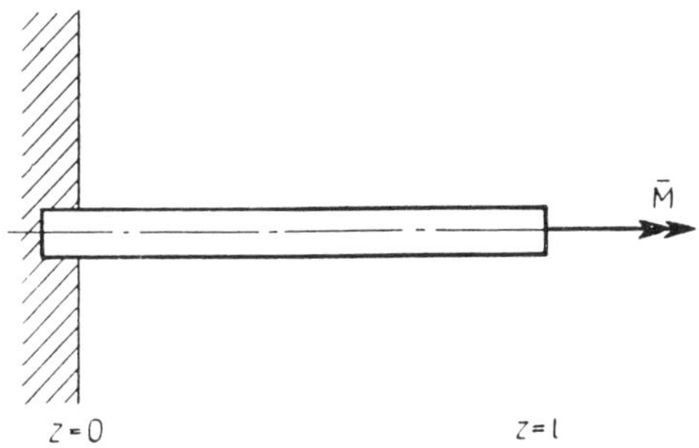

FIG. H 12.

cross-section. Following Reissner's papers (Ref. 6.4), and using the principle of virtual work or the principle of minimum potential energy, derive the following relations:

(1) Assuming

$$u = -\vartheta(z)\,y, \quad v = \vartheta(z)\,x, \quad w = \vartheta'(z)\,\varphi(x, y) \tag{i}$$

show that the governing equation and boundary conditions for $\vartheta(z)$ are given by

$$GJ\vartheta' - E\Gamma\vartheta''' = \bar{M}, \tag{ii}$$

and

$$\vartheta(0) = \vartheta'(0) = \vartheta''(l) = 0, \tag{iii}$$

respectively, and the strain energy stored in the bar is given by

$$\tfrac{1}{2}\int_0^l [GJ(\vartheta')^2 + E\Gamma(\vartheta'')^2]\,dz, \tag{iv}$$

where $\varphi(x, y)$ is the Saint-Venant warping function of the cross-section, $(\ )' = d(\ )/dz$ and

$$\Gamma = \iint \varphi^2\,dx\,dy. \tag{v}$$

The function $\varphi(x, y)$ and the x- and y-axes are chosen as in Problem 3. Show also that the present formulation would not close to the exact solution around $z = 0$, since equations (i) and (iii) combined with the stress–strain relations provide $\tau_{xz} = \tau_{yz} = 0$ at $z = 0$.

(2) Assuming

$$u = -\vartheta(z)\,y, \quad v = \vartheta(z)\,x, \quad w = \alpha(z)\,\varphi(x, y), \tag{vi}$$

show that the governing equations and boundary conditions for ϑ and α are given by

$$\begin{aligned} GJ\vartheta' - GD(\alpha - \vartheta') &= \bar{M}, \\ E\Gamma\alpha'' - GD(\alpha - \vartheta') &= 0, \end{aligned} \tag{vii}$$

and

$$\vartheta(0) = \alpha(0) = \alpha'(l) = 0, \tag{viii}$$

respectively, and the strain energy stored in the bar is given by

$$\tfrac{1}{2}\int_0^l [GJ(\vartheta')^2 + GD(\alpha - \vartheta')^2 + E\Gamma(\alpha')^2]\,dz, \tag{ix}$$

where

$$D = \iint [(\varphi_{,x})^2 + (\varphi_{,y})^2]\,dx\,dy.$$

Show also that the present formulation provides an approximate solution only.

14. We consider a torsional buckling problem of a bar which is clamped at one end ($z = 0$), and is subjected to a critical axial load P_{cr} at the other end ($z = l$) as shown in Fig. H 13. It is assumed that the bar has doubly symmetric cross-section, and the force P_{cr} changes neither its magnitude nor its direction while the buckling occurs.

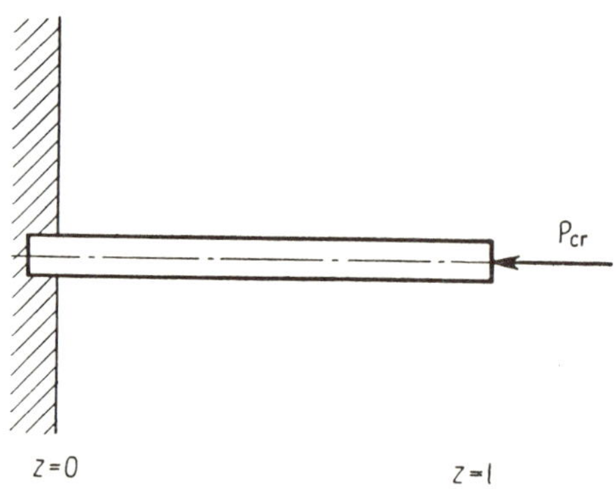

FIG. H 13.

(1) We assume that displacement components are given by

$$\begin{aligned} u &= -x(1 - \cos\vartheta) - y\sin\vartheta, \\ v &= x\sin\vartheta - y(1 - \cos\vartheta), \\ w &= \vartheta'\varphi(x, y), \end{aligned} \tag{i}$$

where u, v, w are measured just prior to the occurrence of the buckling, $\varphi(x, y)$ is the Saint-Venant warping function of the cross-section, ϑ is a function of z only, and $(\ )' = d(\ )/dz$. The function $\varphi(x, y)$ and the x-and y-axes are chosen as in Problem 3. By use of Eq. (5.5) and neglecting terms of higher order, show that the governing equation is finally reduced to

$$GJ\vartheta'' - E\Gamma\vartheta'''' - P_{cr}i_p^2\vartheta'' = 0, \tag{ii}$$

and the boundary conditions to

$$\vartheta = \vartheta' = 0 \quad \text{at} \quad z = 0$$

and

$$E\Gamma\vartheta'' = 0, \quad GJ\vartheta' - E\Gamma\vartheta''' - P_{cr}i_p^2\vartheta' = 0 \quad \text{at} \quad z = l, \tag{iii}$$

where $\Gamma = \int\int \varphi^2\, dx\, dy$, $I_p = \int\int (x^2 + y^2)\, dx\, dy$, $A_0 = \int\int dx\, dy$, $i_p^2 = I_p/A_0$. Note: The strain–displacement relations to be used in the above formulation are

$$e_{zz} = \vartheta''\varphi + \tfrac{1}{2}(x^2 + y^2)(\vartheta')^2, \quad \varepsilon_z = \vartheta''\varphi,$$
$$\gamma_{xz} = \vartheta'(\varphi_{,x} - y), \quad \gamma_{yz} = \vartheta'(\varphi_{,y} + x),$$

where the term $\frac{1}{2}(\vartheta''\varphi)^2$ is neglected in the expression of e_{zz} due to its negligible contribution to the final result. See Refs. 6.7, 6.8 and 6.19.

(2) Next, we assume that displacement components are given by

$$\begin{aligned} u &= -x(1 - \cos\vartheta) - y\sin\vartheta, \\ v &= x\sin\vartheta - y(1 - \cos\vartheta), \\ w &= \alpha\varphi(x, y), \end{aligned} \tag{iv}$$

where ϑ and α are functions of z only. Show that we have the governing equations and boundary conditions as follows:

$$\begin{aligned} GJ\vartheta'' - GD(\alpha' - \vartheta'') - P_{cr}i_p^2\vartheta'' &= 0, \\ E\Gamma\alpha'' - GD(\alpha - \vartheta') &= 0 \end{aligned} \tag{v}$$

and

$$\begin{aligned} &\alpha = \vartheta = 0 \quad \text{at} \quad z = 0 \\ &\alpha' = 0,\ GJ\vartheta' - GD(\alpha - \vartheta') - P_{cr}i_p^2\vartheta' = 0 \quad \text{at} \quad z = l. \end{aligned} \tag{vi}$$

•

CHAPTER 7

Problems Related to Section 7.4

1. We consider free lateral vibration of a beam clamped at one end ($x = 0$) and supported at the other end ($x = l$) with a spring of stiffness k. Show that the functional for the principle of stationary potential energy of this problem is given by

$$\Pi = \frac{1}{2}\int_0^l EI(w'')^2\,dx + \frac{1}{2}k[w(l)]^2 - \frac{1}{2}\omega^2\int_0^l mw^2\,dx$$

with subsidiary conditions $w(0) = w'(0) = 0$, and derive the governing equation and boundary conditions. Show also that the Rayleigh quotient for this problem is given by

$$\omega^2 = \frac{\frac{1}{2}\int_0^l EI(w'')^2\,dx + \frac{1}{2}k\,[w(l)]^2}{\frac{1}{2}\int_0^l mw^2\,dx}.$$

2. We consider lateral vibration of a beam with n constraint conditions:

$$\int_0^l mw\phi_i\,dx = 0 \quad (i = 1, 2, 3, \ldots, n), \tag{i}$$

where $\phi_i(x)$; $i = 1, 2, \ldots, n$ are prescribed functions. Show that the functional for the principle of stationary potential energy of this problem is given by

$$\Pi = \frac{1}{2}\int_0^l EI\,(w'')^2\,dx - \frac{1}{2}\omega^2\int_0^l mw^2\,dx + \sum_{i=1}^n \mu_i\int_0^l mw\phi_i\,dx, \tag{ii}$$

where μ_i; $i = 1, 2, \ldots, n$ are Lagrange multipliers, and derive the stationary conditions of the functional (ii). Show also that if a constraint is given by

$$w(a) = 0, \quad 0 < a < l, \tag{iii}$$

we have

$$\Pi = \frac{1}{2}\int_0^l EI\,(w'')^2\,dx - \frac{1}{2}\omega^2\int_0^l mw^2\,dx + \mu w(a), \tag{iv}$$

where μ is a Lagrange multiplier, and derive the stationary conditions of the functional (iv).

3. We consider free lateral vibration of a cantilever beam which is rotating with constant angular velocity Ω as shown in Fig. H 14. Show that the functional for the principle of stationary potential energy of this problem is given by

$$\Pi = \frac{1}{2}\int_0^l EI(w'')^2\,dx + \frac{1}{2}\int_0^l \sigma_x^{(0)}\,A_0(w')^2\,dx - \frac{1}{2}\omega^2\int_0^l mw^2\,dx,$$

with subsidiary conditions $w(0) = w'(0) = 0$, where $\sigma_x^{(0)}$ is the initial stress caused in the beam by the centrifugal force:

$$\sigma_x^{(0)} A_0 = \int_x^l m\Omega^2 x \, dx$$

and A_0 is the area of the cross-section. Show also that the governing equation and boundary conditions are obtainable from the principle as follows:

$$(EIw'')'' - (\sigma_x^{(0)} A_0 w')' - m\omega^2 w = 0,$$

and

$$w = w' = 0 \quad \text{at} \quad x = 0,$$

$$EIw'' = (EIw'')' = 0 \quad \text{at} \quad x = l.$$

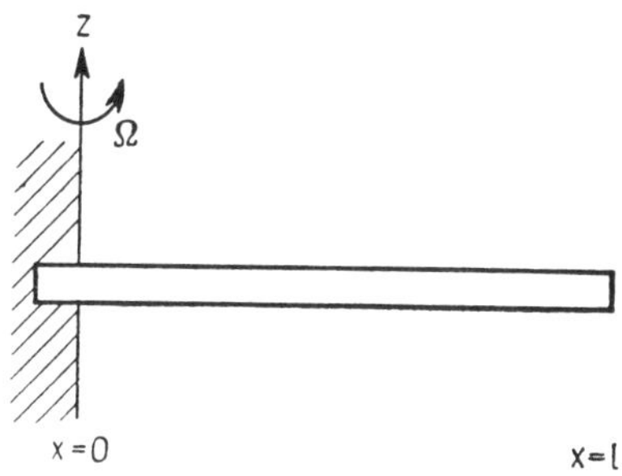

FIG. H 14.

Problems Related to Section 7.5

4. Using Eqs. (3.19), (7.11) and (7.12), show that the strain e_{xx} of a beam in the finite displacement theory based on the Bernoulli–Euler hypothesis is given by

$$e_{xx} = u' + \tfrac{1}{2}[(u')^2 + (w')^2] - z(1 + \varepsilon_0)\,\theta' + \tfrac{1}{2} z^2(\theta')^2, \tag{i}$$

where $1 + \varepsilon_0 = \sqrt{(1 + u')^2 + (w')^2}$ and $\mathbf{n} = -\sin\theta \mathbf{i}_1 + \cos\theta \mathbf{i}_3$.

5. Using the principle of virtual work and equation (i) of Problem 4, show that equilibrium equations of the beam in the finite displacement theory are given by

$$\left\{N_x(1 + u') - M_x\theta' \cos\theta - \frac{\sin\theta}{1 + \varepsilon_0}[M_x(1 + \varepsilon_0) - M_{xx}\theta']'\right\}' + \bar{X} = 0,$$

$$\left\{N_x w' - M_x\theta' \sin\theta + \frac{\cos\theta}{1 + \varepsilon_0}[M_x(1 + \varepsilon_0) - M_{xx}\theta']'\right\}' + \bar{Z} = 0,$$

where $N_x = \iint \sigma_x \, dy\, dz$, $M_x = \iint \sigma_x z \, dy \, dz$, $M_{xx} = \iint \sigma_x z^2 \, dy \, dz$, and where $\bar{X}$ and $\bar{Z}$ are the external loads per unit length of the undeformed centroid locus in the directions of the x- and z-axes, respectively. Next, show that the same equations are obtainable from a consideration of equilibrium conditions of a beam element. Note: The internal force normal to the cross-section of the beam is $\sigma_x(\partial \mathbf{r}/\partial x)$ per unit undeformed area.

6. Show that if the centroid locus is assumed inextensional, namely, $\varepsilon_0 = 0$, and the term containing z^2 is neglected, equation (i) of Problem 4 reduces to

$$e_{xx} = -\frac{w''}{\sqrt{1-(w')^2}}\, z.$$

Show also that using the above equation and the principle of virtual work, we can derive the beam equation known as Euler's elastica. Note: see Ref. 3.1, pp. 347–51 and Ref. 3.21, pp. 183–6.

Problems Related to Section 7.6

7. We consider a beam shown in Fig. 7.6 and find that the total potential energy of the system is given by

$$\Pi = \frac{1}{2}\int_0^l \{EI\,[u' + \tfrac{1}{2}(w')^2]^2 + EI(w'')^2\}\, dx + P_{cr}\, u(l)$$

for post buckling configuration, where u and w are measured from the undeformed state. Applying the results of Section 3.10, especially Eq. (3.85), to the present problem, derive the governing equation and boundary conditions for the buckling and compare them with those obtained in Section 7.6. Note: see Ref. 3.1, pp. 358–60.

8. A cantilever beam is executing a small disturbed motion under a follower force $\bar{P}$ as shown in Fig. H 15, where $\theta = w'(l)$ and α is a specified

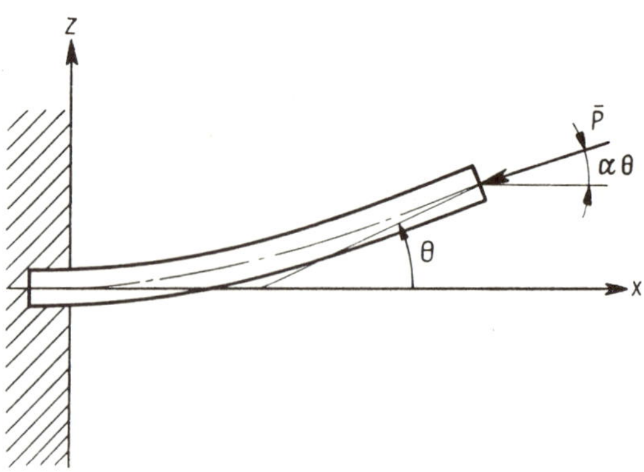

FIG. H 15.

constant. Show, by use of the principle of virtual work, that the equation of motion and boundary conditions are

$$(EIw'')'' + \bar{P}w'' + m\ddot{w} = 0,$$

and

$$w = 0, \quad w' = 0 \quad \text{at} \quad x = 0$$
$$EIw'' = 0, \quad (EIw'')' + \bar{P}(1 - \alpha)\, w' = 0 \quad \text{at} \quad x = l$$

respectively, where a dot denotes differentiation with respect to time. Discuss also whether or not variational principles can be formulated for this problem. Note: see Ref. 3.23.

9. Show that if the effect of transverse shear deformation is taken into account, the functional (7.87) is to be replaced by

$$\Pi = \tfrac{1}{2}\int_0^l [EI\,(u_1')^2 + GkA_0\,(w' + u_1)^2]\, dx - \tfrac{1}{2} P_{cr} \int_0^l (w')^2\, dx. \qquad \text{(i)}$$

where u_1 and w are defined in Section 7.7. Show also that by use of functional (i), we obtain the governing equations and boundary conditions of the problem treated in Section 7.6 as follows:

$$\left.\begin{aligned} [EIu_1']' - GkA_0(w' + u_1) &= 0, \\ [GkA_0(w' + u_1)]' - P_{cr}w'' &= 0 \end{aligned}\right\} \qquad \text{(ii)}$$

and

$$\left.\begin{aligned} u_1 = 0, \quad w = 0 \quad \text{at} \quad x = 0; \\ EIu_1' = 0, \quad w = 0 \quad \text{at} \quad x = l. \end{aligned}\right\} \qquad \text{(iii)}$$

Problems Related to Coupling of Bending and Torsion†

10. Following Trefftz (Refs. 7.3 and 7.4), show that the point (y_s, z_s) defined by

$$y_s = -\frac{1}{I_y}\iint z\varphi\, dy\, dz, \qquad z_s = \frac{1}{I_z}\iint y\varphi\, dy\, dz \qquad \text{(i)}$$

coincides with the center of shear and center of twist of the cross-section of a uniform bar, where the y- and z-axes are taken to coincide with the principal axes through the centroid, $I_y = \iint z^2\, dy\, dz$ and $I_z = \iint y^2\, dy\, dz$. The function $\varphi(y, z)$ is the Saint-Venant warping function of the cross-section with the x-axis as the axis of rotation and is chosen so that $\iint \varphi\, dy\, dz = 0$. Show also that if $\varphi_s(y, z)$ is the Saint-Venant warping function with the locus of the point (y_s, z_s) as the axis of rotation and is so chosen that $\iint \varphi_s(y, z)\, dy\, dz = 0$, we have

$$\varphi_s(y, z) = \varphi(y, z) - z_s y + y_s z,$$
$$\Gamma_s = \Gamma - y_s^2 I_y - z_s^2 I_z,$$

where $\Gamma_s = \iint \varphi_s^2(y, z)\, dy\, dz$ and $\Gamma = \iint \varphi^2(y, z)\, dy\, dz$.

† Beams are assumed to have uniform cross-section along the x-axis in Problems 10 through 15.

11. We can calculate the point (y_s, z_s) of a thin-walled open section by the combined use of equation (vii) of Problem 4 of Chapter 6 and equation (i) of Problem 10 of Chapter 7 as follows:

$$y_s = \frac{1}{I_y} \int_0^l \left(\int_0^s r_s \, ds \right) zt \, ds,$$

$$z_s = -\frac{1}{I_z} \int_0^l \left(\int_0^s r_s \, ds \right) yt \, ds,$$

where the term $-\int_0^\zeta r_n \, d\zeta$ has been neglected due to its small contribution. Show that the point (y_s, z_s) thus obtained is in coincidence with the center of shear derived from the shearing stress distribution due to torsion-free bending. Note: see Ref. 7.32, p. 210 for the shearing stress distribution due to torsion-free bending and the center of shear of a thin-walled open section.

12. We can calculate the point (y_s, z_s) of the thin-walled closed section shown in Fig. H 8 by the combined use of equation (i) of Problem 8 of Chapter 6 and equation (i) of Problem 10 of Chapter 7 as follows:

$$y_s = -\frac{2A_0}{I_y} \frac{\oint_C \left(\int_0^s \frac{ds}{t} \right) zt \, ds}{\oint_C \frac{ds}{t}} + \frac{1}{I_y} \oint_C \left(\int_0^s r_s \, ds \right) zt \, ds,$$

$$z_s = \frac{2A_0}{I_z} \frac{\oint_C \left(\int_0^s \frac{ds}{t} \right) yt \, ds}{\oint_C \frac{ds}{t}} - \frac{1}{I_z} \oint_C \left(\int_0^s r_s \, ds \right) yt \, ds,$$

where the term $-\int_0^\zeta r_n \, d\zeta$ has been neglected due to its small contribution. Show that the point (y_s, z_s) thus obtained is in coincidence with the center of shear derived from the shearing stress distribution due to torsion-free bending. Note: see Ref. 7.7, p. 474 for the shearing stress distribution of a thin-walled closed section due to torsion-free bending.

13. We consider an approximate formulation of a bending-torsion problem of a cantilever beam which is fixed at one end $(x = 0)$, and at the other end $(x = l)$ is subjected to terminal loads:

$$X_\nu = 0, \quad Y_\nu = \bar{Y}_x, \quad Z_\nu = \bar{Z}_x. \tag{i}$$

We assume the displacement components to be given by

$$\begin{aligned} U &= u - yv' - zw' + \vartheta'\varphi \\ V &= v - z\vartheta \\ W &= w + y\vartheta \end{aligned} \tag{ii}$$

where u, v, w and ϑ are functions of x only. The y- and z-axes are taken in coincidence with the principal axes through the centroid of the cross-section. The function $\varphi(y, z)$ is the Saint-Venent warping function of the cross-section and is chosen as mentioned in Problem 10. Show the following relations:

(1) By using Eq. (1.32) and equations (ii), the principle of virtual work can be written as

$$\int_0^l [N\,\delta u' + M_z\,\delta v'' - M_y\,\delta w'' + H\,\delta\vartheta'' + M_T\,\delta\vartheta']\,dx - \bar{P}_y\,\delta v(l) - \bar{P}_z\,\delta w(l) - \bar{M}\,\delta\vartheta(l) = 0 \tag{iii}$$

where

$$\begin{aligned} &N = \iint \sigma_x\,dy\,dz, \quad M_z = -\iint \sigma_x y\,dy\,dz, \\ &M_y = \iint \sigma_x z\,dy\,dz, \quad H = \iint \sigma_x\varphi\,dy\,dz, \\ &M_T = \iint [\tau_{xy}(\varphi_{,y} - z) + \tau_{xz}(\varphi_{,z} + y)]\,dy\,dz, \end{aligned} \tag{iv}$$

and

$$\begin{aligned} &\bar{P}_y = \iint \bar{Y}_x\,dy\,dz, \quad \bar{P}_z = \iint \bar{Z}_x\,dy\,dz, \\ &\bar{M} = \iint (\bar{Z}_x y - \bar{Y}_x z)\,dy\,dz. \end{aligned} \tag{v}$$

Note: since the term $\iint[\tau_{xy}\varphi_{,y} + \tau_{xz}\varphi_{,z}]\,dy\,dz$ in equations (iv) above is finally found to vanish, we have

$$M_T = \iint [\tau_{xz}y - \tau_{xy}z]\,dy\,dz.$$

(2) The governing equations are obtained from equation (iii) as

$$\begin{aligned} &N' = 0, \quad M_z'' = 0, \quad M_y'' = 0, \\ &M_T' - H'' = 0, \end{aligned} \tag{vi}$$

together with

$$\begin{aligned} &u = v = v' = w = w' = \vartheta = \vartheta' = 0 \quad \text{at} \quad x = 0; \\ &N = 0, \quad M_z = 0, \quad M_z' = -\bar{P}_y, \quad M_y = 0, \\ &M_y' = \bar{P}_z, \quad H = 0, \quad M_T - H' = \bar{M} \quad \text{at} \quad x = l. \end{aligned} \tag{vii}$$

(3) By using Eqs. (7.2) and equations (ii) and (iv), the stress resultant–displacement relations are given as follows:

$$
\begin{aligned}
N &= EA_0 u', \\
M_z &= EI_z(v'' - z_s\vartheta''), \\
M_y &= -EI_y(w'' + y_s\vartheta''), \\
H &= E(\Gamma\vartheta'' - z_sI_zv'' + y_sI_yw''), \\
M_T &= GJ\vartheta',
\end{aligned}
\qquad \text{(viii)}
$$

where I_y, I_z, y_s, z_s and Γ are defined in the same manner as in Problem 10.

(4) Consequently, the problem is reduced to solving the differential equations:

$$
\begin{aligned}
&EI_z(v - z_s\vartheta)''' + \bar{P}_y = 0, \\
&EI_y(w + y_s\vartheta)''' + \bar{P}_z = 0, \\
&E(\Gamma\vartheta - z_sI_zv + y_sI_yw)''' - GJ\vartheta' + \bar{M} = 0,
\end{aligned}
\qquad \text{(ix)}
$$

under the boundary conditions:

$$
\begin{aligned}
&v = v' = w = w' = \vartheta = \vartheta' = 0 \quad \text{at} \quad x = 0; \\
&EI_z(v - z_s\vartheta)'' = 0, \quad EI_y(w + y_s\vartheta)'' = 0, \\
&E(\Gamma\vartheta - z_sI_zv + y_sI_yw)'' = 0 \quad \text{at} \quad x = l.
\end{aligned}
\qquad \text{(x)}
$$

(5) Equations (ix) can be transformed into

$$
\begin{aligned}
-EI_zv_s''' &= \bar{P}_y, \\
-EI_yw_s''' &= \bar{P}_z, \\
-E\Gamma_s\vartheta''' + GJ\vartheta' &= \bar{M} + z_s\bar{P}_y - y_s\bar{P}_z,
\end{aligned}
\qquad \text{(xi)}
$$

where

$$
v_s = v - z_s\vartheta, \quad w_s = w + y_s\vartheta \qquad \text{(xii)}
$$

and Γ_s is defined in Problem 10. Equations (xi) indicate the physical meaning of the point (y_s, z_s): the choice of the locus of the point (y_s, z_s) as the reference axis decouples the governing equations into two groups and allows us to treat bending and torsion of the beam separately. Note: see Sections 35 through 38 of Ref. 7.33. See also Ref. 7.28.

Next, derive another approximate formulation of the problem by assuming that

$$
\begin{aligned}
U &= u - yv' - zw' + \alpha\varphi \\
V &= v - z\vartheta \\
W &= w + y\vartheta
\end{aligned}
\qquad \text{(xiii)}
$$

where u, v, w, α and ϑ are functions of x only.

14. We consider an approximate formulation of a torsional–flexural buckling problem of a beam which is clamped at one end ($x = 0$) and is subjected to an axial force P_{cr} at the other end ($x = l$), as shown in Fig. H 13. The symmetry of the cross-section is no longer assumed. The displacement components measured from the state just prior to the occurrence of the buckling are assumed to be

$$\begin{aligned} U &= u - yv' - zw' + \vartheta'\varphi, \\ V &= v - y(1 - \cos\vartheta) - z\sin\vartheta, \\ W &= w + y\sin\vartheta - z(1 - \cos\vartheta), \end{aligned} \tag{i}$$

where u, v, w and ϑ are functions of x only. The y- and z-axes are taken to coincide with the principal axes through the centroid of the cross-section. The function $\varphi(y, z)$ is the Saint-Venant warping function of the cross-section and is chosen as in Problem 10. By use of Eq. (5.5) and neglecting terms of higher order, show that the governing equations of the buckling problem are given by

$$\begin{aligned} &[EI_z(v'' - z_s\vartheta'') + P_{cr}v]'' = 0, \\ &[EI_y(w'' + y_s\vartheta'') + P_{cr}w]'' = 0, \\ &E(\Gamma\vartheta - z_sI_zv + y_sI_yw)'''' - GJ\vartheta'' + P_{cr}i^2\vartheta'' = 0, \end{aligned} \tag{ii}$$

and boundary conditions by

$$\begin{aligned} &v = v' = w = w' = \vartheta = \vartheta' = 0 \quad \text{at} \quad x = 0; \\ &EI_z(v''' - z_s\vartheta''') + P_{cr}v' = 0, \quad EI_z(v'' - z_s\vartheta'') = 0, \\ &EI_y(w''' + y_s\vartheta''') + P_{cr}w' = 0, \quad EI_y(w'' + y_s\vartheta'') = 0, \\ &E(\Gamma\vartheta - z_sI_zv + y_sI_yw)''' - GJ\vartheta' + P_{cr}i^2\vartheta' = 0, \\ &E(\Gamma\vartheta - z_sI_zv + y_sI_yw)'' = 0 \quad \text{at} \quad x = l, \end{aligned} \tag{iii}$$

where $\Gamma = \iint \varphi^2\,dy\,dz$, $A_0 = \iint dy\,dz$, $I_p = \iint (y^2 + z^2)\,dy\,dz$, $i_p = I_p/A_0$, and y_s, z_s, I_y, I_z are defined in Problem 10. Note: see Refs. 7.30 through 7.33.

Next, derive another approximate formulation of the problem by assuming that

$$\begin{aligned} U &= u - yv' - zw' + \alpha\varphi, \\ V &= v - y(1 - \cos\vartheta) - z\sin\vartheta, \\ W &= w + y\sin\vartheta - z(1 - \cos\vartheta), \end{aligned} \tag{iv}$$

where u, v, w, α and ϑ are functions of x only.

15. We consider the lateral buckling of a cantilever beam with rectangular cross-section which is clamped at one end ($x = 0$) and is under a concentrated load P_{cr} at the other end ($x = l$) as shown in Fig. H 16. The y- and z-axes are taken to coincide with the principal axes through the centroid

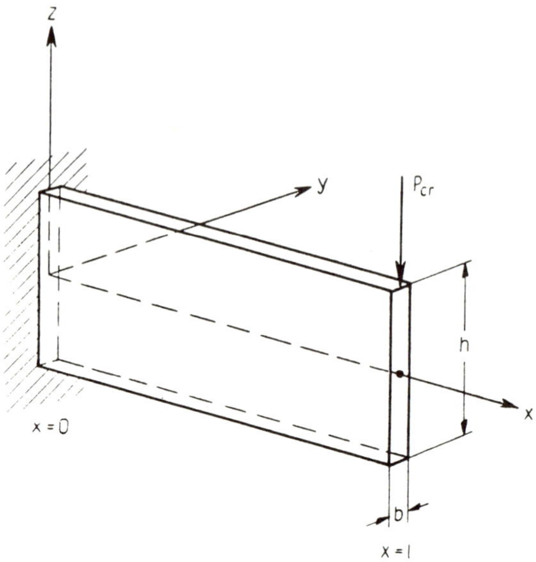

FIG. H 16.

of the cross-section. The stresses caused in the beam by P_{cr} are given by

$$\begin{aligned} \sigma_x^{(0)} &= (P_{cr}/I_y)(l - x)\, z, \\ \tau_{xz}^{(0)} &= (P_{cr}/2I_y)[z^2 - (h/2)^2], \end{aligned} \tag{i}$$

where $I_y = \iint z^2\, dy\, dz = \frac{1}{12} bh^3$. The force P_{cr}, which is acting at the middle point of the upper side of the end section, is assumed to vary neither its magnitude nor its direction while buckling occurs. We assume

$$\begin{aligned} U &= u - yv' - zw' + \vartheta'\varphi, \\ V &= v - y(1 - \cos\vartheta) - z \sin\vartheta, \\ W &= w + y \sin\vartheta - z(1 - \cos\vartheta), \end{aligned} \tag{ii}$$

where u, v, w, ϑ are functions of x only. The function $\varphi(y, z)$ is the Saint-Venant warping function and is chosen as in Problem 10. Using equations (i), (ii) and Eq. (5.5), show that the governing equations for v and ϑ are given by

$$\begin{aligned} [EI_z v'' + P_{cr}(l - x)\vartheta]'' &= 0, \\ EI_\varphi \vartheta'''' - GJ\vartheta'' + P_{cr}(l - x) v'' &= 0 \end{aligned} \tag{iii}$$

and boundary conditions by

$$v = v' = \vartheta = \vartheta' = 0 \quad \text{at} \quad x = 0;$$
$$EI_z v''' - P_{cr}\vartheta = 0, \quad EI_z v'' = 0, \tag{iv}$$
$$E\Gamma\vartheta''' - GJ\vartheta' + \tfrac{1}{2} hP_{cr}\vartheta = 0, \quad E\Gamma\vartheta'' = 0 \quad \text{at} \quad x = l.$$

Note: see Ref. 7.16. Next, derive another approximate formulation of the problem by assuming that

$$U = u - yv' - zw' + \alpha\varphi,$$
$$V = v - y(1 - \cos\vartheta) - z\sin\vartheta,$$
$$W = w + y\sin\vartheta - z(1 - \cos\vartheta), \tag{v}$$

where u, v, w, α and ϑ are functions of x only.

Problems Related to a Beam with Small Initial Deflection

16. Confining our problem to torsion-free bending in the (x, z) plane, we consider a beam, the locus of the centroid of which has a small initial deflection

$$z = z(x), \tag{i}$$

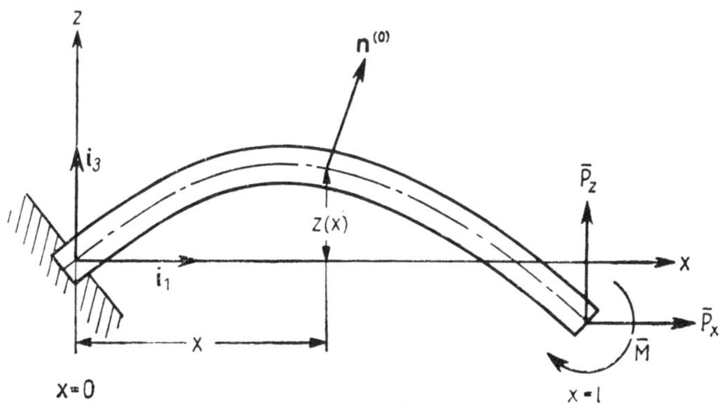

FIG. H 17.

as shown in Fig. H 17. We represent the position vector of an arbitrary point of the undeformed locus by

$$\mathbf{r}_0^{(0)} = x\mathbf{i}_1 + z(x)\,\mathbf{i}_3, \tag{ii}$$

and that of an arbitrary point of the undeformed beam by

$$\mathbf{r}^{(0)} = \mathbf{r}_0^{(0)} + y\mathbf{i}_2 + \zeta\mathbf{n}^{(0)}, \tag{iii}$$

where $\mathbf{i}_1$, $\mathbf{i}_2$ and $\mathbf{i}_3$ are unit vectors in the directions of the x-, y- and z-axes respectively. In equation (iii), $\mathbf{n}^{(0)}$ is a unit normal drawn perpendicular to the undeformed locus and is calculated by

$$\mathbf{n}^{(0)} = \mathbf{r}_0^{(0)\prime} \times \mathbf{i}_2 / |\mathbf{r}_0^{(0)\prime}| \tag{iv}$$

where $(\ )' = d(\ )/dx$. Equation (iii) suggests that the beam is specified by the coordinates (x, y, ζ) which form an orthogonal curvilinear coordinate system. Consequently, by taking $\alpha^1 = x$, $\alpha^2 = y$ and $\alpha^3 = \zeta$, we may apply the formulation developed in Chapter 4. Next, we define the displacement vector of the centroid by

$$\mathbf{u}_0 = u\mathbf{i}_1 + w\mathbf{i}_3 \tag{v}$$

and employ the Bernoulli–Euler hypothesis to obtain

$$\mathbf{r} = \mathbf{r}_0 + y\mathbf{i}_2 + \zeta\mathbf{n} \tag{vi}$$

where

$$\mathbf{r}_0 = \mathbf{r}_0^{(0)} + \mathbf{u}_0 \tag{vii}$$

and

$$\mathbf{n} = \mathbf{r}_0' \times \mathbf{i}_2 / |\mathbf{r}_0'|. \tag{viii}$$

With these preliminaries, derive the following relations:

(1) It is assumed hereafter that the beam is slender and the initial deflection is so small that

$$(z')^2 \ll 1. \tag{ix}$$

Then we have

$$\mathbf{n}^{(0)} = -z'\mathbf{i}_1 + \mathbf{i}_3 \tag{x}$$

and observe that the (x, y, ζ) coordinate system can be taken approximately to be locally rectangular Cartesian.

(2) The displacements are assumed to be so small that

$$u' \sim (w')^2 \ll 1. \tag{xi}$$

Then, we have

$$\mathbf{n} = -(z' + w')\,\mathbf{i}_1 + \mathbf{i}_3 \tag{xii}$$

and

$$\begin{aligned} f_{11} &= \tfrac{1}{2}(\mathbf{r}' \cdot \mathbf{r}' - \mathbf{r}^{(0)\prime} \cdot \mathbf{r}^{(0)\prime}) \\ &= u' + z'w' + \tfrac{1}{2}(w')^2 - \zeta w''. \end{aligned} \tag{xiii}$$

Higher order terms have been neglected in deriving equations (xii) and (xiii) as well as equation (x).

(3) By use of Eq. (4.80) and equation (xiii), equations of equilibrium are obtained as follows:

$$N' + \bar{p}_x = 0, \quad M'' + [(z' + w')\,N]' + \bar{p}_z = 0, \tag{xiv}$$

where it is defined that

$$N = \iint \tau^{11}\, dy\, d\zeta, \quad M = \iint \tau^{11} \zeta\, dy\, d\zeta \tag{xv}$$

and where $\bar{p}_x$ and $\bar{p}_z$ are distributed external loads per unit length of the x-axis in the directions of the x- and z-axes, respectively.

(4) If mechanical boundary conditions are specified at $x = l$, we have

$$\begin{aligned} &N = \bar{P}_x, \quad (z' + w')\, N + M' = \bar{P}_z, \\ &M = \bar{M} \qquad\qquad\qquad \text{at} \quad x = l \end{aligned} \tag{xvi}$$

where $\bar{P}_x$ and $\bar{P}_z$ are concentrated external forces in the directions of the x- and z-axes respectively.

(5) Since the (x, y, ζ) system is taken to be approximately a locally rectangular Cartesian system, we may take

$$\tau^{11} = E f_{11} \tag{xvii}$$

and obtain stress resultant–displacement relations as follows:

$$\begin{aligned} N &= EA_0\, [u' + z'w' + \tfrac{1}{2}(w')^2], \\ M &= -EIw''. \end{aligned} \tag{xviii}$$

Note: see Section 8.9 for a similar development applied to thin shallow shells.

17. We apply the results of Problem 16 to a snap-through problem as shown in Fig. H 18. The total potential energy of the system is given by

$$\Pi = \frac{EA_0}{2} \int_0^l \left\{ \left[u' + z'w' + \frac{1}{2}(w')^2 \right]^2 + i^2 (w'')^2 \right\} dx + \bar{P} w(l/2),$$

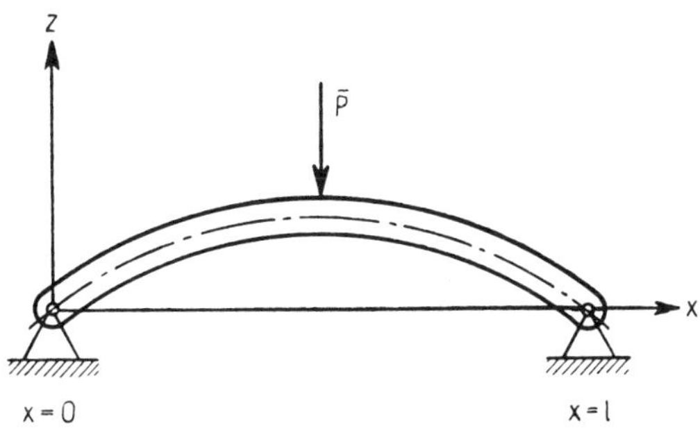

FIG. H 18.

where $z = z(x)$ is the small initial deflection of the beam, l is the span of the beam, $i^2 = I/A_0$, and $\bar{P}$ is the external force applied at $x = l/2$ in the negative direction of the z-axis.

(1) Derive the stationary conditions of the functional Π, where the independent functions subject to variation are u and w under the boundary conditions

$$u = w = 0 \quad \text{at} \quad x = 0 \quad \text{and} \quad x = l.$$

(2) Derive an approximate solution by assuming that

$$z = f_0 \sin \frac{\pi x}{l}, \quad w = -f_1 \sin \frac{\pi x}{l} - f_2 \sin \frac{2\pi x}{l}$$

and noticing that

$$\Pi = \frac{\pi^4 i^4 E A_0}{4 l^3} \left[\lambda_1^2 + 16\lambda_2^2 + \frac{1}{2}\left(\lambda_1\lambda_0 - \frac{1}{2}\lambda_1^2 - 2\lambda_2^2\right)^2 - \bar{p}\lambda_0\lambda_1 \right],$$

where $\lambda_0 = f_0/i$, $\lambda_1 = f_1/i$, $\lambda_2 = f_2/i$, $\bar{p} = (4l/\pi^2 f_0)(\bar{P}/P_e)$, $P_e = \pi^2 EI/l^2$. Show also that we have the following critical load for the snap-through problem:

$$\bar{P}_{cr} = P_e \frac{\pi^2 f_0}{4l} \left[2 + \frac{1}{\lambda_0} \left(\sqrt{\frac{\lambda_0^2 - 4}{3}} \right)^3 \right]; \quad \lambda_0 \leqq \sqrt{22},$$

$$\bar{P}_{cr} = P_e \frac{\pi^2 f_0}{4l} \left[2 + \frac{6}{\lambda_0} \sqrt{\lambda_0^2 - 16} \right]; \quad \lambda_0 \geqq \sqrt{22}.$$

Note: see Ref. 7.34. See also Refs. 3.19 and 3.20.

CHAPTER 8†

Problems Related to Sections 8.2 and 8.4

1. We consider bending of a square plate with all edges built in and subjected to uniform pressure $\bar{p}$. Looking for an approximate solution, we assume

$$w = c(1 - \xi^2)^2 (1 - \eta^2)^2,$$

where c is an arbitrary constant, $\xi = x/(a/2)$, $\eta = y/(a/2)$ and a is the side length of the square. Using the principle of minimum potential energy and applying the Rayleigh–Ritz method, show that we obtain

$$c = 0.001329\, \bar{p}a^4/D$$

and

		Exact solution
$(w)_{x=0,\ y=0} =$	$0.00133\bar{p}a^4/D$	$0.00126\bar{p}a^4/D$
$(M_x)_{x=a/2,\ y=0} =$	$-0.0425\bar{p}a^2$	$-0.0513\bar{p}a^2$
$(M_x)_{x=0,\ y=0} =$	$0.0276\bar{p}a^2$	$0.0231\bar{p}a^2$

† Unless otherwise stated, plates are assumed to have constant thickness and density in Problems 1 through 11.

where $\nu = 0.3$. The accuracy of the approximate solution is shown in Fig. H 19 for reference by introducing a quantity defined by

$$\Delta = \frac{D}{\bar{p}} \left(\frac{\partial^4 w}{\partial x^4} + 2 \frac{\partial^4 w}{\partial x^2 \, \partial y^2} + \frac{\partial^4 w}{\partial y^4} \right) - 1.$$

Note: for details of the exact solution, see Ref. 8.2, pp. 197–202. See also Ref. 8.45, pp. 413–19.

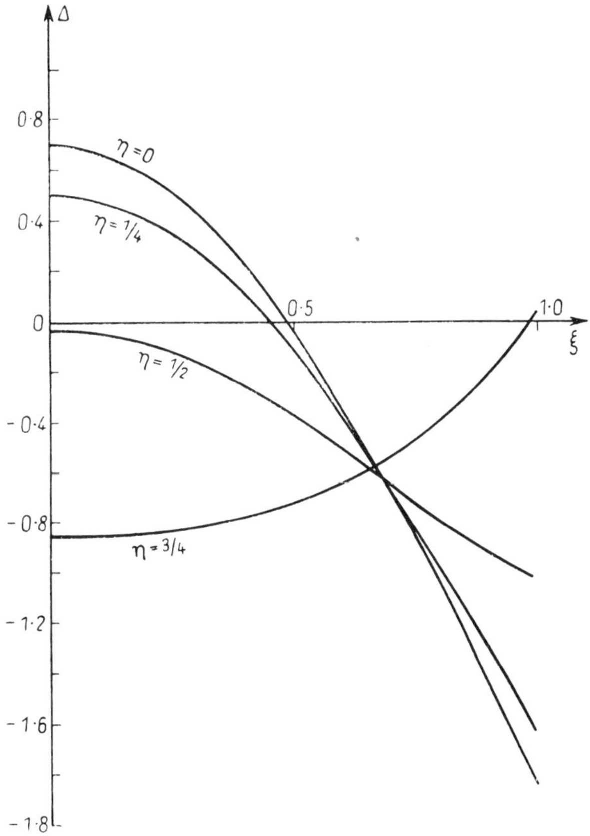

FIG. H 19.

2. We consider bending of a cantilever solid wing of variable thickness under the distributed pressure $\bar{p}(x, y)$. Assuming that

$$w(x, y) = w(y) - \theta(y)\, x,$$

derive governing differential equations and boundary conditions for $w(y)$ and $\theta(y)$ by use of the functional (8.51) and explain the physical meanings of these equations. Note: see Ref. 8.10, pp. 60–6.

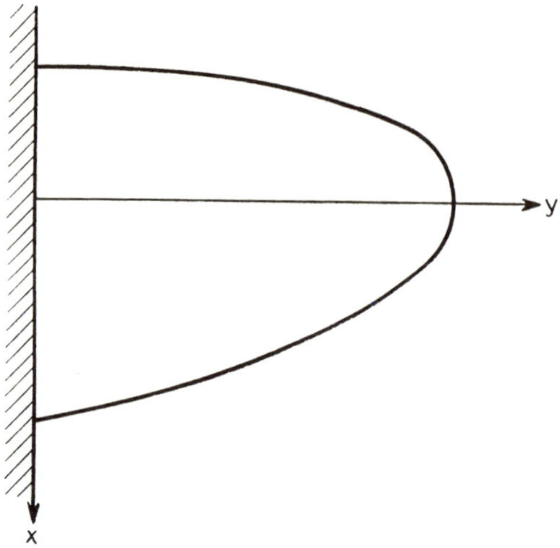

FIG. H 20.

Problems Related to Sections 8.5 and 8.6

3. Show that Eq. (8.71) can be obtained directly from Eqs. (8.67) through the elimination of u and v by use of the identity

$$\frac{\partial^2}{\partial y^2}\left(\frac{\partial u}{\partial x}\right) + \frac{\partial^2}{\partial x^2}\left(\frac{\partial v}{\partial y}\right) = \frac{\partial^2}{\partial x\,\partial y}\left(\frac{\partial u}{\partial y} + \frac{\partial v}{\partial x}\right),$$

and expressing e_{xxo}, e_{yyo} and e_{xyo} in terms of F by use of Eqs. (8.46) and (8.66).

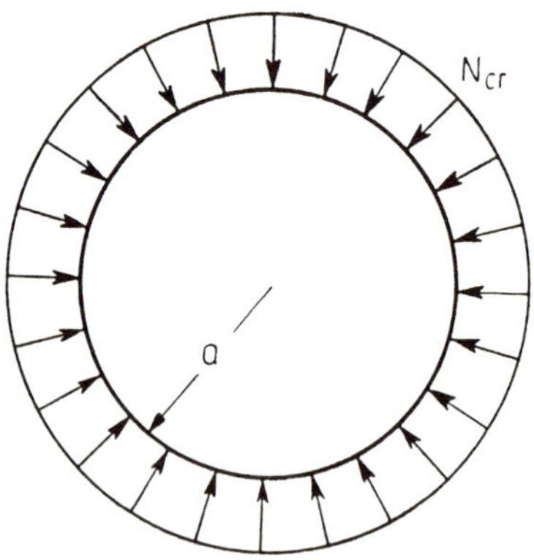

FIG. H 21.

4. We consider the problem of buckling of a uniformly compressed circular plate as shown in Fig. H 21. The plate is simply supported ar $r = a$.

Confining our problem to rotationally symmetric buckling modes, show that the principle of virtual work is finally reduced to

$$\int^{a} [rM_r \delta w'' + M_\theta \delta w' + rN_{cr}\, w' \delta w']\, dr = 0,$$

from which we obtain a differential equation

$$r^2 \Phi'' + r\Phi' + (r^2 \alpha^2 - 1)\, \Phi = 0$$

and boundary conditions

$$\lim_{r \to 0} (r\Phi' + \nu\Phi) = 0, \quad a\Phi'(a) + \nu\Phi(a) = 0$$

for the determination of the critical loads, where $(\ )' = d(\ )/dr$, $\alpha^2 = N_{cr}/D$ and $\Phi = w'$. Note: see Problem 17 of this chapter for the strain–displacement relations expressed in cylindrical coordinates.

5. We consider the problem of buckling of a circular plate with a concentric circular hole which is subjected to internal and external pressures

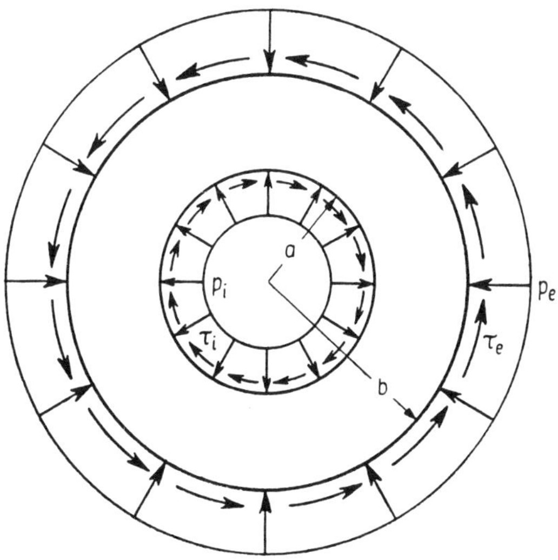

FIG. H 22.

plus shearing forces as shown in Fig. H 22. Show that the governing equation for the buckling is given by

$$D\Delta\Delta w = \frac{1}{r}\frac{\partial}{\partial r}\left(r\sigma_r^{(0)} h \frac{\partial w}{\partial r}\right) + \frac{1}{r}\frac{\partial}{\partial r}\left(\tau_{r\theta}^{(0)} h \frac{\partial w}{\partial \theta}\right)$$
$$+ \frac{1}{r}\frac{\partial}{\partial \theta}\left(\tau_{r\theta}^{(0)} h \frac{\partial w}{\partial r}\right) + \frac{1}{r}\frac{\partial}{\partial \theta}\left(\sigma_\theta^{(0)} h \frac{\partial w}{\partial \theta}\right), \quad \text{(i)}$$

where

$$\Delta(\ ) = \frac{\partial^2(\ )}{\partial r^2} + \frac{1}{r}\frac{\partial(\ )}{\partial r} + \frac{1}{r^2}\frac{\partial^2(\ )}{\partial \theta^2}. \tag{ii}$$

In equation (i), $\sigma_r^{(0)}$, $\sigma_\theta^{(0)}$ and $\tau_\theta^{(0)}$ are the initial stresses caused by the internal pressure p_i, external pressure p_e plus the shearing forces τ_i and τ_e, and are given by

$$\begin{aligned}
\sigma_r^{(0)} &= \frac{a^2b^2(p_e - p_i)}{b^2 - a^2}\frac{1}{r^2} + \frac{p_i a^2 - p_e b^2}{b^2 - a^2,}, \\
\sigma_\theta^{(0)} &= -\frac{a^2b^2(p_e - p_i)}{b^2 - a^2}\frac{1}{r^2} + \frac{p_i a^2 - p_e b^2}{b^2 - a^2}, \\
\tau_{r\theta}^{(0)} &= \frac{b^2\tau_e}{r^2}.
\end{aligned} \tag{iii}$$

The suffixes i and e denote that the quantities are referred to the internal and external boundaries respectively.

Problems Related to Section 8.7

6. A circular plate is subjected to a temperature distribution $\theta(r)$. The surface of the disc is traction free. Derive from the principle of minimum complementary energy the governing equations for σ_r and σ_θ:

$$[r(r\sigma_r)']' - \sigma_r + \alpha E r\theta' = 0, \quad (r\sigma_r)' = \sigma_\theta,$$

and the boundary conditions

$$\lim_{r\to 0}\,[r^3\sigma_r' + (1 - \nu)\, r^2\sigma_r] = 0, \quad \sigma_r(a) = 0,$$

where α is the coefficient of thermal expansion, a is the radius of the plate and $(\ )' = d(\ )/dr$. Show that if $\theta(r)$ is postulated as a polynomial of r, i.e.

$$\theta = \sum_{k=0}^{n} b_k\, r^k,$$

we have

$$\sigma_r = -\alpha E \sum_{k=1}^{n} \frac{b_k}{k+2}(r^k - a^k),$$

$$\sigma_\theta = -\alpha E \sum_{k=1}^{n} \frac{b_k}{k+1}[(k+1)\, r^k - a^k].$$

7. Show that a thermal stress problem for a plate in large deflection can be formulated from the equations developed in the small displacement theory (see Section 8.7) by replacing ε_x, ε_y and γ_{xy} with e_{xx}, e_{yy} and $2e_{xy}$, respectively, and that Eqs. (8.70) and (8.71) are generalized to include thermal effects as follows:

$$D\Delta\Delta w + \frac{1}{1-\nu}\Delta M_T = \bar{p} + F_{,yy}w_{,xx} + F_{,xx}w_{,yy} - 2F_{,xy}w_{,xy}.$$
$$\Delta\Delta F + \Delta N_T = Eh(w_{,xy}^2 - w_{,xx}w_{,yy}).$$

Problems Related to Lateral Vibration of Plates

8. Weinstein's method mentioned in Section 2.8 may be applied to free lateral vibrations of a clamped plate. An intermediate problem can be defined as follows: choose a sequence of linearly independent functions $p_1(x, y)$, $p_2(x, y)$ ... and $p_n(x, y)$, all of which are taken to be plane harmonic functions, and relax the geometrical boundary conditions of the original problem:

$$w = 0, \quad \partial w/\partial \nu = 0 \quad \text{on} \quad C \tag{i}$$

so that they are replaced by

$$w = 0, \quad \int_C p_i(\partial w/\partial \nu)\, ds = 0, \quad i = 1, 2, \ldots, n \quad \text{on} \quad C. \tag{ii}$$

Show that the intermediate problem can be fornulated by the following variational expression:

$$\Pi = \frac{1}{2} D \int_{S_m} (w_{,xx} + w_{,yy})^2\, dx\, dy - \frac{1}{2}\varrho h\omega^2 \iint w^2\, dx\, dy$$
$$- \sum_{i=1}^{n} a_i \int_C p_i\, (\partial w/\partial \nu)\, ds + \int_C qw\, ds, \tag{iii}$$

where a_i; $i = 1, 2, \ldots, n$ and $q(s)$ are Lagrange multipliers. Show also that the following relation is obtained as a natural boundary condition of the functional (iii):

$$D\,\Delta\, w = \sum_{i=1}^{n} a_i\, p_i \tag{iv}$$

Note: see Ref. 2.27.

9. Show that the functional for the principle of stationary potential energy of a free lateral vibration problem for a flat plate with initial membrane stresses $N_x^{(0)}$, $N_y^{(0)}$ and $N_{xy}^{(0)}$ is given by

$$\Pi = \frac{1}{2}\iint_{S_m} \{D\,[(w_{,xx} + w_{,yy})^2 + 2(1-\nu)(w_{,xy}^2 - w_{,xx}\, w_{,yy})]$$
$$+ N_x^{(0)}(w_{,x})^2 + N_y^{(0)}(w_{,y})^2 + 2N_{xy}^{(0)}\, w_{,x}w_{,y} - \varrho h\omega^2 w^2\}\, dx\, dy,$$

where the plate is assumed to be traction-free on the upper and lower surfaces ($z = \pm h/2$) and on the C_1 part of the side boundary, while it is geometrically fixed on the remaining part C_2 of the side boundary. The initial membrane stresses are so chosen that they satisfy

$$N^{(0)}_{x,x} + N^{(0)}_{xy,y} = 0, \quad N^{(0)}_{xy,x} + N^{(0)}_{y,y} = 0 \quad \text{in} \quad S_m,$$

and

$$N^{(0)}_x l + N^{(0)}_{xy} m = 0, \quad N^{(0)}_{xy} l + N^{(0)}_y m = 0 \quad \text{on} \quad C_1.$$

10. We consider the free lateral vibration of a circular plate subject to an initial stress system:

$$N^{(0)}_r = \beta(a^2 - r^2), \quad N_\theta = \beta(a^2 - 3r^2), \quad N^{(0)}_{r\theta} = 0,$$

where β is a constant, and a is the radius of the plate. The plate is assumed traction-free. Confining our problem to rotationally symmetric modes of vibration where the rigid body mode is excluded, show that the functional for the principle of stationary potential energy of this problem is given by

$$\Pi = \pi \int_0^a \left\{ D \left[\left(w'' + \frac{w'}{r} \right)^2 - 2(1 - \nu) \frac{w'w''}{r} \right] + N^{(0)}_r (w')^2 - \varrho h \omega^2 w^2 \right\} r\, dr,$$

where the quantity subject to variation is $w(r)$, under a subsidiary condition

$$\int_0^a wr\, dr = 0.$$

Show also that the governing equation and boundary conditions are given by

$$D\left(\frac{d^2}{dr^2} + \frac{1}{r}\frac{d}{dr}\right)\left(\frac{d^2 w}{dr^2} + \frac{1}{r}\frac{dw}{dr}\right) = \frac{1}{r}\,[r\, N^{(0)}_r\, w']' + \varrho h \omega^2 w$$

and

$$D(rw'' + \nu w') = 0,$$

$$D(rw'' + \nu w')' - D\left(\nu w'' + \frac{w'}{r}\right) - rN^{(0)}_r\, w' = 0,$$

at $r = 0$ and $r = a$, respectively.

11. We consider an approximate solution of Problem 10 by choosing

$$w = c\,(r^2 - \tfrac{1}{2} a^2)$$

and applying the Rayleigh–Ritz method, where c is an arbitrary constant. Show that we obtain the following approximate eigenfrequency of the lowest mode:

(1) From the Rayleigh–Ritz method, we obtain

$$\varrho h \omega^2 = 96(1 + \nu)\,(D/a^4) + 8\beta.$$

(2) From the modified Rayleigh–Ritz method, we obtain

$$\varrho h\omega^2 = \frac{480(1+\nu)}{7+2\nu}\frac{D}{a^4} + 8\beta$$

Note: (1) The values of $\varkappa$ for the eigenfrequency of the lowest mode when the initial stresses are absent are $\varkappa = 8.8896$ for $\nu = \frac{1}{4}$ and $\varkappa = 9.0760$ for $\nu = \frac{1}{3}$, where $\omega = \varkappa\sqrt{D/\varrho h a^4}$.

(2) The value of the critical β which causes buckling of the lowest mode is $\beta = -(3.135)^2 (D/a^4)$ for $\nu = 0.3$ (Ref. 8.46).

Problems Related to the Conditions of Compatibility and Stress Functions

12. We assume the displacement components as given by Eqs. (8.99):

$$U = u + zu_1, \quad V = v + zv_1, \quad W = w, \tag{i}$$

where u, v, w, u_1 and v_1 are functions of (x, y) only. Show that we have

$$\begin{aligned} \varepsilon_x &= u_{,x} + zu_{1,x}, & \gamma_{yz} &= w_{,y} + v_1, \\ \varepsilon_y &= v_{,y} + zv_{1,y}, & \gamma_{xz} &= w_{,x} + u_1, \\ \varepsilon_z &= 0, & \gamma_{xy} &= u_{,y} + v_{,x} + z(u_{1,y} + v_{1,x}). \end{aligned} \tag{ii}$$

Show also that by use of the principle of virtual work

$$\iiint\limits_V (\sigma_x\,\delta\varepsilon_x + \sigma_y\,\delta\varepsilon_y + \tau_{xy}\,\delta\gamma_{xy} + \tau_{xz}\,\delta\gamma_{xz} + \tau_{yz}\,\delta\gamma_{yz})\,dx\,dy\,dz - \iint \bar{p}\,\delta w\,dx\,dy + \cdots = 0 \tag{iii}$$

and equation (ii), the equations of equilibrium for the problem presented in Section 8.2 are given by

$$\begin{aligned} N_{x,x} + N_{xy,y} &= 0, & N_{xy,x} + N_{y,y} &= 0, \\ M_{x,x} + M_{xy,y} - Q_x &= 0, & M_{xy,x} + M_{y,y} - Q_y &= 0, \\ Q_{x,x} + Q_{y,y} + \bar{p} &= 0, \end{aligned} \tag{iv}$$

where N_x, N_y, ..., M_{xy} are defined by Eqs. (8.17), while Q_x and Q_y are defined by

$$Q_x = \int_{-h/2}^{h/2} \tau_{xz}\,dz, \quad Q_y = \int_{-/h}^{h/2} \tau_{yz}\,dz. \tag{v}$$

Compare equations (iv) with Eqs. (8.22) and (8.30).

13. We consider the same problem as Problem 12, and write

$$\begin{aligned} \varepsilon_x &= \varepsilon_{x0} - z\varkappa_x, & \gamma_{yz} &= \gamma_{yz0}, \\ \varepsilon_y &= \varepsilon_{y0} - z\varkappa_y, & \gamma_{xz} &= \gamma_{xz0}, \\ \varepsilon_z &= 0, & \gamma_{xy} &= \gamma_{xy0} - 2z\varkappa_{xy} \end{aligned} \tag{i}$$

in view of equations (ii) of Problem 12, where $\varepsilon_{x0}, \varepsilon_{y0}, \ldots$, and $\varkappa_{xy}$ are functions of (x, y) only. Then, Eqs. (1.16) are written as follows for the present problem:

$$\begin{aligned} R_x &= R_y = U_z = 0, \\ R_z &= \varepsilon_{y0,xx} + \varepsilon_{x0,yy} - \gamma_{xy0,xy} - z(\varkappa_{y,xx} + \varkappa_{x,yy} - 2\varkappa_{xy,xy}), \\ U_x &= \varkappa_{x,y} + \tfrac{1}{2}(-\gamma_{yz0,xx} + \gamma_{xz0,yx} - 2\varkappa_{xy,x}), \\ U_y &= \varkappa_{y,x} + \tfrac{1}{2}(\gamma_{yz0,xy} - \gamma_{xz0,yy} - 2\varkappa_{xy,y}). \end{aligned} \quad \text{(ii)}$$

Show that by use of Lagrange multipliers χ_3, ψ_1 and ψ_2, the principle of virtual work can be written as

$$\begin{aligned} \iint \{N_x\,\delta\varepsilon_{x0} + N_y\,\delta\varepsilon_{y0} + N_{xy}\,\delta\gamma_{xy0} + Q_x\,\delta\gamma_{xz0} + Q_y\,\delta\gamma_{yz0} \\ - M_x\,\delta\varkappa_x - M_y\,\delta\varkappa_y - 2M_{xy}\,\delta\varkappa_{xy}\}\,dx\,dy \\ - \iiint \{\chi_3\,\delta R_z + \psi_1\,\delta U_x + \psi_2\,\delta U_y\}\,dx\,dy\,dz + \cdots = 0 \end{aligned} \quad \text{(iii)}$$

and that from the requirement that coefficients of $\delta\varepsilon_{x0}, \delta\varepsilon_{y0}, \ldots$ must vanish in equation (iii), we obtain

$$\begin{aligned} N_x &= F_{,yy}, \quad N_y = F_{,xx}, \quad N_{xy} = -F_{,xy}, \\ M_x &= \Psi_{1,y}, \quad M_y = \Psi_{2,x}, \quad M_{xy} = -\tfrac{1}{2}(\Psi_{1,x} + \Psi_{2,y}), \\ Q_x &= \tfrac{1}{2}(\Psi_{1,xy} - \Psi_{2,yy}), \quad Q_y = \tfrac{1}{2}(-\Psi_{1,xx} + \Psi_{2,xy}), \end{aligned} \quad \text{(iv)}$$

where

$$F(x, y) = \int \chi_3\,dz, \quad \Psi_1(x, y) = \int \psi_1\,dz, \quad \Psi_2(x, y) = \int \psi_2\,dz. \quad \text{(v)}$$

Next, by substituting equations (iv) into equations (iv) of Problem 12, show that the functions F, Ψ_1 and Ψ_2 thus introduced play the role of stress functions. Discuss also the role played by the function F^* in equation (iii), where $F^*(x, y) = \int \chi_3 z\,dz$.

Problems Related to Curvilinear Coordinates

14. We represent the middle surface of the plate by a nonorthogonal curvilinear coordinate system (α^1, α^2) such that

$$(ds^{(0)})^2 = g_{ij}\,d\alpha^i\,d\alpha^j + (dz)^2, \quad \text{(i)}\dagger$$

where g_{ij}; $i, j = 1, 2$ are functions of (α^1, α^2) only. Using the formulations developed in Sections 4.1, 4.2 and 8.1, show that we obtain a plate theory based on Kirchhoff's hypothesis as follows:

† It is noted here that a roman letter is used in place of (1, 2) in Problems 14 and 18. The summation convention is employed. Thus, a twice-appearing roman letter means summation with respect to (1, 2).

(1) The displacement vector is given by

$$\mathbf{u} = \mathbf{u}_0 + (\mathbf{n} - \mathbf{i}_3)\, z. \tag{ii}$$

By use of the relations

$$\mathbf{u}_0 = v^1\mathbf{g}_1 + v^2\mathbf{g}_2 + w\mathbf{i}_3, \tag{iii}$$

$$\mathbf{r}_0 = \mathbf{r}_0^{(0)} + \mathbf{u}_0, \tag{iv}$$

$$\mathbf{n} = \frac{\partial \mathbf{r}_0}{\partial \alpha^1} \times \frac{\partial \mathbf{r}_0}{\partial \alpha^2} \Bigg/ \left| \frac{\partial \mathbf{r}_0}{\partial \alpha^1} \times \frac{\partial \mathbf{r}_0}{\partial \alpha^2} \right|, \tag{v}$$

where $\mathbf{g}_1 = \partial \mathbf{r}_0^{(0)}/\partial\alpha^1$, $\mathbf{g}_2 = \partial \mathbf{r}_0^{(0)}/\partial\alpha^2$, and v^1, v^2, w are functions of (α^1, α^2) only, we obtain

$$\mathbf{u} = (v^1 + l^1 z)\,\mathbf{g}_1 + (v^2 + l^2 z)\,\mathbf{g}_2 + w\mathbf{i}_3, \tag{vi}$$

$$l^i = -g^{ij} w_{,j}; \quad i = 1, 2 \tag{vii}$$

and $g^{11} = g_{22}/g$, $g^{22} = g_{11}/g$, $g^{12} = g^{21} = -g_{12}/g$, $(\ )_{,i} = \partial(\ )/\partial\alpha^i$.

(2) The strain–displacement relations for the Kármán theory are given by

$$\begin{aligned} f_{11} &= g_{1k}v^k{}_{;1} + \underline{\tfrac{1}{2}(w_{,1})^2} + g_{1k}l^k{}_{;1}z, \\ f_{22} &= g_{2k}v^k{}_{;2} + \underline{\tfrac{1}{2}(w_{,2})^2} + g_{2k}l^k{}_{;2}z, \\ 2f_{12} &= g_{1k}v^k{}_{;2} + g_{2k}v^k{}_{;1} + \underline{w_{,1}w_{,2}} \\ &\qquad + (g_{1k}l^k{}_{;2} + g_{2k}l^k{}_{;1})\, z; \end{aligned} \tag{viii}$$

where $v^k{}_{;j} = v^k{}_{,j} + \left\{ {k \atop ij} \right\} v^i$. The strain–displacement relations for the small displacement theory are obtainable by neglecting the underlined terms in equations (viii).

(3) A plate theory based on the Kirchhoff hypothesis is obtainable by the use of these relations and the principle of virtual work, Eq. (4.80).

15. We represent the middle surface of the plate by a skew coordinate system (ξ, η) such that

$$(ds^{(0)})^2 = (d\xi)^2 + 2\cos\alpha\, d\xi\, d\eta + (d\eta)^2 + (dz)^2, \tag{i}$$

where α is a constant. The displacement vector of the middle surface is denoted by

$$\mathbf{u}_0 = u\mathbf{g}_1 + v\mathbf{g}_2 + w\mathbf{i}_3, \tag{ii}$$

where $\mathbf{g}_1$ and $\mathbf{g}_2$ are unit vectors in the directions of the ξ- and η-axes, respectively, and u, v, w are functions of (ξ, η) only. Using the results of Problem 11 of Chapter 4 and Problem 14 of Chapter 8, and confining our problem to the small displacement theory, show that we obtain the following relations based on the Kirchhoff hypothesis:

(1) We have the displacement and strain–displacement relations as follows:

$$\mathbf{u} = [u - (\operatorname{cosec}^2 \alpha\, w_{,\xi} - \cot\alpha \operatorname{cosec}\alpha\, w_{,\eta})\, z]\, \mathbf{g}_1 + [v - (-\cot\alpha \operatorname{cosec}\alpha\, w_{,\xi} + \operatorname{cosec}^2\alpha\, w_{,\eta})\, z]\, \mathbf{g}_2 + w\mathbf{i}_3, \tag{iii}$$

$$f_{11} = (u + v\cos\alpha)_{,\xi} - w_{,\xi\xi}z,$$
$$f_{22} = (u\cos\alpha + v)_{,\eta} - w_{,\eta\eta}z,$$
$$2f_{12} = (u + v\cos\alpha)_{,\eta} + (u\cos\alpha + v)_{,\xi} - 2w_{,\xi\eta}z. \tag{iv}$$

(2) By use of Eq. (4.80), the equations of equilibrium are obtained as follows:

$$N^{11}{}_{,\xi} + N^{21}{}_{,\eta} = 0, \quad N^{12}{}_{,\xi} + N^{22}{}_{,\eta} = 0,$$
$$M^{11}{}_{,\xi\xi} + 2M^{12}{}_{,\xi\eta} + M^{22}{}_{,\eta\eta} + \bar{p} = 0, \tag{v}$$

where

$$[N^{11}, N^{22}, N^{12}, N^{21}) = \int [\tau^{11}, \tau^{22}, \tau^{12}, \tau^{21}]\, dz,$$

and

$$[M^{11}, M^{22}, M^{12}, M^{21}] = \int [\tau^{11}, \tau^{22}, \tau^{12}, \tau^{21}]\, z\, dz.$$

(3) The equations which correspond to Eqs. (8.49) and (8.34) are obtained in the (ξ, η) coordinate system as follows:

$$\Delta_{(\xi\eta)}\Delta_{(\xi\eta)} F = 0, \tag{vi}$$

and

$$D\Delta_{(\xi\eta)}\Delta_{(\xi\eta)} w = \bar{p}\sin^4\alpha, \tag{vii}$$

respectively, where

$$\Delta_{(\xi\eta)} = \frac{\partial^2}{\partial\xi^2} - 2\cos\alpha\frac{\partial^2}{\partial\xi\,\partial\eta} + \frac{\partial^2}{\partial\eta^2},$$

$$\Delta_{(\xi\eta)}\Delta_{(\xi\eta)} = \frac{\partial^4}{\partial\xi^4} - 4\cos\alpha\frac{\partial^4}{\partial\xi^3\,\partial\eta} + 2(1 + 2\cos^2\alpha)\frac{\partial^4}{\partial\xi^2\,\partial\eta^2} - 4\cos\alpha\frac{\partial^4}{\partial\xi\,\partial\eta^3} + \frac{\partial^4}{\partial\eta^4}.$$

Note: see Ref. 8.47.

16. We represent the middle surface of the plate by an orthogonal curvilinear coordinate system (α, β) such that

$$(ds^{(0)})^2 = A^2(d\alpha)^2 + B^2(d\beta)^2 + (dz)^2, \tag{i}$$

where A and B are functions of (α, β) only. Using the results of Problem 14, and denoting the unit vectors in the α- and β-coordinates by $\mathbf{a}^{(0)}$ and $\mathbf{b}^{(0)}$, respectively, show that we have the following relations based on the Kirchhoff hypothesis:

(1) We denote the displacement vector of the middle surface by

$$\mathbf{u}_0 = u\mathbf{a}^{(0)} + v\mathbf{b}^{(0)} + w\mathbf{n}^{(0)} \tag{ii}$$

where u, v, w are functions of (α, β) only, and obtain

$$\mathbf{u} = \left(u - \frac{1}{A}\frac{\partial w}{\partial \alpha} z\right)\mathbf{a}^{(0)} + \left(v - \frac{1}{B}\frac{\partial w}{\partial \beta} z\right)\mathbf{b}^{(0)} + w\mathbf{n}^{(0)}. \tag{iii}$$

(2) The strain–displacement relations for Kármán's theory are given by

$$e_{\alpha\alpha} = \frac{1}{A}\frac{\partial u}{\partial \alpha} + \frac{v}{AB}\frac{\partial A}{\partial \beta} + \underline{\frac{1}{2A^2}\left(\frac{\partial w}{\partial \alpha}\right)^2} - z\left[\frac{1}{A}\frac{\partial}{\partial \alpha}\left(\frac{1}{A}\frac{\partial w}{\partial \alpha}\right) + \frac{1}{AB^2}\frac{\partial A}{\partial \beta}\frac{\partial w}{\partial \beta}\right],$$

$$e_{\beta\beta} = \frac{1}{B}\frac{\partial v}{\partial \beta} + \frac{u}{AB}\frac{\partial B}{\partial \alpha} + \underline{\frac{1}{2B^2}\left(\frac{\partial w}{\partial \beta}\right)^2} - z\left[\frac{1}{B}\frac{\partial}{\partial \beta}\left(\frac{1}{B}\frac{\partial w}{\partial \beta}\right) + \frac{1}{A^2B}\frac{\partial B}{\partial \alpha}\frac{\partial w}{\partial \alpha}\right],$$

$$\begin{aligned} 2e_{\alpha\beta} = {} & \frac{1}{A}\frac{\partial v}{\partial \alpha} - \frac{u}{AB}\frac{\partial A}{\partial \beta} + \frac{1}{B}\frac{\partial u}{\partial \beta} - \frac{v}{AB}\frac{\partial B}{\partial \alpha} + \underline{\frac{1}{AB}\frac{\partial w}{\partial \alpha}\frac{\partial w}{\partial \beta}} \\ & - z\left[\frac{1}{A}\frac{\partial}{\partial \alpha}\left(\frac{1}{B}\frac{\partial w}{\partial \beta}\right) - \frac{1}{A^2B}\frac{\partial A}{\partial \beta}\frac{\partial w}{\partial \alpha}\right. \\ & \left. + \frac{1}{B}\frac{\partial}{\partial \beta}\left(\frac{1}{A}\frac{\partial w}{\partial \alpha}\right) - \frac{1}{AB^2}\frac{\partial B}{\partial \alpha}\frac{\partial w}{\partial \beta}\right]. \end{aligned} \tag{iv}$$

The strain–displacement relations for the small displacement theory are obtainable by neglecting the underlined terms. Note: it is obvious that these relations coincide with those obtained from the equations of Problem 9 of Chapter 9 by setting $1/R_\alpha = 1/R_\beta = 0$.

17. Show that the displacement and strain–displacement relations expressed in a cylindrical coordinate system:

$$x = r\cos\theta, \quad y = r\sin\theta, \quad z = z \tag{i}$$

are given as follows for Kármán's theory:

$$\mathbf{u} = \left(u - \frac{\partial w}{\partial r} z\right)\mathbf{a}^{(0)} + \left(v - \frac{1}{r}\frac{\partial w}{\partial \theta} z\right)\mathbf{b}^{(0)} + w\mathbf{i}_3,$$

$$e_{rr} = \frac{\partial u}{\partial r} + \underline{\frac{1}{2}\left(\frac{\partial w}{\partial r}\right)^2} - z\frac{\partial^2 w}{\partial r^2},$$

$$e_{\theta\theta} = \frac{1}{r}\frac{\partial v}{\partial \theta} + \frac{u}{r} + \underline{\frac{1}{2r^2}\left(\frac{\partial w}{\partial \theta}\right)^2} - z\left[\frac{1}{r^2}\frac{\partial^2 w}{\partial \theta^2} + \frac{1}{r}\frac{\partial w}{\partial r}\right],$$

$$\begin{aligned} 2e_{r\theta} = {} & \frac{\partial v}{\partial r} + \frac{1}{r}\frac{\partial u}{\partial \theta} - \frac{v}{r} + \underline{\frac{1}{r}\frac{\partial w}{\partial r}\frac{\partial w}{\partial \theta}} \\ & - z\left[\frac{\partial}{\partial r}\left(\frac{1}{r}\frac{\partial w}{\partial \theta}\right) + \frac{1}{r}\frac{\partial}{\partial \theta}\left(\frac{\partial w}{\partial r}\right) - \frac{1}{r^2}\frac{\partial w}{\partial \theta}\right], \end{aligned}$$

where $\mathbf{a}^{(0)}$ and $\mathbf{b}^{(0)}$ are unit vectors in the r- and θ-coordinates, respectively. It is obvious that strain–displacement relations for the small displacement theory are obtainable by neglecting the underlined terms.

18. We represent the middle surface of the plate by a nonorthogonal curvilinear coordinate system (α^1, α^2) such that that

$$(ds^{(0)})^2 = g_{ij}\, d\alpha^i\, d\alpha^j + (dz)^2 \tag{i}$$

and formulate a plate theory including the effect of transverse shear deformation by assuming that

$$\mathbf{u} = (v_0^1 + \overset{1}{v}_{\textcircled{1}}\, z)\, \mathbf{g}_1 + (v_0^2 + \overset{2}{v}_{\textcircled{1}}\, z)\, \mathbf{g}_2 + w\mathbf{i}_3, \tag{ii}$$

where v_0^1, v_0^2, w, $\overset{1}{v}_{\textcircled{1}}$, $\overset{2}{v}_{\textcircled{1}}$ are functions of (α^1, α^2) only, and $\mathbf{g}_1$, $\mathbf{g}_2$ are defined as in Problem 14. Confining our problem to the small displacement theory, show that strain–displacement relations are given by

$$\begin{aligned}
f_{11} &= g_{1i}\, v^i_{0;1} + g_{1i}\, \overset{i}{v}_{\textcircled{1};1}\, z, \\
f_{22} &= g_{2i}\, v^i_{0;2} + g_{2i}\overset{i}{v}_{\textcircled{1};2}\, z, \\
f_{33} &= 0, \\
2f_{12} &= (g_{1i}\, v^i_{0;2} + g_{2i} v^i_{0;1}) + (g_{1i}\overset{i}{v}_{\textcircled{1};2} + g_{2i}\overset{i}{v}_{\textcircled{1};1})\, z, \\
2f_{13} &= g_{1i}\, \overset{i}{v}_{\textcircled{1}} + w_{,1}, \\
2f_{23} &= g^2{}_i\, \overset{i}{v}_{\textcircled{1}} + w_{,2}.
\end{aligned} \tag{iii}$$

Show also that a plate theory including the effect of transverse shear deformation is obtainable in the nonorthogonal curvilinear coordinate system by use of these relations and the principle of virtual work, Eq. (4.80).

CHAPTER 9

Problem Related to Section 9.1

1. Show that Eqs. (9.7) and (9.8) may be derived from the conditions

$$R^{(0)\lambda}{}_{\cdot\mu\nu\omega} = 0$$

which have been introduced in Problem 7 of Chapter 4.

Problems Related to Sections 9.2, 9.3 and 9.4

2. We assume the displacement components

$$U = u + \zeta u_1, \quad V = v + \zeta v_1, \quad W = w \tag{i}$$

as given by Eqs. (9.30). Confining our problem to the small displacement theory, show the following relations:

(1) The strain–displacement relations are given by

$$\begin{aligned}
f_{11} &= A^2(1 - \zeta/R_\alpha)(l_{11} + \zeta m_{11}), \\
f_{22} &= B^2(1 - \zeta/R_\beta)(l_{22} + \zeta m_{22}), \\
f_{33} &= 0, \\
2f_{12} &= AB[(1 - \zeta/R_\alpha)(l_{12} + \zeta m_{12}) + (1 - \zeta/R_\beta)(l_{21} + \zeta m_{21})] \\
2f_{13} &= A(u_1 + l_{31}), \quad 2f_{23} = B(v_1 + l_{32}).
\end{aligned} \tag{ii}$$

(2) By use of equations (ii), we have

$$\begin{aligned}
&\iiint \tau^{\lambda\mu}\, \delta f_{\lambda\mu} \sqrt{g}\, d\alpha^1\, d\alpha^2\, d\alpha^3 \\
&\quad = \iint [N_\alpha\, \delta l_{11} + N_\beta\, \delta l_{22} + N_{\alpha\beta}\, \delta l_{21} + N_{\beta\alpha}\, \delta l_{12} \\
&\qquad + M_\alpha\, \delta m_{11} + M_\beta\, \delta m_{22} + M_{\alpha\beta}\, \delta m_{21} + M_{\beta\alpha}\, \delta m_{12} \\
&\qquad + Q_\alpha \delta(u_1 + l_{31}) + Q_\beta \delta(v_1 + l_{32})]\, AB\, d\alpha\, d\beta,
\end{aligned} \tag{iii}$$

where N_α, N_β, $N_{\alpha\beta}$, $N_{\beta\alpha}$, M_α, M_β, $M_{\alpha\beta}$, $M_{\beta\alpha}$ are defined by Eqs. (9.58) and (9.59), while Q_α and Q_β are defined by

$$Q_\alpha = \int \tau_{\alpha\zeta}\left(1 - \frac{\zeta}{R_\beta}\right) d\zeta, \quad Q_\beta = \int \tau_{\beta\zeta}\left(1 - \frac{\zeta}{R_\alpha}\right) d\zeta. \tag{iv}$$

(3) By use of equation (iii) and the principle of virtual work, we may derive the equations of equilibrium for the problem presented in Section 9.4 in the following form:

$$\begin{aligned}
&\frac{\partial}{\partial\alpha}(BN_\alpha) + \frac{\partial}{\partial\beta}(AN_{\beta\alpha}) + \frac{\partial A}{\partial\beta} N_{\alpha\beta} - \frac{\partial B}{\partial\alpha} N_\beta - \frac{AB}{R_\alpha} Q_\alpha + \bar{Y}_\alpha AB = 0, \\
&\frac{\partial}{\partial\beta}(AN_\beta) + \frac{\partial}{\partial\alpha}(BN_{\alpha\beta}) + \frac{\partial B}{\partial\alpha} N_{\beta\alpha} - \frac{\partial A}{\partial\beta} N_\alpha - \frac{AB}{R_\beta} Q_\beta + \bar{Y}_\beta AB = 0, \\
&\frac{\partial}{\partial\alpha}(BQ_\alpha) + \frac{\partial}{\partial\beta}(AQ_\beta) + AB\left(\frac{N_\alpha}{R_\alpha} + \frac{N_\beta}{R_\beta}\right) + \bar{Y}_n AB = 0, \\
&\frac{\partial}{\partial\alpha}(BM_\alpha) + \frac{\partial}{\partial\beta}(AM_{\beta\alpha}) + \frac{\partial A}{\partial\beta} M_{\alpha\beta} - \frac{\partial B}{\partial\alpha} M_\beta - ABQ_\alpha = 0, \\
&\frac{\partial}{\partial\beta}(AM_\beta) + \frac{\partial}{\partial\alpha}(BM_{\alpha\beta}) + \frac{\partial B}{\partial\alpha} M_{\beta\alpha} - \frac{\partial A}{\partial\beta} M_\alpha - ABQ_\beta = 0.
\end{aligned} \tag{v}$$

Compare these equations with Eqs. (9.60) and (9.68).

3. Show that equations (v) of Problem 2 are obtainable in a different way from the following vectorial equations:

$$\begin{aligned}&\frac{\partial}{\partial\alpha}\,[(N_\alpha \mathbf{a}^{(0)} + N_{\alpha\beta}\mathbf{b}^{(0)} + Q_\alpha \mathbf{n}^{(0)})\,B\,d\beta]\,d\alpha \\ &\quad + \frac{\partial}{\partial\beta}\,[(N_{\beta\alpha}\mathbf{a}^{(0)} + N_\beta \mathbf{b}^{(0)} + Q_\beta \mathbf{n}^{(0)})\,A\,d\alpha]\,d\beta \\ &\quad + [\bar{Y}_\alpha \mathbf{a}^{(0)} + \bar{Y}_\beta \mathbf{b}^{(0)} + \bar{Y}_n \mathbf{n}^{(0)}]\,AB\,d\alpha\,d\beta = 0, \end{aligned} \tag{i}$$

and

$$\begin{aligned}&\frac{\partial \mathbf{r}_0^{(0)}}{\partial\alpha}\,d\alpha \times [N_\alpha \mathbf{a}^{(0)} + N_{\alpha\beta}\mathbf{b}^{(0)} + Q_\alpha \mathbf{n}^{(0)}]\,B\,d\beta \\ &\quad + \frac{\partial \mathbf{r}_0^{(0)}}{\partial\beta}\,d\beta \times [N_{\beta\alpha}\mathbf{a}^{(0)} + N_\beta \mathbf{b}^{(0)} + Q_\beta \mathbf{n}^{(0)}]\,A\,d\alpha \\ &\quad + \frac{\partial}{\partial\alpha}\,[(-M_{\alpha\beta}\mathbf{a}^{(0)} + M_\alpha \mathbf{b}^{(0)})\,B\,d\beta]\,d\alpha \\ &\quad + \frac{\partial}{\partial\beta}\,[(-M_\beta \mathbf{a}^{(0)} + M_{\beta\alpha}\mathbf{b}^{(0)})\,A\,d\alpha]\,d\beta = 0, \end{aligned} \tag{ii}$$

which are derived by considering the equilibrium conditions of the shell element shown in Fig. 9.5 with respect to forces and moments.

4. We represent the unit vectors in the directions of the α- and β-coordinates, and the unit vector normal to the middle surface after deformation, by

$$\mathbf{a},\ \mathbf{b},\ \mathbf{n} \tag{i}$$

respectively:

$$\mathbf{a} = \frac{\mathbf{r}_{0,\alpha}}{|\mathbf{r}_{0,\alpha}|},\quad \mathbf{b} = \frac{\mathbf{r}_{0,\beta}}{|\mathbf{r}_{0,\beta}|},\quad \mathbf{n} = \frac{\mathbf{a}\times\mathbf{b}}{|\mathbf{a}\times\mathbf{b}|}. \tag{ii}$$

Show that linearization of equations (ii) leads to

$$\begin{bmatrix}\mathbf{a}\\ \mathbf{b}\\ \mathbf{n}\end{bmatrix} = \begin{bmatrix}1 & l_{21} & l_{31}\\ l_{12} & 1 & l_{32}\\ -l_{31} & -l_{32} & 1\end{bmatrix}\begin{bmatrix}\mathbf{a}^{(0)}\\ \mathbf{b}^{(0)}\\ \mathbf{n}^{(0)}\end{bmatrix} \tag{iii}$$

where terms higher than the second order with respect to the displacement components (u, v, w) have been neglected, assuming the displacements of the shell to be small. Show also that differentiation of equations (iii) with respect to α and β leads to

$$\frac{\partial}{\partial\alpha}\begin{bmatrix}\mathbf{a}\\ \mathbf{b}\\ \mathbf{n}\end{bmatrix} = \begin{bmatrix}\dfrac{l_{21}}{B}\dfrac{\partial A}{\partial\beta} - \dfrac{Al_{31}}{R_\alpha}, & -\dfrac{1}{B}\dfrac{\partial A}{\partial\beta} + \dfrac{\partial l_{21}}{\partial\alpha}, & \dfrac{A}{R_\alpha} + \dfrac{\partial l_{31}}{\partial\alpha}\\ \dfrac{1}{B}\dfrac{\partial A}{\partial\beta} + \dfrac{\partial l_{12}}{\partial\alpha} - \dfrac{Al_{32}}{R_\alpha}, & -\dfrac{l_{12}}{B}\dfrac{\partial A}{\partial\beta}, & \dfrac{\partial l_{32}}{\partial\alpha} + \dfrac{Al_{12}}{R_\alpha}\\ -\dfrac{A}{R_\alpha} - \dfrac{\partial l_{31}}{\partial\alpha} - \dfrac{l_{32}}{B}\dfrac{\partial A}{\partial\beta}, & -\dfrac{\partial l_{32}}{\partial\alpha} + \dfrac{l_{31}}{B}\dfrac{\partial A}{\partial\beta}, & -\dfrac{Al_{31}}{R_\alpha}\end{bmatrix}\begin{bmatrix}\mathbf{a}^{(0)}\\ \mathbf{b}^{(0)}\\ \mathbf{n}^{(0)}\end{bmatrix} \tag{iv}$$

and

$$\frac{\partial}{\partial\beta}\begin{bmatrix}\mathbf{a}\\ \mathbf{b}\\ \mathbf{n}\end{bmatrix} = \begin{bmatrix} -\dfrac{l_{21}}{A}\dfrac{\partial B}{\partial\alpha}, & \dfrac{1}{A}\dfrac{\partial B}{\partial\alpha}+\dfrac{\partial l_{21}}{\partial\beta}-\dfrac{Bl_{31}}{R_\beta}, & \dfrac{\partial l_{31}}{\partial\beta}+\dfrac{Bl_{21}}{R_\beta} \\ -\dfrac{1}{A}\dfrac{\partial B}{\partial\alpha}+\dfrac{\partial l_{12}}{\partial\beta}, & \dfrac{l_{12}}{A}\dfrac{\partial B}{\partial\alpha}-\dfrac{Bl_{32}}{R_\beta}, & \dfrac{B}{R_\beta}+\dfrac{\partial l_{32}}{\partial\beta} \\ -\dfrac{\partial l_{31}}{\partial\beta}+\dfrac{l_{32}}{A}\dfrac{\partial B}{\partial\alpha}, & -\dfrac{B}{R_\beta}-\dfrac{\partial l_{32}}{\partial\beta}-\dfrac{l_{31}}{A}\dfrac{\partial B}{\partial\alpha}, & -\dfrac{Bl_{32}}{R_\beta} \end{bmatrix}\begin{bmatrix}\mathbf{a}^{(0)}\\ \mathbf{b}^{(0)}\\ \mathbf{n}^{(0)}\end{bmatrix}.$$

Problems Related to the Conditions of Compatibility and Stress Functions

5. Using the results of Problem 9 of Chapter 4 and equations (ii) of Problem 2, write the Riemann–Christoffel curvature tensor

$$R_{2323} \quad R_{3131} \quad R_{1212}$$
$$R_{1231} \quad R_{2312} \quad R_{3123}$$

in terms of l_{11}, l_{22}, l_{12}, l_{21}, m_{11}, m_{22}, m_{12}, m_{21}, $u_1 + l_{31}$ and $v_1 + l_{32}$. After having noticed that the Riemann–Christoffel curvature tensor thus obtained can be expanded into power series with respect to ζ, obtain the expressions of these tensor components at $\zeta = 0$.

6. By use of Lagrange multipliers χ_3, ψ_1, ψ_2 and ψ_3, the principle of virtual work can be written as

$$\iint [N_\alpha\,\delta l_{11} + N_\beta\,\delta l_{22} + \cdots + Q_\beta\delta(v_1 + l_{32})]\,AB\,d\alpha\,d\beta$$
$$-\iiint [\chi_3\,\delta R_{1212} + \psi_1\,\delta R_{1231} + \psi_2\,\delta R_{2312} + \psi_3\,\delta R_{3123}]\sqrt{g}\,d\alpha\,d\beta\,d\zeta$$
$$+ \cdots = 0, \qquad \text{(i)}$$

where the expressions of the Riemann–Christoffel curvature tensor obtained in Problem 5 have been substituted. We expand the Christoffel's symbols and $\sqrt{g}$ in power series of ζ and introduce the following notations:

$$F = \int \chi_3\,d\zeta, \quad \Psi_1 = \int \psi_1\,d\zeta, \quad \Psi_2 = \int \psi_2\,d\zeta, \quad \Psi_3 = \int \psi_3\,d\zeta. \qquad \text{(ii)}$$

Show that from the requirement that the coefficients of δl_{11}, δl_{22}, ... must vanish in equation (i), we obtain N_α, N_β, ... and Q_β in terms of F, Ψ_1, Ψ_2 and Ψ_3, thus discovering that the latter play the role of stress functions in the small displacement theory of shells based on equation (i) of Problem 2. Show also that the stress functions thus obtained are equivalent to those derived in pages 33–6 of Ref. 9.2.

Problems Related to Other Theories of Shells

7. An approximate nonlinear theory for a thin shell has been developed in Section 5.2 of Ref. 9.16, which assumes

$$U = u - l_{31}\zeta, \quad V = v - l_{32}\zeta, \quad W = w, \tag{i}$$

$$e_{\alpha\alpha} = e_{\alpha\alpha 0} - \zeta\varkappa_\alpha, \quad e_{\beta\beta} = e_{\beta\beta 0} - \zeta\varkappa_\beta,$$
$$e_{\alpha\beta} = e_{\alpha\beta 0} - \zeta\varkappa_{\alpha\beta}, \tag{ii}$$

where $e_{\alpha\alpha 0}$, $e_{\beta\beta 0}$, $e_{\alpha\beta 0}$; $\varkappa_\alpha$, $\varkappa_\beta$; and $\varkappa_{\alpha\beta}$ are given by Eqs. (9.82); (9.50); and

$$2\varkappa_{\alpha\beta} = -\hat{m}_{21} - \hat{m}_{12}. \tag{iii}$$

The assumptions Eqs. (9.77) and (9.78) are also employed. Derive the equations of equilibrium, mechanical boundary conditions and stress resultant–displacement relations for the present approximate theory and compare them with those derived in Section 9.7.

8. An approximate small displacement theory for a thin shell is obtained by assuming that

$$U = u - \frac{1}{A}\frac{\partial w}{\partial \alpha}\zeta, \quad V = v - \frac{1}{B}\frac{\partial w}{\partial \beta}\zeta, \quad W = w, \tag{i}$$

$$\varepsilon_\alpha = \varepsilon_{\alpha 0} - \zeta\varkappa_\alpha, \quad \varepsilon_\beta = \varepsilon_{\beta 0} - \zeta\varkappa_\beta,$$
$$\gamma_{\alpha\beta} = \gamma_{\alpha\beta 0} - 2\zeta\varkappa_{\alpha\beta}, \tag{ii}$$

where $\varepsilon_{\alpha 0}$, $\varepsilon_{\beta 0}$, $\gamma_{\alpha\beta 0}$; $\varkappa_\alpha$, $\varkappa_\beta$ and $\varkappa_{\alpha\beta}$ are given by Eqs. (9.36); and

$$\varkappa_\alpha = \frac{1}{A}\frac{\partial}{\partial \alpha}\left(\frac{1}{A}\frac{\partial w}{\partial \alpha}\right) + \frac{1}{AB^2}\frac{\partial A}{\partial \beta}\frac{\partial w}{\partial \beta},$$
$$\varkappa_\beta = \frac{1}{B}\frac{\partial}{\partial \beta}\left(\frac{1}{B}\frac{\partial w}{\partial \beta}\right) + \frac{1}{A^2B}\frac{\partial B}{\partial \alpha}\frac{\partial w}{\partial \alpha},$$
$$2\varkappa_{\alpha\beta} = \frac{1}{B}\frac{\partial}{\partial \beta}\left(\frac{1}{A}\frac{\partial w}{\partial \alpha}\right) - \frac{1}{AB^2}\frac{\partial B}{\partial \alpha}\frac{\partial w}{\partial \beta} \tag{iii}$$
$$+ \frac{1}{A}\frac{\partial}{\partial \alpha}\left(\frac{1}{B}\frac{\partial w}{\partial \beta}\right) - \frac{1}{A^2B}\frac{\partial A}{\partial \beta}\frac{\partial w}{\partial \alpha}.$$

The assumptions Eqs. (9.77) and (9.78) are also employed. By use of these equations and the principle of virtual work, show that the equations of equilibrium are given by

$$[BN_\alpha]_{,\alpha} + [AN_{\beta\alpha}]_{,\beta} + \frac{\partial A}{\partial \beta}N_{\alpha\beta} - \frac{\partial B}{\partial \alpha}N_\beta + \bar{Y}_\alpha AB = 0,$$

$$[AN_\beta]_{,\beta} + [BN_{\alpha\beta}]_{,\alpha} + \frac{\partial B}{\partial \alpha}N_{\beta\alpha} - \frac{\partial A}{\partial \beta}N_\alpha + \bar{Y}_\beta AB = 0, \tag{iv}$$

$$[B\tilde{Q}_\alpha]_{,\alpha} + [A\tilde{Q}_\beta]_{,\beta} + AB\left(\frac{N_\alpha}{R_\alpha} + \frac{N_\beta}{R_\beta}\right) + \bar{Y}_n AB = 0,$$

mechanical boundary conditions by

$$N_{\alpha\nu} = \bar{N}_{\alpha\nu}, \quad N_{\beta\nu} = \bar{N}_{\beta\nu},$$
$$\tilde{Q}_\alpha l + \tilde{Q}_\beta m + \frac{\partial M_{\nu s}}{\partial s} = \bar{V}_n + \frac{\partial \bar{M}_{\nu s}}{\partial s}, \quad M_\nu = \bar{M}_\nu \tag{v}$$

and stress resultant–displacement relations by

$$N_\alpha = \frac{Eh}{(1-\nu^2)}[\varepsilon_{\alpha 0} + \nu\varepsilon_{\beta 0}], \quad N_\beta = \frac{Eh}{(1-\nu^2)}[\nu\varepsilon_{\alpha 0} + \varepsilon_{\beta 0}],$$
$$N_{\alpha\beta} = N_{\beta\alpha} = Gh\gamma_{\alpha\beta 0}, \tag{vi}$$
$$M_\alpha = -D[\varkappa_\alpha + \nu\varkappa_\beta], \quad M_\beta = -D[\nu\varkappa_\alpha + \varkappa_\beta],$$
$$M_{\alpha\beta} = M_{\beta\alpha} = -D(1-\nu)\varkappa_{\alpha\beta} \tag{vii}$$

for the present approximate theory, which is equivalent to the theory of Mushtari-Vlassov (see Ref. 9.1).

9. An approximate nonlinear theory for a thin shell is obtained by assuming

$$U = u - \frac{1}{A}\frac{\partial w}{\partial \alpha}\zeta, \quad V = v - \frac{1}{B}\frac{\partial w}{\partial \beta}\zeta, \quad W = w, \tag{i}$$

$$e_{\alpha\alpha} = e_{\alpha\alpha 0} - \zeta\varkappa_\alpha, \quad e_{\beta\beta} = e_{\beta\beta 0} - \zeta\varkappa_\beta,$$
$$e_{\alpha\beta} = e_{\alpha\beta 0} - \zeta\varkappa_{\alpha\beta} \tag{ii}$$

where $e_{\alpha\alpha 0}$, $e_{\beta\beta 0}$, $e_{\alpha\beta 0}$; $\varkappa_\alpha$, $\varkappa_\beta$, $\varkappa_{\alpha\beta}$ are given by

$$e_{\alpha\alpha 0} = \frac{1}{A}\frac{\partial u}{\partial \alpha} + \frac{v}{AB}\frac{\partial A}{\partial \beta} - \frac{w}{R_\alpha} + \frac{1}{2A^2}\left(\frac{\partial w}{\partial \alpha}\right)^2,$$
$$e_{\beta\beta 0} = \frac{1}{B}\frac{\partial v}{\partial \beta} + \frac{u}{AB}\frac{\partial B}{\partial \alpha} - \frac{w}{R_\beta} + \frac{1}{2B^2}\left(\frac{\partial w}{\partial \beta}\right)^2,$$
$$2e_{\alpha\beta 0} = \frac{1}{A}\frac{\partial v}{\partial \alpha} - \frac{u}{AB}\frac{\partial A}{\partial \beta} + \frac{1}{B}\frac{\partial u}{\partial \beta} - \frac{v}{AB}\frac{\partial B}{\partial \alpha} + \frac{1}{AB}\frac{\partial w}{\partial \alpha}\frac{\partial w}{\partial \beta}; \tag{iii}$$

and equations (iii) of Problem 8, respectively. The assumptions Eqs. (9.77) and (9.78) are also employed. With the aid of the principle of virtual work, show that the equations of equilibrium and mechanical boundary conditions for the present approximate theory are given by equations (iv) and (v) of Problem 8, if $\tilde{Q}_\alpha$ and $\tilde{Q}_\beta$ are replaced by

$$\tilde{Q}_\alpha + N_\alpha \frac{1}{A}\frac{\partial w}{\partial \alpha} + N_{\alpha\beta}\frac{1}{B}\frac{\partial w}{\partial \beta}, \tag{iv}$$

and

$$\tilde{Q}_\beta + N_{\beta\alpha}\frac{1}{A}\frac{\partial w}{\partial \alpha} + N_\beta \frac{1}{B}\frac{\partial w}{\partial \beta}, \tag{v}$$

respectively, and that stress resultant–displacement relations are obtainable from equations (vi) and (vii) of Problem 8 by replacing $\varepsilon_{\alpha 0}, \varepsilon_{\beta 0}, \gamma_{\alpha\beta 0}$ with $e_{\alpha\alpha 0}$, $e_{\beta\beta 0}$, $2e_{\alpha\beta 0}$, respectively. Note: see Ref. 9.29, p. 189.

10. We consider equilibrium conditions of the shell element shown in Fig. 9.5 to obtain the following vectorial equations:

$$\frac{\partial}{\partial\alpha}[(N_\alpha \mathbf{a} + N_{\alpha\beta}\mathbf{b} + Q_\alpha \mathbf{n})\, B\, d\beta]\, d\alpha$$

$$+ \frac{\partial}{\partial\beta}[(N_{\beta\alpha}\mathbf{a} + N_\beta \mathbf{b} + Q_\beta \mathbf{n})\, A\, d\alpha]\, d\beta$$

$$+ [\bar{Y}_\alpha \mathbf{a}^{(0)} + \bar{Y}_\beta \mathbf{b}^{(0)} + \bar{Y}_n \mathbf{n}^{(0)}]\, AB\, d\alpha\, d\beta = 0, \qquad \text{(i)}$$

and

$$\frac{\partial \mathbf{r}_0}{\partial\alpha} d\alpha \times [N_\alpha \mathbf{a} + N_{\alpha\beta}\mathbf{b} + Q_\alpha \mathbf{n}]\, B\, d\beta$$

$$+ \frac{\partial \mathbf{r}_0}{\partial\beta} d\beta \times [N_{\beta\alpha}\mathbf{a} + N_\beta \mathbf{b} + Q_\beta \mathbf{n}]\, A\, d\alpha$$

$$+ \frac{\partial}{\partial\alpha}[(-M_{\alpha\beta}\mathbf{a} + M_\alpha \mathbf{b})\, B\, d\beta]\, d\alpha$$

$$+ \frac{\partial}{\partial\beta}[(-M_\beta \mathbf{a} + M_{\beta\alpha}\mathbf{b})\, A\, d\alpha]\, d\beta = 0 \qquad \text{(ii)}$$

where **a**, **b** and **n** are defined in Problem 4 (compare these equations with equations (i) and (ii) of Problem 3). By use of equations (i) and (ii) thus obtained, plus equations (iii) and (iv) of Problem 4, derive equations of equilibrium in scalar forms for an approximate nonlinear theory of a thin shell based on the Kirchhoff hypothesis and compare them with Eqs. (9.94).

Problems Related to Nonorthogonal Curvilinear Coordinates† ‡

11. We represent the middle surface of the shell before deformation by a pair of parameters (α^1, α^2) so that

$$\mathbf{r}_0^{(0)} = \mathbf{r}_0^{(0)}(\alpha^1, \alpha^2) \qquad \text{(i)}$$

† See Refs. 10.2, 10.4, 10.8 and 10.14.

‡ It is noted here that a Greek letter will be assigned in place of (1, 2, 3) and a roman letter in place of (1, 2) in Problems 11 and 12. The summation convention will be employed. Thus, a twice-appearing Greek or roman letter means summation with respect to (1, 2, 3) or (1, 2), respectively.

and define the two base vectors in the middle surface and the unit vector normal to the middle surface by

$$\mathbf{g}_1 = \frac{\partial \mathbf{r}_0^{(0)}}{\partial \alpha^1}, \quad \mathbf{g}_2 = \frac{\partial \mathbf{r}_0^{(0)}}{\partial \alpha^2}, \quad \mathbf{g}_3. \tag{ii)†$$

By use of these vectors, we define $g_{\lambda\mu}$ and $g^{\lambda\mu}$ by

$$g_{\lambda\mu} = \mathbf{g}_\lambda \cdot \mathbf{g}_\mu = g_{\mu\lambda}, \quad g^{\lambda\nu} g_{\nu\mu} = \delta^\lambda_\mu. \tag{iii)†$$

Next, we represent the position vector of an arbitrary point of the shell before deformation by

$$\mathbf{r}^{(0)} = \mathbf{r}_0^{(0)} + \zeta \mathbf{g}_3, \tag{iv}$$

and employ the set of the three parameters $(\alpha^1, \alpha^2, \zeta)$ as a system of curvilinear coordinates, writing $\zeta = \alpha^3$ whenever convenient. With these preliminaries, show the following geometrical relations:

(1) Concerning the derivatives of the vectors $\mathbf{g}_\lambda$, we have the well-known formulae of Gauss and Weingarten in the theory of differential geometry:

$$\frac{\partial \mathbf{g}_j}{\partial \alpha^k} = \left\{ \begin{matrix} i \\ jk \end{matrix} \right\}_0 \mathbf{g}_i + H_{jk} \mathbf{g}_3, \tag{v}$$

$$\frac{\partial \mathbf{g}_3}{\partial \alpha^k} = -g^{ij} H_{jk} \mathbf{g}_i \tag{vi}$$

where

$$\left\{ \begin{matrix} i \\ jk \end{matrix} \right\}_0 = \frac{1}{2} g^{ia} (g_{aj,k} + g_{ak,j} - g_{jk,a}), \tag{vii}$$

$$H_{ij} = \frac{\partial^2 \mathbf{r}_0^{(0)}}{\partial \alpha^i \, \partial \alpha^j} \cdot \mathbf{g}_3 = H_{ji}. \tag{viii}$$

It is obvious that in the orthogonal curvilinear coordinate system introduced in Section 9.1, we have

$$H_{11} = \frac{1}{R_\alpha}, \quad H_{22} = \frac{1}{R_\beta}, \quad H_{12} = H_{21} = 0. \tag{ix}$$

(2) The distance between two neighboring points $(\alpha^1, \alpha^2, \alpha^3)$ and $(\alpha^1 + d\alpha^1, \alpha^2 + d\alpha^2, \alpha^3 + d\alpha^3)$ is given by

$$(ds^{(0)})^2 = (g_{ij} - 2H_{ij}\zeta + g^{pq} H_{pi} H_{qj} \zeta^2) \, d\alpha^i \, d\alpha^j + (d\zeta)^2. \tag{x}$$

† It is noted that $\mathbf{g}_\lambda$, $g_{\lambda\mu}$ and $g^{\lambda\mu}$ defined here are different from those introduced in Chapter 4. For the purpose of consistency, it may be better to write $\mathbf{g}_{0\lambda}$, $g_{0\lambda\mu}$ and $g_0^{\lambda\mu}$ instead of $\mathbf{g}_\lambda$, $g_{\lambda\mu}$ and $g^{\lambda\mu}$, respectively. However, we prefer simpler notations as far as Problems 11 and 12 are concerned.

(3) We calculate Christoffel's symbol $\left\{ \begin{matrix} \lambda \\ \mu\nu \end{matrix} \right\}$ in the space defined by equation (x), and denote its value at $\zeta = 0$ by $\left\{ \begin{matrix} \lambda \\ \mu\nu \end{matrix} \right\}_0$. We have

$$\left\{ \begin{matrix} i \\ jk \end{matrix} \right\}_0 = \frac{1}{2} g^{ia} (g_{aj,k} + g_{ak,j} - g_{jk,a}),$$

$$\left\{ \begin{matrix} i \\ j3 \end{matrix} \right\}_0 = \left\{ \begin{matrix} i \\ 3j \end{matrix} \right\}_0 = -g^{ia} H_{aj},$$

$$\left\{ \begin{matrix} 3 \\ ij \end{matrix} \right\}_0 = \left\{ \begin{matrix} 3 \\ ji \end{matrix} \right\}_0 = H_{ij}, \tag{xi}$$

$$\left\{ \begin{matrix} 1 \\ 33 \end{matrix} \right\}_0 = \left\{ \begin{matrix} 2 \\ 33 \end{matrix} \right\}_0 = \left\{ \begin{matrix} 3 \\ 33 \end{matrix} \right\}_0 = \left\{ \begin{matrix} 3 \\ 13 \end{matrix} \right\}_0 = \left\{ \begin{matrix} 3 \\ 31 \end{matrix} \right\}_0 = \left\{ \begin{matrix} 3 \\ 23 \end{matrix} \right\}_0 = \left\{ \begin{matrix} 3 \\ 32 \end{matrix} \right\}_0 = 0.$$

(4) Introducing a convention that $H_{\varkappa 3} = H_{3\varkappa} = 0;\ \varkappa = 1, 2, 3$, we can write

$$\left\{ \begin{matrix} \lambda \\ 3\mu \end{matrix} \right\}_0 = -g^{\lambda\varkappa} H_{\varkappa\mu}, \tag{xii}$$

$$\mathbf{g}_{\lambda,k} = \left\{ \begin{matrix} \mu \\ \lambda k \end{matrix} \right\}_0 \mathbf{g}_\mu, \tag{xiii}$$

$$\mathbf{r}^{(0)}{}_{,\lambda} = \mathbf{g}_\lambda + \zeta \left\{ \begin{matrix} \mu \\ 3\lambda \end{matrix} \right\}_0 \mathbf{g}_\mu. \tag{xiv}$$

(5) We define components of a vector $\mathbf{u}(\alpha^1, \alpha^2, \alpha^3)$ as follows:

$$\mathbf{u} = v^\lambda \mathbf{g}_\lambda. \tag{xv}$$

Then we have

$$\mathbf{u}_{,k} = v^\lambda{}_{;k}\, \mathbf{g}_\lambda \tag{xvi}$$

where and throughout Problems 11 and 12 it is defined that

$$v^\lambda{}_{;k} = \frac{\partial v^\lambda}{\partial \alpha^k} + \left\{ \begin{matrix} \lambda \\ \mu k \end{matrix} \right\}_0 v^\mu. \tag{xvii}$$

12. We assume the displacement vector $\mathbf{u}$ in the power series of ζ to be such that

$$\mathbf{u} = \sum_{n=0}^{m} v^\lambda_{\textcircled{n}} \zeta^n \mathbf{g}_\lambda, \tag{i}$$

where $v^\lambda_{\textcircled{n}}$ is a function of (α^1, α^2) only. By use of equation (i), an approximate theory of a shell of moderate thickness can be formulated by use of the principle of virtual work. For example, we may assume

$$\mathbf{u} = (v^1_0 + \zeta v^1_{\textcircled{1}})\, \mathbf{g}_1 + (v^2_0 + \zeta v^2_{\textcircled{1}})\mathbf{g}_2 + v^3_0 \mathbf{g}_3 \tag{ii}$$

and obtain

$$f_{ij} = {}_0f_{ij} + \zeta_{①}\, f_{ij} + \zeta^2_{②}\, f_{ij},$$
$$f_{i3} = {}_0f_{i3}, \quad f_{33} = 0 \tag{iii}$$

where

$$2_0 f_{ij} = g_{\lambda j} v^{\lambda}_{0;\,i} + g_{\lambda i} v^{\lambda}_{0;\,j},$$
$$2_{①} f_{ij} = g_{\lambda j} \overset{\lambda}{v}_{①;\,i} + g_{\lambda i}\, \overset{\lambda}{v}_{①;\,j},$$
$$+ g_{\lambda\mu}\left[\begin{Bmatrix}\mu\\3j\end{Bmatrix}_0 v^{\lambda}_{0;\,i} + \begin{Bmatrix}\mu\\3i\end{Bmatrix}_0 v^{\lambda}_{0;\,j}\right],$$
$$2_{②} f_{ij} = g_{\lambda\mu}\left[\begin{Bmatrix}\mu\\3j\end{Bmatrix}_0 \overset{\lambda}{v}_{①;\,i} + \begin{Bmatrix}\mu\\3i\end{Bmatrix}_0 v^{\lambda}_{0;\,j}\right],$$
$$2_0 f_{i3} = g_{i\lambda}\, \overset{\lambda}{v}_{①} + g_{3\lambda} v^{\lambda}_{0;\,i} \tag{iv}$$

and $\overset{3}{v}_{①} = 0$. In deriving equations (iv), only linear terms are retained, confining our problem to the small displacement theory. By use of these equations and the principle of virtual work, we may derive the small displacement theory of a thin shell which includes the effect of transverse shear deformation and is expressed in the non-orthogonal curvilinear coordinate system.

CHAPTER 10

Problems Related to Sections 10.2, 10.3 and 10.4

1. Show that combination of Eqs. (10.36), (10.39) and (10.41) yields:

$$\{u\} = [[K] - [H]\,[G]^{-1}\,[H]]\,\{\bar{X}\},$$

which provides the deflection influence coefficients, where

$$[K] = [\alpha]'\,[C]\,[\alpha].$$

2. We have considered in Section 10.2 a truss problem where the geometrical boundary conditions are prescribed by Eqs. (10.17), and obtained the conditions of compatibility (10.34) and the relations (10.40). Show that if the geometrical boundary conditions are prescribed such that

$$u_i = \bar{u}_i, \quad v_i = \bar{v}_i, \quad w_i = \bar{w}_i, \quad (i = k+1, \ldots, n)$$

the conditions of compatibility are given by

$$\sum_{(ij)} a_{ijp}\,\delta_{ij} - \sum_{i=k+1}^{J}\left[\sum_j (\bar{u}_i\lambda_{ij} + \bar{v}_i\mu_{ij} + \bar{w}_i\nu_{ij})\, a_{ijp}\right] = 0,$$
$$p = 1, 2, \ldots, R$$

and Castigliano's theorem provides:

$$\sum_{(ij)} \alpha_{ijl}\,\delta_{ij} - \sum_{i=k+1}^{J}\left[\bar{u}_i \sum_j (\lambda_{ij}\alpha_{ijl}) + \bar{v}_i \sum_j (\mu_{ij}\alpha_{ijl}) + \bar{w}_i \sum_j (\nu_{ij}\alpha_{ijl})\right] = u_l,$$

$$\sum_{(ij)} \beta_{ijl}\,\delta_{ij} - \sum_{i=k+1}^{J}\left[\bar{u}_i \sum_j (\lambda_{ij}\beta_{ijl}) + \bar{v}_i \sum_j (\mu_{ij}\beta_{ijl}) + \bar{w}_i \sum_j (\nu_{ij}\beta_{ijl})\right] = v_l,$$

$$\sum_{(ij)} \gamma_{ijl}\,\delta_{ij} - \sum_{i=k+1}^{J} \left[\bar{u}_i \sum_j (\lambda_{ij}\gamma_{ijl}) + \bar{v}_i \sum_j (\mu_{ij}\gamma_{ijl}) + \bar{w}_i \sum_j (\nu_{ij}\gamma_{ijl})\right] = w_l,$$
$$l = 1, 2, \ldots, k.$$

Problems Related to Section 10.7

3. We consider the beam element shown in Fig. 10.5, and assume that the force $\bar{P}$ is absent. Show that the strain energy stored in the beam is given by

$$U_{12} = \frac{EA_0}{2l}(u_2 - u_1)^2 + \frac{6EI}{l^3}\Big[(v_2 + \theta_1 l - v_1)^2 + l(v_2 + \theta_1 l - v_1)(\theta_2 - \theta_1) + \frac{1}{3} l^2(\theta_2 - \theta_1)^2\Big]. \qquad \text{(i)}$$

4. We consider the frame structure shown in Fig. 10.6, and cut it fictitiously into four members ⑭, ⑮, ㊾, and ㉓, where the section ⑤ is chosen perpendicular to the centroid locus of the frame at the point of application of the external force $\bar{P}$. Show that the expression of Π for the principle of minimum potential energy for the present problem is given by

$$\Pi = U_{15} + U_{52} + U_{14} + U_{23} + \bar{P}v_5, \qquad \text{(i)}$$

where $U_{15}(u_1, v_1, \theta_1, u_5, v_5, \theta_5)$, $U_{52}(u_5, v_5, \theta_5, u_2, v_2, \theta_2)$, $U_{14}(u_1, v_1, \theta_1)$ and $U_{23}(u_2, v_2, \theta_2)$ are the strain energies stored in the beam elements, respectively, and their expressions are obtainable by the use of equation (i) of Problem 3. The physical meanings of u_1, v_1, θ_1; u_2, v_2, θ_2; u_5, v_5, θ_5 are the displacement components in the x- and y-directions, and the rotational angle in the clockwise direction of the joints ①, ② and section ⑤, respectively.

Show also that by the use of the stationary conditions of Π with respect to u_5, v_5 and θ_5:

$$\frac{\partial \Pi}{\partial u_5} = 0, \quad \frac{\partial \Pi}{\partial v_5} = 0, \quad \frac{\partial \Pi}{\partial \theta_5} = 0, \qquad \text{(ii)}$$

we may eliminate u_5, v_5 and θ_5 from Π to obtain

$$\Pi^* = U_{12} + U_{14} + U_{23} - \frac{1}{384}\frac{\bar{P}^2 l^3}{EI} + \bar{P}\Big[\frac{1}{2}(v_1 + v_2) + \frac{l}{8}(\theta_2 - \theta_1)\Big] \qquad \text{(iii)}$$

and that the function Π^* thus obtained coincides with Eq. (10.68).

5. We consider a frame structure subject to concentrated forces and moments as shown in Fig. H 23. Show that we have

$$\begin{aligned}\Pi_c^* = {} & V_{12} + V_{23} + V_{14} + (N_{12} + Q_{14} + \bar{X}_1)\,u_1 \\ & + (Q_{12} - N_{14} - \bar{P} + \bar{Y}_1)\,v_1 \\ & + (M_{12} - M_{14} - Q_{12}l + \tfrac{1}{2}\bar{P}l + \bar{M}_1)\,\theta_1 \\ & + (-N_{12} + Q_{23} + \bar{X}_2)\,u_2 \\ & + (-Q_{12} - N_{23} + \bar{Y}_2)\,v_2 \\ & + (-M_{12} - M_{23} + \bar{M}_2)\,\theta_2\end{aligned}$$

instead of Eqs. (10.66) for the present frame problem.

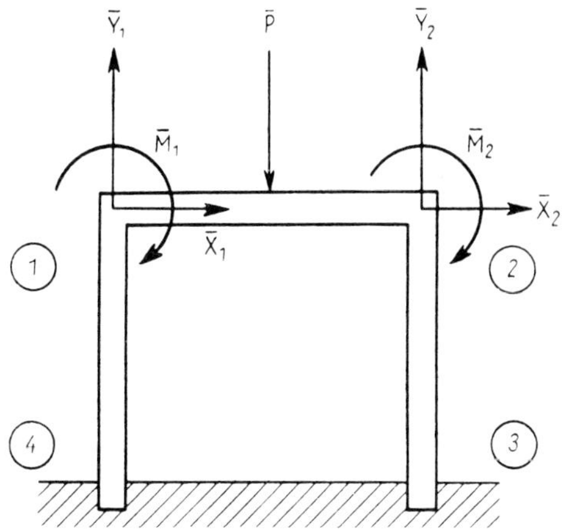

FIG. H 23.

6. We consider a plane frame structure as shown in Fig. H 24. Remembering that, from elementary beam theory, the complementary energy stored in the curved beam is given by

$$V = \int \left(\frac{N^2}{2A_0E} + \frac{M^2}{2EI} \right) ds,$$

where s is measured along the centroid of the curved beam, show that the force method is applicable to the present problem in a manner similar to the development of Section 10.6.

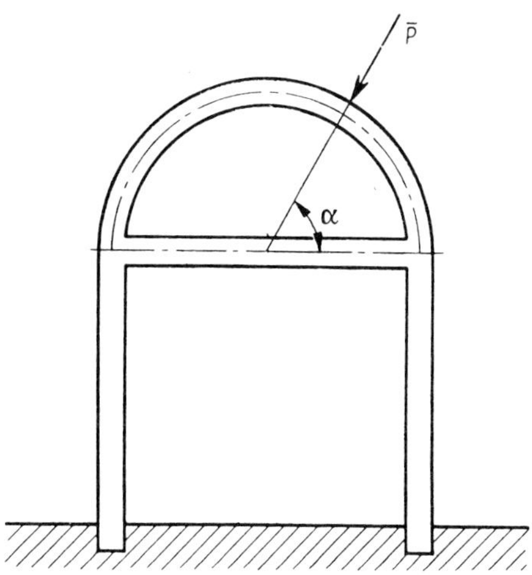

FIG. H 24.

Problem Related to Section 10.8

7. We consider a semi-monocoque plane structure consisting of panels and stringers as shown in Fig. H 25, and assume internal forces in these members as shown in Figs. 10.8 and 10.9. Show that by use of the principle of minimum complementary energy, into which the equilibrium conditions between members are introduced by use of Lagrange multipliers, we have the following compatibility condition:

$$\gamma_{\text{panel}} = \frac{1}{l_1}\left(\frac{1}{l_2}\int_0^{l_2} v_{12}\,dy - \frac{1}{l_2}\int_0^{l_2} v_{34}\,dy\right) + \frac{1}{l_2}\left(\frac{1}{l_1}\int_0^{l_1} u_{14}\,dx - \frac{1}{l_1}\int_0^{l_1} u_{23}\,dx\right),$$

where γ_{panel} is the shearing strain of the panel ⓵, while $u_{14}(x)$, $u_{23}(x)$ and $v_{12}(y)$, $v_{34}(y)$ are the displacements of the four stringers in the directions of the x- and y-axes, respectively.

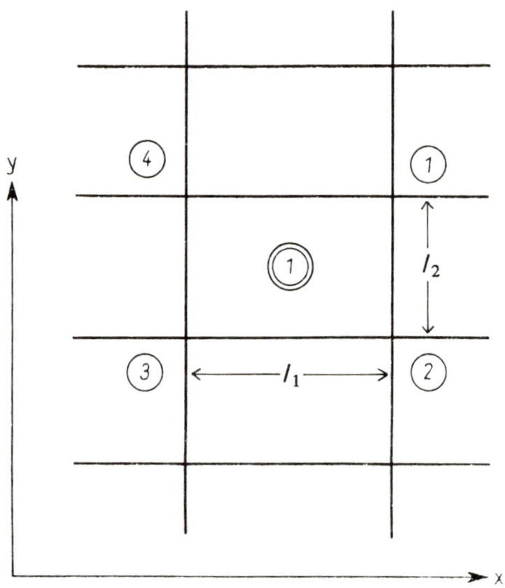

FIG. H 25.

Problems Related to the Unit Load Method

8. We consider a truss problem and denote the actual solution of the internal force and elongation of the ij-th member by P_{ij} and δ_{ij}, respectively. Show that if p_{ij} denotes the internal force in the ij-th member due to a unit virtual load acting at the joint for which the deflection is required (the load acts in the direction of the deflection), then the unit load theorem provides:

$$\delta = \sum_{(ij)} \left(\frac{pPl}{A_0 E}\right)_{ij}. \tag{i}$$

Next, show that if we substitute the relation

$$\delta_{ij} = (Pl/A_0E)_{ij} \tag{ii}$$

into equation (i), and write

$$\delta = \sum_{(ij)} (p\delta)_{ij}, \tag{iii}$$

equation (iii) holds irrespectively of the load–elongation relations. Namely, equation (iii) is applicable to problems of plastic as well as elastic truss problems. Show also that Eqs. (10.47) are obtainable by an application of the unit load method.

9. We consider a plane frame structure consisting of straight members, and employ for the analysis of the structure the elementary beam theory plus the assumption that the deformation due to axial force is negligible compared to that due to bending moment. The actual solutions of the curvature and bending moments of the *ij*-th member are denoted by $\varkappa_{ij}(x)$ and $M_{ij}(x)$ respectively, where x is measured along the centroid locus. Show that if $m_{ij}(x)$ denotes the bending moment in the *ij*-th member due to a unit virtual load acting at the point for which the deflection δ is required (the load acts in the direction of the deflection), then the unit load theorem provides:

$$\delta = \sum_{(ij)} \int_{ij} \frac{mM}{EI}\, dx. \tag{i}$$

Show also that if $m_{ij}(x)$ denotes the bending moment in the *ij*-th member due to unit virtual external moment acting at the point for which the rotation θ is required (the moment acts in the direction of the rotation), then the unit load theorem provides:

$$\theta = \sum_{(ij)} \int_{ij} \frac{mM}{EI}\, dx. \tag{ii}$$

Next, show that if we substitute the relation

$$M_{ij} = -(EI\varkappa)_{ij} \tag{iii}$$

into equations (i) and (ii), and write

$$\delta = -\sum_{(ij)} \int_{ij} m\varkappa\, dx \tag{iv}$$

and

$$\theta = -\sum_{(ij)} \int_{ij} m\varkappa\, dx, \tag{v}$$

respectively, equations (iv) and (v) hold irrespectively of the moment–curvature relation. Namely, these equations are applicable to problems of plastic as well as elastic frames. Show also that Eq. (10.65) is obtainable by an application of the unit load method.

10. Show that the unit load method can be applied also to a three-dimensional frame structure with naturally curved elements. The expression of the complementary energy

$$\int\left(\frac{N^2}{2EA_0}+\frac{M_x^2}{2EI_x}+\frac{M_y^2}{2EI_y}+\frac{T^2}{2GJ}+\frac{Q_x^2}{2Gk_xA_0}+\frac{Q_y^2}{2Gk_yA_0}\right)ds$$

may be helpful for the derivation of the unit load method formulae, where N is the axial force, M_x and M_y are bending moments about the two principal axes, T is the torsional moment, Q_x and Q_y are shearing forces, and s is measured along the centroid locus. Show also that by substitutions such as those mentioned in Problems 8 and 9, we obtain unit load formulae which are applicable to problems of plastic frame problems as well. Note: see Refs. 10.1, 10.2, 10.3 and 10.9 for numerical examples and further applications of the unit load method.

Problems Related to Initial Strains

11. We consider a truss problem and assume that the ij-th member has an initial excess of length of $\delta_{ij}^{(0)}$. Show that in applying the force method to the problem, the complementary energy

$$\left(\frac{P^2l}{2A_0E}\right)_{ij}$$

is to be replaced by

$$\left(\frac{P^2l}{2A_0E}+P\delta^{(0)}\right)_{ij}.$$

12. We consider a plane frame problem and assume that the ij-th member has initial strain $\varepsilon_x^{(0)}(z)$, where the (x, z) coordinate system is taken in a manner similar to Chapter 7. Show that in applying the force method to the problem, the complementary energy

$$\frac{N^2}{2A_0E}+\frac{M^2}{2EI}$$

is to be replaced by

$$\frac{N^2}{2A_0E}+\frac{N}{A_0}\iint\varepsilon_x^{(0)}\,dy\,dz$$
$$+\frac{M^2}{2EI}+\frac{M}{I}\iint\varepsilon_x^{(0)}\,z\,dy\,dz.$$

APPENDIX I

VARIATIONAL PRINCIPLES AS A BASIS FOR THE FINITE ELEMENT METHOD

Section 1. Introduction

Mathematical formulation of a problem for a continuous body is usually made by the use of differential equations, as exemplified by the finite displacement theory of elasticity introduced in Chapter 3, where mechanical or physical quantities of the continuous body, such as displacement, stress, strain and so forth, are assumed to be continuous functions of the space coordinates x_i; $i = 1, 2, 3$ and the continuous body is treated as an assembly of fictitious elements of *infinitesimal magnitude* as shown in Fig. 3.1.

On the other hand, the continuous body is divided into a number of fictitious elements of *finite magnitude*, ("finite elements"), and is treated as an assembly of these elements in the formulation of the *finite element method* (often abbreviated FEM). The continuous functions for the mechanical or physical quantities are now replaced by approximate functions which are smooth in each element, but are continuous and piecewise smooth in the whole body. These approximate functions are constructed by the use of unknown parameters such as values of the quantities at the so-called nodal points combined with the use of interpolation functions, in such a way that distributions of the quantities in each element may be determined uniquely once the values of the unknown parameters have been specified. Thus, we are replacing the original differential equations by a number of algebraic equations which govern the unknown parameters. Consequently, our next problem is how to obtain the governing equations for the unknown parameters.

It has been well established that the variational method provides a powerful and systematic tool for derivation of the governing equations for these unknowns. We remember that some mention has been made already in Chapter 10 of interrelations between variational methods and FEM. There, the generalized Galerkin method based on the principle of virtual work, and the Rayleigh–Ritz method based on the principles of minimum potential energy, of minimum complementary energy and so forth are shown to be applicable to the structural analysis of various assemblies of finite elements such as truss, frame and semi-monocoque structures. It is noted here that the terminology *finite element method* may include those techniques based on the generalized Galerkin method as well as the Rayleigh–Ritz method.†

It is now widely recognized that Courant was one of the pioneering mathematicians in the development of FEM. He presented an approximate solution of the Saint-Venant torsion problem formulated by the use of the principle of minimum complementary energy, assuming a linear distribution of the stress function in each of the assemblage of triangular elements.[1] On the other hand, the paper by Turner, Clough, Martin and Topp[2] and the work by Argyris and Kelsey[3] have been regarded as the most important and historical contributions among pioneering works in FEM in the field of structure. Since the appearance of these literatures, the variational method has been used extensively in the mathematical formulation of FEM. Conversely, the remarkable development of FEM has given great stimulus to the advancement of the variational methods: new variational principles such as variational principles with relaxed continuity requirements,[4–8] Herrmann's principles for incompressible and nearly incompressible materials[9, 10] and also for bending of plates[11, 12] and so forth have been established during the last ten years. The objective of this new appendix is to present a brief survey of recent developments of variational principles which provide a basis for the formulation of FEM in elasticity and plasticity. For practical applications of these principles to the formulation of FEM, the reader is directed to papers such as Refs. 5, 6 and 7.

As the contents of Refs. 2 and 3 show, the primary purpose of these works was to develop a numerical method of analysing the rigidity and stress of an elastic airplane. Since the appearance of these pioneering works, numerous papers have been published concerning applications of FEM in a broad field of engineering science.[13–16] FEM is now widely used not only for numerical analysis of stresses and displacements of elasto-plastic structures, but also for a variety of non-structural problems such as hydrodynamics, heat transfer, seepage and so forth. Thus, the FEM technology, aided by amazing advances of the digital computer, has been making a great contribution to practical applications and will be much more developed and much more in use in the future.

The bibliography of this short appendix is not intended to be complete. The author is satisfied with referring only to a very limited number of papers which are directly and closely related to mathematical formulations introduced in this new appendix. The reader is directed to Refs. 17 and 18, for example, for a complete bibliography for FEM.

† For the sake of simplicity, the generalized Galerkin method will be called the Galerkin method in this appendix.

Section 2. Conventional Variational Principles for the Small Displacement Theory of Elastostatics

The first topic of this appendix will be a review of conventional variational principles for the small displacement theory of elastostatics, the governing equations of which may be given as follows.†

(1) *Equations of equilibrium:*

$$\sigma_{ij,j} + \bar{f}_i = 0. \tag{I-2.1}$$‡

(2) *Strain–displacement relations:*

$$\varepsilon_{ij} = \tfrac{1}{2}(u_{i,j} + u_{j,i}). \tag{I-2.2}$$

(3) *Stress–strain relations:*

$$\sigma_{ij} = a_{ijkl}\varepsilon_{kl}, \tag{I-2.3}$$

or conversely

$$\varepsilon_{ij} = b_{ijkl}\sigma_{kl}. \tag{I-2.4}$$

(4) *Mechanical boundary conditions:*

$$T_i = \bar{T}_i \qquad \text{on } S_\sigma, \tag{I-2.5}$$§||

where

$$T_i = \sigma_{ij}n_j. \tag{I-2.6}$$¶

(5) *Geometrical boundary conditions:*

$$u_i = \bar{u}_i \qquad \text{on } S_u. \tag{I-2.7}$$||

In the above, Eqs. (I-2.3) and (I-2.4) are equivalent to Eqs. (1.6) and (1.8), respectively. For later convenience, Eqs. (1.6) and (1.8) are expressed in matrix forms as follows:

$$\{\sigma\} = [A]\{\varepsilon\} \tag{I-2.8}$$

$$\{\varepsilon\} = [B]\{\sigma\} \tag{I-2.9}$$

where

$$\{\sigma\}^T = [\sigma_x, \sigma_y, \sigma_z, \tau_{yz}, \tau_{zx}, \tau_{xy}],$$
$$\{\varepsilon\}^T = [\varepsilon_x, \varepsilon_y, \varepsilon_z, \gamma_{yz}, \gamma_{zx}, \gamma_{xy}],$$

† The summation convention is employed in this appendix unless otherwise stated. See footnote on page 231 for the convention.

‡ Notations $\bar{f}_i$; $i = 1, 2, 3$ are used instead of $\bar{X}$, $\bar{Y}$, $\bar{Z}$, respectively.

§ Notations T_i and $\bar{T}_i$; $i = 1, 2, 3$ are used instead of X_ν, Y_ν, Z_ν and $\bar{X}_\nu$, $\bar{Y}_\nu$, $\bar{Z}_\nu$, respectively.

¶ Notations n_i; $i = 1, 2, 3$ are used instead of l, m, n, respectively.

|| Notations S_σ and S_u are used instead of S_1 and S_2, respectively.

and where $[A]$ and $[B]$ are positive definite symmetric matrices which obey the relation

$$[B] = [A]^{-1}. \tag{I-2.10}$$

The strain energy function A and the complementary energy function B may be written as either

$$A(\varepsilon_{mn}) = \tfrac{1}{2}a_{ijkl}\varepsilon_{ij}\varepsilon_{kl}, \tag{I-2.11}$$

$$B(\sigma_{mn}) = \tfrac{1}{2}b_{ijkl}\sigma_{ij}\sigma_{kl}, \tag{I-2.12}$$

or

$$A(\varepsilon_{mn}) = \tfrac{1}{2}\{\varepsilon\}^T[A]\{\varepsilon\}, \tag{I-2.13†}$$

$$B(\sigma_{mn}) = \tfrac{1}{2}\{\sigma\}^T[B]\{\sigma\}. \tag{I-2.14†}$$

For later convenience, a notation $A(u_i)$ is introduced here. It is obtainable by substituting Eq. (I-2.2) into Eq. (I-2.11) and expressing the strain energy function in terms of displacement components:

$$A(u_i) = \tfrac{1}{8}a_{klmn}(u_{k,l} + u_{l,k})(u_{m,n} + u_{n,m}). \tag{I-2.15}$$

With these preliminaries, the conventional variational principles mentioned in Chapter 2 may be summarized as follows.

2.1. Principle of Virtual Work

The principle of virtual work may be written as

$$\iiint_V \sigma_{ij}\delta\varepsilon_{ij}dV - \iiint_V \bar{f}_i\delta u_i dV - \iint_{S_\sigma} \bar{T}_i\delta u_i dS = 0, \tag{I-2.16‡}$$

where the subsidiary conditions are

$$\delta\varepsilon_{ij} = \tfrac{1}{2}(\delta u_{i,j} + \delta u_{j,i}), \tag{I-2.17}$$

and

$$\delta u_i = 0 \quad \text{on} \quad S_u. \tag{I-2.18}$$

† Compare with Eqs. (2.2) and (2.20), respectively.
‡ Compare with Eq. (1.32).

2.2. Principle of Minimum Potential Energy

The functional for the principle of minimum potential energy may be written as

$$\Pi_P = \iiint_V [A(u_i) - \bar{f}_i u_i] dV - \iint_{S_\sigma} \bar{T}_i u_i dS \qquad \text{(I-2.19)}\dagger$$

where the subsidiary conditions are

$$u_i = \bar{u}_i \quad \text{on} \quad S_u. \qquad \text{(I-2.20)}$$

2.3. Generalized Principle

The functional for the generalized principle may be written as

$$\Pi_{G1} = \iiint_V \{A(\varepsilon_{ij}) - \bar{f}_i u_i - \sigma_{ij}[\varepsilon_{ij} - \tfrac{1}{2}(u_{i,j} + u_{j,i})]\} dV$$
$$- \iint_{S_\sigma} \bar{T}_i u_i dS - \iint_{S_u} p_i(u_i - \bar{u}_i) dS, \qquad \text{(I-2.21)}\ddagger$$

with no subsidiary conditions, where σ_{ij} and p_i are Lagrange multipliers, of which the physical meanings are given by Eqs. (2.28) and (2.33), respectively.

An alternate expression of the generalized principle may be written in the following form:

$$\Pi_{G2} = \iiint_V \{A(\varepsilon_{ij}) - \bar{f}_i u_i - \sigma_{ij}[\varepsilon_{ij} - \tfrac{1}{2}(u_{i,j} + u_{j,i})]\} dV$$
$$- \iint_{S_\sigma} \bar{T}_i u_i dS - \iint_{S_u} T_i(u_i - \bar{u}_i) dS, \qquad \text{(I-2.22)}\S$$

with no subsidiary conditions. The generalized principle, expressed by Eqs. (I-2.21) and (I-2.22), is sometimes called the Hu–Washizu principle.

2.4. Hellinger–Reissner principle

The functional for the Hellinger–Reissner principle may be written as

$$\Pi_R = \iiint_V [\tfrac{1}{2}\sigma_{ij}(u_{i,j} + u_{j,i}) - B(\sigma_{ij}) - \bar{f}_i u_i] dV$$
$$- \iint_{S_\sigma} \bar{T}_i u_i dS - \iint_{S_u} T_i(u_i - \bar{u}_i) dS, \qquad \text{(I-2.23)}\P$$

† Compare with Eq. (2.12). The notation Π_P is now widely used for expressing the functional for the principle of potential energy.

‡ Compare with Eq. (2.26). The notation Π_G will be used in this appendix to express the functional for the generalized principle.

§ Compare with Eq. (2.34).

¶ Compare with Eq. (2.37). We note that p_i in Eq. (2.37) have been replaced by T_i in Eq. (I-2.23).

with no subsidiary conditions. Integrations by parts lead to the following alternate expression for the principle:

$$-\Pi_R^* = \iiint_V [B(\sigma_{ij}) + (\sigma_{ij,j} + \bar{f}_i)u_i]dV - \iint_{S_\sigma} (T_i - \bar{T}_i)u_i dS - \iint_{S_u} T_i \bar{u}_i dS, \qquad \text{(I-2.24)†}$$

where no subsidiary conditions are imposed.

2.5. Principle of Minimum Complementary Energy

The functional for the principle of minimum complementary energy may be written as follows:

$$\Pi_C = \iiint_V B(\sigma_{ij})dV - \iint_{S_u} T_i \bar{u}_i dS, \qquad \text{(I-2.25)‡}$$

where the subsidiary conditions are

$$\sigma_{ij,j} + \bar{f}_i = 0 \quad \text{in} \quad V, \qquad \text{(I-2.26)}$$

and

$$T_i = \bar{T}_i \quad \text{on} \quad S_\sigma. \qquad \text{(I-2.27)}$$

2.6. Principle of Complementary Virtual Work

This principle may be written as

$$\iiint_V \varepsilon_{ij} \delta\sigma_{ij} dV - \iint_{S_u} \delta T_i \bar{u}_i dS = 0, \qquad \text{(I-2.28)§}$$

where the subsidiary conditions are given by

$$(\delta\sigma_{ij})_{,j} = 0 \quad \text{in} \quad V, \qquad \text{(I-2.29)}$$

and

$$\delta T_i = 0 \quad \text{on} \quad S_\sigma. \qquad \text{(I-2.30)}$$

These variational principles are represented in the left-hand column of Fig. I-1 in the form of a flow diagram.

† Compare with Eq. (2.41).
‡ Compare with Eq. (2.23).
§ Compare with Eq. (1.50).

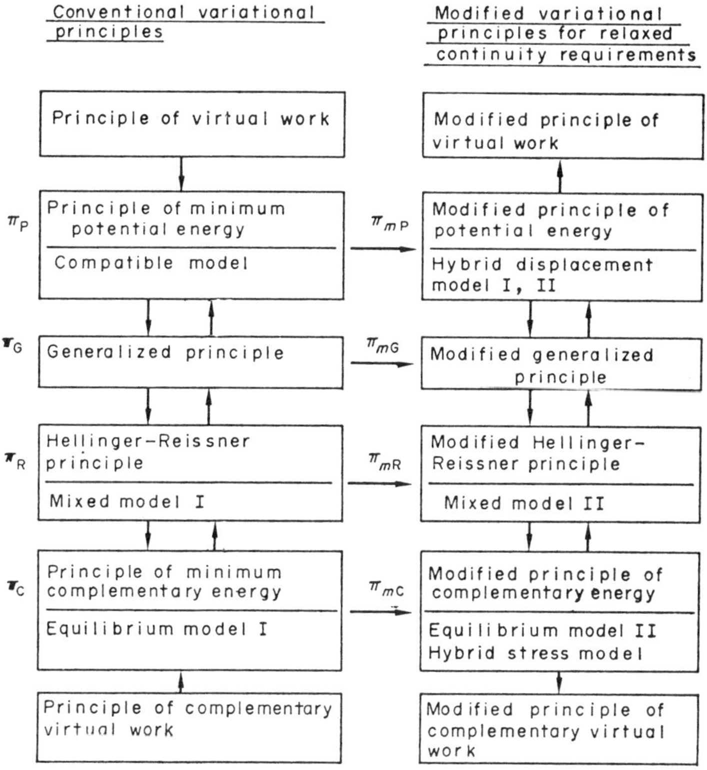

FIG. I-1. A flow diagram for the small displacement theory of elastostatics.

Section 3. Derivation of Modified Variational Principles from the Principle of Minimum Potential Energy

The purpose of the present section is to follow in Fig. I-1 an avenue which starts from the principle of minimum potential energy and leads to the modified principle of potential energy, the modified generalized principle and finally to the modified Hellinger–Reissner principle. We shall treat a solid body problem which is the same as defined in the preceding section, except that the region V is now subdivided fictitiously into a finite number of elements: V_1, V_2, V_3, ..., V_N. For later convenience, we denote two arbitrary adjacent elements by V_a and V_b, and the interelement boundary between V_a and V_b by S_{ab} as shown in Fig. I-2, where tetrahedral elements are used for the purpose of illustration. Two symbols S_{ab}^* and S_{ba}^* will be used whenever necessary to distinguish the interelement boundary S_{ab} belonging to ∂V_a and ∂V_b, respectively.†

† ∂V_a denote the entire boundary of V_a.

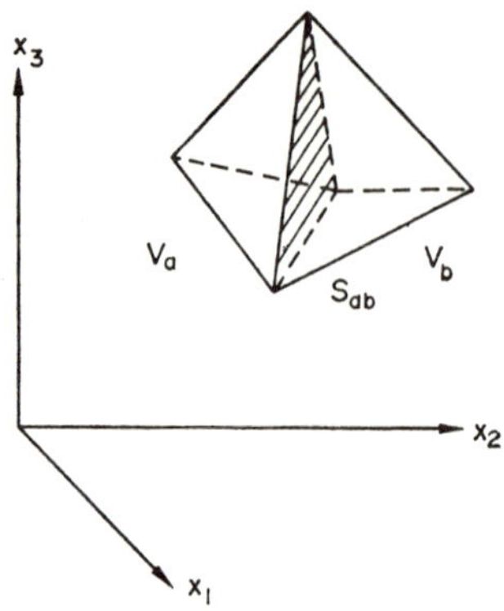

FIG. I-2. V_a, V_b and S_{ab}.

3.1. Principle of Minimum Potential Energy

We shall denote displacements u_i in each element by

$$u_i^{(1)}, u_i^{(2)}, \ldots, u_i^{(a)}, u_i^{(b)}, \ldots u_i^{(N)}; \; i = 1, 2, 3,$$

each of which will be called displacement functions. Then, the assembly of these displacement functions may be taken as admissible functions for the functional of the principle of minimum potential energy, if they satisfy following requirements:

(i) They are continuous and single-valued in each element.
(ii) They are conforming on interelement boundaries:†

$$u_i^{(a)} = u_i^{(b)} \quad \text{on} \quad S_{ab} \tag{I-3.1}$$

(iii) Those belonging to an element containing S_u satisfy Eq. (I-2.7).
Consequently, if the displacement functions are so chosen as to satisfy the requirements (i), (ii) and (iii), the functional for the principle of minimum potential energy is given by

$$\Pi_P = \Sigma \iiint_{V_a} [A(u_i) - \bar{f}_i u_i] dV - \iint_{S_\sigma} \bar{T}_i u_i dS, \tag{I-3.2‡}$$

where the notation Σ means summation over all the elements. The independent quantities subject to variation in Π_P are $u_i^{(1)}, u_i^{(2)}, \ldots, u_i^{(a)}, u_i^{(b)}, \ldots, u_i^{(N)}$ (abridged as $u_i^{(a)}$ hereafter).

† See Refs. 14 and 19 for the definition of the terminology *conforming*.
‡ Compare with Eq. (I-2.19).

3.2. Modified Principles of Potential Energy

Next, we shall formulate a variational principle in which the subsidiary conditions (I-3.1) are introduced into the framework of the variational expression. By the use of Lagrange multipliers λ_i defined on S_{ab}, we obtain the functional for a modified principle as follows:

$$\Pi_{mP1} = \Pi_P - \Sigma H_{ab1} \tag{I-3.3}$$

where Π_P is given by Eq. (I-3.2), and

$$H_{ab1} = \iint\limits_{S_{ab}} \lambda_i(u_i^{(a)} - u_i^{(b)})dS. \tag{I-3.4}$$

In Eq. (I-3.3), the notation Σ in front of H_{ab1} means summation over all the interelement boundaries. The independent quantities subject to variation in Π_{mP1} are $u_i^{(a)}$ and λ_i under the subsidiary conditions, Eq. (I-2.7). The principle for the functional Π_{mP1} will be called *the first modified principle of potential energy with relaxed continuity requirements* because the requirement (ii) is relaxed in Π_{mP1} and the displacement functions in each element may be chosen independently without any concern about the conformity requirement. After some manipulation including integrations by parts, the first variation of Π_{mP1} on S_{ab} is shown to be

$$\begin{aligned}\delta\Pi_{mP1} = & \text{-----------} \\ & + \iint\limits_{S_{ab}} \{[T_i^{(a)}(u_j^{(a)}) - \lambda_i]\delta u_i^{(a)} + [T_i^{(b)}(u_j^{(b)}) + \lambda_i]\delta u_i^{(b)} \\ & \qquad - (u_i^{(a)} - u_j^{(b)})\delta\lambda_i\}dS + \dots,\end{aligned} \tag{I-3.5}$$

and we obtain the following stationary conditions on S_{ab}:

$$T_i^{(a)}(u_j^{(a)}) = -T_i^{(b)}(u_j^{(b)}) = \lambda_i, \tag{I-3.6}$$

$$u_i^{(a)} = u_i^{(b)} \tag{I-3.7}$$

where $T_i^{(a)}(u_j^{(a)})$ and $T_i^{(b)}(u_j^{(b)})$ are obtainable from

$$T_i^{(a)} = \sigma_{ij}^{(a)} n_j^{(a)}, \; T_i^{(b)} = \sigma_{ij}^{(b)} n_j^{(b)}, \tag{I-3.8}$$

by substituting Eqs. (I-2.2) and (I-2.3) to express $\sigma_{ij}^{(a)}$ and $\sigma_{ij}^{(b)}$ in terms of $u_i^{(a)}$ and $u_i^{(b)}$, respectively. Needless to say, $n_i^{(a)}$ and $n_i^{(b)}$ are the direction cosines of the outward-drawn normals on S_{ab}^* and S_{ba}^*, respectively, and we have

$$n_i^{(a)} = -n_i^{(b)}. \tag{I-3.9}$$

The stationary conditions (I-3.6) indicate the physical meaning of the Lagrange multiplier: λ_i is equal to $T_i^{(a)}(u_j^{(a)})$ on S_{ab}. It is noted here that the modified principle is no longer a minimum principle, but keeps its stationary property only. The functional Π_{mP1} was originally proposed by Jones [20] and later developed further by Yamamoto.[21]

The functional Π_{mP1} will be modified slightly. We introduce two functions $\lambda_i^{(a)}$ and $\lambda_i^{(b)}$, which are defined on S_{ab}^* and S_{ba}^*, respectively, and obey the following relation:

$$\lambda_i^{(a)} + \lambda_i^{(b)} = 0. \tag{I-3.10}$$

Then, by writing

$$\lambda_i = \lambda_i^{(a)}, \quad -\lambda_i = \lambda_i^{(b)}, \tag{I-3.11}$$

the integrand of Eq. (I-3.4) can be expressed by

$$\lambda_i^{(a)} u_i^{(a)} + \lambda_i^{(b)} u_i^{(b)}$$

under the subsidiary condition (I-3.10). Consequently, introducing a new Lagrange multiplier μ_i defined on S_{ab}, we may write Eq. (I-3.4) in an equivalent form denoted by H_{ab2} as follows:

$$H_{ab2} = \iint_{S_{ab}} [\lambda_i^{(a)} u_i^{(a)} + \lambda_i^{(b)} u_i^{(b)} - \mu_i(\lambda_i^{(a)} + \lambda_i^{(b)})] dS, \tag{I-3.12}$$

or

$$H_{ab2} = \iint_{S_{ab}^*} \lambda_i^{(a)}(u_i^{(a)} - \mu_i) dS + \iint_{S_{ba}^*} \lambda_i^{(b)}(u_i^{(b)} - \mu_i) dS. \tag{I-3.13}$$

By the use of H_{ab2} thus defined, Eq. (I-3.3) may be written in another form as follows:

$$\Pi_{mP2} = \Pi_P - \Sigma H_{ab2}. \tag{I-3.14}$$

This principle will be called *the second modified principle of potential energy with relaxed continuity requirements*, where the independent quantities subject to variation are $u_i^{(a)}$, $\lambda_i^{(a)}$ and μ_i under the subsidiary conditions Eq. (I-2.7). Among these quantities, $u_i^{(a)}$ in V_a, and $\lambda_i^{(a)}$ on S_{ab}^* may be chosen independently of $u_i^{(b)}$ in V_b, and $\lambda_i^{(b)}$ on S_{ba}^*, respectively, while μ_i defined on S_{ab} must be common to S_{ab}^* and S_{ba}^*. After some manipulation including partial integrations, the first variation of Π_{mP2} on S_{ab} is shown to be

$$\begin{aligned} \delta\Pi_{mP2} = \ldots & \\ + \iint_{S_{ab}} \{ & [T_i^{(a)}(u_j^{(a)}) - \lambda_i^{(a)}]\delta u_i^{(a)} + [T_i^{(b)}(u_j^{(b)}) - \lambda_i^{(b)}]\delta u_i^{(b)} \\ & + (u_i^{(a)} - \mu_i)\delta\lambda_i^{(a)} + (u_i^{(b)} - \mu_i)\delta\lambda_i^{(b)} \\ & - (\lambda_i^{(a)} + \lambda_i^{(b)})\delta\mu_i \} dS + \ldots \end{aligned} \tag{I-3.15}$$

and we obtain the following stationary conditions on S_{ab}:

$$\lambda_i^{(a)} = T_i^{(a)}(u_j^{(a)}), \quad \lambda_i^{(b)} = T_i^{(b)}(u_j^{(b)}), \tag{I-3.16}$$

$$u_i^{(a)} = \mu_i, \quad u_i^{(b)} = \mu_i, \tag{I-3.17}$$

$$\lambda_i^{(a)} + \lambda_i^{(b)} = 0. \tag{I-3.18}$$

The stationary conditions (I-3.16) and (I.3.17) indicate the physical meaning of the Lagrange multipliers: $\lambda_i^{(a)}$, $\lambda_i^{(b)}$ and μ_i are equal to $T_i^{(a)}(u_j^{(a)})$ on S_{ab}^*, $T_i^{(b)}(u_j^{(b)})$ on S_{ba}^* and u_i on S_{ab}, respectively.

If we employ the stationary conditions (I-3.16) in order to eliminate $\lambda_i^{(a)}$ and $\lambda_i^{(b)}$, we may write H_{ab2} in another form as follows:

$$H_{ab3} = \iint_{S_{ab}^*} T_i^{(a)}(u_j^{(a)})(u_i^{(a)} - \mu_i)dS$$
$$+ \iint_{S_{ba}^*} T_i^{(b)}(u_j^{(b)})(u_i^{(b)} - \mu_i)dS, \tag{I-3.19}$$

and we obtain

$$\Pi_{mP3} = \Pi_P - \Sigma H_{ab3}. \tag{I-3.20}$$

This principle will be called *the third modified principle of potential energy with relaxed continuity requirements*, where the independent quantities subject to variation are $u_i^{(a)}$ and μ_i under the subsidiary conditions Eq. (I-2.7). Among these quantities subject to variation, $u_i^{(a)}$ in $V^{(a)}$ may be chosen independently of $u_i^{(b)}$ in V_b, while μ_i should be common to S_{ab}^* and S_{ba}^*. The functional Π_{mP2} and Π_{mP3} are equivalent to those derived originally by Tong.[22] The modified principles with relaxed continuity requirements will be called modified principles hereafter for the sake of brevity.

3.3. Modified Generalized Principle

The modified principles of potential energy thus derived may be generalized in a familiar manner. We shall start from the functional Π_{mP2} to obtain the functional for a generalized principle as follows:

$$\Pi_{mG1} = \Sigma \iiint_{V_a} \{A(\varepsilon_{ij}) - \bar{f}_i u_i$$
$$- \sigma_{ij}[\varepsilon_{ij} - \tfrac{1}{2}(u_{i,j} + u_{j,i})]\} dV - \Sigma H_{ab2}$$
$$- \iint_{S_\sigma} \bar{T}_i u_i dS - \iint_{S_u} p_i(u_i - \bar{u}_i)dS, \tag{I-3.21†}$$

where the independent quantities subject to variation are $\varepsilon_{ij}^{(a)}$, $\sigma_{ij}^{(a)}$, $u_i^{(a)}$, $\lambda_i^{(a)}$ and μ_i with no subsidiary conditions. It can be shown that stationary conditions of Π_{mG1} on S_{ab} provide,

$$\lambda_i^{(a)} = T_i^{(a)}, \ \lambda_i^{(b)} = T_i^{(b)}, \tag{I-3.22}$$

† Compare with Eq. (I-2.21).

together with Eqs. (I-3.17) and (I-3.18). Consequently, we may write the functional for the generalized principle in another equivalent form as follows:

$$\Pi_{mG2} = \Sigma \iiint_{V_a} \{A(\varepsilon_{ij}) - \bar{f}_i u_i - \sigma_{ij}[\varepsilon_{ij} - \tfrac{1}{2}(u_{i,j} + u_{j,i})]\} dV - \Sigma H_{ab4} - \iint_{S_\sigma} \bar{T}_i u_i dS - \iint_{S_u} T_i(u_i - \bar{u}_i) dS, \qquad \text{(I-3.23)}\dagger$$

where

$$H_{ab4} = \iint_{S_{ab}} [T_i^{(a)} u_i^{(a)} + T_i^{(b)} u_i^{(b)} - \mu_i (T_i^{(a)} + T_i^{(b)})] dS, \qquad \text{(I-3.24)}$$

or

$$H_{ab4} = \iint_{S_{ab}^*} T_i^{(a)} (u_i^{(a)} - \mu_i) dS + \iint_{S_{ba}^*} T_i^{(b)} (u_i^{(b)} - \mu_i) dS. \qquad \text{(I-3.25)}$$

In Eq. (I-3.23), the independent quantities subject to variation are $\varepsilon_{ij}^{(a)}$, $\sigma_{ij}^{(a)}$, $u_i^{(a)}$ and μ_i with no subsidiary conditions.

3.4. Modified Hellinger–Reissner Principle

Elimination of ε_{ij} from the functional Π_{mG2} by the use of the stationary conditions (I-2.4) leads to the modified Hellinger–Reissner functional:

$$\Pi_{mR} = \Sigma \iiint_{V_a} [-B(\sigma_{ij}) + \tfrac{1}{2}\sigma_{ij}(u_{i,j} + u_{j,i}) - \bar{f}_i u_i] dV - \Sigma \iint_{S_{ab}} [T_i^{(a)} u_i^{(a)} + T_i^{(b)} u_i^{(b)} - \mu_i (T_i^{(a)} + T_i^{(b)})] dS - \iint_{S_\sigma} \bar{T}_i u_i dS - \iint_{S_u} T_i(u_i - \bar{u}_i) dS, \qquad \text{(I-3.26)}\ddagger$$

where the independent quantities subject to variation are $\sigma_{ij}^{(a)}$, $u_i^{(a)}$ and μ_i with no subsidiary conditions. Through integrations by parts, we may obtain another expression for the modified Hellinger–Reissner functional:

$$-\Pi_{mR}^* = \Sigma \iiint_{V_a} [B(\sigma_{ij}) + (\sigma_{ij,j} + \bar{f}_i) u_i] dV - \Sigma G_{ab} - \iint_{S_\sigma} (T_i - \bar{T}_i) u_i dS - \iint_{S_a} T_i \bar{u}_i dS, \qquad \text{(I-3.27)}\S$$

where

$$G_{ab} = \iint_{S_{ab}} \mu_i (T_i^{(a)} + T_i^{(b)}) dS, \qquad \text{(I-3.28)}$$

and the independent quantities subject to variations are $\sigma_{ij}^{(a)}$, $u_i^{(a)}$ and μ_i with no subsidiary conditions.

† Compare with Eq. (I-2.22).
‡ Compare with Eq. (I-2.23).
§ Compare with Eq. (I-2.24).

Section 4. Derivation of Modified Variational Principles from the Principle of Minimum Complementary Energy

The purpose of this section is to follow in Fig. I-1 an avenue which starts from the principle of minimum complementary energy, leading to the modified principle of complementary energy and finally to the modified Hellinger–Reissner principle. We shall treat the same problem as defined in the beginning of Section 3, and proceed to a formulation of the principle of minimum complementary energy for the assembly of the finite elements.

4.1. Principle of Minimum Complementary Energy

We shall denote stresses in each element by

$$\sigma_{ij}^{(1)}, \sigma_{ij}^{(2)}, \ldots, \sigma_{ij}^{(a)}, \sigma_{ij}^{(b)}, \ldots, \sigma_{ij}^{(N)}; \ i, j = 1, 2, 3$$

each of which will be called a function for stresses. The assembly of these functions for stresses may be taken as admissible functions for the functional of the principle of minimum complementary energy, if they satisfy following requirements:

(i) They are continuous, single-valued and satisfy Eq. (I-2.1) in each element.

(ii) They satisfy equilibrium conditions on inter-element boundaries:

$$T_i^{(a)} + T_i^{(b)} = 0 \quad \text{on} \quad S_{ab}, \tag{I-4.1}$$

where $T_i^{(a)}$ and $T_i^{(b)}$ are defined by Eqs. (I-3.8).

(iii) Those belonging to an element containing S_σ satisfy Eq. (I-2.5).

Consequently, if the functions for stresses are so chosen as to satisfy the requirements (i), (ii) and (iii), the functional for the principle of minimum complementary energy is given by

$$\Pi_C = \Sigma \iiint_{V_a} B(\sigma_{ij}) dV - \iint_{S_u} T_i \bar{u}_i dS, \tag{I-4.2}$$ †

where the independent quantities subject to variation are $\sigma_{ij}^{(a)}$.

Next, we shall formulate a variational principle in which the subsidiary conditions (I-4.1) are introduced into the framework of the variational expression. By the use of Lagrange multipliers μ_i defined on S_{ab}, we obtain the functional for a modified principle as follows:

$$\Pi_{mC} = \Pi_C - \Sigma G_{ab}, \tag{I-4.3}$$

† Compare with Eq. (I-2.25).

where Π_C is given by Eq. (I-4.2) and where it is defined that

$$G_{ab} = \iint_{S_{ab}} \mu_i (T_i^{(a)} + T_i^{(b)}) dS \tag{I-4.4}$$

and the independent quantities subjected to variation are $\sigma_{ij}^{(a)}$ and μ_i under the subsidiary conditions, Eq. (I-2.1) and (I-2.5). The principle for the functional Π_{mC} will be called *the modified principle of complementary energy with relaxed continuity requirements*, because the requirement (ii) is relaxed in Π_{mC} and the functions for stresses in each element may be chosen independently without any concern about the equilibrium requirements on the interelement boundaries. It is noted here that the modified principle is no longer a minimum principle, but keeps its stationary property only. The functional Π_{mC} was originally formulated by Pian.[4, 5]

4.2. Modified Hellinger–Reissner Principle

Next, we shall introduce the subsidiary conditions, (I-2.1) and (I-2.5), into the framework of the variational expression Π_{mC} by the use of Lagrange multipliers u_i. Then, we may have a functional which is the same as that of the modified Hellinger–Reissner principle $-\Pi_{mR}^*$ given by Eq. (I-3.27). Needless to say, it is a simple matter to transform $-\Pi_{mR}^*$ thus obtained into Π_{mR} defined by Eq. (I-3.26) through integration by parts.

Thus far, two avenues in the flow diagram of Fig. I-1 have been traced. Arrows in the diagram show conventional avenues leading from one principle to another. The reader is advised to follow these arrows and familiarize himself with these transformations.

Several typical finite element models are also listed in the flow diagram, together with the variational principles on which the models are based. A detailed description of interrelations between these variational principles and related finite element models are beyond the intended scope of this appendix. Therefore only a brief mention will be made of finite element models based on the principle of virtual work. For details of these interrelations, the reader is directed to Refs. 5 through 8 and 23, for example.

As mentioned in the Introduction of this book, an approximate method of solution based on the principle of virtual work is called the Galerkin method, which may be considered as an application of the method of weighted residuals. As far as the elastostatic problem in the small displacement theory is concerned, this method provides a finite element formulation which is equivalent to that obtained by the use of the compatible model. However, the principle of virtual work or its equivalent provides a basis which is broader than variational principles when applied to problems outside of the small displacement elasticity problem. Similar observations may be made concerning finite element formulations based on the principle of complementary virtual work, the modified principle of virtual work, and the modified principle of complementary virtual work.

PROBLEM 1. Show that the modified principle of virtual work is given as follows:

$$\Sigma\iiint_{V_a}[\sigma_{ij}\delta\varepsilon_{ij} - \bar{f}_i\delta u_i]dV - \Sigma\delta\iint_{S_{ab}}\lambda_i(u_i^{(a)} - u_i^{(b)})dS - \iint_{S_\sigma}\bar{T}_i\delta u_i dS = 0, \tag{I-4.5}$$

where the subsidiary conditions are given by Eqs. (I-2.17) and (I-2.18).

PROBLEM 2. Show that the modified principle of complementary virtual work is given as follows:

$$\Sigma\iiint_{V_a}\varepsilon_{ij}\delta\sigma_{ij}dV - \Sigma\delta\iint_{S_{ab}}\mu_i(T_i^{(a)} + T_i^{(b)})dS - \iint_{S_u}\delta T_i\bar{u}_i dS = 0, \tag{I-4.6}$$

where the subsidiary conditions are given by Eqs. (I-2.29) and (I-2.30).

PROBLEM 3. Read Refs. 5, 6, 7 and 22, and show that:

(a) Displacements $\mu_i(=u_i)$ are assumed along all the interelement boundaries, and the stiffness matrix of each element is to be obtained by the use of the principle of minimum potential energy in the hybrid displacement model II based on Π_{mP2} or Π_{mP3}.

(b) Displacements $\mu_i(=u_i)$ are assumed along all the interelement boundaries, and the stiffness matrix of each element is to be obtained by the use of the principle of minimum complementary energy in the hybrid stress model based on Π_{mC}.

PROBLEM 4. Show that by the introduction of a new quantity e defined by

$$3e = \varepsilon_x + \varepsilon_y + \varepsilon_z,$$

the strain energy function A given by Eq. (2.3) can be generalized as follows:

$$\begin{aligned} A(\varepsilon_x, \varepsilon_y, \ldots, \gamma_{xy}; e, H) &= \frac{G\nu}{(1-2\nu)}(3e)^2 + G(\varepsilon_x^2 + \varepsilon_y^2 + \varepsilon_z^2) \\ &\quad + \tfrac{1}{2}G(\gamma_{yz}^2 + \gamma_{zx}^2 + \gamma_{xy}^2) \\ &\quad - 2\nu GH[3e - (\varepsilon_x + \varepsilon_y + \varepsilon_z)], \end{aligned}$$

where H is a Lagrange multiplier, which is multiplied by $2\nu G$ for the sake of convenience.

Next, show that $A(\varepsilon_x, \varepsilon_y, \ldots, \gamma_{xy}; e, H)$ is transformed into

$$\begin{aligned} &A(\varepsilon_x, \varepsilon_y, \ldots, \gamma_{xy}; H) \\ &= G[\varepsilon_x^2 + \varepsilon_y^2 + \varepsilon_z^2 + \tfrac{1}{2}(\gamma_{yz}^2 + \gamma_{zx}^2 + \gamma_{xy}^2) \\ &\quad + 2\nu H(\varepsilon_x + \varepsilon_y + \varepsilon_z) - \nu(1 - 2\nu)H^2] \end{aligned}$$

through elimination of e by the use of the stationary condition of $A(\varepsilon_x, \varepsilon_y, \ldots, \gamma_{xy}; e, H)$ with respect to e, namely,

$$3e = (1 - 2\nu)H.$$

Finally, indicate that $A(\varepsilon_x, \varepsilon_y, \ldots, \gamma_{xy}; H)$ is equivalent to the strain energy function derived by Herrmann for nearly incompressible materials.(9, 10)

Section 5. Conventional Variational Principles for the Bending of a Thin Plate

We shall devote the present and next sections to the derivation of conventional and modified variational principles for the problem defined in Section 8.1, namely, the bending of a thin plate based on the Kirchhoff hypotheses, because problems of plate bending are frequently treated in numerical examples of various finite element models. We shall first review some fundamental relations of the problem. Unless otherwise stated, we shall employ the same notation as used in Chapter 8.

We remember that in the bending theory of a thin plate, the stress–strain relations are given by Eq. (8-2), and the strain energy function A and complementary energy function B are given by

$$A = \frac{E}{2(1 - \nu^2)}(\varepsilon_x + \varepsilon_y)^2 + \frac{G}{2}(\gamma_{yz}^2 + \gamma_{zx}^2 + \gamma_{xy}^2 - 4\varepsilon_x\varepsilon_y), \qquad \text{(I-5.1)†}$$

and

$$B = \frac{1}{2E}[(\sigma_x + \sigma_y)^2 + 2(1 + \nu)(\tau_{yz}^2 + \tau_{zx}^2 + \tau_{xy}^2 - \sigma_x\sigma_y)] \qquad \text{(I-5.2)†}$$

respectively. We remember also that the Kirchhoff hypothesis imposes geometrical constraints such that

$$U = -zw_{,x}, \; V = -zw_{,y}, \; W = w, \qquad \text{(I-5.3)‡}$$

and consequently,

$$\varepsilon_x = -zw_{,xx}, \; \varepsilon_y = -zw_{,yy}, \; \gamma_{xy} = -2zw_{,xy}, \; \varepsilon_z = \gamma_{xz} = \gamma_{yz} = 0, \qquad \text{(I-5.4)§}$$

† Refer to Eqs. (3) and (4) of Appendix B.
‡ Refer to Eqs. (8.14).
§ Refer to Eq. (8.15).

where $w(x, y)$ is the displacement of the middle surface in the direction of the z-axis. Two relations are noted here, since they are frequently used in subsequent formulations:

$$\begin{aligned} (\ )_{,x} &= l(\ )_{,\nu} - m(\ )_{,s}, \\ (\ )_{,y} &= m(\ )_{,\nu} + l(\ )_{,s}, \end{aligned} \qquad \text{(I-5.5)†}$$

which hold on the boundary C and

$$\begin{aligned} &\iint_{S_m} [M_x w_{,xx} + 2M_{xy} w_{,xy} + M_y w_{,yy}] dxdy \\ &= \int_{C_\sigma + C_u} [-V_z w + M_\nu w_{,\nu} + M_{\nu s} w_{,s}] ds \\ &+ \iint_{S_m} [M_{x,xx} + 2M_{xy,xy} + M_{y,yy}] w dxdy, \end{aligned} \qquad \text{(I-5.6)‡}$$

where V_z, M_ν and $M_{\nu s}$ are defined by Eqs. (8.24) and (8.25) by the use of M_x, M_y and M_{xy}.

5.1. Principle of Minimum Potential Energy

The functional for the principle of minimum potential energy for the plate bending problem is given as follows:

$$\begin{aligned} \Pi_P &= \iint_{S_m} [\hat{A}(w) - \bar{p}w] dxdy \\ &+ \int_{C_\sigma} [-\bar{V}_z w + \bar{M}_\nu w_{,\nu} + \bar{M}_{\nu s} w_{,s}] ds, \end{aligned} \qquad \text{(I-5.7)}$$

where

$$\hat{A}(w) = \frac{D}{2} [(w_{,xx} + w_{,yy})^2 + 2(1 - \nu)(w^2_{,xy} - w_{,xx} w_{,yy})], \qquad \text{(I-5.8)}$$

and where the subsidiary conditions are

$$w = \bar{w}, \quad w_{,\nu} = \overline{w_{,\nu}} \quad \text{on} \quad C_u. \qquad \text{(I-5.9)}$$

The functional (I-5.7) can be derived from Eq. (I-2.19) in a manner similar to the development in Section 8.2, by first substituting Eqs. (I-5.1), (I-5.3) and (I-5.4) into Eq. (I-2.19) and then performing integrations with respect to z, noticing that

$$dS = dzds. \qquad \text{(I-5.10)}$$

A further partial integration was performed in Section 8.2 to obtain the

† Refer to Eq. (8.20).

‡ Refer to Eqs. (8.19), (8.20) and (8.21). Notations C_σ and C_u are used instead of C_1 and C_2, respectively.

mechanical boundary conditions in a manner as given by Eq. (8.31). However, it is preferable for later formulations to write the integration on C_σ as it is in Eq. (I-5.7).

5.2. Generalized Principle

The functional (I-5.7) may be transformed through familiar procedure to obtain the functional for a generalized principle:

$$\begin{aligned}\Pi_{G1} = \iint_{S_m} [\hat{A}(\varkappa_x, \varkappa_y, \varkappa_{xy}) + (\varkappa_x - w_{,xx})M_x \\ + (\varkappa_y - w_{,yy})M_y + 2(\varkappa_{xy} - w_{,xy})M_{xy} - \bar{p}w]dxdy \\ + \int_{C_\sigma} [-\bar{V}_z w + \bar{M}_\nu w_{,\nu} + \bar{M}_{\nu s} w_{,s}]ds \\ + \int_{C_u} [-(w - \bar{w})P_3 + (w_{,\nu} - \overline{w_{,\nu}})P_1 \\ + (w_{,s} - \bar{w}_{,s})P_2]ds,\end{aligned} \qquad \text{(I-5.11)}$$

where

$$\hat{A}(\varkappa_x, \varkappa_y, \varkappa_{xy}) = \frac{D}{2}[(\varkappa_x + \varkappa_y)^2 + 2(1-\nu)(\varkappa_{xy}^2 - \varkappa_x \varkappa_y)], \qquad \text{(I-5.12)}$$

and where P_1, P_2 and P_3 are Lagrange multipliers on C_u defined later by Eqs. (I-5.17). The last term on the right-hand side of Eq. (I-5.11) is obtainable from the last term on the right-hand side of Eq. (2.26) of Chapter 2, which is written for the present problem as follows:

$$-\int_{C_u} \int_{-h/2}^{h/2} [(U - \bar{U})p_x + (V - \bar{V})p_y + (W - \bar{W})p_z]dzds, \qquad \text{(I-5.13)}$$

where p_x, p_y and p_z are Lagrange multipliers which introduce the geometrical boundary conditions into the variational expression. By the use of Eq. (I-5.3), (I-5.5) and (I-5.9), we may derive following geometrical relations on the boundary C_u:

$$\begin{aligned} U &= -zw_{,x} = -z(lw_{,\nu} - mw_{,s}), \\ V &= -zw_{,y} = -z(mw_{,\nu} + lw_{,s}), \\ W &= w, \end{aligned} \qquad \text{(I-5.14)}$$

and

$$\begin{aligned} \bar{U} &= -z(l\overline{w_{,\nu}} - m\bar{w}_{,s}), \\ \bar{V} &= -z(m\overline{w_{,\nu}} + l\bar{w}_{,s}), \\ \bar{W} &= \bar{w}. \end{aligned} \qquad \text{(I-5.15)}$$

Substitution of Eqs. (I-5.14) and (I-5.15) into the integral (I-5.13) and integrations with respect to z transform the integral into

$$\int_{C_u} [-(w - \bar{w})P_3 + (w_{,\nu} - \overline{w_{,\nu}})P_1 + (w_{,s} - \bar{w}_{,s})P_2]ds, \qquad \text{(I-5.16)}$$

where

$$P_1 = l\int p_x z dz + m\int p_y z dz,$$
$$P_2 = -m\int p_x z dz + l\int p_y z dz,$$
$$P_3 = \int p_z dz, \tag{I-5.17}$$

and we obtain the last term in Eq. (I-5.11). It is noted here that the Lagrange multipliers P_2 and P_3 in Eq. (I-5.11) cannot be assumed independently, because w and $w_{,s}$ are not independent on C_u.†

We may obtain another expression of the generalized variational principle in which the Lagrange multipliers P_1, P_2 and P_3 have been eliminated. For this purpose, we may require the coefficients of δw and $\delta w_{,\nu}$ on C_u of $\delta\Pi_{G1}$ to vanish. After some manipulations including integrations by parts and the use of Eq. (I-5.6), we find that the first variation of Π_{G1} on C_u takes the following form:

$$\int_{C_u}[(V_z - P_3)\delta w - (M_\nu - P_1)\delta w_{,\nu} - (M_{\nu s} - P_2)\delta w_{,s}]ds + \ldots$$
$$= -(M_{\nu s} - P_2)\delta w|_{C_u} + \int_{C_u}\{[(V_z + M_{\nu s,s}) - (P_3 + P_{2,s})]\delta w$$
$$- (M_\nu - P_1)\delta w_{,\nu}\}ds + \ldots \quad . \tag{I-5.18}$$

Consequently, the requirement that the coefficients of δw and $\delta w_{,\nu}$ on C_u must vanish provides:

$$V_z + M_{\nu s,s} = P_3 + P_{2,s}, \quad M_\nu = P_1 \quad \text{on} \quad C_u, \tag{I-5.19}$$

and

$$M_{\nu s} = P_2 \quad \text{at the ends of} \quad C_u. \tag{I-5.20}$$

We find from Eqs. (I-5.19) and (I-5.20) that P_1, P_2 and P_3 may be replaced by M_ν, $M_{\nu s}$ and V_z, respectively, in the integral (I-5.16). Thus, we may transform Π_{G1} into

$$\Pi_{G2} = \iint_{S_m}[\hat{A}(\varkappa_x, \varkappa_y, \varkappa_{xy}) + (\varkappa_x - w_{,xx})M_x$$
$$+ (\varkappa_y - w_{,yy})M_y + 2(\varkappa_{xy} - w_{,xy})M_{xy} - \bar{p}w]dxdy$$
$$+ \int_{C_\sigma}[-\bar{V}_z w + \bar{M}_\nu w_{,\nu} + \bar{M}_{\nu s} w_{,s}]ds$$
$$+ \int_{C_u}[- V_z(w - \bar{w}) + M_\nu(w_{,\nu} - \overline{w_{,\nu}}) + M_{\nu s}(w_{,s} - \bar{w}_{,s})]ds, \tag{I-5.21}$$

where the independent quantities subject to variation are $\varkappa_x$, $\varkappa_y$, $\varkappa_{xy}$, w, M_x, M_y and M_{xy} with subsidiary conditions that $w = \bar{w}$ at the ends of C_u.‡

† See Section 8.2.

‡ The functional Π_{G2} with no subsidiary conditions is obtainable by eliminating P_1, P_2 and P_3 by the use of the stationary conditions of Π_{G1} on $C_\sigma + C_u$. It is given by adding $-\Sigma(w - \bar{w})(M_{\nu s} - \bar{M}_{\nu s})|_{C_u}$ terms to the R.H.S. of Eq. (I-5.21), where Σ means summations over all the C_us.

5.3. Hellinger-Reissner Principle

We may eliminate $\varkappa_x$, $\varkappa_y$ and $\varkappa_{xy}$ from the functional Π_{G2} through the use of the stationary conditions, Eq. (8.54), to obtain the functional for the Hellinger–Reissner principle:

$$\Pi_R = \iint_{S_m} [-M_x w_{,xx} - M_y w_{,yy} - 2M_{xy} w_{,xy} - \hat{B}(M_x, M_y, M_{xy}) - \bar{p}w]dxdy + \int_{C_\sigma} [-\bar{V}_z w + \bar{M}_\nu w_{,\nu} + \bar{M}_{\nu s} w_{,s}]ds + \int_{C_u} [-V_z(w - \bar{w}) + M_\nu(w_{,\nu} - \overline{w_{,\nu}}) + M_{\nu s}(w_{,s} - \bar{w}_{,s})]ds, \quad \text{(I-5.22)}$$

where

$$\hat{B}(M_x, M_y, M_{xy}) = \frac{1}{2}\frac{12}{Eh^3}[(M_x + M_y)^2 + 2(1+\nu)(M_{xy}^2 - M_x M_y)]. \quad \text{(I-5.23)}$$

By the use of Eq. (I-5.6), the functional (I-5.22) may be transformed into another expression of the functional for the Hellinger–Reissner principle:

$$-\Pi_R^* = \iint_{S_m} [\hat{B}(M_x, M_y, M_{xy}) + (M_{x,xx} + 2M_{xy,xy} + M_{y,yy} + \bar{p})w]dxdy + \int_{C_\sigma} [-(V_z - \bar{V}_z)w + (M_\nu - \bar{M}_\nu)w_{,\nu} + (M_{\nu s} - \bar{M}_{\nu s})w_{,s}]ds + \int_{C_u} [-V_z \bar{w} + M_\nu \overline{w_{,\nu}} + M_{\nu s}\bar{w}_{,s}]ds. \quad \text{(I-5.24)}$$

It is obvious that the functional for the principle of minimum complementary energy may be derived from Eq. (I-5.24). We repeat here that special care must be taken for formulating the mechanical boundary conditions for plate bending problems under the Kirchhoff hypothesis.

Section 6. Derivation of Modified Variational Principles for the Bending of a Thin Plate

We shall continue to treat the problem defined in the preceding section, except that the region S_m is now divided into a number of finite elements: $S_1, S_2, \ldots, S_N$, and the whole region is treated as an assembly of these elements. For later convenience, we denote two arbitrary adjacent elements by S_a and S_b, and the interelement boundary between S_a and S_b by C_{ab} as shown in Fig. I-3. Two symbols C^*_{ab} and C^*_{ba} will be used whenever necessary to distinguish the interelement boundary C_{ab} belonging to ∂S_a and ∂S_b, respectively. Arrows labelled with s_a and s_b in the same figure denote the

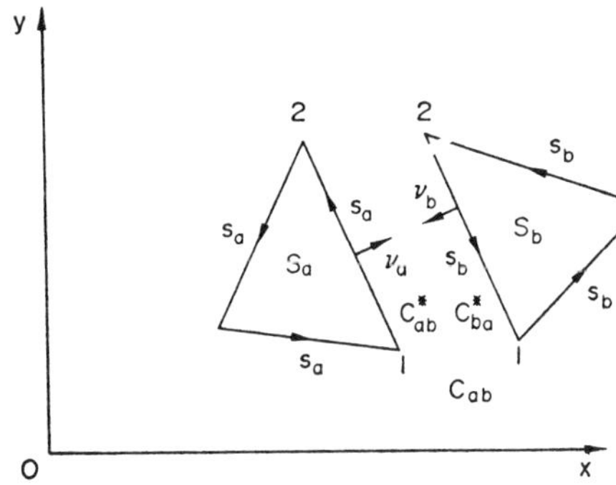

FIG. I-3. S_a, S_b and C_{ab}.

directions of measuring s along the boundaries of ∂S_a and ∂S_b, respectively. Moreover, two arrows labelled with ν_a and ν_b denote the outward normals on C^*_{ab} and C^*_{ba}, respectively.

6.1. Principle of Minimum Potential Energy

We shall denote the deflection $w(x, y)$ in each element by

$$w^{(1)}, w^{(2)}, \ldots, w^{(a)}, w^{(b)}, \ldots, w^{(N)}.$$

The assembly of these displacement functions may be taken as admissible functions for the functional of the principle of minimum potential energy, if they satisfy following requirements:

(i) They are continuous and single-valued in each element.
(ii) They are conforming on interelement boundaries:

$$w^{(a)} = w^{(b)}, \quad w^{(a)}_{,\nu_a} = -w^{(b)}_{,\nu_b} \quad \text{on} \quad C_{ab}. \tag{I-6.1}$$

(iii) Those belonging to an element containing C_u satisfy Eq. (I-5.9).

Consequently, if the displacement functions are chosen to satisfy the requirements (i), (ii) and (iii), the functional for the principle of minimum potential energy is given by

$$\Pi_P = \Sigma \iint_{S_a} [\hat{A}(w) - \bar{p}w]dxdy$$
$$+\int_{C_\sigma} [-\bar{V}_z w + \bar{M}_\nu w_{,\nu} + \bar{M}_{\nu s} w_{,s}]ds, \qquad \text{(I-6.2)†}$$

where the notation Σ means summation over the entire elements. The independent quantities subject to variation in Π_P are $w^{(a)}$ under the subsidiary conditions (ii) and (iii).

6.2. Modified Principles of Potential Energy

Next, we shall formulate a variational principle in which the subsidiary conditions (I-6.1) are introduced into the framework of the variational expression. By the use of H_{ab1} defined by Eq. (I-3.4) and remembering that

$$U^{(a)} = -zw^{(a)}_{,x} \qquad U^{(b)} = -zw^{(b)}_{,x},$$
$$V^{(a)} = -zw^{(a)}_{,y} \qquad V^{(b)} = -zw^{(b)}_{,y},$$
$$W^{(a)} = w^{(a)}, \qquad W^{(b)} = w^{(b)},$$

and

$$(\ )_{,x} = l(\ )_{,\nu_a} - m(\ )_{,s_a},$$
$$(\ )_{,y} = m(\ )_{,\nu_a} + l(\ )_{,s_a},$$
$$(\ )_{,x} = -l(\ )_{,\nu_b} + m(\ )_{,s_b},$$
$$(\ )_{,y} = -m(\ )_{,\nu_b} - l(\ )_{,s_b},$$

where l and m are direction cosines of the normal ν_a, we may transform H_{ab1} into:

$$H_{ab1} = \int_{C^*_{ab}} [\Lambda_3 w^{(a)} - \Lambda_1 w^{(a)}_{,\nu_a} - \Lambda_2 w^{(a)}_{,s_a}]ds_a$$
$$+ \int_{C^*_{ba}} [-\Lambda_3 w^{(b)} - \Lambda_1 w^{(b)}_{,\nu_b} - \Lambda_2 w^{(b)}_{,s_b}]ds_b, \qquad \text{(I-6.3)}$$

where

$$\Lambda_1 = l\int \lambda_1 z dz + m \int \lambda_2 z dz,$$
$$\Lambda_2 = -m\int \lambda_1 z dz + l \int \lambda_2 z dz,$$
$$\Lambda_3 = \int \lambda_3 dz \qquad \text{(I-6.4)}$$

are Lagrange multipliers derived from $(\lambda_1, \lambda_2, \lambda_3)$. Consequently, we have the following functional for the modified principle of potential energy:

$$\Pi_{mP1} = \Pi_P - \Sigma H_{ab1}, \qquad \text{(I-6.5)}$$

† Compare with Eq. (I-5.7).

where Π_P and H_{ab1} are given by Eqs. (I-6.2) and (I-6.3), respectively. It is noted here that the Lagrange multipliers Λ_2 and Λ_3 in H_{ab1} cannot be assumed independently, because $w^{(a)}$ and $w^{(a)}_{,s_a}$ and also $w^{(b)}$ and $w^{(b)}_{,s_b}$ are not independent on C_{ab}.

PROBLEM. Show that H_{ab1} may be transformed through integrations by parts into the following form:

$$H_{ab1} = \int_{C^*_{ab}} [(\Lambda_3 + \Lambda_{2,s_a})w^{(a)} - \Lambda_1 w^{(a)}_{,\nu_a}]ds_a$$

$$+ \int_{C^*_{ba}} [-(\Lambda_3 + \Lambda_{2,s_a})w^{(b)} - \Lambda_1 w^{(b)}_{,\nu_b}]ds_b - \Lambda_2(w^{(a)} - w^{(b)})|_1^2, \qquad \text{(I-6.6)}$$

where the notation $|_1^2$ means that values at the ends of C_{ab} are taken. *Note:* See Eq. (8.27).

6.3. Modified Generalized Principle

We note (omitting the algebra details) that the following functional for the modified generalized principle may be derived from Eq. (I-6.5):

$$\Pi_{mG} = \Sigma \iint_{S_a} [\hat{A}(\varkappa_x, \varkappa_y, \varkappa_{xy}) + (\varkappa_x - w_{,xx})M_x$$

$$+ (\varkappa_y - w_{,yy})M_y + 2(\varkappa_{xy} - w_{,xy})M_{xy} - \bar{p}w]dxdy$$

$$-\Sigma H_{ab1} + \int_{C_\sigma} [-\bar{V}_z w + \bar{M}_\nu w_{,\nu} + \bar{M}_{\nu s} w_{,s}]ds$$

$$+ \int_{C_u} [-V_z(w - \bar{w}) + M_\nu(w_{,\nu} - \overline{w_{,\nu}}) + M_{\nu s}(w_{,s} - \bar{w}_{,s})]ds. \qquad \text{(I-6.7)†}$$

6.4. Modified Hellinger–Reissner Principle

By the use of Eqs. (8.54), we may eliminate $\varkappa_x$, $\varkappa_y$ and $\varkappa_{xy}$ from Eq. (I-6.7) to obtain a functional for the modified Hellinger–Reissner principle:

$$\Pi_{mR} = \Sigma \iint_{S_a} [-M_x w_{,xx} - M_y w_{,yy}$$

$$- 2M_{xy} w_{,xy} - \hat{B}(M_x, M_y, M_{xy}) - \bar{p}w]dxdy$$

$$- \Sigma H_{ab1} + \int_{C_\sigma} [-\bar{V}_z w + \bar{M}_\nu w_{,\nu} + \bar{M}_{\nu s} w_{,s}]ds$$

$$+ \int_{C_u} [-V_z(w - \bar{w}) + M_\nu(w_{,\nu} - \overline{w_{,\nu}})$$

$$+ M_{\nu s}(w_{,s} - \overline{w}_{,s})]ds. \qquad \text{(I-6.8)‡}$$

† Compare with Eq. (I-5.21).
‡ Compare with Eq. (5.22).

PROBLEM 1. Show that by the introduction of new functions:

$$\Lambda_1^{(a)}, \Lambda_2^{(a)}, \Lambda_3^{(a)} \quad \text{defined on} \quad C^*_{ab}$$
$$\Lambda_1^{(b)}, \Lambda_2^{(b)}, \Lambda_3^{(b)} \quad \text{defined on} \quad C^*_{ba}$$

and

$$\mu_1, \mu_2 \quad \text{defined on} \quad C_{ab},$$

the expression (I-6.3) may be written in an equivalent form denoted by H_{ab2} as follows:

$$H_{ab2} = \int_{C^*_{ab}} [\Lambda_3^{(a)}(w^{(a)} - \mu_1) - \Lambda_1^{(a)}(w^{(a)}_{,\nu_a} - \mu_2) - \Lambda_2^{(a)}(w^{(a)} - \mu_1)_{,s_a}]ds_a$$
$$+ \int_{C^*_{ba}} [\Lambda_3^{(b)}(w^{(b)} - \mu_1) - \Lambda_1^{(b)}(w^{(b)}_{,\nu_b} + \mu_2) - \Lambda_2^{(b)}(w^{(b)} - \mu_1)_{,s_b}]ds_b. \quad \text{(I-6.9)}$$

PROBLEM 2. By replacing H_{ab1} in Eq. (I-6.8) with H_{ab2} of Eq. (I-6.9), show that the stationary conditions of Eq. (I-6.8) on C_{ab} with respect to $w^{(a)}$ and $w^{(b)}$ allow us to set:

$$\Lambda_3^{(a)} = V_z^{(a)}, \Lambda_1^{(a)} = M_\nu^{(a)}, \Lambda_2^{(a)} = M_{\nu s}^{(a)},$$
$$\Lambda_3^{(b)} = V_z^{(b)}, \Lambda_1^{(b)} = M_\nu^{(b)}, \Lambda_2^{(b)} = M_{\nu s}^{(b)}, \quad \text{(I-6.10)}$$

in the Eq. (I-6.9) and consequently, we may write H_{ab2} in an equivalent form denoted by H_{ab4} as follows:

$$H_{ab4} = \int_{C^*_{ab}} [V_z^{(a)}(w^{(a)} - \mu_1) - M_\nu^{(a)}(w^{(a)}_{,\nu_a} - \mu_2) - M_{\nu s}^{(a)}(w^{(a)} - \mu_1)_{,s_a}]ds_a$$
$$+ \int_{C^*_{ba}} [V_z^{(b)}(w^{(b)} - \mu_1) - M_\nu^{(b)}(w^{(b)}_{,\nu_b} + \mu_2) - M_{\nu s}^{(b)}(w^{(b)} - \mu_1)_{,s_b}]ds_b. \quad \text{(I-6.11)}$$

6.5. Another Derivation of the Modified Hellinger–Reissner Principle

Thus far, we have formulated the modified Hellinger–Reissner principle from the modified principle of potential energy. Now, we shall trace another avenue and derive the modified Hellinger–Reissner principle from Eq. (I-3.27), where the term G_{ab} is given as follows:

$$G_{ab} = \iint_{S_{ab}} \mu_i(T_i^{(a)} + T_i^{(b)})dS. \quad \text{(I-6.12)}$$

Remembering that the function μ_1, μ_2 and μ_3 in Eq. (I-6.12) correspond to U, V and W on S_{ab}, respectively, and the U, V and W are expressible as given by Eq. (I-5.3), we may write Eq. (I-6.12) as follows:

$$G_{ab} = \int_{C_{ab}} \int_{-h/2}^{h/2} [(X_\nu^{(a)} + X_\nu^{(b)})(-zw_{,x}) + (Y_\nu^{(a)} + Y_\nu^{(b)})(-zw_{,y})$$
$$+ (Z_\nu^{(a)} + Z_\nu^{(b)})w]dzds. \quad \text{(I-6.13)}$$

After some manipulation, we obtain:

$$G_{ab} = \int_{C^*_{ab}} [V_z^{(a)}\mu_1 - M_v^{(a)}\mu_2 - M_{vs}^{(a)}\mu_{1,s_a}]ds_a$$
$$+ \int_{C^*_{ba}} [V_z^{(b)}\mu_1 + M_v^{(b)}\mu_2 - M_{vs}^{(b)}\mu_{1,s_b}]ds_b, \tag{I-6.14}$$

where we set

$$w = \mu_1, \qquad w_{,v_a} = \mu_2. \tag{I-6.15}$$

In Eq. (I-6.14), μ_1 and μ_2 should be taken as Lagrange multipliers defined on C_{ab}. Thus, we obtain an expression of the modified Hellinger–Reissner functional as follows:

$$-\Pi^*_{mR} = \sum \iint_{S_a} [\hat{B}(M_x, M_y, M_{xy})$$
$$+ (M_{x,xx} + 2M_{xy,xy} + M_{y,yy} + \bar{p})w]dxdy - \Sigma G_{ab}$$
$$+ \int_{C_\sigma} [-(V_z - \bar{V}_z)w + (M_v - \bar{M}_v)w_{,v} + (M_{vs} - \bar{M}_{vs})w_{,s}]ds$$
$$+ \int_{C_u} [-V_z\bar{w} + M_v\overline{w_{,v}} + M_{vs}\bar{w}_{,s}]ds, \tag{I-6.16}$$

where G_{ab} is given by Eq. (I-6.14).

Performing partial integrations, we may transform the functional given in Eq. (I-6.16) in another form:

$$\Pi_{mR} = \sum \iint_{S_a} [-M_x w_{,xx} - M_y w_{,yy} - 2M_{xy} w_{,xy}$$
$$- \hat{B}(M_x, M_y, M_{xy}) - \bar{p}w]dxdy$$
$$- \sum \Big\{ \int_{C^*_{ab}} [V_z^{(a)}(w^{(a)} - \mu_1) - M_v^{(a)}(w^{(a)}_{,v_a} - \mu_2) - M_{vs}^{(a)}(w^{(a)}_{,s_a} - \mu_{1,s_a})]ds_a$$
$$+ \int_{C^*_{ba}} [V_z^{(b)}(w^{(b)} - \mu_1) - M_v^{(b)}(w^{(b)}_{,v_b} + \mu_2) - M_{vs}^{(b)}(w^{(b)}_{,s_b} - \mu_{1,s_b})]ds_b \Big\}$$
$$+ \int_{C_\sigma} [-\bar{V}_z w + \bar{M}_v w_v + \bar{M}_{vs} w_{,s}]ds$$
$$+ \int_{C_u} [-V_z(w - \bar{w}) + M_v(w_{,v} - \overline{w_{,v}}) + M_{vs}(w_{\ s} - \bar{w}_{,s})]ds. \tag{I-6.17}$$

The functional (I-6.17) is equivalent to that in Eq. (I-6.8) if H_{ab1} is replaced by H_{ab4} given in Eq. (I-6.11).

6.6. A Special Case of the Modified Variational Principles for the Bending of a Thin Plate

As the last topic of this section, we shall consider a special case of the modified variational principles when the displacement functions are so chosen that they are continuous along the entire interelement boundaries:

$$w^{(a)} = w^{(b)} \quad \text{on} \quad C_{ab}. \tag{I-6.18}$$

Then, Eq. (I-6.3) reduces to

$$H_{ab1} = -\int_{C_{ab}} \Lambda_1(w^{(a)}_{,\nu_a} + w^{(b)}_{,\nu_b})ds. \tag{I-6.19}$$

By the introduction of new functions $\Lambda_1^{(a)}$ and $\Lambda_1^{(b)}$ defined on C^*_{ab} and C^*_{ba}, respectively, together with a new Lagrange multiplier μ, Eq. (I-6.19) may be written in an equivalent form as follows:

$$H_{ab2} = -\int_{C_{ab}} [\Lambda_1^{(a)} w^{(a)}_{,\nu_a} + \Lambda_2^{(b)} w^{(b)}_{,\nu_b} - \mu(\Lambda_1^{(a)} - \Lambda_1^{(b)})]ds, \tag{I-6.20}$$

or

$$H_{ab2} = -\int_{C^*_{ab}} \Lambda_1^{(a)}(w^{(a)}_{,\nu_a} - \mu)ds_a - \int_{C^*_{ba}} \Lambda_1^{(b)}(w^{(b)}_{,\nu_b} + \mu)ds. \tag{I-6.21}$$

By the use of Eq. (I-6.21), the functional for the modified principle of potential energy, Eq. (I-6.5), can be written as follows:

$$\Pi_{mP2} = \Pi_P - \Sigma H_{ab2}, \tag{I-6.22}$$

where the independent quantities subject to variation are $w^{(a)}$, $\Lambda_1^{(a)}$ and μ under the subsidiary conditions, Eq. (I-5.9). Taking variations with respect to these quantities, we find that the stationary conditions of Π_{mP2} on C_{ab} provide

$$\Lambda_1^{(a)} = M_\nu^{(a)}(w^{(a)}), \; \Lambda_1^{(b)} = M_\nu^{(b)}(w^{(b)}), \tag{I-6.23}$$

$$w^{(a)}_{,\nu_a} = -w^{(b)}_{,\nu_b} = \mu, \tag{I-6.24}$$

where $M_\nu^{(a)}(w^{(a)})$ and $M_\nu^{(b)}(w^{(b)})$ are obtained by substituting the stress-resultant and displacement relations, Eq. (8.33), into $M_\nu^{(a)}$ and $M_\nu^{(b)}$, respectively, to express them in terms of the displacements only. Eqs. (I-6.23) and (I-6.24) indicate the physical meaning of the Lagrange multipliers $\Lambda_1^{(a)}$, $\Lambda_1^{(b)}$ and μ. We also find that by the use of the stationary conditions, Eq. (I-6.23), we may eliminate $\Lambda_1^{(a)}$ and $\Lambda_1^{(b)}$ to reduce the functional Π_{mP2} to

$$\Pi_{mP3} = \Pi_P - \Sigma H_{ab3}, \tag{I-6.25}$$

where

$$H_{ab3} = -\int_{C^*_{ab}} M_\nu^{(a)}(w^{(a)})(w^{(a)}_{,\nu_a} - \mu)ds_a - \int_{C^*_{ba}} M_\nu^{(b)}(w^{(b)})(w_{,\nu b}^{(b)} + \mu)ds_b. \tag{I-6.26}$$

We may obtain the modified generalized principle and the modified Hellinger–Reissner principle for this special case by substituting Eq. (I-6.20) or Eq. (I-6.21) in place of H_{ab1} into Eq. (I-6.7) and (I-6.8), respectively. We find that the stationary conditions on C_{ab} of the functionals derived through these substitutions provide

$$\Lambda_1^{(a)} = M_\nu^{(a)},\ \Lambda_1^{(b)} = M_\nu^{(b)}. \tag{I-6.27}$$

Consequently, we find that we may replace H_{ab1} in Eqs. (I-6.7) and (I-6.8) by H_{ab4} defined by the following equation:

$$H_{ab4} = -\int_{C_{ab}^*} M_\nu^{(a)}(w_{,\nu_a}^{(a)} - \mu)ds_a - \int_{C_{ba}^*} M_\nu^{(b)}(w_{,\nu_b}^{(b)} + \mu)ds_b, \tag{I-6.28}$$

to obtain alternate expressions of the modified generalized principle and the modified Hellinger–Reissner principle for the special case specified by Eq. (I-6.18).

We shall specialize our problem further by assuming that not only w, but also M_ν are continuous along the entire interelement boundaries:

$$w^{(a)} = w^{(b)},\ M_\nu^{(a)} = M_\nu^{(b)} \quad \text{on} \quad C_{ab}. \tag{I-6.29}$$

Then, Eq. (I-6.28) is reduced to

$$H_{ab5} = -\int_{C_{ab}^*} M_\nu^{(a)} w_{,\nu_a}^{(a)} ds_a - \int_{C_{ba}^*} M_\nu^{(b)} w_{,\nu_b}^{(b)} ds_b, \tag{I-6.30}$$

and we have an expression of the functional for the modified Hellinger–Reissner principle as follows:

$$\begin{aligned} \Pi_{mR} = \sum \iint_{S_a} [& -M_x w_{,xx} - M_y w_{,yy} - 2M_{xy} w_{,xy} \\ & - \hat{B}(M_x, M_y, M_{xy}) - \bar{p}w]dxdy \\ & + \sum \Big\{ \int_{C_{ab}^*} M_\nu^{(a)} w_{,\nu_a}^{(a)} ds_a + \int_{C_{ba}^*} M_\nu^{(b)} w_{,\nu_b}^{(b)} ds_b \Big\} \\ & + \int_{C_\sigma} [-\bar{V}_z w + \bar{M}_\nu w_{,\nu} + \bar{M}_{\nu s} w_{,s}] ds \\ + \int_{C_u} [-V_z(w - \bar{w}) & + M_\nu(w_{,\nu} - \overline{w_{,\nu}}) + M_{\nu s}(w_{,s} - \bar{w}_{,s})]ds. \end{aligned} \tag{I-6.31}$$

Through integrations by parts, we may transform Eq. (I-6.31) into another form:

$$\begin{aligned} \Pi_{mR}^* = \sum \iint_{S_a} [& -\hat{B}(M_x, M_y, M_{xy}) + w_{,x}(M_{x,x} + M_{xy,y}) \\ & + w_{,y}(M_{xy,x} + M_{y,y}) - \bar{p}w]dxdy \\ & - \sum \Big\{ \int_{C_{ab}^*} M_{\nu s}^{(a)} w_{,s_a}^{(a)} ds_a + \int_{C_{ba}^*} M_{\nu s}^{(b)} w_{,s_b}^{(b)} ds_b \Big\} \\ + \int_{C_\sigma} [& -\bar{V}_z w + (\bar{M}_\nu - M_\nu) w_{,\nu} + (\bar{M}_{\nu s} - M_{\nu s}) w_{,s}] ds \\ & + \int_{C_u} [-V_z(w - \bar{w}) - M_\nu \overline{w_{,\nu}} - M_{\nu s} \bar{w}_{,s}] ds. \end{aligned} \tag{I-6.32}$$

The functional (I-6.32) was originally formulated and applied to a finite element analysis of the plate bending problem by Herrmann.[11, 12]

Section 7. Variational Principles for the Small Displacement Theory of Elastodynamics

Our next topic will be variational principles for the small displacement of theory of elastodynamics, for which the governing equations may be given as follows:

(1) *Equations of motion:*

$$\sigma_{ij,j} + \bar{f}_i = \rho \ddot{u}_i \tag{I-7.1}$$

(2) *Strain–displacement relations:*

$$\varepsilon_{ij} = \tfrac{1}{2}(u_{i,j} + u_{j,i}) \tag{I-7.2}$$

(3) *Stress–strain relations:*

$$\sigma_{ij} = a_{ijkl}\varepsilon_{kl} \tag{I-7.3}$$

or conversely

$$\varepsilon_{ij} = b_{ijkl}\sigma_{kl} \tag{I-7.4}$$

(4) *Mechanical boundary conditions:*

$$T_i = \bar{T}_i \quad \text{on} \quad S_\sigma \tag{I-7.5}$$

(5) *Geometrical boundary conditions:*

$$u_i = \bar{u}_i \quad \text{on} \quad S_u, \tag{I-7.6}$$

where the quantities appearing in these equations, namely, σ_{ij}, ε_{ij}, u_i, $\bar{f}_i$,$\bar{T}_i$ and $\bar{u}_i$ are functions of time t as well as space coordinates x_i; $i = 1, 2, 3$. For a complete definition of the elastodynamic problem, the following initial conditions should be added to the above equations:

$$\begin{aligned} u_i(x_1, x_2, x_3, 0) &= \overline{u_i(0)}, \\ \dot{u}_i(x_1, x_2, x_3, 0) &= \overline{\dot{u}_i(0)}, \end{aligned} \tag{I-7.7}$$

where $\overline{u_i(0)}$ and $\overline{\dot{u}_i(0)}$ are prescribed functions of the space coordinates.

Hamilton's principle introduced in Section 5.6 is the best established and most frequently used variational principle among those derived for the elastodynamic problem. Through transformations and generalizations similar to those for the elastostatic problem, we may create a flow diagram for a family related to Hamilton's principle as shown in Fig. I-4. Several papers related to this diagram are listed in the bibliography of this appendix.[24–30]

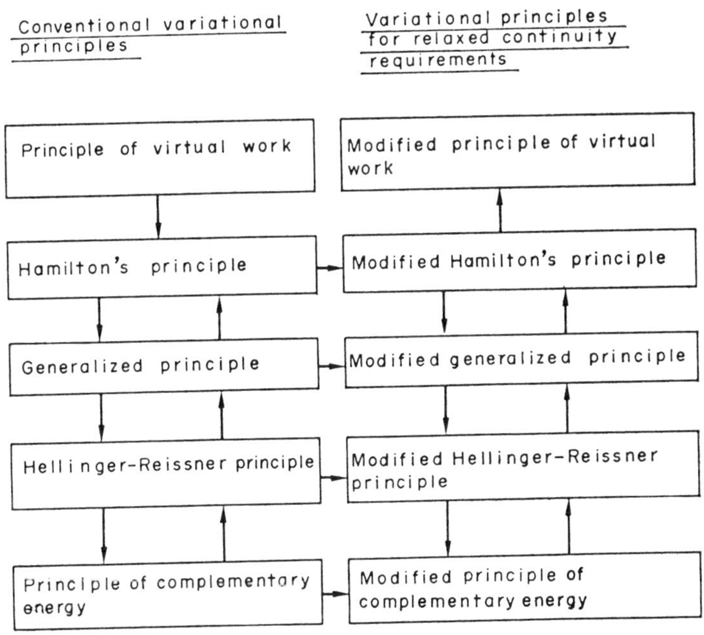

FIG. I-4. A flow diagram for the small displacement theory of elastodynamics.

Here we shall trace only an avenue which leads from the principle of virtual work to the principle of complementary energy. The reader is directed to Refs. 27 and 29 for other routes, including the modified variational principles with relaxed continuity requirements.

7.1. Principle of Virtual Work

Denoting a virtual variation of $u_i(t)$ at time t by $\delta u_i(t)$, we have†

$$-\iiint_V (\sigma_{ij,j} + \bar{f}_i - \rho \ddot{u}_i)\delta u_i dV + \iint_{S_\sigma} (T_i - \bar{T}_i)\delta u_i dS = 0, \qquad \text{(I-7.8)}$$

where the integrations extend over the entire region of V and S_σ at the time t. By integrating Eq. (I-7.8) with respect to time between two limits $t = t_1$ and

† It is repeated that $\delta u_i(t)$ is a virtual variation of $u_i(t)$ at the time t. The reader will find that the function $u_i(t) + \delta u_i(t)$ plays a role of an admissible function in Eq. (1-7.14).

$t = t_2$, and employing the convention that values of u_i at $t = t_1$ and $t = t_2$ are prescribed such that

$$\delta u_i(t_1) = 0, \ \delta u_i(t_2) = 0, \tag{I-7.9†}$$

together with some manipulation including partial integrations with respect to time as well as the space coordinates, we obtain the principle of virtual work for the elastodynamic problem as follows:

$$\int_{t_1}^{t_2} \Big\{ \delta T - \iiint_V \sigma_{ij} \delta \varepsilon_{ij} dV + \iiint_V \bar{f}_i \delta u_i dV + \iint_{S_\sigma} \bar{T}_i \delta u_i dS \Big\} dt = 0, \tag{I-7.10}$$

where

$$T = \tfrac{1}{2} \iiint_V \rho \dot{u}_i \dot{u}_i dV \tag{I-7.11}$$

is the kinetic energy of the elastic body, and where the subsidiary conditions are given by

$$\delta \varepsilon_{ij} = \tfrac{1}{2}(\delta u_{i,j} + \delta u_{j,i}), \tag{I-7.12}$$

and

$$\delta u_i = 0 \quad \text{on} \quad S_u, \tag{I-7.13}$$

together with Eqs. (I-7.9).

7.2. Hamilton's Principle

If the body forces $\bar{f}_i$ and external forces $\bar{T}_i$ on S_σ are assumed to be prescribed in such a way that they are not subjected to variation, we may derive from Eq. (I-7.10), the principle of stationary potential energy, or Hamilton's principle, as follows:

$$\delta \int_{t_1}^{t_2} \{ T - \Pi_P \} dt = 0, \tag{I-7.14‡}$$

where T and Π_P are given by Eqs. (I-7.11) and (I-2.19), respectively, while the subsidiary conditions are given by Eqs. (I-7.6) and (I-7.9).

† This convention means that the initial conditions, namely Eqs. (I-7.7), are not taken into serious consideration in the Hamilton's principle family. It may be said that the primary concern for the family is derivation of the equations of motion and boundary conditions at the time t; the initial conditions are of secondary concern.

‡ Refer to Eq. (5.86).

7.3. Generalized Principle

Next, we shall introduce new functions v_i defined by

$$v_i - \dot{u}_i = 0, \tag{I-7.15}$$

and write the kinetic energy T in a generalized form as follows;

$$T_G = \iiint_V [\tfrac{1}{2}\rho v_i v_i - p_i(v_i - \dot{u}_i)] dV, \tag{I-7.16}$$

where p_i is Lagrange multiplier which introduces the subsidiary condition, Eq. (I-7.15), into the framework of the expression of the kinetic energy. Then, we obtain a generalized principle as follows:

$$\delta \int_{t_1}^{t_2} \{T_G - \Pi_{G2}\} dt = 0, \tag{I-7.17}$$

where T_G and Π_{G2} are given by Eqs. (I-7.16) and (I-2.22), respectively, while the subsidiary conditions are given by Eqs. (I-7.9).

7.4. Hellinger–Reissner Principle

Elimination of v_i and ε_{ij} from Eq. (I-7.17) by the use of the stationary conditions with respect v_i and ε_{ij}; namely

$$\rho v_i = p_i \tag{I-7.18}$$

and Eq. (I-7.3), leads to the Hellinger–Reissner principle:

$$\delta \int_{t_1}^{t_2} \left\{ \iiint_V \left(p_i \dot{u}_i - \frac{1}{2\rho} p_i p_i \right) dV - \Pi_R \right\} dt = 0, \tag{I-7.19}$$

where Π_R is given by Eq. (1-2.23), while the subsidiary conditions are given by Eq. (I-7.9).

Through integrations by parts with respect to time as well as the space coordinates, we obtain another expression for the Hellinger–Reissner principle:

$$\delta \int_{t_1}^{t_2} \left\{ - \iiint_V \left(\dot{p}_i u_i + \frac{1}{2\rho} p_i p_i \right) dV - \Pi_R^* \right\} dt = 0, \tag{I-7.20}$$

where Π_R^* is given by Eq. (I-2.24), and the subsidiary conditions are given by

$$\delta p_i(t_1) = 0, \ \delta p_i(t_2) = 0. \tag{I-7.21}$$

7.5. Principle of Stationary Complementary Energy

The principle of stationary complementary energy is obtainable by taking as subsidiary conditions the stationary conditions with respect to the displacements, namely:

$$\sigma_{ij,j} + \bar{f}_i = \dot{p}_i, \tag{I-7.22}$$

and

$$T_i = \bar{T}_i \quad \text{on} \quad S_\sigma, \tag{I-7.23}$$

and we obtain

$$\delta \int_{t_1}^{t_2} \left\{ -\iiint_V \frac{1}{2\rho} p_i p_i dV + \iiint_V B(\sigma_{ij}) dV - \iint_{S_u} T_i \bar{u}_i dS \right\} dt = 0, \tag{I-7.24}$$

where Eqs. (I-7.21), (I-7.22) and (I-7.23) are taken as subsidiary conditions.

7.6. Another Expression of the Principle of Stationary Complementary Energy

Next, we shall obtain an alternate expression for the principle of stationary complementary energy. First, we introduce the following new notations:

$$\tau_{ij} = \int_0^t \sigma_{ij} dt, \qquad t_i = \tau_{ij} n_j,$$

$$\tilde{\bar{f}}_i = \int_0^t \bar{f}_i dt, \qquad \bar{t}_i = \int_0^t \bar{T}_i dt,$$

$$v_i = \dot{u}_i, \qquad \bar{v}_i = \dot{\bar{u}}_i. \tag{I-7.25}$$ †

Assuming $p_i = 0$ at $t = 0$, we may replace Eqs. (I-7.22) and (I-7.23) by the following equations:

$$\tau_{ij,j} + \tilde{\bar{f}}_i = p_i \tag{I-7.26}$$

and

$$t_i = \bar{t}_i. \tag{I-7.27}$$

We may eliminate p_i from Eq. (I-7.24) by the use of Eq. (I-7.26) and perform partial integrations with respect to time to obtain

† These definitions for τ_{ij}, t_i, $\tilde{\bar{f}}_i$, $\bar{t}_i$, v_i and $\bar{v}_i$ are used in Section 7.6 only.

$$\delta \int_{t_1}^{t_2} \left\{ - \iiint_V \frac{1}{2\rho} (\tau_{ij,j} + \tilde{\bar{f}}_i)(\tau_{ik,k} + \tilde{\bar{f}}_i) dV + \iiint_V B(\dot{\tau}_{ij}) dV + \iint_{S_u} t_i v_i dS \right\} dt = 0, \qquad \text{(I-7.28)}$$

where the subsidiary conditions are Eq. (I-7.27) together with

$$\delta\tau_{ij}(t_1) = 0, \; \delta\tau_{ij}(t_2) = 0. \qquad \text{(I-7.29)}$$

Eq. (I-7.28) is another expression of the principle of stationary complementary energy which is expressed in terms of impulse and velocity instead of force and displacement.[27]

It is noted here that Hamilton's principle and the principle of virtual work have been used frequently in mathematical formulations of the finite element method applied to dynamic response problems. An elastic body under consideration is divided into a number of finite elements and Hamilton's principle is applied to obtain a system of linear algebraic equations which may be written in a matrix form as follows:

$$[M]\{\ddot{q}\} + [C]\{\dot{q}\} + [K]\{q\} = \{\bar{Q}\}, \qquad \text{(I-7.30)}$$

where $[M]$, $[C]$ and $[K]$ are the inertia, damping and stiffness matrices, respectively, while $\{q\}$ is the column vector of nodal displacements, and $\{\bar{Q}\}$ is the external load vector. Eq. (I-7.30) may be solved by either the mode superposition method or a step-by-step integration procedure. The reader is directed for Refs. 31 and 32, for example, for further details. It is also noted here that the principle of stationary complementary energy has been used recently in application to the finite element method.[26–28]

7.7. Gurtin's Principle[33, 34]

We have seen that the initial conditions, Eq. (I-7.7), are not taken into serious consideration in the variational family associated with the Hamilton's principle and, in that sense, none of the family is complete in defining the elastodynamic problem in the form of variational expressions. Gurtin established variational principles which, in contrast to those belonging to the Hamilton family, fully characterize the solution of the elastodynamic problem. His formulation begins by first defining the convolution of two functions $\vartheta(x, t)$ and $\omega(\boldsymbol{x}, t)$ by

$$[\vartheta * \omega](\boldsymbol{x}, t) = \int_0^t \vartheta(\boldsymbol{x}, t - t')\omega(\boldsymbol{x}, t')dt', \qquad \text{(I-7.31)}$$

and then observing that σ_{ij} and u_i satisfy the equations of motion, if and only if

$$g*\sigma_{ij,j} + \hat{\bar{f}}_i = \rho u_i, \tag{I-7.32}$$

where x denotes the space coordinates (x_1, x_2, x_3), and

$$g(t) = t, \tag{I-7.33}$$

$$\hat{\bar{f}}_i(x, t) = [g*\bar{f}_i](x, t) + \rho(x, t)[t\bar{\dot{u}}(x, o) + \bar{u}_i(x, 0)]. \tag{I-7.34}$$

By the use of these relations, Gurtin derived a family of variational principles which have forms similar to those shown in Fig. I-1, except for the presence of g, the use of convolutions, and the appearance of the initial conditions and the term ρ. For details, the reader is directed to Gurtin's original papers. It is noted finally that variational formulations using convolution integrals have been employed recently in the basic theoretical development of the finite element method for time dependent problems.[35–38]

Section 8. Finite Displacement Theory of Elastostatics

In Section 3.5 we defined a problem of the finite displacement theory of elastostatics which is usually called a geometrically nonlinear problem, because the solid body still behaves elastically, although the displacements are finite and no longer small. We formulated the problem by the use of Kirchhoff stress tensor $\sigma^{\lambda\mu}$ and Green strain tensor $e_{\lambda\mu}$ in the first part of Chapter 3.† In the subsequent sections of the chapter, we formulated for the problem the principle of virtual work, the principle of stationary potential energy, the generalized principle, and the Hellinger–Reissner principle, as represented by Eqs. (3.49), (3.68), (3.70) and (3.71), respectively. These variational principles can be modified into those for relaxed continuity requirements and we obtain the flow diagram illustrating interrelations between these variational principles as shown in Fig. I-5.

8.1. Some Remarks on the Flow Diagram

Three comments will be made here with regard to the flow diagram. The first comment concerns the principle of complementary energy for the nonlinear elastostatic problem. It can be shown that by the use of the equations of equilibrium, Eq. (3.27), together with the mechanical boundary conditions, Eq. (3.42), we may reduce the functional of Eq. (3.71) to

$$\Pi_c = \iiint_V [B(\sigma_{ij}) + \tfrac{1}{2}\sigma_{ij}u_{k,i}u_{k,j}]dV - \iint_{S_u} F_i\bar{u}_i dS. \tag{I-8.1‡§}$$

† The stress tensor $\sigma^{\lambda\mu}$ has been named pseudo-stress or generalized stress in the footnote of page 57. It is also called the second Piola–Kirchkoff stress tensor in Ref. 39.

‡ In sections 8 and 9 of this appendix, we shall use subscript Roman letters instead of superscript or subscript Greek letters employed in Chapter 3. Thus we write u_i, σ_{ij}, e_{ij}, F_i, ... instead of u^λ, $\sigma^{\lambda\mu}$, $e_{\lambda\mu}$, F^λ, .., respectively.

§ The body forces and the external forces on S_σ are assumed dead loads.

However, since the coupling of displacements with stress components complicates the expression of Π_c as well as the subsidiary conditions, Eqs. (3.27) and (3.42), there seems to be little merit in deriving the expression for Π_c in the form as shown in Eq. (I-8.1). Consequently, the principle of complementary energy is not listed in the flow diagram of Fig. I-5.†

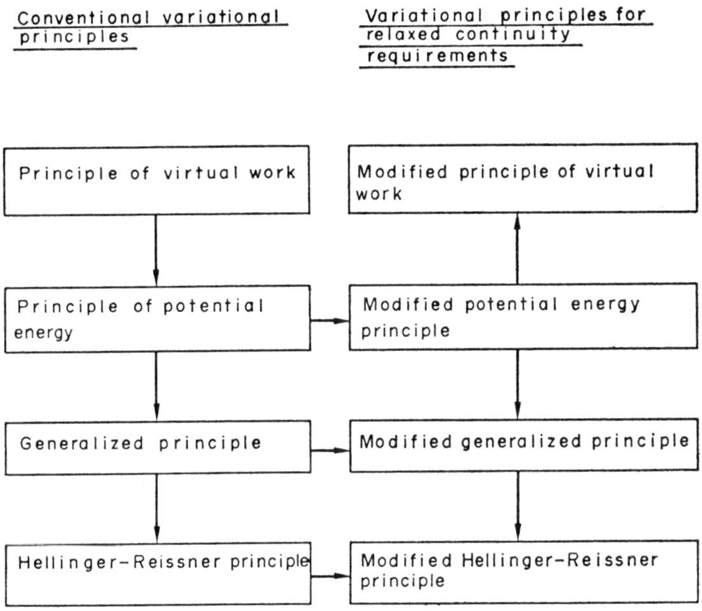

FIG. I-5. A flow diagram for the finite displacement Theories of elastostatics and elastodynamics.

The second observation relates to the variational principles with relaxed continuity requirements. It is easily observed that the functional for the principle of stationary potential energy is given for a finite element formulation as follows:

$$\Pi_P = \sum \iiint_{V_a} [A(u_i) + \Phi(u_i)]dV + \iint_{S_\sigma} \Psi(u_i)dS, \tag{I-8.2}$$

while the functional for the modified principle of potential energy with relaxed continuity requirements is given by

$$\Pi_{mP1} = \Pi_P - \Sigma H_{ab1} \tag{I-8.3}$$

where Π_P is given by Eq. (I-8.2) and

† However, this statement does not reduce the value of the principle of stationary complementary energy which may be formulated for incremental theories of the geometrical nonlinear problem.

$$H_{ab1} = \iint_{S_{ab}} \lambda_i(u_i^{(a)} - u_i^{(b)})dS. \tag{I-8.4}$$

In Eq. (I-8.4), the newly introduced functions λ_i; $i = 1, 2, 3$ are Lagrange multipliers, while $u_i^{(a)}$ and $u_i^{(b)}$ are displacement components belonging to two adjacent elements a and b, respectively.

The functional Π_{mP1} may be transformed into another equivalent functional Π_{mP2} as follows:

$$\Pi_{mP2} = \Pi_P - \Sigma H_{ab2} \tag{I-8.5}$$

where

$$H_{ab2} = \iint_{S_{ab}} [\lambda_i^{(a)}u_i^{(a)} + \lambda_i^{(b)}u_i^{(b)} - u_i(\lambda_i^{(a)} + \lambda_i^{(b)})]dS, \tag{I-8.6}$$

or equivalently

$$H_{ab2} = \iint_{S_{ab}^*} \lambda_i^{(a)}(u_i^{(a)} - u_i)dS + \iint_{S_{ba}^*} \lambda_i^{(b)}(u_i^{(b)} - u_i)dS. \tag{I-8.7}$$

The modified principle of potential energy may be generalized in the usual manner to obtain the functional for the generalized principle:

$$\begin{aligned} \Pi_{mG1} = \Sigma\iiint_{V_a} \{&A(e_{ij}) + \Phi(u_i) \\ &- \sigma_{ij}[e_{ij} - \tfrac{1}{2}(u_{i,j} + u_{j,i} + u_{k,i}u_{kj})]\}dV \\ &- \Sigma H_{ab2} + \iint_{S_\sigma} \Psi(u_i)dS - \iint_{S_u} p_i(u_i - \bar{u}_i)dS. \end{aligned} \tag{I-8.8}$$

Elimination of e_{ij} from Π_{mG1} then leads to the functional for the modified Hellinger–Reissner principle:

$$\begin{aligned} \Pi_{mR} = \Sigma\iiint_{V_a} [&\tfrac{1}{2}(u_{i,j} + u_{j,i} + u_{k,i}u_{k,j})\sigma_{ij} \\ &- B(\sigma_{ij}) + \Phi(u_i)]dV - \Sigma H_{ab2} \\ &+ \iint_{S_\sigma} \Psi(u_i)dS - \iint_{S_u} p_i(u_i - \bar{u}_i)dS. \end{aligned} \tag{I-8.9}$$

PROBLEM. Show that the stationary conditions on S_{ab} of the functional Π_{mG1} of Eq. (I-8.8) provide

$$\lambda_i^{(a)} = F_i^{(a)}, \ \lambda_i^{(b)} = F_i^{(b)}, \tag{I-8.10}$$

where

$$F_i^{(a)} = \sigma_{km}^{(a)}n_k^{(a)}(\delta_{im} + u_{i,m}^{(a)}), \tag{I-8.11}$$

$$F_i^{(b)} = \sigma_{km}^{(b)}n_k^{(b)}(\delta_{im} + u_{i,m}^{(b)}). \tag{I-8.12}$$

Show also that the stationary conditions on S_{ab} of the functional Π_{mP2} of Eq. (I-8.5) provide:

$$\lambda_i^{(a)} = F_i^{(a)}(u_j^{(a)}),\ \lambda_i^{(b)} = F_i^{(b)}(u_j^{(b)}), \tag{I-8.13}$$

where $F_i^{(a)}(u_j^{(a)})$ and $F_i^{(b)}(u_j^{(b)})$ can be obtained from Eqs. (I-8.11) and (I-8.12) by substituting Eqs. (3.33) and (3.18) to express $F_i^{(a)}$ and $F_i^{(b)}$ entirely in terms of the displacement $u_i^{(a)}$ and $u_i^{(b)}$, respectively.

The third observation relates to the problem of the finite displacement theory of elastodynamics defined in Section 5.6. It is apparent that we may obtain for the elastodynamic problem a flow diagram similar to that shown in Fig. I-5, if the inertia term is taken into account.

Thus far, several remarks have been made on the flow diagram of Fig. I-5. It is natural to conclude that we may formulate finite element models corresponding to these variational principles in a manner similar to those for the small displacement elastostatic problem. Among finite element models thus formulated, the most frequently used is the compatible model based on the principle of stationary potential energy. This model will be discussed briefly in the next section.

8.2. A Formulation for the Compatible Model and the Modified Incremental Stiffness Method[40]

A formulation for the compatible model begins by approximating u_i in each element by

$$\{u\} = [S]\{q\}, \tag{I-8.14}$$

with the aid of compatible shape functions, where $\{u\}^T = [u_1, u_2, u_3]$ and $\{q\}$ is a column vector of nodal displacements. If the total strain energy U is expressed in terms of q_i:

$$U = \sum \iiint_{V_c} A(q_i) dV, \tag{I-8.15}$$

we may obtain the following equations by the use of the principle of stationary potential energy:

$$\left\{\frac{\partial U}{\partial q}\right\} = \{\bar{Q}\}, \tag{I-8.16}$$

where $\{\bar{Q}\}$ is a column vector of the generalized forces. Since Eq. (I-8.16) are nonlinear, several iterative solution methods have been proposed.

Here, we shall outline an iterative method called the modified incremental stiffness method, assuming for the sake of simplicity that the elastic body is fixed on S_u. We divide the total strain energy in two parts such that

$$U = U_L + U_{NL} \tag{I-8.17}$$

where U_L is a linear term containing all the quadratic terms with respect to displacements, while U_{NL} is a nonlinear term containing all the remaining higher-order products. The stiffness matrix $[K]$ is then derived from

$$\left\{\frac{\partial U_L}{\partial q}\right\} = [K]\{q\}. \tag{I-8.18}$$

We now divide the loading path of the solid body problem into a number of states:

$$\Omega^{(0)}, \Omega^{(1)}, \Omega^{(2)}, \ldots, \Omega^{(N)}, \Omega^{(N+1)}, \ldots, \Omega^{(f)},$$

where $\Omega^{(0)}$ and $\Omega^{(f)}$ are the initial and final states of the deformation, respectively, while $\Omega^{(N)}$ is an arbitrary intermediate state. We shall derive an incremental formulation for the determination of the $\Omega^{(N+1)}$ state assuming that this state is incrementally close to the $\Omega^{(N)}$ state and that the $\Omega^{(N)}$ state is known. Denoting the generalized forces and displacements corresponding to the $\Omega^{(N)}$ and $\Omega^{(N+1)}$ states by $\{\bar{Q}^{(N)}\}$, $\{q^{(N)}\}$ and $\{\bar{Q}^{(N)} + \Delta\bar{Q}\}$, $\{q^{(N)} + \Delta q\}$, respectively, and by the use of Eqs. (I-8.17) and (I-8.18), we may write Eq. (I-8.16) for the $\Omega^{(NH)}$ state as follows:

$$[K](\{q^{(N)}\} + \{\Delta q\}) + \left\{\frac{\partial U_{NL}(q^{(N)} + \Delta q)}{\partial q}\right\} = \{\bar{Q}^{(N)}\} + \{\Delta\bar{Q}\}. \tag{I-8.19}$$

By the use of a Taylor series expansion

$$\frac{\partial U_{NL}(q_k^{(N)} + \Delta q_k)}{\partial q_i} = \frac{\partial U_{NL}(q_k^{(N)})}{\partial q_i} + \sum_j \frac{\partial^2 U_{NL}}{\partial q_i \partial q_j}\Delta q_j + \ldots$$

in which the higher order terms are neglected, we may have

$$\left([K] + \left[\frac{\partial^2 U_{NL}}{\partial q_i \partial q_j}\right]\right)\{\Delta q\}$$

$$= \{\Delta\bar{Q}\} + \{\bar{Q}^{(N)}\} - [K]\{q^{(N)}\} - \left\{\frac{\partial U_{NL}(q^{(N)})}{\partial q}\right\}. \tag{I-8.20}$$

We obtain $\{\Delta q\}$ by solving Eq. (I-8.20), and the displacements corresponding to the $\Omega^{(N+1)}$ state are given by $\{q^{(N)} + \Delta q\}$.

It is a unique characteristic of the modified incremental stiffness method that the term

$$\{\bar{Q}^{(N)}\} - [K]\{q^{(N)}\} - \left\{\frac{\partial U_{NL}(q^{(N)})}{\partial q}\right\} \tag{I-8.21}$$

is retained on the right-hand side of Eq. (I-8.20) for an equilibrium check. It is stated in Ref. 40 that the equilibrium check term plays the essential role of preventing an approximate solution based on this incremental formulation from drifting away from the exact solution.

A review has been given in Ref. 41 on various formulations for solving the geometrically nonlinear problem numerically. These includes the incremental stiffness procedure, self-correcting incremental procedure (modified incremental stiffness method), Newton–Raphson method, perturbation method and initial-value formulation. Distinguished features of each formulation are discussed and recommendations are made as to which procedures are the best suited. It is also stated in the reference that the treatment of the nonlinear problem as an initial-value problem opens the door to a large number of solution procedures. For details of these formulations and their applications to FEM, the reader is directed to Refs. 40 through 44.

8.3. A Generalized Variational Principle by the Use of the Piola Stress Tensor

The last topic of this section will be a derivation of another generalized principle from the principle of stationary potential energy, Eq. (3.69). To begin with, we find that the strain e_{ij} is a function of $u_{m,n}$ and may be written as

$$e_{ij} = \tfrac{1}{2}(\alpha_{ij} + \alpha_{ji} + \alpha_{ki}\alpha_{kj}), \tag{I-8.22}$$

where by definition

$$\alpha_{ij} = u_{i,j} \tag{I-8.23}$$

By the use of Eq. (I-8.22), we may express the strain energy function $A(e_{ij})$ in terms of α_{kl} and write

$$A[e_{ij}(\alpha_{kl})] \equiv A(\alpha_{ij})$$

for the sake of brevity. Then, by the introduction of Lagrange multipliers $\tilde{\sigma}_{ij}$ and $\tilde{p}_i$, we may derive from Eq. (3.69) the following generalized functional:

$$\begin{aligned} \tilde{\Pi}_{G1} = &\iiint_V \{A(\alpha_{ij}) + \Phi(u_i) - \tilde{\sigma}_{ji}(\alpha_{ij} - u_{i,j})\} dV \\ &+ \iint_{S_\sigma} \Psi(u_i) dS - \iint_{S_u} \tilde{p}_i(u_i - \bar{u}_i) dS, \end{aligned} \tag{I-8.24}$$

where the independent quantities subject to variation are α_{ij}, u_i, $\tilde{\sigma}_{ij}$ and $\tilde{p}_i$ with no subsidiary conditions. Taking variations with respect to these quantities, we obtain the following stationary conditions:

$$\frac{\partial A}{\partial e_{mn}} \frac{\partial e_{mn}}{\partial \alpha_{ij}} = \tilde{\sigma}_{ji}, \tag{I-8.25}$$

$$\tilde{\sigma}_{ji,j} + \bar{P}_i = 0, \tag{I-8.26}$$

$$\alpha_{ij} - u_{i,j} = 0, \tag{I-8.27}$$

$$\tilde{\sigma}_{ji} n_j = \bar{F}_i \quad \text{on} \quad S_\sigma, \tag{I-8.28}$$

$$\tilde{\sigma}_{ji} n_j = \tilde{p}_i \quad \text{on} \quad S_\sigma, \tag{I-8.29}$$

$$u_i = \bar{u}_i \quad \text{on} \quad S_u. \tag{I-8.30}$$

These equations indicate the physical meaning of the Lagrange multipliers. It is seen from Eqs. (I-8.25) and (I-8.26) that $\tilde{\sigma}_{ij}$ is the Piola stress tensor.†,‡

If the body forces $\bar{P}_i$ and the external force $\bar{F}_i$ on S_σ are treated as dead loads, we may use Eqs. (I-8.26), (I-8.28) and (I-8.29) for the elimination of u_i to transform Eq. (I-8.24) into

$$\tilde{\Pi}_{G2} = \iiint_V [A(\alpha_{ij}) - \tilde{\sigma}_{ji}\alpha_{ij})dV + \iint_{S_u} \tilde{\sigma}_{ji} n_j \bar{u}_i dS, \tag{I-8.31}$$

where the independent quantities subject to variation are α_{ij} and $\tilde{\sigma}_{ij}$ under the subsidiary conditions of Eqs. (I-8.26) and (I-8.28). Thus, the merit of the use of the Piola stress tensor is that the subsidiary conditions are expressed in terms of $\tilde{\sigma}_{ij}$ only in linear forms.

If it were possible to eliminate α_{ij} from Eq. (I-8.31) by the use of Eq. (I-8.25), we might obtain a functional expressed entirely in terms of $\tilde{\sigma}_{ij}$ and similar in form to that of the principle of minimum complementary energy in the small displacement theory of elasticity. However, this elimination is difficult in general.[46] Consequently, it would seem advantageous, for practical applications to FEM, not to struggle with the elimination to obtain the principle of stationary complementary energy, but to be satisfied with the functional $\tilde{\Pi}_{G2}$, taking $\tilde{\sigma}_{ij}$ and α_{ij} as independent quantities subject to variation under the subsidiary conditions Eqs. (I-8.26) and (I-8.28).

Section 9. Two Incremental Theories

In the present section, we shall formulate two incremental theories for a nonlinear solid body problem with geometrical and material nonlinearity. The deformation of the body is characterized by the features that not only its displacements are finite, but also its strains are no longer small, and the material behavior is elastic–plastic.

The formulation of the incremental theories begins by dividing the loading path of the solid body problem into a number of equilibrium states

$$\Omega^{(0)}, \Omega^{(1)}, \ldots, \Omega^{(N)}, \Omega^{(N+1)}, \ldots, \Omega^{(f)},$$

where $\Omega^{(0)}$ and $\Omega^{(f)}$ are the initial and final states of the deformation, respectively, while $\Omega^{(N)}$ is an arbitrary intermediate state. It is assumed that all

† The Piola stress tensor is also called the Lagrange stress tensor,[45] or the first Piola–Kirchhoff stress tensor.[39] It is defined by $\boldsymbol{\sigma}_i = \tilde{\sigma}_{ij}\mathbf{i}_j$, where $\boldsymbol{\sigma}_i$ and $\mathbf{i}_j$ have the same meaning as $\boldsymbol{\sigma}^\lambda$ and $\mathbf{i}_\mu$ introduced in Section 3.2. Unlike the Kirchhoff stress tensor σ_{ij}, the Piola stress tensor is generally unsymmetric.

‡ By combining $\boldsymbol{\sigma}_i = \tilde{\sigma}_{ij}\mathbf{i}_j$ with $\boldsymbol{\sigma}_i = \sigma_{ij}(\delta_{kj} + \alpha_{kj})\mathbf{i}_k$ given by Eqs. (3.17) and (3.23), we obtain $\tilde{\sigma}_{ij} = \sigma_{ik}(\delta_{jk} + \alpha_{jk})$, which is equivalent to Eq. (I-8.25).

the state variables such as stresses, strains and displacements, together with the loading history, are known up to the $\Omega^{(N)}$ state. Our problem is then to formulate an incremental theory for determining all the state variables in the $\Omega^{(N+1)}$ state, under an assumption that the $\Omega^{(N+1)}$ state is incrementally close to the $\Omega^{(N)}$ state and all the governing equations may be linearized with respect to the incremental quantities. The step characterizing the deformation process from the $\Omega^{(N)}$ state to the $\Omega^{(N+1)}$ state will be referred to as the $(N+1)$-th step.

Let the positions of an arbitrary material point of the body in the $\Omega^{(0)}$, $\Omega^{(N)}$ and $\Omega^{(N+1)}$ states be denoted by $P^{(0)}$, $P^{(N)}$ and $P^{(N+1)}$, respectively, and the position vectors to these points by $\mathbf{r}^{(0)}$, $\mathbf{r}^{(N)}$ and $\mathbf{r}^{(N+1)}$, respectively, as

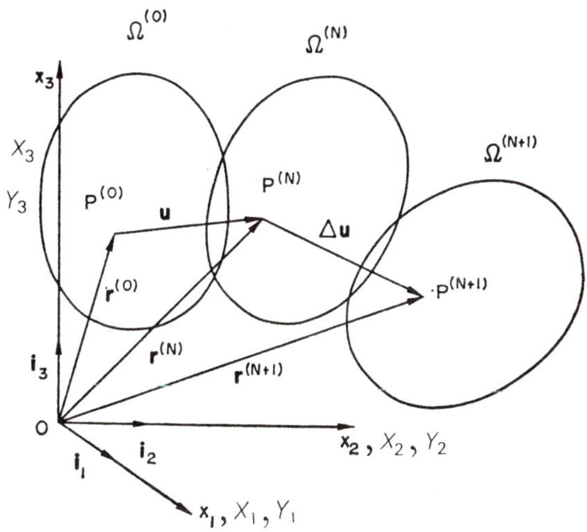

FIG. I-6. $\Omega^{(0)}$, $\Omega^{(N)}$ and $\Omega^{(N+1)}$

shown in Fig. I-6, and let the rectangular Cartesian coordinates of the positions $P^{(0)}$, $P^{(N)}$ and $P^{(N+1)}$ be represented by x_i, X_i and Y_i, respectively. Then, we have

$$\mathbf{r}^{(0)} = x_i\mathbf{i}_i, \tag{I-9.1}$$

$$\mathbf{r}^{(N)} = X_i\mathbf{i}_i = \mathbf{r}^{(0)} + \mathbf{u} = (x_i + u_i)\mathbf{i}_i, \tag{I-9.2}$$

$$\begin{aligned}\mathbf{r}^{(N+1)} &= Y_i\mathbf{i}_i = \mathbf{r}^{(0)} + \mathbf{u} + \Delta\mathbf{u} \\ &= (X_i + \Delta u_i)\mathbf{i}_i = (x_i + u_i + \Delta u_i)\mathbf{i}_i,\end{aligned} \tag{I-9.3}$$

where $\mathbf{i}_i$; $i = 1, 2, 3$ are the base vectors of the rectangular Cartesian coordinates, while $\mathbf{u}$ and $\mathbf{u} + \Delta\mathbf{u}$, and u_i and $u_i + \Delta u_i$; $i = 1, 2, 3$ are the displacement vectors and their components of the point in the $\Omega^{(N)}$ and $\Omega^{(N+1)}$ states, respectively.

9.1. Definitions of Strains

We shall denote the familiar Green strain tensors at the $\Omega^{(N)}$ and $\Omega^{(N+1)}$ states by e_{ij} and $e_{ij} + \Delta e_{ij}$, respectively. These are defined by

$$\begin{aligned} 2e_{ij} &= \mathbf{r}^{(N)}_{,i} \cdot \mathbf{r}^{(N)}_{,j} - \mathbf{r}^{(0)}_{,i} \cdot \mathbf{r}^{(0)}_{,j} \\ &= u_{i,j} + u_{j,i} + u_{k,i} u_{k,j}, \end{aligned} \tag{I-9.4}$$

and

$$\begin{aligned} 2(e_{ij} + \Delta e_{ij}) &= \mathbf{r}^{(N+1)}_{,i} \cdot \mathbf{r}^{(N+1)}_{,j} - \mathbf{r}^{(0)}_{,i} \cdot \mathbf{r}^{(0)}_{,j} \\ &= (u_i + \Delta u_i)_{,j} + (u_j + \Delta u_j)_{,i} + (u_k + \Delta u_k)_{,i}(u_k + \Delta u_k)_{,j} \end{aligned} \tag{I-9.5}$$

respectively, where $(\ )_{,i} = \partial(\ )/\partial x_i$. It is readily obtained from Eqs. (I-9.4) and (I-9.5) that

$$\begin{aligned} 2\Delta e_{ij} &= \mathbf{r}^{(N+1)}_{,i} \cdot \mathbf{r}^{(N+1)}_{,j} - \mathbf{r}^{(N)}_{,i} \cdot \mathbf{r}^{(N)}_{,j} \\ &= (\delta_{kj} + u_{k,j})\Delta u_{k,i} + (\delta_{ki} + u_{k,i})\Delta u_{k,j} + \Delta u_{k,i} \Delta u_{k,j}. \end{aligned} \tag{I-9.6}$$

On the other hand, we may have another definition of the strain increments for the $(N + 1)$-th step, taking the $\Omega^{(N)}$ state as an initial state, and by the use of the rectangular Cartesian coordinates (X_1, X_2, X_3). Denoting the strain increments by $\Delta^* e_{ij}$, we may have

$$\begin{aligned} 2\Delta^* e_{ij} &= \frac{\partial \mathbf{r}^{(N+1)}}{\partial X_i} \cdot \frac{\partial \mathbf{r}^{(N+1)}}{\partial X_j} - \frac{\partial \mathbf{r}^{(N)}}{\partial X_i} \cdot \frac{\partial \mathbf{r}^{(N)}}{\partial X_j} \\ &= \frac{\partial \Delta u_i}{\partial X_j} + \frac{\partial \Delta u_j}{\partial X_i} + \frac{\partial \Delta u_k}{\partial X_i} \frac{\partial \Delta u_k}{\partial X_j}. \end{aligned} \tag{I-9.7}$$

The transformation laws between Δe_{ij} and $\Delta^* e_{ij}$ are as follows:

$$\Delta e_{ij} = \frac{\partial X_m}{\partial x_i} \frac{\partial X_n}{\partial x_j} \Delta^* e_{mn}, \tag{I-9.8}$$

$$\Delta^* e_{ij} = \frac{\partial x_m}{\partial X_i} \frac{\partial x_n}{\partial X_j} \Delta e_{mn}. \tag{I-9.9}$$

If Δe_{ij} and $\Delta^* e_{ij}$ are linearized with respect to Δu_k, we obtain

$$2\Delta \varepsilon_{ij} = (\delta_{kj} + u_{k,j})\Delta u_{k,i} + (\delta_{ki} + u_{k,i})\Delta u_{k,j} \tag{I-9.10}$$

$$2\Delta^* \varepsilon_{ij} = \frac{\partial \Delta u_i}{\partial X_j} + \frac{\partial \Delta u_j}{\partial X_i}. \tag{I-9.11}$$

We note here some of the geometrical relations which are useful in carrying out later formulations. First, we define the Jacobians as follows:

$$D^{(N)} = \frac{\partial(X_1, X_2, X_3)}{\partial(x_1, x_2, x_3)} = |X_{i,j}|,$$
$$D^{(N+1)} = \frac{\partial(Y_1, Y_2, Y_3)}{\partial(x_1, x_2, x_3)} = |Y_{i,j}|, \tag{I-9.12}$$

and we obtain

$$\frac{\partial(Y_1, Y_2, Y_3)}{\partial(X_1, X_2, X_3)} = \frac{\partial(Y_1, Y_2, Y_3)}{\partial(x_1, x_2, x_3)} \bigg/ \frac{\partial(X_1, X_2, X_3)}{\partial(x_1, x_2, x_3)} = \frac{D^{(N+1)}}{D^{(N)}} \tag{I-9.13}$$

Second, the following relations are also worth noting:

$$X_{i,j} = \delta_{ij} + u_{i,j}$$
$$Y_{i,j} = \delta_{ij} + (u_i + \Delta u_i)_{,j} \tag{I-9.14}$$

and

$$\begin{bmatrix} \frac{\partial X_1}{\partial x_1} & \frac{\partial X_1}{\partial x_2} & \frac{\partial X_1}{\partial x_3} \\ \frac{\partial X_2}{\partial x_1} & \frac{\partial X_2}{\partial x_2} & \frac{\partial X_2}{\partial x_3} \\ \frac{\partial X_3}{\partial x_1} & \frac{\partial X_3}{\partial x_2} & \frac{\partial X_3}{\partial x_3} \end{bmatrix} \begin{bmatrix} \frac{\partial x_1}{\partial X_1} & \frac{\partial x_1}{\partial X_2} & \frac{\partial x_1}{\partial X_3} \\ \frac{\partial x_2}{\partial X_1} & \frac{\partial x_2}{\partial X_2} & \frac{\partial x_2}{\partial X_3} \\ \frac{\partial x_3}{\partial X_1} & \frac{\partial x_3}{\partial X_2} & \frac{\partial x_3}{\partial X_3} \end{bmatrix} = [I], \tag{I-9.15}$$

where $[I]$ is a unit matrix. Third, if Δu_i is assumed small, we may write

$$\frac{\partial(Y_1, Y_2, Y_3)}{\partial(X_1, X_2, X_3)} = 1 + \Delta^*\varepsilon_{ii} + O(\Delta u_i \Delta u_j), \tag{I-9.16}$$

where $\Delta^*\varepsilon_{ii} = \sum_{i=1}^{3}[\partial(\Delta u_i)/\partial X_i]$.

9.2. Definitions of Stresses

First, we define the Kirchhoff stress tensors by the use of the (x_1, x_2, x_3) coordinates, and denote those defined at the points $P^{(N)}$ and $P^{(N+1)}$ by σ_{ij} and $\sigma_{ij} + \Delta\sigma_{ij}$, respectively as shown in Fig. I-7.† These stress tensors are defined per unit area of the $\Omega^{(0)}$ state as introduced in Chapter 3.

Second, we define the Euler stress tensors[45] at the point $P^{(N)}$, and denote them by σ_{ij}^E; the Euler stresses are those acting on six surfaces:

$$X_i = \text{const.}, \; X_i + dX_i = \text{const.}; \; i = 1, 2, 3$$

† The stress tensors σ_{ij} and $\sigma_{ij} + \Delta\sigma_{ij}$ defined here are respectively the same as $\sigma^{\lambda\mu}$ and $\sigma^{\lambda\mu} + \sigma_*^{\lambda\mu}$ defined in Section 3.11.

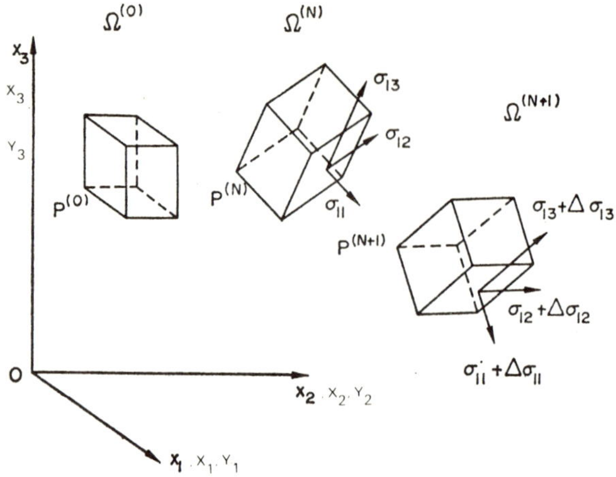

FIG. I-7. Definition of Kirchhoff stress tensors by the use of the (x_1, x_2, x_3) coordinate system.

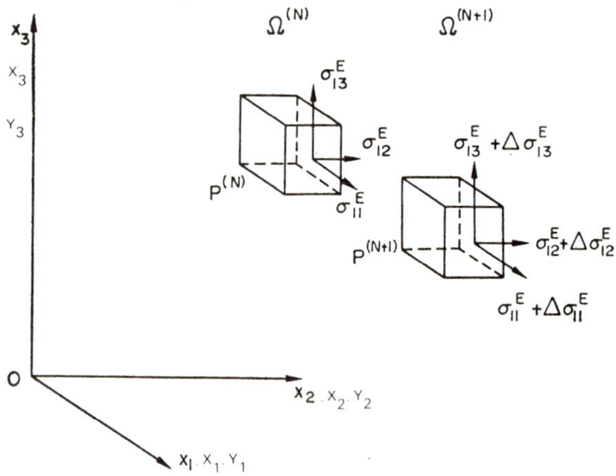

FIG. I-8. Definition of Euler stress tensors.

of an infinitesimal rectangular parallelepiped containing the point $P^{(N)}$ as shown in Fig. I-8. It should be noted that the Euler stress tensors σ_{ij}^E are defined per unit area of the $\Omega^{(N)}$ state and they are taken in the directions of the rectangular Cartesian coordinate axes, namely in the directions of $\mathbf{i}_i$; $i = 1, 2, 3$. Following Ref. 45, we have the transformation law between σ_{ij}^E and σ_{ij} as follows:

$$\sigma_{ij}^E = \frac{1}{D^{(N)}} \frac{\partial X_i}{\partial x_k} \frac{\partial X_j}{\partial x_l} \sigma_{kl}. \tag{I-9.17}$$

Third, we define the Euler stress tensors $\sigma_{ij}^E + \Delta\sigma_{ij}^E$ at the point $P^{(N+1)}$ and obtain the transformation law between $\sigma_{ij}^E + \Delta\sigma_{ij}^E$ and $\sigma_{ij} + \Delta\sigma_{ij}$ as follows:

$$\sigma_{ij}^E + \Delta\sigma_{ij}^E = \frac{1}{D^{(N+1)}} \frac{\partial Y_i}{\partial x_k} \frac{\partial Y_j}{\partial x_l} (\sigma_{kl} + \Delta\sigma_{kl}). \tag{I-9.18}$$

Fourth, we define another set of the Kirchhoff stress tensors at the point $P^{(N+1)}$ by the use of the (X_1, X_2, X_3) coordinates. We denote its components

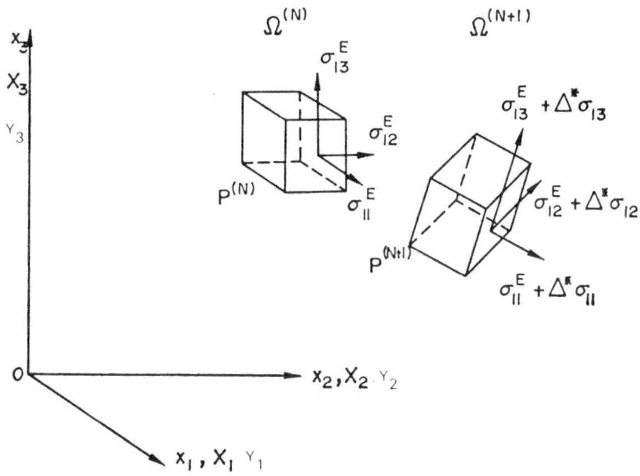

FIG. I-9. Definition of the Kirchhoff stress tensors by the use of the (X_1, X_2, X_3) coordinate system.

by $\sigma_{ij}^E + \Delta^*\sigma_{ij}$, as shown in Fig. I-9.† The transformation law between $\sigma_{ij}^E + \Delta^*\sigma_{ij}$ and $\sigma_{ij}^E + \Delta\sigma_{ij}^E$ is written in the following form:

$$\sigma_{ij}^E + \Delta\sigma_{ij}^E = \frac{1}{\dfrac{\partial(Y_1, Y_2, Y_3)}{\partial(X_1, X_2, X_3)}} \frac{\partial Y_i}{\partial X_k} \frac{\partial Y_j}{\partial X_l} (\sigma_{kl}^E + \Delta^*\sigma_{kl}). \tag{I-9.19}$$

Combining Eqs. (I-9.18) and (I-9.19), and using the relation (I-9.13), we obtain

$$\sigma_{ij}^E + \Delta^*\sigma_{ij} = \frac{1}{D^{(N)}} \frac{\partial X_i}{\partial x_k} \frac{\partial X_j}{\partial x_l} (\sigma_{kl} + \Delta\sigma_{kl}). \tag{I-9.20}$$

Consequently, from Eqs. (I-9.17) and (I-9.20), we obtain

$$\Delta^*\sigma_{ij} = \frac{1}{D^{(N)}} \frac{\partial X_i}{\partial x_k} \frac{\partial X_j}{\partial x_l} \Delta\sigma_{kl}. \tag{I-9.21}$$

† It is seen that σ_{ij}^E and $\Delta^*\sigma_{ij}$ defined here are respectively the same as $\sigma^{(0)\lambda\mu}$ and $\sigma^{\lambda\mu}$ defined in Section 5.1. It is also observed that Δ^*e_{ij} defined by Eq. (I-9.7) are the same as $e_{\lambda\mu}$ defined by Eq. (5.6). $\Delta^*\sigma_{ij}$ is sometimes called the Truesdell stress increment tensor.[47]

We note here the following relation which can be derived from Eq. (I-9.19) neglecting the terms of higher order product of the incremental displacements Δu_i and the incremental stresses $\Delta\sigma^E_{ij}$:

$$\Delta^*\sigma_{ij} = \Delta\sigma^E_{ij} - \sigma^E_{ip}\Delta\omega_{pj} - \sigma^E_{jp}\Delta\omega_{pi} - \sigma^E_{ik}\Delta^*\varepsilon_{jk} - \sigma^E_{jk}\Delta^*\varepsilon_{ik} + \sigma^E_{ij}\Delta^*\varepsilon_{kk}, \tag{I-9.22}$$

where it is defined that

$$\Delta\omega_{ij} = \frac{1}{2}\left(\frac{\partial\Delta u_j}{\partial X_i} - \frac{\partial\Delta u_i}{\partial X_j}\right). \tag{I-9.23}$$

Eq. (I-9.16) together with the relation

$$\frac{\partial\Delta u_i}{\partial X_j} = \Delta^*\varepsilon_{ij} - \Delta\omega_{ij} \tag{I-9.24}$$

have been used in the derivation.

Finally, we define the Jaumann stress increment tensors. We denote the Euler stresses acting on the six surfaces of an infinitesimal rectangular parallelepiped $P^{(N+1)}Q^{(N+1)}R^{(N+1)}S^{(N+1)}$ by $\sigma^E_{ij} + \Delta\sigma^J_{ij}$ as shown in Fig. I-10. The direction cosines of the three sides of the parallelepiped relative to the rectangular Cartesian coordinates (X_1, X_3, X_3) are specified in the following table†:

	X_1	X_2	X_3
$\overrightarrow{P^{(N+1)}Q^{(N+1)}}$	1	$\Delta\omega_{12}$	$-\Delta\omega_{31}$
$\overrightarrow{P^{(N+1)}R^{(N+1)}}$	$-\Delta\omega_{12}$	1	$\Delta\omega_{23}$
$\overrightarrow{P^{(N+1)}S^{(N+1)}}$	$\Delta\omega_{31}$	$-\Delta\omega_{23}$	1

where $\Delta\omega_{ij}$ have been defined by Eq. (I-9.23). The quantities $\Delta\omega_{ij}$ represent the rigid body rotation experienced by the infinitesimal rectangular parallelepiped $P^{(N)}Q^{(N)}R^{(N)}S^{(N)}$ at the $(N + 1)$-th step. The stress increments $\Delta\sigma^J_{ij}$ thus defined are called the Jaumann stress increment tensors.[47]

Next, we shall derive the relations among $\Delta\sigma^J_{ij}$, $\Delta\sigma^E_{ij}$ and $\Delta^*\sigma_{ij}$. We denote the matrices of $\sigma^E_{ij} + \Delta\sigma^J_{ij}$ and $\sigma^E_{ij} + \Delta\sigma^E_{ij}$ at the point $P^{(N+1)}$ by $[\sigma^E + \Delta\sigma^J]$ and $[\sigma^E + \Delta\sigma^E]$ respectively.‡ The transformation law may be written in the following form:

† The table indicates that the direction cosines of the vector $\overrightarrow{P^{(N+1)}Q^{(N+1)}}$ relative to the (X_1, X_2, X_3) coordinate axes are $(1, \Delta\omega_{12} - \Delta\omega_{31})$, and so forth.

‡
$$[\sigma] = \begin{bmatrix} \sigma_{11} & \sigma_{12} & \sigma_{13} \\ \sigma_{21} & \sigma_{22} & \sigma_{23} \\ \sigma_{31} & \sigma_{32} & \sigma_{33} \end{bmatrix}$$

$$[\sigma^E + \Delta\sigma^J] = [L][\sigma^E + \Delta\sigma^E][L]^T, \qquad \text{(I-9.25)}$$

where

$$[L] = \begin{bmatrix} 1 & \Delta\omega_{12} & -\Delta\omega_{31} \\ -\Delta\omega_{12} & 1 & \Delta\omega_{23} \\ \Delta\omega_{31} & -\Delta\omega_{23} & 1 \end{bmatrix} \qquad \text{(I-9.26)}$$

which may be decomposed into the form

$$[L] = [I] + [\Delta\omega], \qquad \text{(I-9.27)}$$

where

$$[\Delta\omega] = \begin{bmatrix} 0 & \Delta\omega_{12} & -\Delta\omega_{31} \\ -\Delta\omega_{12} & 0 & \Delta\omega_{23} \\ \Delta\omega_{31} & -\Delta\omega_{23} & 0 \end{bmatrix} \qquad \text{(I-9.28)}$$

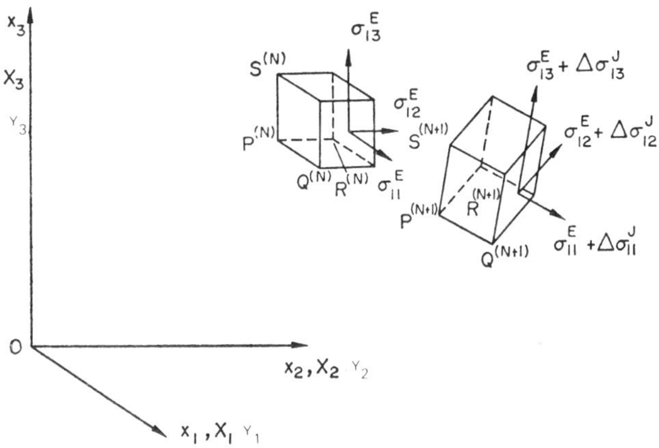

FIG. I-10. Definition of the Jaumann stress increment tensors.

Neglecting terms of higher order product, we obtain from Eq. (I-9.25) the following relation:

$$[\Delta\sigma^J] = [\Delta\sigma^E] + [\sigma^E][\Delta\omega]^T + [\Delta\omega][\sigma^E], \qquad \text{(I-9.29)}$$

or

$$\Delta\sigma^J_{ij} = \Delta\sigma^E_{ij} - \sigma^E_{ip}\Delta\omega_{pj} - \sigma^E_{jp}\Delta\omega_{pi}. \qquad \text{(I-9.30)}$$

By combining Eq. (I-9.22) with Eq. (I-9.30), we obtain

$$\Delta^*\sigma_{ij} = \Delta\sigma^J_{ij} - \sigma^E_{ik}\Delta^*\varepsilon_{jk} - \sigma^E_{jk}\Delta^*\varepsilon_{ik} + \sigma^E_{ij}\Delta^*\varepsilon_{kk}. \qquad \text{(I-9.31)}$$

Eqs. (I-9.22), (I-9.30) and (I-9.31) show the relations among $\Delta\sigma^J_{ij}$, $\Delta\sigma^E_{ij}$ and $\Delta^*\sigma_{ij}$. It is seen that $\Delta^*\sigma_{ij}$ reduces to $\Delta\sigma^J_{ij}$ if the strains $\Delta^*\varepsilon_{kl}$ may be assumed small quantities.

9.3. Relations between Stress and Strain Increments

The next step in the incremental formulation is to assume relations between the stress increments and the strain increments. One of the most natural assumptions may be to postulate the relations between $\Delta^*\sigma_{ij}$ and Δ^*e_{kl} in the following form:

$$\Delta^*\sigma_{ij} = C^*_{ijkl}\Delta^*e_{kl}, \tag{I-9.32}$$

or in a linearized form

$$\Delta^*\sigma_{ij} = C^*_{ijkl}\Delta^*\varepsilon_{kl}. \tag{I-9.33}$$†

In these equations, C^*_{ijkl} may include the effect of past history as mentioned in the flow theory of plasticity introduced in Chapter 12. It is noted here that since C^*_{ijkl} may be multi-valued in the flow theory of plasticity, some technique is required to choose proper values of C^*_{ijkl} for an element under consideration, if this incremental theory is applied to a finite element analysis.[48] We can derive the relations between $\Delta\sigma_{ij}$ and Δe_{kl} with the aid of Eqs. (I-9.9), (I-9.21) and (I-9.32). The result may be written in the following form:

$$\Delta\sigma_{ij} = C_{ijkl}\Delta e_{kl}, \tag{I-9.34}$$

or in a linearized form

$$\Delta\sigma_{ij} = C_{ijkl}\Delta\varepsilon_{kl}, \tag{I-9.35}$$

where

$$C_{ijkl} = D^{(N)}C^*_{pqrs}\frac{\partial x_i}{\partial X_p}\frac{\partial x_j}{\partial X_q}\frac{\partial x_k}{\partial X_r}\frac{\partial x_l}{\partial X_s}. \tag{I-9.36}$$

An alternate natural assumption may be to postulate the relations between $\Delta\sigma^J_{ij}$ and $\Delta^*\varepsilon_{kl}$ in the following form:

$$\Delta\sigma^J_{ij} = C^J_{ijkl}\Delta^*\varepsilon_{kl}. \tag{I-9.37}$$

Eqs. (I-9.37) have been used frequently in the theoretical development and analysis of elastic–plastic problems.

If Eqs. (I-9.37) are postulated, we can derive the relations between $\Delta^*\sigma_{ij}$ and $\Delta^*\varepsilon_{kl}$ by the use of Eqs. (I-9.31) and (I-9.37) and obtain C^*_{ijkl} as follows:

$$C^*_{ijkl} = C^J_{ijkl} - \sigma^E_{ik}\,\delta_{jl} - \sigma^E_{jk}\,\delta_{il} + \sigma^E_{ij}\delta_{kl}, \tag{I-9.38}$$

which are to be used for Eq. (I-9.33). We can now determine C_{ijkl} with the aid of Eqs. (I-9.36) and (I-9.38). The result is

† The linear relations between $d\sigma_{ij}$ and $d\varepsilon_{kl}$ derived for the flow theory of plasticity in Chapter 12 may be interpreted as the relations either between $\Delta^*\sigma_{ij}$ and $\Delta^*\varepsilon_{kl}$ of Eq. (I-9.33) or between $\Delta\sigma^J_{ij}$ and $\Delta^*\varepsilon_{kl}$ of Eq. (I-9.37).

$$C_{ijkl} = D^{(N)} \frac{\partial x_i}{\partial X_p} \frac{\partial x_j}{\partial X_q} \frac{\partial x_k}{\partial X_r} \frac{\partial x_l}{\partial X_s} [C^J_{pqrs} - \sigma^E_{pr}\delta_{qs} - \sigma^E_{qr}\delta_{ps} + \sigma^E_{pq}\delta_{rs}], \qquad \text{(I-9.39)}$$

which are to be used for Eq. (I-9.35). With these preliminaries, we shall now proceed to formulating the incremental theories.

9.4. An Incremental Theory by the Lagrangian Approach

First, we shall formulate an incremental theory by the Lagrangian approach. We begin by defining the stresses, strains, displacements, body forces, external forces acting on S_σ and the displacements prescribed on S_u in the $\Omega^{(N)}$ and $\Omega^{(N+1)}$ states by

$$\sigma_{ij},\ e_{ij},\ u_i,\ \bar{P}_i,\ \bar{F}_i,\ \bar{u}_i,$$

and

$$\sigma_{ij} + \Delta\sigma_{ij},\ e_{ij} + \Delta e_{ij},\ u_i + \Delta u_i,$$
$$\bar{P}_i + \Delta\bar{P}_i,\ \bar{F}_i + \Delta\bar{F}_i,\ \bar{u}_i + \Delta\bar{u}_i,$$

respectively. Then, in a manner similar to the development of Section 3.11, the principle of virtual work for the $\Omega^{(N+1)}$ state is expressed by

$$\iiint_V [(\sigma_{ij} + \Delta\sigma_{ij})\delta(e_{ij} + \Delta e_{ij}) - (\bar{P}_i + \Delta\bar{P}_i)\delta\Delta u_i]dV^{(0)} - \iint_{S_\sigma}(\bar{F}_i + \Delta\bar{F}_i)\delta\Delta u_i dS^{(0)} = 0, \qquad \text{(I-9.40)}$$

where

$$\Delta u_i = \Delta\bar{u}_i \quad \text{on} \quad S_u, \qquad \text{(I-9.41)}$$

and where $e_{ij} + \Delta e_{ij}$ is given by Eq. (I-9.5). We repeat here that the body forces and the surface forces on S_σ are defined per unit volume and per unit area in the $\Omega^{(0)}$ state, and that $dV^{(0)} = dx_1dx_2dx_3$ and $dS^{(0)}$ are respectively elementary volume and elementary surface area in the $\Omega^{(0)}$ state. Neglecting the terms of higher order product of the incremental displacements, we obtain after some manipulations

$$\iiint_V \{\Delta\sigma_{ij}\delta\Delta\varepsilon_{ij} + \sigma_{ij}\delta(\tfrac{1}{2}\Delta u_{k,i}\Delta u_{k,j}) - \Delta\bar{P}_i\delta\Delta u_i + [\sigma_{ij}\delta\Delta\varepsilon_{ij} - \bar{P}_i\delta\Delta u_i]\}dV^{(0)} - \iint_{S_\sigma} (\Delta\bar{F}_i + \bar{F}_i)\delta\Delta u_i dS^{(0)} = 0. \qquad \text{(I-9.42)}$$

If it is assured that the $\Omega^{(N)}$ state is in equilibrium, then the terms

$$\iiint_V [\sigma_{ij}\delta\Delta\varepsilon_{ij} - \bar{P}_i\delta\Delta u_i]dV^{(0)} - \iint_{S_\sigma} \bar{F}_i\delta\Delta u_i dS^{(0)} \qquad \text{(I-9.43)}$$

will vanish in Eq. (I-9.42). However, the $\Omega^{(N)}$ state may not be in complete equilibrium in this kind of incremental theory due to neglect of the higher order terms and computational inaccuracies. Consequently, it is essential to retain these terms in Eq. (I-9.42) for an equilibrium check, as mentioned in the preceding section of this appendix. The principle of virtual work thus established holds irrespective of the incremental stress–strain relations.

An application of Eq. (I-9.42) to a finite element formulation will be discussed briefly at this point. To begin with, Δu_i within each finite element is approximated by

$$\Delta u_i = \sum_k \phi_{ik} \Delta q_k, \tag{I-9.44}$$

where $\phi_{ik}(x_1, x_2, x_3)$ are the shape functions and Δq_k are incremental nodal point displacements. We assume that these shape functions are chosen so that the Δu_i given by Eq. (I-9.44) are compatible with those of the adjacent elements. Substituting Eqs. (I-9.10), (I-9.35) and (I-9.44) into Eq. (I-9.42), we find that the terms representing the contribution from an arbitrary finite element to the left-hand side of Eq. (I-9.42) can be expressed in the following form:

$$\sum_i \{\sum_j (k_{ij}^{(0)} + k_{ij}^{(1)} + k_{ij}^{(2)}) \Delta q_j - \Delta \bar{Q}_i - \Delta \epsilon_i\} \delta \Delta q_i,$$

which may also be expressed in a matrix form as

$$\{\delta \Delta q\}^T [([k^{(0)}] + [k^{(1)}] + [k^{(2)}])\{\Delta q\} - \{\Delta \bar{Q}\} - \{\Delta \epsilon\}],$$

where

$$\begin{aligned}
k_{ij}^{(0)} &= \iiint_{V_n} C_{klpq} \phi_{ki,l} \phi_{pj,q} dV^{(0)}, \\
k_{ij}^{(1)} &= \iiint_{V_n} C_{klpq} (u_{r,k} \delta_{sp} + u_{s,p} \delta_{rk} \\
&\quad + u_{r,k} u_{s,p}) \phi_{ri,s} \phi_{sj,q} dV^{(0)}, \\
k_{ij}^{(2)} &= \iiint_{V_n} \sigma_{kl} \phi_{pi,k} \phi_{pj,l} dV^{(0)}, \\
\Delta \bar{Q}_i &= \iiint_{V_n} \Delta \bar{P}_k \phi_{ki} dV^{(0)} + \iint_{S_{\sigma_n}} \Delta \bar{F}_k \phi_{ki} dS^{(0)}, \\
\Delta \epsilon_i &= \iiint_{V_n} [-\sigma_{kl}(\delta_{pl} + u_{p,l}) \phi_{pi,k} + \bar{P}_k \phi_{ki}] dV^{(0)} \\
&\quad + \iint_{S_{\sigma_n}} \bar{F}_k \phi_{ki} dS^{(0)},
\end{aligned} \tag{I-9.45}$$

and where V_n and S_{σ_n} are respectively the region and the portion of S_σ belonging to the element under consideration. The matrix $[k^{(0)}]$ is the incremental stiffness matrix. The matrices $[k^{(1)}]$ and $[k^{(2)}]$ are called the initial displacement stiffness matrix and the initial stress stiffness matrix, respectively.[49] The matrix $\{\Delta \epsilon\}$ may be called the residual matrix. It is a common practice to assemble the terms representing the contributions from all the elements to obtain a system of linear incremental equilibrium equations for the entire structure, which are subsequently solved to determine the state variables in the $\Omega^{(N+1)}$ state such as the stresses $\sigma_{ij} + \Delta\sigma_{ij}$, the increments of the nodal point displacements Δu_i and so forth.

PROBLEM 1. Show that if applied to the geometrically nonlinear problem for which the principle of stationary potential energy holds, the formulation developed here is equivalent to that of the modified incremental stiffness method treated in Section 8 of this Appendix.

PROBLEM 2. Compare the present method with the Euler method for the stability problem introduced in Section 3.11 of Chapter 3.

9.5. Another Incremental Theory by Combined Use of the Eulerian and Lagrangian Approaches

We shall formulate a second incremental theory by combining the Eulerian and Lagrangian approaches.[50] As mentioned in 9.2, we introduce the (X_1, X_2, X_3) coordinates in the $\Omega^{(N)}$ state, and denote the Euler stress tensors by σ^E_{ij}, body forces by $\bar{P}_i$, and the surface forces on S_σ by $\bar{F}_i$ in the $\Omega^{(N)}$ state. It is noted here that σ^E_{ij} and $\bar{F}_i$ are defined per unit area and $\bar{P}_i$ are defined per unit volume of the $\Omega^{(N)}$ state. On the other hand, we define the Kirchhoff stress tensors $\sigma^E_{ij} + \Delta^*\sigma_{ij}$, the body forces $\bar{P}_i + \Delta\bar{P}_i$ and the surface forces $\bar{F}_i + \Delta\bar{F}_i$ on S_σ in the $\Omega^{(N+1)}$ state, where it is understood that all these quantities are defined per unit area and per unit volume of the $\Omega^{(N)}$ state. Then, we may write the principle of virtual work for the $\Omega^{(N+1)}$ state as follows:

$$\iiint_V [(\sigma^E_{ij} + \Delta^*\sigma_{ij})\delta\Delta^*e_{ij} - (\bar{P}_i + \Delta\bar{P}_i)\delta\Delta u_i]dV^{(N)} - \iint_{S_\sigma}(\bar{F}_i + \Delta\bar{F}_i)\delta\Delta u_i dS^{(N)} = 0, \tag{I-9.46†}$$

where

$$\Delta u_i = \Delta\bar{u}_i \quad \text{on} \quad S_u, \tag{I-9.47}$$

and where $dV^{(N)} = dX_1 dX_2 dX_3$ and $dS^{(N)}$ are respectively elementary volume and elementary surface area of the $\Omega^{(N)}$ state. Neglecting the terms of higher order product of the incremental displacements, we obtain after some manipulation

$$\iiint_V \{\Delta^*\sigma_{ij}\delta\Delta^*\varepsilon_{ij} + \tfrac{1}{2}\sigma^E_{ij}\,\delta\left(\frac{\partial\Delta u_k}{\partial X_i}\frac{\partial\Delta u_k}{\partial X_j}\right) - \Delta\bar{P}_i\delta\Delta u_i + [\sigma^E_{ij}\delta\Delta^*\varepsilon_{ij} - \bar{P}_i\delta\Delta u_i]\}dV^{(N)} - \iint_{S_\sigma}(\Delta\bar{F}_i + \bar{F}_i)\delta\Delta u_i dS^{(N)} = 0. \tag{I-9.48}$$

† Eq. (I-9.46) is equivalent to Eq. (5.5) of Chapter 5.

By the use of Eq. (I-9.48), we can establish a finite element formulation in a manner similar to the development in 9.4. By approximating Δu_i in each element by

$$\Delta u_i = \sum_k \phi_{ik} \Delta q_k, \tag{I-9.49}$$

where $\phi_{ik}(X_1, X_2, X_3)$ are compatible shape functions, and using Eqs. (I-9.11) and (I-9.33), we find that the terms representing the contribution from an arbitrary finite element to the left hand side of Eq. (I-9.48) can be expressed in the following form:

$$\sum_i \{\sum_j (k_{ij} + k_{ij}^{(G)}) \Delta q_j - \Delta \bar{Q}_i - \Delta \epsilon_i\} \delta \Delta q_i,$$

which may also be expressed in a matrix form as

$$\{\delta \Delta q\}^T [([k] + [k^{(G)}]) \{\Delta q\} - \{\Delta \bar{Q}\} - \{\Delta \epsilon\}],$$

where

$$k_{ij} = \iiint_{V_n} C^*_{klpq} \frac{\partial \phi_{ki}}{\partial X_l} \frac{\partial \phi_{pj}}{\partial X_q} dV^{(N)}$$

$$k_{ij}^{(G)} = \iiint_{V_n} \sigma_{kl}^E \frac{\partial \phi_{pi}}{\partial X_k} \frac{\partial \phi_{pj}}{\partial X_l} dV^{(N)}$$

$$\Delta \bar{Q}_i = \iiint_{V_n} \Delta \bar{P}_k \phi_{ki} dV^{(N)} + \iint_{S_{\sigma_n}} \Delta \bar{F}_k \phi_{ki} dS^{(N)},$$

$$\Delta \epsilon_i = \iiint_{V_n} [-\sigma_{kl}^E \phi_{ki,l} + \bar{P}_k \phi_{ki}] dV^{(N)} + \iint_{S_{\sigma_n}} \bar{F}_k \phi_{ki} dS^{(N)}, \tag{I-9.50}$$

and where $[k^{(G)}]$ is called the incremental geometric stiffness matrix.

By assembling the terms representing the contributions from all the elements to obtain a system of linear equations for the entire structure and solving these equations, we can obtain the state variables in the $\Omega^{(N+1)}$ state. The stresses $\sigma_{ij}^E + \Delta^* \sigma_{ij}$ thus obtained are now transformed by the use of Eq. (I-9.19) into $\sigma_{ij}^E + \Delta \sigma_{ij}^E$, which provide the initial stress for the $(N + 2)$th step. It should be noted here that after each succeeding step, total displacements are computed by adding all incremental contributions to update nodal point coordinates, and the stiffness matrices $[k]$ and $[k^{(G)}]$ are recomputed for each step.

The above is an outline of the incremental theory developed in Ref. 50. It is stated in the reference that if the structural response is highly nonlinear, even the above procedure may lead to computed results which are in error.

It is also suggested that for this class of problems, Newton–Raphson iteration procedures can be employed to reduce the error in the nodal point equilibrium to any desired degree. The reader is directed to Refs. 48 through 51, for further details of the incremental theories and other formulations, together with their practical applications to geometrical and material nonlinear problems.

PROBLEM 1. The two incremental theories formulated in Section 9 have been made with reference to the rectangular Cartesian coordinate system. Extend the above theories and develop them in the general curvilinear coordinate system which have been introduced in Chapter 4.

PROBLEM 2. Show that if the structural response is highly nonlinear, the relations between stress and strain increments as given by Eq. (I-9.32), or given in a more general form by

$$\Delta^*\sigma_{ij} = f_{ij}(\Delta^* e_{kl})$$

must be employed, and the principles of virtual work, Eq. (I-9.40) and Eq. (I-9.46), must be used without neglecting the terms of higher order product.

PROBLEM 3. Show that Eq. (I-9.46) is equivalent to Eq. (I-9.40). *Note*: Relations such as Eqs. (I-9.9), and (I-9.20) and $dV^{(N)} = D^{(N)} dV^{(0)}$ are useful for the proof.

PROBLEM 4. Compare the incremental theory formulated in 9.5 with the flow theory of plasticity introduced in Chapter 12.

Section 10. Some Remarks on Discrete Analysis

The term *discrete analysis* seems to cover a wide spectrum of numerical analysis methods wherein a system having an infinite number of degrees of freedom is approximated by a system having a finite number of degrees of freedom. Thus, differential or integral equations established for a continuous body problem are reduced to a finite number of algebraic equations in discrete analysis.† As is well known, the method of weighted residuals (abreviated MWR)[56–59] and the finite difference method (abreviated FDM) are two major discrete analysis methods.‡ As the last topic of this appendix, we shall examine the MWR because it provides a broader and more flexible basis to the formulation of FEM than the variational method.§

Following Ref. 58, we shall take, as an example, a two dimensional heat conduction problem defined by the following differential equation:

$$\frac{\partial}{\partial x}\left(K\frac{\partial\theta}{\partial x}\right) + \frac{\partial}{\partial y}\left(K\frac{\partial\theta}{\partial y}\right) + \bar{Q} = 0, \text{ in } S, \tag{I-10.1}$$

† For discrete analysis applied to integral equations, see Refs. 52 through 55, for example.

‡ It is stated in Ref. 58 that FDM, which originally appeared to be a different process, has recently been formulated on variational basis and can be identified in FEM terminology.

§ I wish to express my gratitude to Professor O. C. Zienkiewicz for his permission of my frequent reference to Ref. 58 in the writing of Section 10.

together with prescribed boundary conditions:

$$K\frac{\partial\theta}{\partial n} = \bar{q} \quad \text{on} \quad C_1, \tag{I-10.2}$$

$$\theta = \bar{\theta} \quad \text{on} \quad C_2, \tag{I-10.3}$$

where θ, K and $\bar{Q}$ are the temperature, the heat conductivity and the heat source intensity respectively, while n is the normal drawn outwards on the boundary, and $\bar{q}$ and $\bar{\theta}$ are prescribed functions of the space coordinates.

10.1. A Variational Principle

A variational principle will be derived here for this problem for later reference. In a manner similar to the development for the linear elasticity problem, we begin by writing the following equation:

$$-\iint_S \left[\frac{\partial}{\partial x}\left(\frac{K\partial\theta}{\partial x}\right) + \frac{\partial}{\partial y}\left(K\frac{\partial\theta}{\partial y}\right) + \bar{Q}\right]\delta\theta\, dxdy$$
$$+\int_{C_1}\left(K\frac{\partial\theta}{\partial n} - \bar{q}\right)\delta\theta\, ds = 0, \tag{I-10.4}$$

where $\delta\theta$ is a virtual variation of θ, and Eq. (I-10.3) is taken as a subsidiary condition. If it is assumed that $\delta\theta$ is a continuous function in S, integrations by parts transform Eq. (I-10.4) into

$$\iint_S \left[K\left(\frac{\partial\theta}{\partial x}\frac{\partial\delta\theta}{\partial x} + \frac{\partial\theta}{\partial y}\frac{\partial\delta\theta}{\partial y}\right) - \bar{Q}\delta\theta\right]dxdy$$
$$-\int_{C_1}\bar{q}\delta\theta ds = 0. \tag{I-10.5}$$

If it is further assumed that K, $\bar{Q}$ and $\bar{q}$ are not subjected to variation, we have from Eq. (I-10.5) the following variational principle:

$$\delta\Pi = 0, \tag{I-10.6}$$

where

$$\Pi = \iint_S \left\{\tfrac{1}{2}K\left[\left(\frac{\partial\theta}{\partial x}\right)^2 + \left(\frac{\partial\theta}{\partial y}\right)^2\right] - \bar{Q}\theta\right\}dxdy$$
$$-\int_{C_1}\bar{q}\theta ds. \tag{I-10.7}$$

10.2. Method of Weighted Residuals

Returning to the topic of MWR, let us denote an approximate solution for θ by $\tilde{\theta}$ and express it as follows:

$$\tilde{\theta} = \sum_{i=1}^{N} a_i \phi_i(x, y) + \phi_0(x, y), \tag{I-10.8}$$

where $\phi_i(x, y)$; $i = 1, 2, \ldots, N$ are coordinate functions defined in the domain S, and a_i; $i = 1, 2, \ldots, N$ are parameters to be determined. The function $\phi_0(x, y)$ is included in Eq. (I-10.8) to take care of some inhomogeneous terms appearing in Eqs. (I-10.1), (I-10.2) and (I-10.3). Introducing Eq. (I-10.8) into Eqs. (I-10.1), (I-10.2) and (I-10.3), we have the so-called residuals defined as follows:

$$-R_S \equiv \frac{\partial}{\partial x}\left(K\frac{\partial \tilde{\theta}}{\partial x}\right) + \frac{\partial}{\partial y}\left(K\frac{\partial \tilde{\theta}}{\partial y}\right) + \bar{Q} \quad \text{in} \quad S, \tag{I-10.9}$$

$$R_{C_1} \equiv K\frac{\partial \tilde{\theta}}{\partial n} - \bar{q} \quad \text{on} \quad C_1, \tag{I-10.10}$$

$$R_{C_2} \equiv \tilde{\theta} - \bar{\theta} \quad \text{on} \quad C_2. \tag{I-10.11}$$

Unless $\tilde{\theta}$ is an exact solution by chance, these residuals never vanish. The method of weighted residuals proposes to determine values of a_i in such a way that the residuals are reduced to zero in the sense of weighted mean, namely,

$$\iint_S R_S W_i dxdy + \int_{C_1} R_{C_1} W_i ds + \int_{C_2} R_{C_2} W_i ds = 0, \quad i = 1, 2, \ldots, N, \tag{I-10.12}$$

where W_i; $i = 1, 2, \ldots, N$ are the so-called weighting functions. They may be any functions and have no continuity requirements. They may be discontinuous functions including the delta function. There are several ways of choosing the weighting functions. A different choice leads to a different formulation. Some special cases of these choices will be shown in the following.

10.3. Point Collocation and Subdomain Collocation

If the weighting functions are chosen to be delta functions in such a way that

$$W_i = \delta(x - x_i, y - y_i); \; i = 1, 2, \ldots, N, \tag{I-10.13}$$

where δ is the delta function, and (x_i, y_i) is the coordinates of a point in S, or on C_1, or on C_2, we have a formulation called point collocation.

Next, we divide the region S, and the boundaries C_1 and C_2 into a number of subdomains $\Omega_1, \Omega_2, \ldots$ and choose the weighting functions in such a way that

$$W_i = 1 \quad \text{in the subdomain } \Omega_i,$$
$$W_i = 0 \quad \text{elsewhere,}$$

to obtain a formulation called subdomain collocation.

10.4. Galerkin Method

Now, we shall choose $\tilde{\theta}$ of Eq. (I-10.8) in such a way that it satisfies Eq. (I-10.3) and write Eq. (I-10.12) as follows:

$$\iint_S R_S W_i dxdy + \int_{C_1} R_{C_1} W_i ds = 0,$$
$$i = 1, 2, \ldots, N. \qquad \text{(I-10.14)†}$$

If the continuity of the weighting functions is assumed, integrations by parts transform Eq. (I-10.14) into

$$\iint_S \left[K\left(\frac{\partial\tilde{\theta}}{\partial x}\frac{\partial W_i}{\partial x} + \frac{\partial\tilde{\theta}}{\partial y}\frac{\partial W_i}{\partial y} \right) - \bar{Q} W_i \right] dxdy$$
$$-\int_{C_1} \bar{q} W_i ds = 0. \qquad \text{(I-10.15)‡}$$

The Galerkin method proposes to employ either Eq.(I-10.14) or Eq. (I-10.15) and to take

$$W_i = \phi_i(x, y);\ i = 1, 2, \ldots, N, \qquad \text{(I-10.16)}$$

for the determination of the unknown parameters a_i. In other words, the weighting functions are taken in coincidence with the coordinate functions in the Galerkin method.

10.5. Rayleigh–Ritz Method

Needless to say, the Rayleigh–Ritz method asserts that

$$\frac{\partial \Pi}{\partial a_i} = 0;\ i = 1, 2, \ldots, N, \qquad \text{(I-10.17)}$$

where

$$\Pi = \iint_S \left\{ \frac{1}{2} K\left[\left(\frac{\partial\tilde{\theta}}{\partial x} \right)^2 + \left(\frac{\partial\tilde{\theta}}{\partial y} \right)^2 \right] - \bar{Q}\tilde{\theta} \right\} dxdy$$
$$- \int_{C_1} \bar{q}\tilde{\theta} ds, \qquad \text{(I-10.18)}$$

† Compare with Eq. (I-10.4).
‡ Compare with Eq. (I-10.5).

and $\tilde{\theta}$ is given by Eq. (I-10.8), while Eq. (I-10.3) is taken as a subsidiary condition. As expected, the equations obtained from Eq. (I-10.17) are equivalent to those obtained by the use of the Galerkin method based on Eq. (I-10.15).

As mentioned before, the weighting functions have no continuity requirements as far as Eq. (I-10.12) are concerned. However, the continuity on the weighting functions is required in transforming Eq. (I-10.14) into Eq. (I-10.15). It is stated in Ref. 58 that integrations of Eq. (I-10.14) by parts into Eq. (I-10.15) reduce the continuity requirements on the coordinate functions, but increase those on the weighting functions.

Thus far, we have seen that MWR includes several methods such as collocation methods, Galerkin method and variational method, providing a broad basis for the discrete analysis technique and elucidating the features of individual methods. MWR can be formulated for almost any problem in engineering science and consequently has universality in applications to practical problems. For further details, the reader is directed to Refs. 56 through 59, for example.

Bibliography

1. R. Courant, Variational Methods for the Solution of Problems of Equilibrium and Vibrations, *Bulletin of the American Mathematical Society*, Vol. 49, pp. 1–23, January 1943.
2. M. J. Turner, R. W. Clough, H. C. Martin and J. L. Topp, Stiffness and Deflection Analysis of Complex Structures, *Journal of Aeronautical Sciences*, Vol. 23, No. 9, pp. 805–824, 1956.
3. J. H. Argyris, Energy Theorems and Structural Analysis. Part I. General Theory, *Aircraft Engineering*, Vol. 26, pp. 347–356, October 1954; pp. 383–387, 394 November 1954; and Vol. 27, pp. 42–58, February 1955; pp. 80–94, March 1955; pp. 125–134, April 1955; pp. 145–158, May 1955. J. H. Argyris and S. Kelsey, Energy Theorems and Structural Analysis. Part II. Applications to Thermal Stress Analysis and to Upper and Lower Limits of Saint-Venant Torsion Constant, *Aircraft Engineering*, Vol. 26, pp. 410–422, December 1954 (reprinted as Ref. 2.15).
4. T. H. H. Pian, Derivation of Element Stiffness Matrices by Assumed Stress Distribution, *AIAA Journal*, Vol. 2, No. 7, pp. 1333–1336, July 1964.
5. T. H. H. Pian and P. Tong, Basis of Finite Element Methods for Solid Continua, *International Journal for Numerical Methods in Engineering*, Vol. 1, No. 1, pp. 3–28, January–March 1969.
6. T. H. H. Pian, Formulation of Finite Element Methods for Solid Continua, in *Recent Advances in Matrix Methods of Structural Analysis and Design*, edited by R. H. Gallagher, Y. Yamada and J. T. Oden, The University of Alabama in Huntsville Press, pp. 49–83, 1971.
7. T. H. H. Pian, Finite Element Methods by Variational Principles with Relaxed Continuity Requirements, in *Variational Methods in Engineering*, edited by C. A. Brebbia and H. Tottenham, Southampton University Press, pp. 3/1–3/24, 1973.
8. T. H. H. Pian and P. Tong, Finite Element Methods in Continuum Mechanics, in *Advances in Applied Mechanics*, edited by C. S. Yih, Academic Press, Vol. 12, pp. 1–58, 1972.

9. L. R. HERRMANN and R. M. TOMS, A Reformulation of the Elastic Field Equations, in Terms of Displacements, Valid for All Admissible Value of Poisson's Ratio, *Transactions of the ASME, Journal of Applied Mechanics*, Vol. 86, Ser. E, pp. 140–141, 1964.
10. L. R. HERRMANN, Elasticity Equations for Incompressible and Nearly Incompressible Materials by a Variational Theorem, *AIAA Journal*, Vol. 3, No. 10, pp. 1896–1900, October 1965.
11. L. R. HERRMANN, A Bending Analysis for Plates, *Proceedings of the Conference on Matrix Methods in Structural Mechanics*, AFFDL-TR-66-80, pp. 577–601, 1965.
12. L. R. HERRMANN, Finite Element Bending Analysis for Plates, *Journal of Engineering Mechanics Division, Proceedings of the American Society of Civil Engineers*, Vol. EM5, pp. 13–26, October 1967.
13. O. C. ZIENKIEWICZ, The Finite Element Method: From Intuition to Generality, *Applied Mechanics Reviews*, Vol. 23, No. 3, pp. 249–256, March 1970.
14. O. C. ZIENKIEWICZ and Y. K. CHEUNG, *The Finite Element Method in Structural and Continuum Mechanics*, McGraw-Hill, 1967.
15. O. C. ZIENKIEWICZ, *The Finite Element Method in Engineering Science*, McGraw-Hill, 1971.
16. J. H. ARGYRIS, The Impact of the Digital Computer on Engineering Sciences, *Aeronautical Journal of the Royal Society*, Vol. 74, pp. 13–41, 1970 and Vol. 74, pp. 111–127, 1970.
17. J. R. WHITEMAN, *A Bibliography for Finite Element Methods*, Department of Mathematics, Brunel University, TR/9, March 1972.
18. J. E. AKIN, D. L. Fenton and W. C. T. Stoddart, *The Finite Element Method, A Bibliography of its Theory and Applications*, Department of Engineering Mechanics, the University of Tennesse, Knoxville, Report EM 72-1, February 1972.
19. G. STRANG and G. FIX, *An Analysis of the Finite Element Method*, Prentice Hall, 1973.
20. R. E. Jones, A Generalization of the Discrete-Stiffness Method of Structural Analysis, *AIAA Journal*, Vol. 2, No. 5, pp. 821–826, May 1964.
21. Y. YAMAMOTO, *A Formulation of Matrix Displacement Method*, Department of Aeronautics and Astronautics, Massachusetts Institute of Technology, 1966.
22. P. TONG, New Displacement Hybrid Finite Element Models for Solid Continua, *International Journal for Numerical Methods in Engineering*, Vol. 2, No. 1, pp. 73–83, January–March 1970.
23. K. WASHIZU, Outline of Variational Principles in Elasticity (in Japanese), in *Series in Computer-Oriented Structural Engineering*, Vol. II-3-A, Baifukan Publishing Co., Tokyo, 1972.
24. R. A. TOUPIN, A Variational Principle for the Mesh-Type Analysis of Mechanical Systems, *Transactions of ASME, Journal of Applied Mechanics*, Vol. 74, pp. 151–152, 1952.
25. K. WASHIZU, *On the Variational Principles Applied to Dynamic Problems of Elastic Bodies*, Aeroelastic and Structures Research Laboratory, Massachusetts Institute of Technology, March 1957.
26. R. L. SAKAGUCHI and B. TABARROK, Calculation of Plate Frequencies from Complementary Energy Formulation, *International Journal for Numerical Methods in Engineering*, Vol. 2, No. 2, pp. 283–293, April–June 1970.
27. B. TABARROK, Complementary Energy Method in Elastodynamics, in *High Speed Computing of Elastic Structures*, edited by B. Fraeijs de Veubeke, University of Liège, Belgium, pp. 625–662, 1971.
28. M. GERADIN, Computation Efficiency of Equilibrium Models in Eigenvalue Analysis, in *High Speed Computing of Elastic Structures*, edited by B. Fraeijs de Veubeke, University of Liège, Belgium, pp. 589–623, 1971.
29. K. WASHIZU, Some Considerations on Basic Theory for the Finite Element Method, in *Advances in Computational Methods in Structural Mechanics and Design*, edited by R. W. Clough, Y. Yamamoto and J. T. Oden, The University of Alabama in Huntsville Press, pp. 39–53, 1972.
30. B. FRAEIJS DE VEUBEKE, The Duality Principles of Elastodynamics Finite Element Applications, in *Lectures on Finite Element Methods in Continuum Mechanics*, edited

by J. T. Oden and E. R. de Arantes e Oliveira. The University of Alabama in Huntsville Press, pp. 357–377, 1973.
31. R. W. Clough and K. J. Bathe, Finite Element Analysis of Dynamic Response, in *Advances in Computational Methods in Structural Mechanics and Design*, edited by R. W. Clough, Y. Yamamoto and J. T. Oden, The University of Alabama in Huntsville Press, pp. 153–179, 1972.
32. R. W. Clough, Basic Principles of Structural Dynamics, pp. 495–511; Vibration Analysis of Finite Element Systems, pp. 513–523; Numerical Integration of the Equations of Motion, pp. 525–533. in *Lectures on Finite Element Methods in Continuum Mechanics*, edited by J. T. Oden and E. R. de Arantes e Oliveira. The University of Alabama in Huntsville Press, 1973.
33. M. E. Gurtin, Variational Principles for Linear Elastodynamics, *Archiv for Rational Mechanics and Analysis*, Vol. 16, pp. 34–50, 1964.
34. M. E. Gurtin, Variational Principles for the Linear Theory of Viscoelasticity, *Archiv for Rational Mechanics and Analysis*, Vol. 13, pp. 179–191, 1963.
35. E. L. Wilson and R. E. Nickel, Application of the Finite Element Method to Heat Conduction Analysis, *Nuclear Engineering and Design*, Vol. 4, pp. 276–286, North-Holland Publishing Co., Amsterdam 1966.
36. R. S. Dunham, R. E. Nickel and D. C. Strickler, Integration Operators for Transient Structural Response, *Computers and Structures*, Vol. 2, pp. 1–15, 1972.
37. J. Ghaboussi and E. L. Wilson, Variational Formulation of Dynamics of Fluid-Saturated Porous Elastic Solids, *Proceedings of the American Society of Civil Engineers, Journal of the Engineering Mechanics Division*, Vol. EM4, pp. 947–963, August 1972.
38. S. Atluri, An Assumed Stress Hybrid Finite Element Model for Linear Elastodynamic Analysis, *AIAA Journal*, Vol. 11, No. 7, pp. 1028–1031, July 1973.
39. C. Truesdell and W. Noll, The Non-Linear Field Theories of Mechanics, in *Handbuch der Physik Band* 111/3 edited by S. Flügge, Springer Verlag, 1965.
40. J. A. Stricklin, W. E. Haisler and W. A. von Riesemann, Geometrically Nonlinear Structural Analysis by Direct Stiffness Method, *Journal of the Structural Division, ASCE*, vol. 97, No. ST9, pp. 2299–2314, Sept. 1971.
41. W. Haisler, J. A. Stricklin and F. J. Stebbins, Development and Evaluation of Solution Procedures for Geometrically Nonlinear Structural Analysis, *AIAA Journal*, Vol. 10, No. 3, pp. 264–272, March 1972.
42. J. T. Oden, *Finite Elements of Nonlinear Continua*, McGraw-Hill, 1972.
43. R. H. Gallagher, Finite Element Analysis of Geometrically Nonlinear Problems, in *Theory and Practice in Finite Element Structural Analysis*, edited by Y. Yamada and R. H. Gallagher. The University of Tokyo Press, pp. 109–124, 1973.
44. H. C. Martin and G. F. Carey, *Introduction to Finite Element Analysis. Theory and Application*, McGraw-Hill Company, 1973.
45. Y. C. Fung, *Foundations of Solid Mechanics*, Prentice-Hall, 1965.
46. W. T. Koiter, *On the Principle of Stationary Complementary Energy in the Nonlinear Theory of Elasticity*, Report No. 488, Laboratory of Engineering, Delft University of Technology, the Netherland, January 1973, and also *SIAM Journal on Applied Mathematics*, Vol. 25, No. 3, pp. 424–434, November 1973.
47. W. Prager *Introduction to Mechanics of Continua*, Ginn and Company, 1961.
48. Yamada, T. Kawai., N. Yoshimura and T. Sakurai, Analysis of the Elastic-Plastic Problem by the Matrix Displacement Method, *Proceedings of the Second Conference on Matrix Methods in Structural Mechanics*, Wright-Patterson Air Force Base, Ohio, Oct. 15–17, 1968, AFFDL-68-150, Dec. 1969, pp. 1271–1299.
49. P. V. Marcal, Large Strain, Large Displacement Analysis, pp. 535–543; Instability Analysis Using the Incremental Stiffness Matrices, pp. 545–561. In *Lectures on Finite Element Methods in Continuum Mechanics*, edited by J. T. Oden and E. R. de Arantes e Oliveira, The University of Alabama in Huntsville Press, 1973.
50. L. D. Hofmeister, G. A. Greenbaum and D. A. Evensen, Large Strain, Elasto-Plastic Finite Element Analysis, *AIAA Journal*, Vol. 9, No. 7, pp. 1248–1254, July 1971.
51. J. A. Stricklin, W. S. von Riesmann, J. R. Tillerson and W. E. Haisler, Static Geometric and Material Nonlinear Analysis, in *Advances in Computational Methods in*

Structural Mechanics and Design, edited by R. W. Clough, Y. Yamamoto and J. T. Oden, The University of Alabama in Huntsville Press, pp. 301–324, 1972.
52. R. L. Bisplinghoff and H. Ashley, *Principles of Aeroelasticity*, John Wiley & Sons, New York, 1962.
53. H. Ashley, S. Widnall and M. T. Landahl, New Directions in Lifting Surface Theory, *AIAA Journal*, Vol. 3, No. 1, pp. 3–16, January 1965.
54. H. Ashley, Some Considerations Relative to the Predictions of Unsteady Airloads in Lifting Configuration, *Journal of Aircraft*, Vol. 8, No. 10, pp. 747–756, October 1971.
55. J. L. Hess and A. M. O. Smith, Calculation of Potential Flow about Arbitrary Bodies, in *Progress in Aeronautical Sciences* Vol. 8, edited by D. Küchemann, Pergamon Press, 1967, pp. 1–138.
56. S. H. Crandall, *Engineering Analysis*, McGraw-Hill, 1956.
57. B. A. Finlayson, *The Method of Weighted Residuals and Variational Principles*, Academic Press, 1972.
58. O. C. Zienkiewicz, *Note on the Finite Element Method and Its Applications*, Industrial Center of Technology, Japan, 1972.
59. O. C. Zienkiewicz, Weighted Residual Processes in Finite Element with Particular Reference to Some Transient and Coupled Problems, in *Lectures on Finite Element Methods in Continuum Mechanics*, edited by J. T. Oden and E. R. de Arantes e Oliveira, The University of Alabama in Huntsville Press, pp. 415–458, 1973.

APPENDIX J

NOTES ON THE PRINCIPLE OF VIRTUAL WORK

WE shall make two short notes here on the principle of virtual work expressed by Eq. (1.32). The first note is concerned with Eq. (1.28). That is: the terms such as

$$\left(\frac{\partial \sigma_x}{\partial x} + \frac{\partial \tau_{xy}}{\partial y} + \frac{\partial \tau_{zx}}{\partial z} + \bar{X}\right), (\ldots), (\ldots);$$
$$(X_\nu - \bar{X}_\nu), (\ldots), (\ldots)$$

appearing in Eq. (1.28) are taken from Eqs. (1.26) and (1.27), namely, the equations of equilibrium in V and the mechanical boundary conditions on S_1 of the solid body *before the execution of the virtual displacements* δu, δv *and* δw. In other words, $(\sigma_x, \sigma_y, \ldots, \tau_{xy})$ and $(\delta u, \delta v, \delta w)$ are independent of each other.

The second note is that Eq. (1.32) does not state the first law of thermodynamics, but state merely a kind of divergence theorem which is a special case of Eq. (1.76). A physical interpretation of Eq. (1.32) may be given as follows: We consider an infinitesimal rectangular parallelepiped enclosed by the following six surfaces:

$$x = \text{const.}, \quad x + dx = \text{const.};$$
$$y = \text{const.}, \quad y + dy = \text{const.};$$
$$z = \text{const.}, \quad z + dz = \text{const.},$$

in the body V before the execution of the infinitesimal virtual displacements

$$\delta \mathbf{u} = \delta u \mathbf{i} + \delta v \mathbf{j} + \delta w \mathbf{k}, \tag{J-1}$$

and denote stresses acting on these six surfaces by

$$-(\sigma_x \mathbf{i} + \tau_{xy} \mathbf{j} + \tau_{xz} \mathbf{k}),$$
$$\sigma_x \mathbf{i} + \tau_{xy} \mathbf{j} + \tau_{xz} \mathbf{k} + \frac{\partial}{\partial x}(\sigma_x \mathbf{i} + \tau_{xy} \mathbf{j} + \tau_{xz} \mathbf{k}) dx; \ldots . \tag{J-2}$$

Then, the virtual work done, during the virtual displacements, *by these stresses and the body forces acting on this infinitesimal rectangular parallelepiped* is given by

$$-(\sigma_x\mathbf{i} + \tau_{xy}\mathbf{j} + \tau_{xz}\mathbf{k}) \,.\, \delta\mathbf{u}dydz + \left[\sigma_x\mathbf{i} + \tau_{xy}\mathbf{j} + \tau_{xz}\mathbf{k}\right.$$

$$\left. + \frac{\partial}{\partial x}(\sigma_x\mathbf{i} + \tau_{xy}\mathbf{j} + \tau_{xz}\mathbf{k})dx\right] \,.\, \left[\delta\mathbf{u} + \frac{\partial \delta\mathbf{u}}{\partial x}dx\right] dydz$$

$$+ \ldots$$

$$+ (\bar{X}\delta u + \bar{Y}\delta v + \bar{Z}\delta w)dxdydz,$$

$$= \left[\sigma_x \frac{\partial \delta u}{\partial x} + \sigma_y \frac{\partial \delta v}{\partial y} + \ldots + \tau_{xy}\left(\frac{\partial \delta u}{\partial y} + \frac{\partial \delta v}{\partial x}\right)\right] dxdydz, \qquad \text{(J.3)}$$

where higher order terms are neglected and Eqs. (1.2) have been substituted.

Now, we divide the solid body fictitiously into a large number of infinitesimal rectangular parallelepipeds and write relations such as given by Eq. (J-3) for all the parallelepipeds. If we sum up these relations over all the parallelepipeds, we find that the terms representing the contribution from the virtual work done by the stresses acting on the interfaces between the adjacent parallelepipeds are cancelled out. Consequently, by the use of the four relations which hold on the boundary, namely, the relation

$$(X_\nu \delta u + Y_\nu \delta v + Z_\nu \delta w)dS$$
$$= (\sigma_x \delta u + \tau_{xy}\delta v + \tau_{xz}\delta w)ldS + (\tau_{yx}\delta u + \sigma_y \delta v + \tau_{yz}\delta w)mdS$$
$$+ (\tau_{zx}\delta u + \tau_{zy}\delta v + \sigma_z \delta w)ndS, \qquad \text{(J-4)}$$

together with Eqs. (1.29), (1.30) and (1.12), we finally find that the sums of the terms appearing in the left hand side of Eq. (J-3) is equal to the virtual work done, during the virtual displacements, by the entire body forces and the prescribed surface forces on S_1. Thus we obtain:

$$\iiint_V (\sigma_x \delta\varepsilon_x + \sigma_y \delta\varepsilon_y + \ldots + \tau_{xy}\delta\gamma_{xy})dxdydz$$
$$= \iiint_V (\bar{X}\delta u + \bar{Y}\delta v + \bar{Z}\delta w)dxdydz$$
$$+ \iint_{S_1} (\bar{X}_\nu \delta u + \bar{Y}_\nu\, \delta v + \bar{Z}_\nu\, \delta w)dS, \qquad \text{(J-5)}$$

where $\delta\varepsilon_x$, $\delta\varepsilon_y$, ... and $\delta\gamma_{xy}$ are given in terms of δu, δv and δw as shown in Eq. (1.33). Eq. (J-5) states: *The virtual work done by the internal forces is equal to the virtual work done by the external forces in arbitrary infinitesimal virtual displacements satisfying the prescribed geometrical boundary conditions.* This is an interpretation of the principle of virtual work expressed by Eq. (1.32).

PROBLEM 1. Show that the above interpretation is similar to that of the divergence theorem of Gauss introduced in the footnote of page 14.

PROBLEM 2. Show that the integrand $\boldsymbol{\sigma}^\lambda \,.\, (\delta\mathbf{r})_{,\lambda}dV$ in Eq. (3.47) may be interpreted as the virtual work done, during the infinitesimal virtual displacements, by the body forces and the surface forces acting on the deformed infinitesimal parallelepiped.

Note:

$$-\boldsymbol{\sigma}^1 . \delta\mathbf{r}dx^2dx^3 + (\boldsymbol{\sigma}^1 + \boldsymbol{\sigma}^1_{,1}dx^1) . [\delta\mathbf{r} + (\delta\mathbf{r})_{,1}dx^1]dx^2dx^3$$
$$+ \ldots + \bar{\mathbf{P}} . \delta\mathbf{r}dx^1dx^2dx^3$$
$$= \boldsymbol{\sigma}^\lambda . (\delta\mathbf{r})_{,\lambda}dx^1dx^2dx^3 + (\text{higher order terms}).$$

INDEX